INTRODUCTION TO NONLINEAR DIFFERENTIAL AND INTEGRAL EQUATIONS

HAROLD T. DAVIS

DOVER PUBLICATIONS, INC., NEW YORK

Published in Canada by General Publishing Company, Ltd., 30 Lesmill Road, Don Mills, Toronto, Ontario.

Published in the United Kingdom by Constable and Company, Ltd., 10 Orange Street, London WC 2.

International Standard Book Number: 0-486-60971-5
Manufactured in the United States of America
Dover Publications, Inc.
180 Varick Street
New York, N.Y. 10014

Preface

The work presented in the ensuing pages began some 17 years ago with the direction of a dissertation on the nonlinear integro-differential equation introduced by Vito Volterra in his theory of population growth. Although the limited progress made at that time was achieved mainly by the use of linear processes, it was an auspicious moment to begin the study of such nonlinear problems. The great computing devices of the present day were then being developed. These gave great promise of being able to unlock secrets in nonlinear problems which had hitherto baffled the most skillful analysts.

It is a matter of considerable historical interest to observe that by the middle of the 20th century the astronomers had been struggling for more than 250 years with the nonlinear system of equations which describes the motion of the planets. The first edition of Newton's *Principia* was published in 1687. Euler's *Theoria motuum planetarum et cometarum* appeared in 1744. Since the publication of these critical works, incredible ingenuity has been exhibited by a sequence of some of the most brilliant analysts in the history of mathematics— Lagrange, Laplace, Gauss, and their distinguished colleagues. Equations which required in excess of 170 pages to print one of them had been produced by C. Delaunay in his attempt to mathematize the motion of the moon. G. W. Hill had introduced an infinite determinant in his theory of the lunar perigee. The problem of the stability of the solar system presents baffling difficulties even to the present day. For the only tools then available were those that belonged to the linear algorithm. But the problems to be solved were those that belonged to nonlinear mathematics. Substantial progress could be made only when clever transformations had reduced the nonlinear problems to linear ones, or to problems asymptotic to some linear algorithm.

But the advent of the machines has altered the picture. These great tools have made it possible to achieve progress where none was possible before and in the development of the present volume generous use has been made of them. But the powers of the analytical method are not to be disregarded in favor of the powers of the machine. The solutions of nonlinear equations still possess singularities, which only the analytical method can discover and describe.

In preparing the present volume the author has made full use of the achievements of a number of distinguished analysts, who, in the last years of the 19th century and in the early years of the 20th, discovered much about the stability of solutions of systems of equations, explored the mysteries of limit cycles, and made attacks upon the problems of periodicity. Most prominent among these was H. Poincaré and A. Liapounoff, whose achievements remain a cornerstone in the history of the subject.

From another and quite different point of view we find the important discoveries of P. Painlevé, B. Gambier, and their colleagues with respect to a class of differential equations of second order with fixed critical points. Their work was perhaps the most important advance made during the early years of the present century in the resolution of some of the mysteries of nonlinear equations and the problem of the classification of equations.

In his admiration of these studies, however, one should not disregard the history of the development of existence theorems, which were largely the product of that same period. Although the problem finds its origin earlier in the work of Cauchy, the modern form for these theorems was achieved principally by the investigations of E. Goursat and E. Picard, and their colleagues, who made fundamental use of conditions introduced in 1876 by R. Lipschitz.

During the last 20 years there has been an explosive interest in the problem of nonlinear equations. An abundance of memoirs and treatises has appeared in the current literature of mathematics. By referring to the appended *Bibliography* the reader can see for himself the direction in which the subject has moved and note those who have contributed to its development.

The present work has made an attempt to attain a modest measure of completeness in defining the field of nonlinear problems and to indicate within its scope something of the progress that has been attained. The needs of those who are required to apply such equations to the problems of the physical world have been kept constantly in view. Physicists and engineers, in fact all who work in the natural sciences, are continually challenged by these equations. Something must be done to show them how to attack their problems.

The contents of the present volume can be summarized as follows. After a general survey of the problem presented by nonlinear equations, the differential equation of first order is studied. Classical theories of integration, the integrating factor, particular equations, and the problem of singular solutions are discussed. This is followed by a systematic study of the Riccati equation. This equation was specifically chosen because of its wide application, the interest which it has had for many mathematicians, its connection with the linear problem, and the fact that it illustrates in a simple way many of the differences between the linear and the nonlinear problem. It forms a kind of bridge between the two domains.

Following this introduction, existence theorems are presented and the algorithms, which they contain, are critically examined. Cauchy's *calculus of limits*, in spite of the restrictions which its assumptions of analyticity impose, is found to suggest the method which the author discovered was his most useful tool in probing the great problems of the subject.

Before proceeding to more general matters, the book considers two particular problems, which serve to illustrate much of the content of later theorems. The first of these is Volterra's theory of the growth of conflicting populations, which is an example both of à vortex cycle and of a periodic solution defined by nonlinear equations. The second, the problem of pursuit, illustrates on the one hand the usefulness of graphical methods in a solution of certain types of nonlinear equations, and on the other, provides an excellent example of a limit cycle. Both problems are later analyzed with respect to more general theories of such phenomena.

It was found necessary to introduce a chapter on elliptic integrals, elliptic functions, and theta functions, since these are fundamental to any understanding of the problems of nonlinear equations. They provide, for example, the background for such a significant domain as that of the Painlevé transcendents. They are also essential in exploring the complexities of the celebrated *Duffing problem*, that is to say, the problem of the pendulum subjected to the influence of a periodic forcing function.

A general classification of nonlinear problems was found useful in the demarcation of fields, which have been more or less explored. This leads first to a description of methods useful in studying the elliptic equation. The more general problem of second order equations of polynomial class follows as a natural consequence and thus to a study of the Painlevé transcendents.

Although the book contains an introduction to various classical methods of numerical integration of nonlinear equations, these were abandoned in favor of a new tool designated as the method of *continuous analytic continuation*. This is an iterative process based upon Cauchy's method of limits. Basically the method is not new since it is founded upon analytic continuation, which has been used for many years in the solution of difficult problems. One noteworthy example is that of Emden's equation, which is described in Chapter 12. The classical application is to develop a Taylor's series using as large a number of derivatives as possible. These computations are always arduous and a practical limit to the number of terms is soon reached. The series thus attained is then employed over as large a range as possible, this range being limited by the requirements of accuracy. At the end of the interval a new series is then constructed from the derivatives of the first and the solution is analytically extended over a new range.

The method of continuous analytic continuation is a variation of this, since the new center of the series is constructed at each infini-

tesimal step.   Like a turtle, the continuation carries its house with it.
The method was found to have astonishing efficiency and many of the
computations in this book were made by introducing terms involving
derivatives of not more than fourth order.   Upon comparing the
digital results with those obtained by use of a differential analyzer,
it was found that the errors were of the same order.   In fact, the
method of continuous analytic continuation appears to be the mathematical analogue of the analogue computer.   But it has the advantage
that the error can be reduced to any desired size by the introduction
of higher derivatives, while the computer's accuracy is limited by its
mechanical parts.

A systematic study of the errors has been made and these are found
to increase very slowly in the case of periodic functions.   Probes as to
the efficiency of the method in computing values into the very heart of
a polar singularity showed a remarkably small error even in this ex-
treme case.   The method is also readily adaptable to computations
around a singular point in the complex plane.   In fact, both its versa-
tility and its ready adaptation to high speed computers, make the
method superior, in the opinion of the author, to the classical ones
which he has applied in the numerical integration of nonlinear equations.

The phenomena of the phase plane (the plane of y and y') is studied
as an introduction to the problems of nonlinear mechanics.   Curiously
enough the essential phenomena were found to be produced by the
solutions of a linear differential equation with constant coefficients to
which has been added a forcing function.   This greatly simplified the
presentation of difficult matters.   But the mystery of such behavior
is readily explained if we write

$$L(y)=z,\ M(z)=0, \tag{1}$$

where $L(y)$ is a linear operator and $M(z)$ is a nonlinear equation defining the forcing function z.   The phenomena of the phase plane are
thus after all determined by what is essentially a nonlinear system.

The contents of what has been called nonlinear mechanics is given
in Chapter 11, together with numerous illustrative examples.   One
of the principal contributions of this chapter is a study of the system

$$y'=P(x,y),\ x'=Q(x,y), \tag{2}$$

where $P(x,y)$ and $Q(x,y)$ are polynomials of second degree.   Although
no claim to an exhaustive treatment of this complex problem is made,
considerable understanding of the phenomena of such a system may
be gained, especially with respect to the question of when the solutions are periodic.

In order to show the application of various techniques, a number of
classical equations are introduced and their solutions discussed.

Lord Rayleigh in 1883 presented an equation which, rediscovered in another form by B. Van der Pol in 1923, has been a classical example of nonlinear oscillatory phenomena. This equation has been thoroughly examined and a set of tables provided for the limit cycle which it defines. Perhaps the most important contribution of this chapter, however, relates to the Duffing problem and the "jump phenomena" which belong to it. Nearly 150 solutions of this equation have been made from which it has been possible to separate the regions of resonance from those of stability. Numerous examples are given to illustrate the phenomena presented by the solutions of this equation.

No work on nonlinear operators can pretend to completeness which does not survey the problems presented by nonlinear integral equations. Unfortunately, progress in this field has been limited. Enough examples exist, however, to show the manner in which nonlinear problems differ from the linear; and existence theorems have been given for general equations of both the Fredholm and the Volterra type. A preliminary study has been made of the integro-differential equation, which Volterra introduced as an essential part of his theory of hereditary mechanics, and the case where the hereditary factor is constant has been solved.

Since the calculus of variations has been a large contributor of problems in the field of nonlinear differential equations, a chapter showing this connection is given. The Euler equation in its various forms now assumes primary importance and we are interested more in the first variation, which produces the problems, than with the second variation, which is of such fundamental importance in answering the question of whether or not the integral does, indeed, attain its extremal value. For this reason only a superficial treatment of the complex problems associated with the second variation has been given.

In conclusion we wish to make certain acknowledgements of the help received in preparing this volume. No work of this magnitude can be brought to completion without the assistance of a number of people. We mention first Dr. Zenon Szatrowski, who, as a courageous graduate student, assailed the heights of the Volterra integro-differential equation and thus began the attack upon the nonlinear problem. Several grants in aid were made by the Graduate School of Northwestern University to assist in the study of the Painlevé problem and these investigations were finally brought to a successful conclusion under an Ordnance Research and Development Project carried out at the University from November 24, 1953, to March 1955. This long study of the Painlevé problem was initiated by the investigations of Richard C. Paxman and Arthur Pancoe. The arduous computations of the fractional linear transformation were made by Mrs. Hugh (Elizabeth) Rowlinson and Mr. Mykola Marchenko and the final construction of tables by IBM calculators was supervised by Robert D. Lowe, who had previously worked on the project.

During a summer course given at the University of California in Berkeley in 1952, the author was assisted in the application of the differential analyzer to nonlinear problems by Dr. John Killeen. This method of analysis by analogue computers was continued at Northwestern University by Dr. Endrick Noges and the study of both the Duffing problem and the problem of system (1) above was initiated at that time.

In the summer of 1957 the author was invited by Dr. Mark Mills to deliver a series of lectures on the nonlinear problem at the Lawrence Radiation Laboratory of the University of California (Livermore branch) and the unparalleled facilities of that Laboratory became available. He is deeply indebted for this opportunity to Dr. Ivan Weeks, at that time a group leader in the Theoretical Division, who has maintained a deep interest in the project. This contact with the Laboratory was resumed for 15 months from June 1958 to September 1959, and as a member of the Theoretical Division, at that time under the administration of Dr. Sidney Fernbach, the author was able to bring the work to its present state of completion.

Among those to whom he is especially indebted is Dr. James E. Faulkner, who took great interest in the system defined by (2) above. The theorem on vortices given in Chapter 11 was proved by him and he contrived many of the examples presented there.

Devoted interest was given to the project by Norriss Hetherington, an expert in the operation of the differential analyzer, who solved problems too numerous to mention. Among his special achievements were the trajectories for the Volterra integro-differential equation, the solutions for special cases of system (1), the analytic continuation around a singular point of the first Painlevé transcendent, and trajectories and curves for the Duffing problem. This work would never have been completed without his constant interest and help.

Exceptional assistance was also given at various stages of the work by C. Douglas Gardner, an expert in the operation of the IBM machines. He computed various tables for the Van der Vol equation, but his principal achievement was his application of continuous analytic continuation to the computation of the solution of a nonlinear differential equation around a singular point in the complex plane.

Among others who assisted in the work was Dr. Roger L. Fulton, who undertook a systematic computation of the solutions of the Duffing problem for small values of the parameter in the forcing function and represented these solutions graphically. He also fitted least-square polynomials to the boundaries between stable and unstable solutions. Robert E. Shafer and Alfred E. Villaire also assisted in various parts of the project, the latter supplying the graphical representations given in Chapter 12 of the various cases of the generalized Emden equation. The author is also indebted to H. Wayne

Hudson, who produced copies of the manuscript and assisted its completion by numerous other services.

And finally the author is indebted to some of his colleagues at Northwestern University, who assisted him with council and information at various stages of the project, particularly, Dr. Walter Scott, Dr. W. T. Reid, and Dr. Ralph Boas. And last, but by no means least, is Miss Vera Fisher. Her ability to make arduous computations, to produce excellent graphs of complicated material, and to manufacture a manuscript better than any one else, was utilized at every step of the work.

Table of Contents

Chapter 1

Introduction

1. Nonlinear Operators

IN ANOTHER VOLUME the author has developed a theory of linear operators, which contains within its scope a considerable domain of analysis. That such a work should include within its limits a large area of mathematics is readily understood from the fact that the assumption of linearity in operational processes underlies most applications of analysis to the problems of the natural world. It is for this reason that a theory of linear operators, in contrast to a theory of nonlinear operators, is comparatively easy to develop. The latter is beset by many difficulties. There are relatively few algorithms which can be applied and the powerful existence theorems of the linear case must be replaced too often by those of special application.

But in spite of the difficulties of the general problem, there exists need for a systematic treatment of nonlinear equations. Nature, with scant regard for the desires of the mathematician, often seems to delight in formulating her mysteries in terms of nonlinear systems of equations. The theories of elasticity and hydrodynamics are especially rich in such systems. Mechanics, relying as it does upon the calculus of variations, Euler's equation, and Hamilton's principle, provides a wealth of other examples. The mathematician, however, with his rich store of linear algorithms, must usually attack these mysteries from the point of view of linear operators. His problem thus becomes that of reducing the equations through various analytical devices to a linear system. Failing in this, he must then try to approximate the solution by some asymptotic process which brings it within the scope of functions which have been defined and studied by linear methods.

The purpose of this work is to set forth some aspects of the problem of nonlinear equations, to exhibit transformations which lead to equations that can be solved by classical methods, to collect the results obtained in certain special cases, and to attempt some useful generalizations of a few problems.

By a *nonlinear operator* we shall mean one that is not linear, and by a *linear operator* L we shall mean one that has the following properties:

$$L(\phi+\psi)=L(\phi)+L(\psi), \tag{1}$$

$$L(k\phi)=kL(\phi),$$

where ϕ and ψ are arbitrary functions and k is a scalar quantity. The difference

$$\Delta(\phi,\psi)=L(\phi+\psi)-\{L(\phi)+L(\psi)\}, \tag{2}$$

we shall call the *linear deficiency* of the operator L.

For example, $L=d^2/dx^2$ is a linear operator, but $L=(d/dx)^2$ is a nonlinear operator, since we have

$$\left[\frac{d(\phi+\psi)}{dx}\right]^2=\left(\frac{d\phi}{dx}+\frac{d\psi}{dx}\right)^2=\left(\frac{d\phi}{dx}\right)^2+2\frac{d\phi}{dx}\frac{d\psi}{dx}+\left(\frac{d\psi}{dx}\right)^2.$$

The linear deficiency of $(d/dx)^2$ is seen to be

$$\Delta(\phi,\psi)=2\frac{d\phi}{dx}\frac{d\psi}{dx}.$$

Some common forms of nonlinear operators in one variable are the following:

(1) $L(u)=\dfrac{du}{dx}+Q(x)u+R(x)u^2;$

(2) $L(u)=\left(\dfrac{du}{dx}\right)^2-a(x)-b(x)u-c(x)u^2-d(x)u^3;$

(3) $L(u)=\dfrac{d^2u}{dx^2}-\dfrac{1}{u}\left(\dfrac{du}{dx}\right)^2-f(x)\dfrac{du}{dx}-g(x)u;$

(4) $L(u)=\displaystyle\int_a^b K(x,s)u(s)u(s+x)ds;$

(5) $L(u)=\dfrac{1}{u}\dfrac{du}{dx}+A(x)+B(x)u+\displaystyle\int_a^x K(x,s)u(s)ds.$

As in the case of linear operators, we may also have nonlinear operators in more than one variable. The following are examples:

(6) $L(u)=\left(\dfrac{\partial u}{\partial x}\right)^2+\left(\dfrac{\partial u}{\partial x}\right)^2;$

(7) $L(u)=u(x,y)+\displaystyle\int_a^b\int_a^b K(x,y;s,t)u^2(s,t)dsdt;$

(8) $L(u)=\dfrac{\partial^2 u}{\partial x^2}+\dfrac{\partial^2 u}{\partial y^2}+\dfrac{\partial^2 u}{\partial z^2}+Ke^u$

2. Nonlinear Equations

When a nonlinear operator is equated to zero or to a given function, we have a *nonlinear equation*. Thus, confining our attention to operators of a single variable, we see that the equation

$$L(u)=f(x), \tag{1}$$

is a nonlinear equation if $L(u)$ is a nonlinear operator.

In the case where $L(u)$ is a linear operator it is customary to distinguish between the equations: $L(u)=0$ and $L(u)=f(x)$, by referring to the first as a *homogeneous equation* and to the second as a *nonhomogeneous equation*. In the case of nonlinear equations the term homogeneous no longer applies. We shall thus introduce arbitrarily the term *null-equation* to refer to $L(u)=0$, when $L(u)$ is a nonlinear operator, and the term *complete equation* when we have $L(u)=f(x)$.

If there exists a function $u(x)$ which satisfies equation (1), then we say that the equation has a *particular solution*. If one or more functions satisfy the equation, then the equation has several solutions. The *general solution* is the totality of the particular solutions.

For example, the equation

$$(x^2+y^2)\frac{dy}{dx}=xy, \tag{2}$$

is a nonlinear equation, which has as its general solution the function y defined by the following implicit function:

$$2y^2 \log cy - x^2 = 0, \tag{3}$$

where c is an arbitrary constant.

On the other hand, the equation

$$\left[\left(\frac{dy}{dx}\right)^2+1\right]^3 = r^2\left(\frac{d^2y}{dx^2}\right)^2, \tag{4}$$

has as a *particular solution* the two-parameter family of circles,

$$(x-a)^2+(y-b)^2=r^2, \tag{5}$$

and also the *singular solutions:* $y=\pm ix$. These functions comprise the *general solution* of the differential equation.

3. The Solution of Nonlinear Equations

In the preceding section we have given two particular nonlinear equations and have exhibited their solutions, implicitly in one case and essentially explicitly in the second, in terms of elementary functions. In general, however, it is impossible to attain such results.

When it is possible to obtain such solutions, however, the advantages are obvious.

In the first place, it is then often a relatively simple matter to define the points of singularity in the solution and to characterize them. In the second place, it is usually possible without undue effort to exhibit the solutions numerically by means of a table of values, or graphically. Such a table or graph can be computed by some one of the many methods available for such problems and the use of one or more of the ever-increasing number of tables of special functions. In the third place, the arbitrary parameters of the solution appear explicitly and their relationship to the variables is thus immediately observed.

It is thus apparent that the first objective in the study of a nonlinear equation is to ascertain whether or not a solution can be obtained either explicitly or implicitly in terms of classical functions. The procedure in such a study is to discover a transformation which will reduce the equation to some type that is known to have a solution of the desired kind. Failing this, one seeks a transformation which will reduce the equation to one that is *asymptotic* to a form solvable by known functions.

A simple example of what is meant by the last statement is furnished by the linear differential equation:

$$x^2y'' + xy' + (x^2 - n^2)y = 0, \tag{1}$$

which has as its general solution the function

$$y = A J_n(x) + B Y_n(x),$$

where $J_n(x)$ and $Y_n(x)$ are Bessel functions of first and second kind respectively.

By means of the transformation:

$$y = \frac{1}{\sqrt{x}} z, \tag{2}$$

equation (1) is reduced to the form

$$z'' + \left(1 - \frac{4n^2 - 1}{4x^2}\right) z = 0. \tag{3}$$

But as x becomes large, the coefficient of z approaches 1 as a limit and the differential equation (3) approaches the equation:

$$z'' + z = 0, \tag{4}$$

as its limiting form. This equation has the general solution:

$$z = K \cos (x + b), \tag{5}$$

where K and b are arbitrary constants.

It is reasonable to assume that any solution of (1) will be asymptotic to the proper specialization of (5). This is, indeed, the case, but the proof is not readily given. Some of the difficulties are exhibited by considering the case where we write $z=\sqrt{x}J_n(x)$. It can than be shown that z has the following formal representation:

$$z=\sqrt{\frac{2}{\pi}}\left\{P_n(x)\cos\left(x-\frac{1}{2}n\pi-\frac{1}{4}\pi\right)-Q_n(x)\sin\left(x-\frac{1}{2}n\pi-\frac{1}{4}\pi\right)\right\},\quad (6)$$

where $P_n(x)$ and $Q_n(x)$ have the following expansions:

$$P_n(x)=1-\frac{(4n^2-1^2)(4n^2-3^2)}{2!(8x)^2}+\cdots,$$

$$Q_n(x)=\frac{(4n^2-1^2)}{1!(8x)}-\frac{(4n^2-1^2)(4n^2-3^2)(4n^2-5^2)}{3!(8x)^3}+\cdots.\quad (7)$$

It is an interesting and curious fact that neither $P_n(x)$ nor $Q_n(x)$ converges for any finite value of x. Nevertheless, it can be proved that z, as represented by (6), is the solution of the differential equation (3) in the sense of *semiconvergent*, or *asymptotic*, series.

A second example is furnished by the following nonlinear equation:

$$\frac{dy}{dx}=y^2+x,\quad (8)$$

upon which we make the following transformation of both the dependent and the independent variables:

$$x=\left(\frac{3}{2}t\right)^{\frac{2}{3}},\qquad y=\sqrt{x}w.\quad (9)$$

Equation (8) is then reduced to the following:

$$\frac{dw}{dt}+\frac{1}{3}\frac{w}{t}=w^2+1,\quad (10)$$

which, as t increases, is asymptotic to the equation:

$$\frac{dw}{dt}=w^2+1.\quad (11)$$

The solution of equation (11) is the function $w=\tan(t-t_0)$ and we can infer, therefore, that w, the solution of equation (10), is asymptotic to this function, that is,

$$w\sim\tan(t-t_0),\quad (12)$$

where the symbol $\sim$ means "is asymptotic to."

Although it may not appear so, the example which we have just presented is equivalent to that given in the first case. To see this, let us make in equation (10) the transformation:

$$w=-\frac{u'}{u},\tag{13}$$

from which we obtain the equation:

$$t\frac{d^2u}{dt^2}+\frac{1}{3}\frac{du}{dt}+tu=0.\tag{14}$$

By means of a second transformation,

$$u=t^{1/3}y(t),\tag{15}$$

equation (14) becomes

$$t^2\frac{d^2y}{dt^2}+t\frac{dy}{dt}+\left(t^2-\frac{1}{9}\right)y=0,\tag{16}$$

which is equivalent to (1) in which $n=1/3$.

But since we obtain by simple computation

$$\frac{u'}{u}=\frac{y'}{y}+\frac{1}{3t},\qquad \frac{y'}{y}=\frac{z'}{z}-\frac{1}{2t^2},$$

and since, by the arguments in the first example, we have the asymptotic values:

$$z\sim K\cos(t-t_0),\qquad z'\sim -K\sin(t-t_0),$$

we thus see that

$$w\sim\frac{\sin(t-t_0)}{\cos(t-t_0)}=\tan(t-t_0).$$

One of the major difficulties encountered in the solution of nonlinear equations, in contrast to the solution of linear equations, is the manner in which the arbitrary parameters enter. The solution of a linear equation appears in the form of a function linear in the arbitrary constants. But this is not the case for nonlinear equations as one sees from the following example:

$$xyy''=yy'+xy'^2$$

which has the general solution: $y=A\exp(Bx^2)$. The arbitrary parameters A and B enter nonlinearly in the solution.

In the domain of linear equations an essentially complete theory exists for differential equations which possess solutions that have at most singularities which are poles and branch points. Although this is not the case for linear equations for which the solutions possess essential singularities, even here considerable progress has been made and a large body of information has been assembled for equations of this class. But in the category of nonlinear differential equations, the situation is very different. Satisfactory information exists in general only for certain restricted types of equations and for a limited number of special cases. Part of the reason for this is to be found in the present status of the theory of functions, which has been developed largely around classes of functions in which the linearity property is an essential factor. A further examination of this point may prove instructive.

As we have already said above, given a nonlinear equation, the first step in its solution is to attempt to find a function,

$$u(x,y)=0,$$

involving one or more arbitrary constants, which satisfies the equation and which can be expressed in terms of the classical functions. These functions have been divided into *algebraic* and *transcendental* categories. The classical transcendental functions, although now a numerous collection, are still inadequate for the solution of most nonlinear equations. They embrace the exponential, hyperbolic, and circular functions with their inverses; the families of functions derived from linear differential equations of second order such as the Legendrian functions, the Bessel functions, the Laguerre, Hermite, Chebyscheff, and Jacobi functions, with their numerous relatives; the elliptic integrals, the elliptic functions, the elliptic modular functions, and the theta functions; the gamma, beta, psi, and Riemann zeta functions, together with a growing class of similar functions defined by integral transforms. Numerous other functions have been defined in various ways in recent years and their properties partially explored, so that an impressive collection now exists from which one can attempt to construct the solutions of nonlinear equations either directly or by the asymptotic method.

But for the most part these functions are derived in one way or another from linear properties. The most notable exception is found, of course, in the elliptic, elliptic modular, and theta functions, which, as we shall show later, are derived from nonlinear equations. They are very useful, therefore, as spearheads into the unknown region of functions, which provide solutions for a certain class of nonlinear equations.

The classical transcendentals are conspicuous for the tractable character of their singularities. These singularities, in general, are either isolated branch points or poles in the finite plane. Essential singularities, such as the point $x=0$ in the function $\sin(1/x)$, are usually referred to the point at infinity. Natural boundaries, such as the unit circle for the function

$$f(x)=1+x+x^2+x^4+x^8+x^{16}+ \ldots ,$$

are rarely found in the classical transcendentals, although such boundaries exist for the elliptic modular functions.*

For the most part the solutions of nonlinear equations do not possess the functional simplicity of the solutions of linear equations. Functions with essential singularities are the rule rather than the exception. Highly restrictive conditions must be imposed if such singularities are to be excluded. Thus the general solution of the equation:

$$(1-y^2)(\text{arcsin } y)^2 = x^2 y'^2,$$

is the function: $y = \sin(c/x)$ together with the two singular solutions: $y = \pm 1$.

In contrast to the linear case, the solutions of nonlinear equations will sometimes have *movable singular points*. For example, the equation

$$y' + y^2 = 0$$

has the general solution: $y = 1/(x-C)$, where C is arbitrary. Thus any point in the plane, by proper specialization of C, can be a pole of the solution, a phenomenon which is absent from the solution of linear equations.

PROBLEMS

1. Show that $x^3 y^2 + 7x = C$ is a solution of the differential equation

$$2x^3 yy' + 3x^2 y^2 + 7 = 0.$$

2. Prove that the general solution of the equation

$$(x^2 + y^2)y' + 2x(y + 2x) = 0,$$

is the function: $y^3 + 4x^3 + 3x^2 y = C$.

3. Show that the general solution of

$$y' + 3y + y^2 = 4$$

is the function: $y = (e^{5x} - 4k)/(e^{5x} + k)$.

4. Show that for the transformation: $u(x) = y' + y^2$, the equation

$$xy'' + (2xy + 2)y' + 2y^2 + 2 = 0, \tag{17}$$

reduces to: $xu' + 2u = -2$.

5. Referring to Problem 4, solve the equation: $xu' + 2u = -2$, and thus obtain the function: $u = c/x^2 - 1$, where c is arbitrary. Setting $c = 2$, verify that the function

$$y = \frac{(x^2 - 1)\cos(x+k) - x\sin(x+k)}{x^2\sin(x+k) + x\cos(x+k)},$$

where k is arbitrary, provides a particular solution of (17).

---

* For a discussion of this example see Whittaker and Watson: *Modern Analysis*, p. 98.

6. Show that for the transformation: $t = 4x^{5/4}/5$, $y = \sqrt{x}z$, the equation

$$\frac{d^2y}{dx^2} = 6y^2 - 6x \qquad (18)$$

becomes

$$\frac{d^2z}{dt^2} + \frac{1}{t}\frac{dz}{dt} - \frac{4}{25t^2} z = 6z^2 - 6.$$

Knowing that the equation: $z'' = 6z^2 - 6$ has a real solution with a real period, what might one conclude about the solution of (18) for large values of x?

7. Show that the transformation

$$wz = 1/(2x+1), \quad 4w^2 = 2x^2 - x + 1/y,$$

where z is the dependent and w the independent variable, leaves the following equation unchanged:

$$y' + (8x^3 - 2x)y^3 + 3y^2 = 0.$$

4. The Origin of Nonlinear Equations.

As we shall see in the subsequent development of our subject, nonlinear equations originate naturally in many different ways. The mathematical description of natural phenomena, such as elastic systems, problems in stress and strain, optical systems, and the like, lead to such equations.

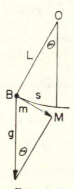

FIGURE 1.

As a simple example, let us consider the vibration of the pendulum shown in Figure 1. The force acting in the direction BM is seen to equal $-mg \sin \theta$. Since this force is also

$$m\frac{d^2s}{dt^2} = mL\frac{d^2\theta}{dt^2},$$

where L is the length of the pendulum, we obtain the differential equation:

$$\frac{d^2\theta}{dt^2} + \frac{g}{L}\sin \theta = 0. \qquad (1)$$

This is a nonlinear equation, which, when displacements of B are small, is approximated by the linear equation:

$$\frac{d^2\theta}{dt^2}+\frac{g}{L}\,\theta=0. \tag{2}$$

The general solution of (1) is attainable only through the use of elliptic functions.

Nonlinear equations are also frequently obtained when parameters are eliminated from a set of parametric curves. Let us consider the doubly infinite family of curves represented by the function

$$f(x,y,\alpha,\beta)=0, \tag{3}$$

where α and β are independent parameters.

A second equation is obtained by differentiation, namely,

$$\frac{\partial f}{\partial x}+\frac{\partial f}{\partial y}\frac{dy}{dx}=0, \tag{4}$$

and a third equation, by means of a second differentiation, as follows:

$$\frac{\partial^2 f}{\partial x^2}+2\,\frac{\partial^2 f}{\partial x\,\partial y}\frac{dy}{dx}+\frac{\partial^2 f}{\partial y^2}\left(\frac{dy}{dx}\right)^2+\frac{\partial f}{\partial y}\frac{d^2 y}{dx^2}=0. \tag{5}$$

If the parameters α and β can be eliminated between equations (3), (4), and (5), a second order differential equation is obtained, which is, in general, nonlinear.

The following examples will illustrate this method of obtaining nonlinear equations:

Example 1.   Find the differential equation of the central conics,

$$x^2+2hxy+by^2=K, \tag{6}$$

where b and h are variable parameters.

Solution: Taking two derivatives of this function, we obtain, together with the original equation, the following set for the elimination of b and h:

$$x^2+2hxy+by^2=K,$$

$$x+hy+(hx+by)\frac{dy}{dx}=0,$$

$$1+2h\frac{dy}{dx}+b\left(\frac{dy}{dx}\right)^2+(hx+by)\frac{d^2 y}{dx^2}=0. \tag{7}$$

Eliminating b and h from these equations, we obtain the following nonlinear equation:

$$Ky^2y''+Ky'^2(xy'-y)+(y-xy')^3=0. \tag{8}$$

Example 2. Find the differential equation for y, where we write

$$y = \frac{a(x) + kb(x)}{c(x) + kd(x)}, \tag{9}$$

in which $a(x)$, $b(x)$, $c(x)$, and $d(x)$ are given functions and k is an arbitrary constant.

Solution: Differentiating y, we get

$$y' = \frac{(c+kd)(a'+kb')}{(c+kd)^2} - \frac{(a+kb)(c'+kd')}{(c+kd)^2},$$

which can also be written

$$y' = \frac{a'+kb'}{c+kd} - \left(\frac{a+kb}{c+kd}\right)\left(\frac{c'+kd'}{c+kd}\right),$$
$$= \frac{a'+kb'}{c+kd} - y\left(\frac{c'+kd'}{c+kd}\right). \tag{10}$$

Solving for k from (9), we get

$$k = -\frac{a-yc}{b-yd}; \tag{11}$$

and solving also for k from (10), we find

$$k = -\frac{a'-yc'-y'c}{b'-yd'-y'd}. \tag{12}$$

Equating these two values of k and clearing fractions, we obtain

$$(a-yc)(b'-yd'-y'd) = (b-yd)(a'-yc'-y'c).$$

From this we derive the equation:

$$(bc-ad)y' + (a'd-ad'+bc'-b'c)y + (cd'-c'd)y^2 = a'b-ab', \tag{13}$$

which can be written in the following form:

$$y' + Q(x)y + R(x)y^2 = P(x), \tag{14}$$

where we use the abbreviations:

$$Q(x) = \frac{a'd-ad'+bc'-b'c}{D}, \ R(x) = \frac{cd'-c'd}{D}, \ P(x) = \frac{a'b-ab'}{D}, \ D = bc-ad.$$

The assumption is now made that D is not identically zero, for in this case the coefficient of y' in equation (14) would be zero and the equation would reduce to a quadratic in y. Equation (14) is called a *Riccati equation*, which will be discussed in some detail in Chapter 3.

PROBLEMS

1. Find the differential equation which has as its general solution the function

$$y = (1 + kx)/(x + 2kx^2).$$

2. Show by the elimination of a and b that the function

$$y = \frac{1}{x-a} + \frac{1}{x-b},$$

is the general solution of the equation

$$y'' + 3yy' + y^3 = 0.$$

3. Determine the differential equation which has the solution

$$(x-a)^2 + (y-b)^2 = r^2,$$

where a and b are arbitrary constants and r is a given quantity.

4. Find the differential equation which has the general solution:

$$(y+a)^3 = (x+b)^2,$$

where a and b are arbitrary constants.

5. Given the functions: $z = x^2 + 2hxy + by^2$, where b and h are arbitrary, form the partial derivatives z_x and z_y and eliminate the parameters. Show that in this manner one obtains the partial differential equation: $2z = xz_x + yz_y$.

5. The Problem of Nonlinear Equations

Having now observed some of the difficulties which are presented by nonlinear equations, we shall survey in a general way a few of the areas in which special progress has been made in developing this field of analysis.

In the first place, it will be necessary to examine the general differential equation of first order, namely,

$$\frac{dy}{dx} = f(x, y), \tag{1}$$

where $f(x,y)$ is a function which must be properly defined. This equation has been the subject of much study. Existence theorems have been given and general algorithms have been provided by means of which solutions can be approximated. A number of particular cases have been investigated and for some of these, special methods of solution have been discovered.

Upon the threshold of this subject appears the Riccati equation,

$$\frac{dy}{dx} + Q(x) \, y + R(x) \, y^2 = P(x), \tag{2}$$

to the discussion of which a surprising number of papers have been devoted. This equation appears in numerous problems. Since the cross-ratio of any four of its solutions is a constant, it has interested the geometers. It appears in the theory of Bessel functions, and it has applications in mechanics. For this reason it has received much attention from analysts.

In one sense a general solution has been achieved for the Riccati. It is customary to regard a linear differential equation as solved, if its solution can be reduced to the quadrature of a known function, even through the quadrature cannot be expressed simply in terms of the classical algebraic or transcendental functions. In the same sense we shall regard a nonlinear equation as solved, if it can be reduced to the solution of a linear equation, even though the solution is not explicitly reducible to the classical functions. Since the transformation:

$$y = \frac{u'}{Ru},$$

(3)

leads to a linear equation of second order in u, equation (2) can thus be regarded as solved in the sense just mentioned. The Riccati equation, because of its intrinsic interest and its importance as an example of a nonlinear equation for which many special results have been attained, will be discussed in some detail in Chapter 3.

Difficulties increase greatly when one considers the problem of the nonlinear differential equation of second order, namely,

$$\frac{d^2y}{dx^2} = f(x,y,y').$$

(4)

We shall not attempt at this place to enumerate the various classes of this equation which have proved to have special interest, but will merely describe a few typical examples. One of these is found in the theory of biological and population growth, to which V. Volterra (1860–1940) devoted major attention. In developing a theory of the conflict of species, he was led to the following system:

$$\frac{dx}{dt} = a(x-xy), \qquad \frac{dy}{dt} = -c(y-xy),$$

(5)

where a and c are positive constants.

When x is eliminated, a nonlinear differential equation of second order in y is obtained, which has a periodic solution. Since system (5) provides an interesting example of a problem in the domain of *nonlinear mechanics*, it serves as an instructive introduction into the phenomena of this subject and will be discussed in considerable detail in a later chapter.

In connection with his study of the problem of individual growth,

Volterra also introduced nonlinear integro-differential equations of which the following is a typical example:

$$\frac{1}{y}\frac{dy}{dt}=A+By+\int_0^t K(t,s)y(s)ds. \tag{6}$$

Another classical problem, which illustrates some of the difficulties encountered in the solution of nonlinear differential equations of second order, is that of the *curve of pursuit*. One is required to determine the path of an object A, which pursues an object B, moving along some prescribed curve, such that A's direction of motion is always toward B. This problem leads to a differential equation of type (4) which, in general, is not integrable in terms of classical functions. But this problem has one conspicuous advantage over other nonlinear problems in the fact that the curve of pursuit can be represented approximately by graphical methods and the more formal analysis can thus be guided.

Notable among the classical functions from the point of view of nonlinear equations are the elliptic integrals and their inverses, the elliptic functions, since these provide solutions of the general equation:

$$\frac{d^2y}{dx^2}=A+By+Cy^2+Dy^3, \tag{7}$$

where A, B, C, and D are constants. Chapter 6 is devoted to the definition of elliptic integrals and elliptic functions and an enumeration of some of their most important properties. Since any knowledge of these functions is imperfect without an introduction to the Theta functions and the elliptic modular functions as well, these are also defined and a few of their properties described.

A natural generalization of the elliptic functions was provided by P. Painlevé (1863–1933) and his collaborators in the early years of the 20th century through the solutions of a certain class of nonlinear differential equations of second order. The unique property of these functions was found in the characterization of their singularities, which admitted *movable poles*, but only *fixed critical points*, that is, branch points and essential singularities. Within this class of equations six were discovered, which defined new transcendental functions. Such, for example, is the equation

$$\frac{d^2y}{dx^2}=6y^2+\lambda x, \tag{8}$$

the solution of which is called the *first Painlevé transcendent*. A large number of memoirs has been devoted to the properties of these interesting equations.

The calculus of variations has contributed its share to the store of nonlinear differential equations. In its simplest form, the problem of this calculus is to determine a function $y(x)$, which will minimize (or maximize) an integral of the form:

$$I = \int_a^b F(x,y,y')dx. \tag{9}$$

The first necessary condition imposed on $y(x)$ is that it shall be a solution of Euler's equation:

$$\frac{\partial F}{\partial y} - \frac{d}{dx}\frac{\partial F}{\partial y'} = 0. \tag{10}$$

In general, for an arbitrary $F(x,y,y')$, this equation is a nonlinear differential equation of second order. The great importance of the problem may be inferred from the fact that the subject of dynamics, from the point of view of the *principle of least action* and its generalization by Hamilton, is founded on the techniques of the calculus of variations.

In addition to these general classes of problems, there exist in the literature of nonlinear equations a number of special equations, which have received considerable attention. Among these, for example, is the following:

$$R(t) = \int_{-\infty}^{\infty} y(s)\, y(s+t)\, ds, \tag{11}$$

which plays an important role in the theory of integral equations.

Numerous special differential equations have been the object of much study, since they have appeared in connection with applications of various kinds. A few of these are listed below as follows:

Emden's differential equation:

$$\frac{d^2y}{dz^2} + \frac{2}{x}\frac{dy}{dx} + y^n = 0; \tag{12}$$

Rayleigh's equation:

$$\frac{d^2y}{dx^2} + K\frac{dy}{dx} + m\left(\frac{dy}{dx}\right)^3 + n^2y = 0; \tag{13}$$

Van der Pol's equation:

$$\frac{d^2y}{dx^2} - \epsilon(1-y^2)\frac{dy}{dx} + ay = 0; \tag{14}$$

Duffing's equation:

$$\frac{d^2y}{dx^2} + ay + by^3 = K\sin \omega t; \tag{15}$$

The Generalized Blasius Equation:

$$\frac{d^3y}{dx^3}+ay\frac{d^2y}{dx^2}=\beta\left[\left(\frac{dy}{dx}\right)^2-1\right]; \tag{16}$$

The White-dwarf equation:

$$\frac{1}{x^2}\frac{d}{dx}\left(x^2\frac{dy}{dx}\right)+(y^2-C)^{3/2}=0; \tag{17}$$

The Thomas-Fermi equation:

$$\frac{d^2y}{dx^2}=\frac{1}{\sqrt{x}}y^{3/2}; \tag{18}$$

Langmuir's equation:

$$3y\frac{d^2y}{dx^2}+\left(\frac{dy}{dx}\right)^2+4y\frac{dy}{dx}+y^2=1; \tag{19}$$

Kidder's equation:

$$\sqrt{(1-\alpha y)}\frac{d^2y}{dx^2}+2x\frac{dy}{dx}=0,\ 0<\alpha<1. \tag{20}$$

6. Systems of Nonlinear Equations

Stimulated originally by the needs of the astronomers and in more recent years by the development in electrical communication, which has introduced nonlinear elements into its circuits, analysts have devoted extensive study to the problems presented by systems of nonlinear equations. Such a system, for example, is the following:

$$\frac{dx}{dt}=A+Bx+Cy+Dx^2+Exy+Fy^2+Gx^3+\ \ldots,$$

$$\frac{dy}{dt}=A'+B'x+C'y+D'x^2+E'xy+F'y^2+G'x^3+\ \ldots, \tag{1}$$

where the multipliers of the variable terms are constants.

When the coefficients are independent of t, the system can be reduced to an equation of first order in x and y, that is,

$$\frac{dy}{dx}=f(x,y). \tag{2}$$

The problem of finding the integral, $F(x,y)=0$, of this equation has been the object of much study, since many of the problems concerning the stability of dynamical systems are thus formulated. The equation, $F(x,y)=0$, is said to define a system of trajectories in *phase-space*, that is to say, a one-parameter family of curves in the plane.

The methods used to determine the nature of the solution of such a system as (1) constitute the subject matter of what has come to be called *nonlinear mechanics*. H. Poincaré, commenting on the problem thus presented, made the following remarks:*

"Considering x and y as the coordinates of a variable point, and t as the time, one seeks the motion of a point to which one gives the velocity as a function of the coordinates. Thus, in the motion which we have studied, we have sought to answer such questions as these: Does the moving point describe a closed curve? Does it always remain in the interior of a certain portion of the plane? In other words, and speaking in the language of astronomy, we have inquired whether the orbit of this point is stable or unstable."

Differential equations of second order, which do not contain terms in which the independent variable appears explicitly, can be reduced to a system such as (1) by the simple device of replacing dy/dt by x. Thus the equation of Van der Pol,

$$\frac{d^2y}{dt^2} - \epsilon(1-y^2)\frac{dy}{dt} + ay = 0, \tag{3}$$

can be replaced by the system:

$$y' = x, \; x' = \epsilon(1-y^2)x - ay. \tag{4}$$

If one equation is divided by the other, the following equation of first order is obtained:

$$\frac{dy}{dx} = \frac{x}{\epsilon(1-y^2)x - ay}. \tag{5}$$

The integral of this equation, $F(x,y) = 0$, defines the phase trajectories of the original differential equation (3). As we shall see later, if a and ϵ are both positive, these phase curves consist of a series of spirals, which are asymptotic to a fixed closed curve called a *limit cycle*.

System (1) is readily generalized by the following:

$$\frac{dx_i}{dt} = X_i(x_1, x_2, \ldots, x_n), \; i = 1, 2, \ldots, n. \tag{6}$$

Since t appears only as an intrinsic variable, it can be eliminated and the following system then defines a set of phase trajectories in a phase-space of n dimensions:

$$\frac{dx_1}{X_1} = \frac{dx_2}{X_2} = \cdots = \frac{dx_n}{X_n}. \tag{7}$$

---

*See, for example, H. Poincaré: "Sur les courbes définies par les équations différentielles," *Journal des Mathématiques*, Vol. 1 (4), 1885, pp. 167–244. Also E. Picard: *Traité d'analyse*, Vol. 3, Paris, 1896, p. 217 *et seq.*

The $n-1$ integrals of this system of equations,

$$F_i(x_1, x_2, \ldots, x_n) = 0, \; i = 1, 2, \ldots, n-1,$$

define a complex of hypersurfaces in phase-space, which depend upon $n-1$ independent parameters. The phase trajectories in n-space are the intersections of these surfaces.

7. Nonlinear Partial Differential Equations

Although we shall be concerned only superficially with a few nonlinear partial differential equations, the subject should not be entirely neglected. Such equations are the breeders of ordinary differential equations, which often provide the most useful solutions of them. We shall thus give a short introduction to the subject.

By a partial differential equation of first order in two independent variables we shall mean an equation of the form

$$F(x, y, z, p, q) = 0, \tag{1}$$

where we use the customary abbreviations:

$$p = \frac{\partial z}{\partial x}, \; q = \frac{\partial z}{\partial y}. \tag{2}$$

Equation (1) is called *linear* if it can be written in the form

$$f_0(x,y)z + f_1(x,y)p + f_2(x,y)q = R(x,y), \tag{3}$$

and *homogeneous* if $R(x,y) = 0$.

The following form of equation (1) is called *quasilinear:*

$$f_1(x,y,z)p + f_2(x,y,z)q = R(x,y,z), \tag{4}$$

provided it does not reduce to (3).

All forms of (1) which cannot be subsumed under (3) and (4) are called *nonlinear* equations.

Thus the equations

$$ap + bq = c, \; z = x^2 p + y^2 q, \; (x^2 - y^2)z = xyp - 3x^2 y, \tag{5}$$

are linear; the equations

$$(xz + y)^2 = y^2 p + x^2 q, \; (1-z)p + (1+z)q = 0, \tag{6}$$

are quasilinear; and the following are examples of nonlinear equations:

$$p^2 = aq + bz + cx + dy, \; pq + xp + yq = z. \tag{7}$$

A partial differential equation of second order in two independent variables is similarly written as follows:

$$F(x,y,z,p,q,r,s,t)=0, \tag{8}$$

where we use the customary abbreviations:

$$r=\frac{\partial^2 z}{\partial x^2}, \qquad s=\frac{\partial^2 z}{\partial x \partial y}, \qquad t=\frac{\partial^2 z}{\partial y^2}. \tag{9}$$

An equation in which z,p,q,r,s,t are linearly connected is called a linear equation, and all others are nonlinear.

Thus, Poisson's equation,

$$r+t=\rho(x,y), \tag{10}$$

is linear, but the equation

$$r^2+t^2=z, \tag{11}$$

is nonlinear.

The intrinsic difficulty in the solution of partial differential equations is readily seen from the long history of the subject. Since the initiating of the study of such equations near the close of the 18th century by Lagrange and Laplace, there has been an increasing interest in the problems presented by them. The method of Lagrange was followed by the theory of Cauchy, who introduced the idea of characteristic strips associated with partial differential equations.

The "problem of Pfaff", initiated by J. F. Pfaff (1765–1825), was that of determining the integral equivalent of the equation

$$P(x,y,z)dx+Q(x,y,z)dy+R(x,y,z)dz=0, \tag{12}$$

and its generalization to n variables. This problem was investigated by C. G. J. Jacobi (1804–51) and R. F. A. Clebsch (1833–72) among others and has stimulated intensive research in recent years.

A large group of analysts was attracted during the second half of the 19th century to the various problems presented by partial differential equations, especially to those of second order. For these equations were found to be essential to the development of the theory of surfaces on the one hand and to the solution of physical problems on the other. In a certain sense these two domains of mathematics were connected by the bridge of the calculus of variations. Various aspects of the subject were studied in particular by J. G. Darboux, A. M. Ampère, C. F. Gauss, and G. F. B. Riemann and by many of their contemporaries. Some of the earlier work of G. Monge (1746–1818) was found to be illuminated by these later investigations.

During the latter half of the 19th century we find a wealth of papers on the subject. Conspicuous among these was the work of Madame

Sophie Kovalevski (1850–91), which gave an existence proof for a general system of partial differential equations in n independent and p dependent variables.

Among the attractive problems derived from the calculus of variations was that of minimal surfaces, which led to a special type of nonlinear equation which can be reduced to the form

$$(1+q^2)r - 2pqs + (1+p^2)t = 0. \tag{13}$$

This equation has received considerable modern attention. *The problem of Plateau*, so called from the investigations of J. Plateau (1801–1883)., is that of finding a continuous minimal surface, which passes through a continuous curve. We shall return to this problem in Chapter 14.

The general theory of partial differential equations was set forth by A. R. Forsyth in the fifth and sixth volumes of his *Theory of Differential Equations* (1900–02) and by E. J. B. Goursat (1858–1936) in his *Cours d'analyse mathématiques*, Vol. 2, 1918, and in his *Leçons sur l'intégration des équations aux dérivées partielles*, Vol. 1, 1891 (equations of first order), Vol. 2, 1896 (equations of second order). A notable contribution both for its intrinsic merit and for the abundance of research which it has stimulated was E. Cartan's *Leçons sur les invariants intégraux*, published in 1922. The contributions to more recent literature of I. M. Janet, J. Horn, J. M. Thomas, T. Y. Thomas, J. A. Schouten, Harry Bateman, and E. Kamke should be mentioned.

But in spite of this abundance of material, the battlements of the nonlinear partial differential equation still present great difficulties to their surmounting. One of the most cogent remarks of Lamb in his *Hydrodynamics*, a work replete with nonlinear problems, is his simple statement: "The exact equations of steady motion are hardly tractable."

Among the more *tractable* equations, however, one finds the following

$$\frac{\partial^2 z}{\partial x^2} + \frac{\partial^2 z}{\partial y^2} = k e^{az}, \tag{14}$$

which is called *Liouville's equation* after J. Liouville (1809–82), who discovered a general integral for it.

This remarkable solution assumes the following form:

$$e^{az} = 2a^{-k} \left[\left(\frac{\partial u}{\partial x}\right)^2 + \left(\frac{\partial u}{\partial y}\right)^2 \right] \bigg/ (u^2 + v^2 + 1)^2, \tag{15}$$

where u and v are defined by the equation: $u + iv = f(x + iy)$, in which $f(w)$ is an arbitrary analytic function.

The equation of Liouville has a surprising number of applications. It is a special case of the equation derived by Lagrange for the stream function $\psi = \psi(x,y)$ in the case of the two-dimensional steady vortex motion of an incompressible fluid, that is,

$$\frac{\partial^2 \psi}{\partial x^2} + \frac{\partial^2 \psi}{\partial y^2} = F(\psi), \tag{16}$$

where $F(\psi)$ is an arbitrary function of ψ.* Liouville's equation is obtained by the obvious specialization: $F(\psi) = ke^{a\psi}$.

This equation also appears in the theory of thermionic emission and in the problem of the isothermal gas sphere. Both of these applications will be discussed in Section 6 of Chapter 12. Because of the intrinsic mathematical interest in the equation, it has been the object of a number of studies prominent among which are the contributions of H. Poincaré, E. Picard, L. Bieberbach, L. Lichtenstein, and G. W Walker. (See *Bibliography*).

The most prolific source of nonlinear differential equations has been the problems of hydrodynamics. This we can immediately observe from the equations themselves. The general equations of two-dimensional flow are the following known as the Navier-Stokes equations:†

$$\frac{\partial u}{\partial t} + u \frac{\partial u}{\partial x} + v \frac{\partial u}{\partial y} = X - \frac{1}{\rho} \frac{\partial p}{\partial x} + \frac{\mu}{\rho} \left(\frac{\partial^2 u}{\partial x^2} + \frac{\partial^2 u}{\partial y^2} \right),$$

$$\frac{\partial v}{\partial t} + u \frac{\partial v}{\partial x} + v \frac{\partial v}{\partial y} = Y - \frac{1}{\rho} \frac{\partial p}{\partial y} + \frac{\mu}{\rho} \left(\frac{\partial^2 v}{\partial x^2} + \frac{\partial^2 v}{\partial y^2} \right), \tag{17}$$

where u and v are the velocity coefficients and X and Y the body forces per unit volume in the x and y directions, p the pressure per unit area, ρ the mass density, and μ the coefficient of viscosity. To this system must also be added the *equation* of *continuity*:

$$\frac{\partial \rho}{\partial t} = \frac{\partial \rho u}{\partial x} + \frac{\partial \rho v}{\partial y}. \tag{18}$$

No general solution exists for these equations, but many special examples have been studied. One of these we shall examine in Section 10 of Chapter 12.

We have indicated in the preceding how nonlinear partial differential equations are obtained from physical problems. Another fruitful source is found in the more purely mathematical problems of geometry. A few illustrations of this may prove instructive.

Thus, let us consider the following equation of a surface in three variables, which depends upon two parameters α and β:

$$f(x,y,z,\alpha,\beta) = 0. \tag{19}$$

---

*See H. Lamb: *Hydrodynamics*, p. 244. See also H. Bateman: *Partial Differential Equations of Mathematical Physics*, Cambridge, 1932, pp. 166–169.

†Named after L. M. H. Navier (1785–1836) and G. G. Stokes (1819–1903), although they were also discovered in a somewhat simpler form by S. D. Poisson.

Differentiating partially with respect to x and y, we obtain the following equations:

$$f_x+f_z p=0, \quad f_y+f_z q=0. \tag{20}$$

We now have three equations, the two just given and equation (19) from which α and β are to be eliminated. The resulting equation will be a partial differential equation of first order, usually nonlinear. For example, if $z=(x-\alpha)^2+(y-\beta)^2$, the resulting differential equation is

$$4z=p^2+q^2. \tag{21}$$

A second method for obtaining a partial differential equation is found in the elimination of the function symbol ϕ from the equation:

$$\phi(u,v)=0, \tag{22}$$

where u and v are given functions of x, y, z and ϕ is an arbitrary function.

To achieve this we now differentiate (22) with respect to x and y and thus obtain the following equations:

$$\phi_u(u_x+pu_z)+\phi_v(v_x+pv_z)=0,$$
$$\phi_u(u_y+qu_z)+\phi_v(v_y+qv_z)=0. \tag{23}$$

Since these equations form a homogeneous system satisfied by ϕ_u and ϕ_v, it is both necessary and sufficient for the existence of nonzero values of ϕ_u and ϕ_v that the determinant of the system be identically zero.

We thus obtain the following equation:

$$(u_x+pu_z)(v_y+qv_z)=(v_x+pv_z)(u_y+qu_z), \tag{24}$$

which can be arranged in the form

$$Pp+Qp=R \tag{25}$$

where P, Q, and R are the following Jacobians:

$$P=\frac{\partial(u,v)}{\partial(y,z)}, \quad Q=\frac{\partial(u,v)}{\partial(z,x)}, \quad R=\frac{\partial(u,v)}{\partial(x,y)}. \tag{26}$$

Although equation (25) is quasilinear, the attainment of its general integral, that is to say, the function defined by (22), usually leads to the solution of a system of ordinary equations which are nonlinear.

To show this, let us find the system of differential equations which have as solutions the functions:

$$u=a, \ v=b, \tag{27}$$

where a and b are arbitrary constants.

Forming differentials of these functions, we get

$$u_x dx + u_y dy + u_z dz = 0, \ v_x dx + v_y dy + v_z dz = 0, \tag{28}$$

and from these we have

$$\frac{dx}{P} = \frac{dy}{Q} = \frac{dz}{R}, \tag{29}$$

where P, Q, and R are the Jacobians defined by (26).

From this we see that to solve equation (25) we first find solutions u and v of (26). Then any arbitrary function of u and v equated to zero provides a solution of (25). In other words, *the quasilinear problem is equivalent to that presented by the system* (29) *which is, in general, nonlinear.*

PROBLEMS

1. Find the differential equation of which the following function is a solution:

$$z = \alpha x + \beta y + f(\alpha, \beta).$$

Use your results to solve the equation

$$p(5p^2 + x) + (6\sqrt{q} + y)q = z.$$

2. Find the differential equation for which the following function is a solution:

$$xz = 2\sqrt{Ax} + Ay + B.$$

3. Verify that the function

$$z = a\sqrt{x^2 + y^2} + f(y/x),$$

is a solution of the following equation:

$$xp + yq = a\sqrt{x^2 + y^2}$$

4. Compute the functions u and v from the equation: $u + iv = (x + iy)^2$. Substitute these in (15) and find the corresponding particular solution of Liouville's equation. Verify that the function thus obtained is indeed a solution.

5. In his original study of equation (14) Liouville introduced the equation:

$$\frac{\partial^2 \log \phi}{\partial s \, \partial t} = \frac{\phi}{2\alpha^2}.$$

with the corresponding solution:

$$\phi = [4\alpha^2 f'(s)g'(t)e^{f(s)+g(t)}]/[1 + e^{f(s)+g(t)}]^2.$$

Use the following transformations:

$$az = \log \phi, \ s = A(x+iy), \ t = B(x-iy),$$

to transform (14) and determine A and B to obtain Liouville's results.

6. Show that the function

$$z = a + b \cosh^{-1}(\rho/b), \ \rho^2 = x^2 + y^2,$$

is a solution of equation (13).

Chapter 2

The Differential Equation of First Order

1. Introduction

IN THIS CHAPTER we shall consider methods for the solution of the differential equation

$$F(x, y, y') = 0, \tag{1}$$

where $F(x,y,y')$ is a function that is continuous and possesses continuous first derivatives with respect to each of the variables in a certain domain R of values of x, y, and y':

If $P_0 = (x_0, y_0, y_0')$ is a point in R which satisfies (1) and if the derivative $F_{y'}(x,y,y')$ does not vanish at P_0, then by the theory of implicit functions there exists a unique function, y', of the variables x and y, which is continuous in the neighborhood of P_0 and which assumes the value y_0' when $x = x_0$ and $y = y_0$. Let us write this function as follows:

$$\frac{dy}{dx} = f(x, y), \tag{2}$$

which is a convenient representation of the general differential equation of first order.

We shall now assume that this equation has a solution in the domain R, which depends upon an arbitrary parameter c, and which can be written in the form

$$u(x,y,c) = 0. \tag{3}$$

For every point (x,y) in the domain R, equation (2) defines the slope of the tangent to every curve represented by (3). Since the explicit determination of (3) is generally a matter of much difficulty, considerable information about the form of the integral curves which it defines can often be obtained if a field of elementary tangents is constructed in the domain R by drawing short tangent lines at a selected number of points.

This preliminary investigation can be readily understood from an example. For this purpose we shall consider the nonlinear equation

$$\frac{dy}{dx} = xy(y-2), \tag{4}$$

which has the following general solution:

$$y=\frac{2}{1+ce^{x^2}}.\tag{5}$$

If the parameter c is determined in such a manner that the integral curve passes through the point (x_0,y_0), then this solution can be written:

$$y=\frac{2y_0}{y_0+(2-y_0)\exp(x^2-x_0^2)}\tag{6}$$

It is clear from (6) that if y_0 is any real number less than 2, then the solution has no singular point in the finite real plane. If $y_0=2$, the solution reduces to $y=y_0$. If, however, y_0 is any real number greater than 2, the y becomes infinite at the points defined by

$$x^2=x_0^2+\log\left(\frac{y_0}{y_0-2}\right)\tag{7}$$

This situation is shown in Figure 1, where the following six integral curves are represented together with a system of elementary tangents.

I. $(x_0, y_0)=(0, 3)$, $y=\dfrac{6}{3-e^{x^2}}.$

II. $(x_0, y_0)=(0, 2.04)$, $y=\dfrac{4.08}{2.04-0.04\ e^{x^2}}.$

III. $(x_0, y_0)=(0, 1.5)$, $y=\dfrac{3}{1.5+0.5\ e^{x^2}}.$

IV. $(x_0, y_0)=(0, 1)$, $y=\dfrac{2}{1+e^{x^2}}.$

V. $(x_0, y_0)=(0,2)$, $y=2.$

VI. $(x_0, y_0)=(0,-1)$, $y=\dfrac{2}{1-3e^{x^2}}.$

The graphical method which we have just illustrated is often the most practical way to obtain a solution if high numerical accuracy is not desired. The method has been described in some detail with a number of illustrations by S. Brodetsky. He described his procedure as follows:*

"Draw the locus of all points at which the required family of curves are parallel to the axis of x: it is of course $f(x,y)=0$. Draw the locus of points where they are parallel to the axis of y, i.e., $1/f(x,y)=0$.

*S. Brodetsky: "The Graphical Treatment of Differential Equations," *Mathematical Gazette*, Vol. 9, 1917-19, pp. 377-382; Vol. 10, 1920-21, pp. 3-8, 35-38, 49-59.

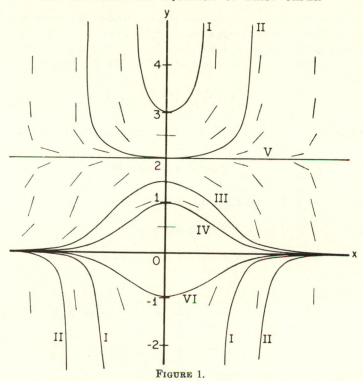

FIGURE 1.

One or other or both of these loci may not exist in the finite part of the plane; but in any case we get the plane divided up into a number of compartments: in some the required curves have positive dy/dx, in others negative dy/dx. Now calculate d^2y/dx^2 from the given differential equation. This can always be done. Draw the locus of points of inflection, i.e., $d^2y/dx^2=0$. We now have a number of compartments, in some of which the curves are concave upward, viz d^2y/dx^2 positive, in others convex downward, viz d^2y/dx^2 negative. We have thus divided up the plane into spaces, in each of which the curves satisfying the differential equation have one of the general forms

$$(1)\ \diagdown,\quad (2)\ \diagup,\quad (3)\ \diagdown,\quad (4)\ \diagup\ \ldots$$

Now draw a number of short tangents at a convenient number of points, and the geometrical solution of the differential equation is obtained.''

The conditions on the first and second derivatives which lead to the four general forms just given may be tabulated as follows:

$$y'<0,\ y''<0,\ \diagdown\ (1)\qquad y'>0,\ y''<0,\ \diagup\ (2),\qquad\qquad (8)$$

$$y''>0,\ \diagdown\ (3)\qquad\qquad y''>0,\ \diagup\ (4).$$

To illustrate this analysis, we shall consider the example given above. We first observe that when $dy/dx=f(x,y)$, the second derivative is readily computed from the following formula provided $f(x,y)$ has first derivatives with respect to each of the variables:

$$\frac{d^2y}{dx^2}=\frac{\partial f}{\partial x}+\frac{\partial f}{\partial y}\frac{dy}{dx}=\frac{\partial f}{\partial x}+\frac{\partial f}{\partial y}f. \tag{9}$$

Since, in the example, we have $dy/dx=xy(y-2)$, we readily compute

$$\frac{d^2y}{dx^2}=y(y-2)(1+2x^2y-2x^2). \tag{10}$$

Setting $xy(y-2)$ equal to zero, we obtain the three lines:

$$x=0,\ y=0,\ y=2,$$

which divide the plane into the six areas shown in Figure 2.

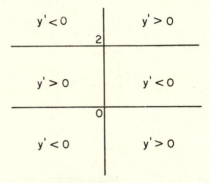

FIGURE 2

Similarly, if we equate $y(y-2)(1+2x^2y-2x^2)$ to zero, we obtain the seven areas shown in Figure 3, along the boundaries of which d^2y/dx^2 is zero. If the areas in these two figures are superimposed

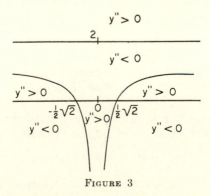

FIGURE 3

and the signs of y' and y'' combined, then from table (8) it is possible to show the characteristic form of the integral in each of the areas. The combined picture is shown in Figure 4.   If this figure is compared with Figure 1, which shows six integral curves, the nature of the

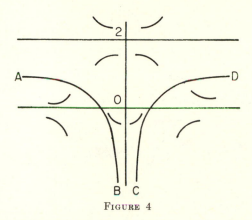

FIGURE 4

analysis and the help which it affords in sketching the solutions of the original differential equation is readily seen.   Along the lines AB and CD the integrals have points of inflection.

2. Isoclines and Curvature

In the preceding section a graphical analysis of the equation,

$$\frac{dy}{dx}=f(x,y),\tag{1}$$

has been described. This analysis can be supplemented, and the actual construction of integral curves aided, by the use of *isoclines* and *circles of curvature*, as we shall now show.

If dy/dx is replaced by a constant m, then the equation,

$$f(x,y)=m,\tag{2}$$

defines a curve along which the derivative is constant and equal to m.   Such a curve is called an *isocline*.   In Section 1 isoclines corresponding to $m=0$ and $m=\infty$ were used in defining areas of the plane within which the derivatives remained of one sign.

A field of isoclines is shown in Figure 5, wherein are graphed ten isoclines for the illustrative equation,

$$\frac{dy}{dx}=xy(y-2).\tag{3}$$

The relationship of these curves to the integrals of (3) is shown in the figure by the three solutions which pass respectively through the points (0,1), (0,1.4), and (0,2.5).

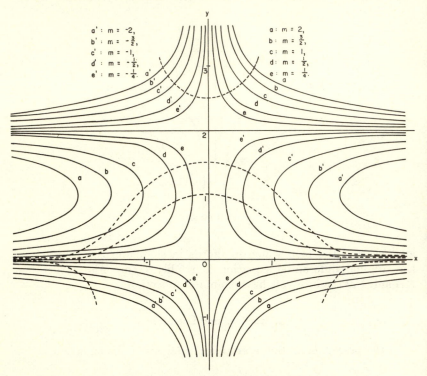

FIGURE 5.—Isoclines corresponding to the equation: $y'=xy(y-2)$.

If, in addition to a pattern of isoclines, we also know something about the curvature at a series of points, an integral curve can be readily drawn. To illustrate, let us assume that we wish to construct a solution of (1) which passes through the point $P=(x_0,y_0)$. We first compute the slope y_0' and the second derivative y_0''. From these two values we can now compute the radius of curvature (R) at P by means of the formula

$$R^2=\frac{(1+y'^2)^3}{y''^2};\tag{4}$$

and, if we so desire, the corresponding center of curvature: $Q=(a,b)$ from the formulas

$$a=x_0-\frac{y_0'R_0}{D},\quad b=y_0+\frac{R_0}{D},\quad D^2=1+y_0'^2.\tag{5}$$

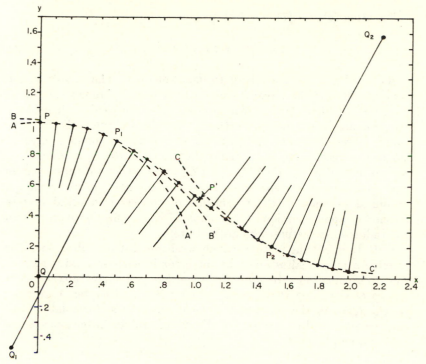

FIGURE 6

Knowing y_0' and R_0, a length R_0 perpendicular to the tangent at P can be laid off, or, if more accuracy is desired, the coordinates of the center of curvature can be computed. An arc of the circle of curvature is now drawn, and at a suitable distance, depending again upon the desired order of approximation, a new point is chosen and the corresponding slope and radius of curvature determined. This process is continued until an approximate graph of the integral curve has been constructed.

In some cases it may be more convenient to use trigonometric functions in computing R.* The procedure in this case can be described as follows: Since $dy/dx = \tan \psi$, where ψ is the slope angle, we have

$$\frac{d^2y}{dx^2} = f(x,y, \tan \psi). \qquad (6)$$

Differentiating $y' = \tan \psi$, we get

$$\frac{d^2y}{dx^2} = \sec^2 \psi \frac{d\psi}{dx} = \sec^2 \psi \frac{d\psi}{ds} \frac{ds}{dx} = \frac{1}{\cos^3 \psi} \frac{1}{R}, \qquad (7)$$

since $R = ds/d\psi$ and $dx/ds = \cos \psi$.

*This is the method suggested by H. Levy and E. A. Baggott: *Numerical Solution of Differential Equations*. American edition, New York, 1950, viii+238 pp. In particular, p. 51.

Equating (6) and (7), we thus obtain

$$\frac{1}{R} = \cos^3 \psi\, f(x,y,\, \tan \psi). \tag{8}$$

As an illustration we return to equation (3). The details of the computation for the integral curve which passes through the initial point: $x_0 = 0$, $y_0 = 1$ are shown in Figure 6. Successive elementary tangents are drawn together with the directions of the radii of curvature. Three centers of curvature are shown at the points Q, Q_1, and Q_2, the corresponding radii being the lengths PQ, P_1Q_1, and P_2Q_2. Arcs of curvature, AA', BB', and CC' are drawn from these centers and their observed departures from the integral curve provide a graphical estimate of the magnitude of the error involved in this method of approximation. The point P' is an inflection point where the radius of curvature becomes infinite.

Unfortunately the method which we have just described does not provide a numerical estimate of the error in each successive step of the approximation. It must be regarded, therefore, merely as a means of sketching the integral curve. But in Chapter 9 we shall return again to the problem. A better analytical formulation will then be given and an estimate of the error involved in each successive step will be provided.

3. The Integrating Factor

In certain special cases the integration of the equation

$$\frac{dy}{dx} = f(x,\, y), \tag{1}$$

can be accomplished by the introduction of what is called an *integrating factor*. To illustrate this method of solution, let us assume that equation (1) has a solution which can be written in the form

$$u(x,\, y) = 0. \tag{2}$$

Forming the differential of $u(x, y)$, we have

$$\frac{\partial u}{\partial x}\, dx + \frac{\partial u}{\partial y}\, dy = 0; \tag{3}$$

from which it follows that we can write

$$f(x,\, y) = -\frac{\dfrac{\partial u}{\partial x}}{\dfrac{\partial u}{\partial y}}. \tag{4}$$

Let us now introduce two functions $P(x, y)$ and $Q(x, y)$ and write $f(x, y)$ as the following quotient:

$$f(x, y) = -\frac{P(x, y)}{Q(x, y)}.$$

In this case equation (1) takes the form

$$P(x, y)\ dx + Q(x, y)\ dy = 0. \tag{5}$$

Upon comparing equation (5) with equation (3) and observing that

$$\frac{\partial}{\partial y}\frac{\partial u}{\partial x} = \frac{\partial}{\partial x}\frac{\partial u}{\partial y},$$

we see that equation (5) is immediately integrable provided P and Q satisfy the following condition:

$$\frac{\partial P}{\partial y} = \frac{\partial Q}{\partial x}. \tag{6}$$

In this case we say that equation (5) is *exact* and the solution is obtained by writing: $\partial u/\partial x = P$ and $\partial u/\partial y = Q$. We thus obtain the function $u(x, y)$ by either of the following integrals:

$$u(x, y) = \int P(x, y)\ \partial x + \phi(y) = \int Q(x, y)\ \partial y + \psi(x), \tag{7}$$

where the symbols ∂x and ∂y are used to indicate that the integration is partial with respect to x in the first case and y in the second.

In order to determine the functions $\phi(y)$ and $\psi(x)$, let us first substitute

$$u(x,y) = \int P(x,y)\partial x + \phi(y)$$

in the equation

$$\frac{\partial u}{\partial y} = Q(x,y).$$

We thus obtain

$$\frac{\partial}{\partial y}\left\{ \int P(x,y)\partial x + \phi(y) \right\} = Q(x,y),$$

from which we have

$$\frac{d\phi(y)}{dy} = Q(x,y) - \frac{\partial}{\partial y}\int P(x,y)\partial x. \tag{8}$$

Since the left side of this equation is a function of y alone, it follows that the right side must also be a function of y alone. We thus have

$$\frac{\partial}{\partial x}\left\{ Q(x,y)-\frac{\partial}{\partial y}\int P(x,y)\,\partial x \right\}=0,$$

which is merely a statement of condition (6) above.

Therefore, upon integrating (8), we obtain $\phi(y)$ from the equation

$$\phi(y)=\int\left\{ Q(x,y)-\frac{\partial}{\partial y}\int P(x,y)\,\partial x \right\}dy. \qquad (9)$$

A similar argument shows that $\psi(x)$ can be obtained from the equation

$$\psi(x)=\int\left\{ P(x,y)-\frac{\partial}{\partial x}\int Q(x,y)\,\partial y \right\}dx. \qquad (10)$$

As an example, consider the equation

$$(3x^2-8xy+6y^2)\,dx+(12xy-4x^2-6y^2)\,dy=0.$$

Writing $P=3x^2-8xy+6y^2$, $Q=12xy-4x^2-6y^2$, we see that the equation is exact since $\partial P/\partial y=-8x+12y=\partial Q/\partial x$.

Making use of (7) we get

$$u(x,y)=\int(3x^2-8xy+6y^2)\,\partial x+\phi(y),$$

$$=x^3-4x^2y+6xy^2+\phi(y).$$

From (9) we then obtain

$$\phi(y)=\int\left\{ 12xy-4x^2-6y^2-\frac{\partial}{\partial y}(x^3-4x^2y+6xy^2) \right\}dy,$$

$$=\int-6y^2\,dy=-2y^3+C.$$

The solution of the equation is thus

$$x^3-4x^2y+6xy^2-2y^3+C=0.$$

The same solution is obtained if we compute $u(x,y)$ from the second equation in (7) and determine $\psi(x)$ by means of (10).

In general, equation (5) is not exact. In this case, however, it is theoretically possible, and in some examples it is practical, to make the equation exact by introducing as a multiplier a function $\mu(x,y)$

called an *integrating factor*. This function must then satisfy the equation:

$$\frac{\partial(\mu P)}{\partial y} = \frac{\partial(\mu Q)}{\partial x},$$

that is to say, the following partial differential equation:

$$Q\frac{\partial \mu}{\partial x} - P\frac{\partial \mu}{\partial y} + \mu\left(\frac{\partial Q}{\partial x} - \frac{\partial P}{\partial y}\right) = 0. \tag{11}$$

That there exists an infinite number of integrating factors is easily seen from the following considerations, provided there exists at least one solution of equation (5). Let μ be any solution of equation (11), let $u(x,y) = 0$ be a solution of equation (5), and let $F(z)$ be a function which possesses a derivative $F'(z)$, but which is otherwise arbitrary. Then the function $\mu F(u)$ is also an integrating factor.

To prove this, we substitute $\mu F(u)$ in equation (11) for μ and thus obtain

$$Q\left(F\frac{\partial \mu}{\partial x} + \mu F'\frac{\partial u}{\partial x}\right) - P\left(F\frac{\partial \mu}{\partial y} + \mu F'\frac{\partial u}{\partial y}\right) + F\mu\left(\frac{\partial Q}{\partial x} - \frac{\partial P}{\partial y}\right)$$

$$= F\left[Q\frac{\partial \mu}{\partial x} - P\frac{\partial \mu}{\partial y} + \mu\left(\frac{\partial Q}{\partial x} - \frac{\partial P}{\partial y}\right)\right] + \mu F'\left(Q\frac{\partial u}{\partial x} - P\frac{\partial u}{\partial y}\right). \tag{12}$$

The expression in the brackets is zero from equation (11). Moreover, since $u(x,y) = 0$ is a solution of equation (5), it follows that

$$\frac{\partial u}{\partial x}\bigg/\frac{\partial u}{\partial y} = \frac{P(x,y)}{Q(x,y)},$$

and thus the multiplier of $\mu F'$ in (12) is also zero. Consequently $\mu F(u)$ is an integrating factor of (5) since it satisfies equation (11).

In spite of the fact that the integrating factors of (5) are thus infinite in number, it is usually impossible to determine even one of them and other methods must be used to solve the equation. Certain special cases exist, however, where the integrating factor can be found and these will now be discussed.

PROBLEMS

1. Show that $(x^2 + y^2)^{-2}$ is an integrating factor for the equation:

$$(x^2 - y^2)\, dx + 2xy\, dy = 0.$$

2. Solve the equation:

$$(x^3 y^4 + xy^2)\, dx + (x^4 y^3 + x^2 y)\, dy = 0.$$

3. Show that $(x^2 + y^2)^{-1}$ is an integrating factor for $y\, dx - x\, dy = 0$.

4. Prove that $\cos x \cos y$ is an integrating factor for the equation:

$$(2x \tan y \sec x + y^2 \sec y)\, dx + (2y \tan x \sec y + x^2 \sec x)\, dy = 0.$$

5. Solve the following equation:

$$\left(2x+\frac{y}{x^2+y^2}\right)dx+\left(2y-\frac{x}{x^2+y^2}\right)dy=0.$$

6. Given the equation

$$[p(x)+yq(x)]dx+y\,dy=0,$$

where $q(x)=k\,p(x)$, k arbitrary, show that the following is an integrating factor:

$$\mu=\exp[-k(y+\int q\,dx].$$

7. Show that

$$\mu=e^{\int p(x)\,dx}$$

is an integrating factor for the equation:

$$[y\,p(x)-q(x)]dx+dy=0.$$

Use this fact to solve the linear differential equation:

$$y'+py=q.$$

8. Prove that $\mu=1/(x^2+y^2)$ is an integrating factor for the equation

$$(y+x\phi)\,dx-(x-y\phi)\,dy=0,$$

where $\phi=\phi(x^2+y^2)$.

4. The Homogenous Case

A function $F(x,y)$ is said to be homogeneous and of degree n provided it satisfies the condition

$$F(\lambda x,\,\lambda y)=\lambda^n\,F(x,y). \tag{1}$$

A fundamental property of such functions is found in *Euler's theorem*, which states that *if $F=F(x,y)$ is a homogeneous function of degree n, then it satisfies the following equation:*

$$nF=x\,\frac{\partial F}{\partial x}+y\,\frac{\partial F}{\partial y}. \tag{2}$$

Euler's *extended theorem* states that if

$$F=F(x,y;\,u,v) \tag{3}$$

is a homogeneous function of degree m in x and y and of degree n in u and v, then F satisfies the equation:

$$(n-m)F=\left(u\,\frac{\partial}{\partial x}+v\,\frac{\partial}{\partial y}\right)\left(x\,\frac{\partial F}{\partial u}+y\,\frac{\partial F}{\partial v}\right)-\left(x\,\frac{\partial}{\partial u}+y\,\frac{\partial}{\partial v}\right)\left(u\,\frac{\partial F}{\partial x}+v\,\frac{\partial F}{\partial y}\right). \tag{4}$$

Returning now to the equation

$$P(x,y)\,dx+Q(x,y)\,dy=0, \tag{5}$$

we shall assume that $P(x,y)$ and $Q(x,y)$ are homogeneous functions of the same degree n. We shall now prove that the function

$$\mu(x,y) = \frac{1}{xP+yQ},\tag{6}$$

is an integration factor for equation (5), provided $xP+yQ \not\equiv 0$.

To establish this, we first compute

$$\frac{\partial(\mu P)}{\partial y} = \mu \frac{\partial P}{\partial y} + P \frac{\partial \mu}{\partial y} = \mu \frac{\partial P}{\partial y} - \mu^2 P \left(x \frac{\partial P}{\partial y} + y \frac{\partial Q}{\partial y} + Q \right),$$

$$\frac{\partial(\mu Q)}{\partial x} = \mu \frac{\partial Q}{\partial x} + Q \frac{\partial \mu}{\partial x} = \mu \frac{\partial Q}{\partial x} - \mu^2 Q \left(x \frac{\partial P}{\partial x} + y \frac{\partial Q}{\partial x} + P \right).$$

The condition that

$$\frac{\partial(\mu P)}{\partial y} = \frac{\partial(\mu Q)}{\partial x}\tag{7}$$

reduces to

$$\frac{\partial P}{\partial y} - \mu P \left(x \frac{\partial P}{\partial y} + y \frac{\partial Q}{\partial y} \right) = \frac{\partial Q}{\partial x} - \mu Q \left(x \frac{\partial P}{\partial x} + y \frac{\partial Q}{\partial x} \right).$$

Multiplying by the reciprocal of μ, that is, by $xP+yQ$, and simplifying the resulting equation, we get

$$Q \left(x \frac{\partial P}{\partial x} + y \frac{\partial P}{\partial y} \right) = P \left(x \frac{\partial Q}{\partial x} + y \frac{\partial Q}{\partial y} \right).\tag{8}$$

Since $P(x,y)$ and $Q(x,y)$ were assumed to be homogeneous functions of degree n, they both satisfy Euler's identity (2); that is, we have

$$x \frac{\partial P}{\partial x} + y \frac{\partial P}{\partial y} = nP, \qquad x \frac{\partial Q}{\partial x} + y \frac{\partial Q}{\partial y} = nQ.$$

When these values are substituted in (8), there results: $Q(nP) = P(nQ)$. This identity carries with it the proof of the equivalence of the two members of equation (7) and thus $\mu(x,y)$ is shown to be the integrating factor for equation (5).

The solution of equation (5) is now readily made by means of the transformation:

$$y = xy.\tag{9}$$

Substituting $dy = x\ dv + v\ dx$ in (5), and observing that both $P(x,y)$ and $Q(x,y)$ are homogeneous functions of degree n, we get

$$P(x,y)dx + Q(x,y)dy = P(x,xv)dx + Q(x,xv)(v\ dx + x\ dv)$$

$$= x^n P(1,v)dx + x^n\ Q(1,v)\ (v\ dx + x\ dv) = 0.$$

This equation reduces to the following:

$$[P(1,v)+v\ Q(1,v)]\ dx+x\ Q(1,v)\ dv=0. \tag{10}$$

If we use the abbreviation: $G(v)=P(1,v)+v\ Q(1,v)$, then equation (10) can be written:

$$\frac{dx}{x}+\frac{Q(1,y)}{G(v)}\ dv=0, \tag{11}$$

and its integral is

$$\log\ kx=-\int\frac{Q(1,v)}{G(v)}\ dv. \tag{12}$$

PROBLEMS

Solve the following equations:

1. $2xy^3\dfrac{dy}{dx}=4x^4-x^2y^2+2y^4.$ 2. $\dfrac{dy}{dx}=\dfrac{y+\sqrt{y^2-x^2}}{x}.$

3. $\dfrac{dy}{dx}=\dfrac{2x^2+y^2}{-2xy+3y^2}.$ 4. $(x^2y-y^3)dx+(xy^2-x^2)dy=0.$

5. Prove Euler's theorem by taking derivatives with respect to λ of both sides of equation (1).

6. If $u=f(a_1,a_2,\ldots,a_n)$ is a linear homogeneous function, that is, $n=1$, prove that

$$\frac{\partial^2u}{\partial a_1{}^2}=-\frac{1}{a_1}\left[a_2\frac{\partial^2u}{\partial a_1\partial a_2}+a_3\frac{\partial^2u}{\partial a_1\partial a_3}+\cdots+a_n\frac{\partial^2u}{\partial a_1\partial a_n}\right].$$

7. Use the proposition of Problem 6 to compute the second partial derivative with respect to x of the function: $u=ax^k\ y^{1-k}.$

5. The Equation: $(Ax+By+E)dx=(Cx+Dy+F)\ dy$

As an application of the theory of the preceding section we shall consider the equation:

$$\frac{dy}{dx}=\frac{Ax+By+E}{Cx+Dy+F}, \tag{1}$$

where the coefficients are constants, subject only to the restriction that: $AD-BC\neq0.$

Equation (1) can be reduced to the form

$$\frac{dy}{dx}=\frac{Ax+By}{Cx+Dy}, \tag{2}$$

by means of the linear transformation:

$$x=x'+h,\ y=y'+k.$$

For if these values are substituted in (1) there results:

$$\frac{dy'}{dx'}=\frac{Ax'+By'+Ah+Bk+E}{Cx'+Dy'+Ch+Dk+F}.$$

Since we have assumed that $AD-BC\neq0$, values of h and k can be determined such that

$$Ah+Bk+E=0,\ Ch+Dk+F=0.$$

Equation (2) will prove to be of great interest to us in connection with the problems of nonlinear mechanics. For this reason we shall give a somewhat detailed description of it. In particular, we shall now describe two methods for its solution, the connection between the two methods, and some special examples.

First Method of Solution. Since the right hand member of equation (2) is a homogeneous equation of degree zero, we can apply the method of Section 4 to solve it. Thus, making the transformation: $y=vx$, and observing, in the notation of Section 4, that we have

$$G(v)=A+(B-C)v-Dv^2,$$

we obtain the solution of (2) in the form:

$$\log kx=\int\frac{C+Dv}{A+(B-C)v-Dv^2}\ dv.\tag{3}$$

Although this solution assumes the nonvanishing of $AD-BC$, this condition can be removed with little difficulty with respect to (1) in the following manner:

The condition $AD=BC$ implies the linear dependence of the terms $(Ax+By)$ and $(Cx+Dy)$ and we readily show that

$$Ax+By=\frac{A}{C}\ (Cx+Dy).$$

Hence, if we introduce the function

$$u=Ax+By,$$

we shall have from equation (1):

$$\frac{du}{dx}=A+B\frac{dy}{dx}=A+B\frac{Ax+By+E}{Cx+Dy+F},$$

$$=A+AB\frac{u+E}{Cu+AF}=\frac{A[(B+C)u+AF+BE]}{Cu+AF}.\tag{4}$$

The variables are thus separated and the equation is integrable.

Second Method of Solution. Introducing an auxiliary variable t, and using the abbreviations: $U=Cx+Dy+F$, $V=Ax+By+E$, we can write

$$\frac{dt}{t}=\frac{dx}{Cx+Dy+F}=\frac{dy}{Ax+By+E}=\frac{dx}{U}=\frac{dy}{V}=\frac{p\,dx+q\,dy}{pU+qV}, \tag{5}$$

where p and q are arbitrary constant parameters.

In order to integrate (5) we now seek to determine p and q so that (5) can be written in the form:

$$\frac{dt}{t}=\frac{p\,dx+q\,dy}{\lambda(px+qy)+r}. \tag{6}$$

For the determination of p and q, we then obtain the equation:

$$pC+qA=\lambda p, \tag{7}$$

$$pD+qB=\lambda q,$$

in terms of which we can then write

$$r=pF+qE. \tag{8}$$

Since (7) is a homogeneous system in p and q, nontrivial solutions will exist for values of λ which satisfy the following quadratic equation:

$$\begin{vmatrix} C-\lambda & A \\ D & B-\lambda \end{vmatrix} \equiv \lambda^2-(B+C)\lambda+BC-AD=0. \tag{9}$$

Let us denote by λ_1 and λ_2 the roots of (9), where λ_1 and λ_2 are assumed to be different from one another, and neither is zero, that is, $BC-AD\neq 0$. Let us also denote by p_1,q_1 and p_2,q_2 the values of p and q which correspond respectively to λ_1 and λ_2 and by r_1 and r_2 the values of r obtained by substituting in (8) the values of p and q.

Returning to (6) we now integrate the equation for each set of parameters and thus obtain:

$$t=K_1[\lambda_1(p_1x+q_1y)+r_1]^{1/\lambda_1},$$

$$t=K_2[\lambda_2(p_2x+q_2y)+r_2]^{1/\lambda_2}. \tag{10}$$

Equating these values of t and simplifying the resulting equation, we obtain the solution of equation (1) in the following form:

$$\lambda_1(p_1x+q_1y)+r_1=K[\lambda_2(p_2x+q_2y)+r_2]^{\lambda_1/\lambda_2}, \tag{11}$$

where K is an arbitrary constant.

If $BC-AD=0$, then one of the roots, let us say λ_1, is zero, and some modification is necessary. In this case equation (6) becomes

$$\frac{dt}{t}=\frac{1}{r}\ (p\ dx+q\ dy).$$

The first equation in (10) is thus replaced by

$$t=K_1e^{(p_1x+q_1y)/r_1}.$$

The resulting solution thus has the form:

$$p_1x+q_1y=(r_1/\lambda_2)\ \log\ [\lambda_2(p_2x+q_2y)+r_2]+K. \tag{12}$$

Relationship Between the Two Methods of Solution. As we shall see later the second method of solution has some advantage over the first in certain applications, since it introduces the characteristic equation (9). This advantage can be restored, however, by making in equation (3) the transformation:

$$w=C+Dv=\frac{Cx+Dy}{x}. \tag{13}$$

Equation (3) then becomes

$$\log kx=-\int\frac{w\ dw}{w^2-(B+C)w+BC-AD},$$

$$=-\int\frac{w\ dw}{(w-\lambda_1)(w-\lambda_2)}, \tag{14}$$

where λ_1 and λ_2 are the roots of equation (9).

Assuming that $BC-AD\neq0$, we first integrate the right-hand member of (14). By substituting the value of w as given by (13) in the resulting function, we obtain, after some simplification, the following function:

$$Cx+Dy-\lambda_2x=K(Cx+Dy-\lambda_1x)^{\lambda_1/\lambda_2}. \tag{15}$$

This function can be shown to be equivalent to (11) in which r_1 and r_2 are equated to zero. Thus, we see by (9) that $\lambda_1+\lambda_2=B+C$, and by the second equation in (7) that

$$p_1=\frac{\lambda_1-B}{D}\ q_1=\frac{C-\lambda_2}{D}\ q_1,\ \text{and}\ p_2=\frac{C-\lambda_1}{D}\ q_2.$$

When these values of p_1 and p_2 are substituted in (11) and the arbitrary constant K is replaced by

$$[D/(\lambda_2q_2)]^{\lambda_1/\lambda_2}(\lambda_1q_1/D)\ K,$$

equation (15) is obtained.

Special Cases.   It will be convenient in a later place in this book to have on record a few special cases for ready reference.   These are as follows:

CASE 1.—*The characteristic roots are pure imaginaries:* $\lambda_1 = \lambda i$, $\lambda_2 = -\lambda i$, $BC - AD > 0$.

Substituting these values of the roots of (9) in (11), and assuming that $r_1 = r_2 = 0$, which means that we are solving equation (2), we obtain after some simplification the following equation:

$$Ax^2 + 2Bxy - Dy^2 = K. \qquad (16)$$

The solution is thus a *conic section*.   But since $B = -C$, and $BC - AD = -B^2 - AD > 0$, the conic is found to be an *ellipse*.

CASE 2.—*The roots are real and equal:* $\lambda_1 = \lambda_2 = \lambda$.
In this case the coefficients of (1) satisfy the equation

$$(B - C)^2 = -4AD. \qquad (17)$$

If $B = C$, then either A or D must be zero, and if $B = -C$, then $BC - AD = 0$.   We shall first consider the solution when neither of these conditions hold.   In this case we obtain from (14)

$$\log kx = -\int \frac{w dw}{(w - \lambda)^2} = -\log (w - \lambda) + \lambda (w - \lambda)^{-1}. \qquad (18)$$

Replacing w by its value from (13), we get

$$\log [k(Cx + Dy - \lambda x)] = \lambda x / (Cx + Dy - \lambda x), \qquad (19)$$

where $\lambda = \frac{1}{2} (B + C)$.

If $B = C$ and $A = 0$, we obtain the solution most readily from (3).   It is found to be

$$Bx = Dy \log ky. \qquad (20)$$

Similarly, if $B = C$ and $D = 0$, the solution becomes

$$By = Ax \log kx. \qquad (21)$$

If $B = C$ and if both A and D are zero, equation (1) becomes

$$\frac{dy}{dx} = \frac{By + E}{Bx + F}, \qquad (22)$$

the solution of which is the *straight line*,

$$By + E = K(Bx + F). \qquad (23)$$

If $B=-C$ and $BC-AD=0$, the solution is reduced to equation (4), which can be written

$$A(AF+BE)dx=(-Bu+AF)du.$$

The solution is found to be

$$A(AF+BE)x=-\tfrac{1}{2}Bu^2+AFu+K,$$

which reduces to the *parabola*

$$(Ax+By)^2+2A(Ex-Fy)=K'. \tag{24}$$

CASE 3.—*The characteristic roots are complex numbers:* $\lambda_1=\lambda+\mu i$, $\lambda_2=\lambda-\mu i$.

It will be convenient in this case to use the form of the solution given by (15), which, however, will be written in the form

$$\lambda_2 \log\,(Cx+Dy-\lambda_2x)=\lambda_1 \log\,(Cx+Dy-\lambda_1x)+k, \tag{25}$$

where $k=\log K$.

Making use of the abbreviations:

$$u=Cx+Dy-\lambda x,\ v=\mu x,\ r^2=u^2+v^2,$$

we observe that we can write (25) as follows:

$$(\lambda-\mu i) \log\,(u+vi)=(\lambda+\mu i) \log\,(u-vi)+k,$$

$$(\lambda-\mu i)\left[\log r+i \arctan \frac{v}{u}\right]=(\lambda+\mu i)\left[\log r-i \arctan \frac{v}{u}\right]+k.$$

This equation readily reduces to the following:

$$2i\left(\lambda \arctan \frac{v}{u}-\mu \log r\right)=k.$$

Since k is arbitrary and may, in particular, contain $2i$ as a factor, we obtain finally the real solution:

$$\mu \log r=\lambda \arctan \frac{v}{u}+k. \tag{26}$$

The following examples will illustrate the theory given above:

Example 1. Solve the equation

$$\frac{dy}{dx}=\frac{2x+2y}{5x-y}. \tag{27}$$

Solution by the First Method. By means of the transformation: $y=vx$, equation (27) reduces to

$$x\,\frac{dv}{dx}=\frac{2-3v+v^2}{5-v},$$

from which we get by (3)

$$\log kx=\int\frac{5-v}{2-3v+v^2}\,dv=\int\left[\frac{-4}{v-1}+\frac{3}{v-2}\right]dv=\log\left[\frac{(v-2)^3}{(v-1)^4}\right].$$

Replacing v by y/x and simplifying, we obtain the solution in the form

$$(y-2x)^3=k(y-x)^4. \tag{28}$$

Solution by the Second Method. Writing equation (27) in the form

$$\frac{dy}{2x+2y}=\frac{dx}{5x-y},$$

we identify the coefficients: $A=2$, $B=2$, $C=5$, $D=-1$, and from them obtain equation (9): $\lambda^2-7\lambda+12=0$, with roots $\lambda_1=3$ and $\lambda_2=4$.

The first equation in (7) becomes: $5p+2q=\lambda p$. Assuming arbitrarily that $p_1=1$, and replacing λ by $\lambda_1=3$, we get $q_1=-1$. Similarly, assuming $p_2=2$ and replacing λ by $\lambda_2=4$, we find $q_2=-1$.

Substituting these values in equation (11), and observing that $r_1=r_2=0$, we obtain the solution in the form

$$3(x-y)=K[4(2x-y)]^{3/4}.$$

This equation can also be written

$$(y-2x)^3=k(y-x)^4,$$

in agreement with (28), if we write $k=-3^4/(K^4 4^3)$.

Example 2. Solve the equation

$$\frac{dx}{x-y}=\frac{dy}{x+y}. \tag{29}$$

Solution: In this case equation (9) becomes: $\lambda^2-2\lambda+2=0$, with roots respectively equal to $1+i$ and $1-i$. The equation is thus included under *Case 3* above, where $\lambda=\mu=1$. Since $u=-y$, $v=x$, we have $r^2=x^2+y^2$.

The solution as given by (26) thus reduces to

$$\log(x^2+y^2)=-2\arctan\frac{x}{y}+k. \tag{30}$$

Introducing polar coordinates, r and θ, where $x=r\cos\theta$, $y=r\sin\theta$, and observing that $x/y=\cot\theta=\tan\left(\frac{1}{2}\pi-\theta\right)$, we reduce (30) to the form

$$\log r^2 = -2\left(\frac{1}{2}\pi-\theta\right)+k=2\theta+k';$$

which can also be written:

$$r^2 = Ke^{2\theta}. \tag{31}$$

PROBLEMS

Solve the following equations:

1. $\dfrac{dy}{dx}=\dfrac{x+2y+3}{3x+2y+4}$.

2. $\dfrac{dy}{dx}=\dfrac{3x+6y+5}{2x-3y-2}$.

3. $\dfrac{dy}{dx}=\dfrac{10x-3y+3}{6x+7y+2}$.

4. $\dfrac{dy}{dx}=\dfrac{2x-y+10}{x+2y+12}$.

5. Show that the solution of the second illustrative example can be written

$$(x^2+y^2)\left(\frac{x+iy}{x-iy}\right)^i=\text{constant}.$$

6. Special Cases for Which the Integrating Factor Is Known

Returning to the general equation:

$$P(x,y)dx+Q(x,y)dy=0, \tag{1}$$

we shall consider certain special cases for which the integrating factor is known.

CASE 1.—The differential equation

$$f(y/x)dx-dy=0, \tag{2}$$

has as an integrating factor the function

$$\mu=1/[y-xf(y/x)]. \tag{3}$$

This is readily verified by direct substitution in the equation

$$\frac{\partial(\mu P)}{\partial y}=\frac{\partial(\mu Q)}{\partial x}, \tag{4}$$

but is more easily proved by observing that $P=f(y/x)$ and $Q=-1$ are homogeneous functions of degree 0. The integrating factor is then obtained from (6) of Section 4.

CASE 2.—The function

$$\mu=\frac{1}{x^2+y^2},\tag{5}$$

is an integrating factor for the equation

$$(y+xF)dx-(x-yF)dy=0,\tag{6}$$

provided $F=F(x^2+y^2)$.

The proof follows by direct substitution in (4).

CASE 3.—If P and Q are functions such that

$$\frac{1}{Q}\left(\frac{\partial P}{\partial y}-\frac{\partial Q}{\partial x}\right)=g(x)\tag{7}$$

where $g(x)$ is a function of x alone, then an integrating factor is furnished by the function

$$\mu=Ke^{\int g(x)dx}.\tag{8}$$

Proof: If μ is assumed to be a function of x alone, then equation (4) reduces to the following:

$$\frac{d\mu}{dx}=\mu\frac{\left(\dfrac{\partial P}{\partial y}-\dfrac{\partial Q}{\partial x}\right)}{Q}.\tag{9}$$

But μ cannot be a function of x alone unless the multiplier of μ in (9) is also a function of x alone, let us say, $g(x)$. In this case (8) is a solution of (9) and the theorem follows as a consequence.

CASE 4.—A similar result is obtained if we assume that

$$\left(\frac{\partial P}{\partial y}-\frac{\partial Q}{\partial x}\right)\Big/\left(Q\frac{\partial\phi}{\partial x}-P\frac{\partial\phi}{\partial y}\right)=g(\phi),\tag{10}$$

where $\phi=\phi(x,y)$ is an arbitrary function of x and y, subject only to the conditions (a) that its first partial derivatives exist and (b) that the denominator of the left-hand member of (10) does not vanish identically.

Under these assumptions the function

$$\mu=Ke^{\int g(\phi)d\phi}\tag{11}$$

is an integrating factor of (1).

Proof: If μ is an integrating factor, then the identity (4) must hold, which in this case reduces to the following equation:

$$\mu\frac{\partial P}{\partial y}+P\frac{d\mu}{d\phi}\frac{\partial\phi}{\partial y}=\mu\frac{\partial Q}{\partial x}+Q\frac{d\mu}{d\phi}\frac{\partial\phi}{\partial x}.$$

This equation can be written:

$$\frac{d\mu}{d\phi}\left[P\,\frac{\partial\phi}{\partial y}-Q\frac{\partial\phi}{\partial x}\right]=\mu\left[\frac{\partial Q}{\partial x}-\frac{\partial P}{\partial y}\right],$$

that is,

$$\frac{d\mu}{d\phi}=\mu g(\phi),$$

the solution of which is (11).

CASE 5.—(Goursat).   The differential equation

$$y(a+\alpha x^m y^n)dx+x(b+\beta x^m y^n)\ dy=0, \tag{12}$$

has as an integrating factor

$$\mu=x^p y^q, \tag{13}$$

provided $(b\alpha-\beta a)\neq 0$.   The values of p and q are determined from the equations

$$bp-aq=a-b,\ \beta p-\alpha q=\alpha(n+1)-\beta(m+1). \tag{14}$$

The proof is immediate if we substitute

$$P=ax^p y^{q+1}+\alpha x^{m+p}y^{n+q+1},\ Q=bx^{p+1}y^q+\beta x^{m+p+1}y^{n+q},$$

in (4) and equate to zero the coefficients of the terms $x^p y^q$ and $x^{m+p}\,y^{n+q}$.

The case where $b\alpha-\beta a=0$ is trivial, since (12) then reduces to

$$\alpha ydx+\beta xdy=0.$$

CASE 6.—If $P=yp(xy)$ and $Q=xq(xy)$, then the function

$$\mu=1/(xP-yQ) \tag{15}$$

is an integrating factor of (1).

Proof: Substituting μP and μQ in (4), we obtain after simplification the following equation which must be identically satisfied:

$$P(xQ_x-yQ_v-Q)=Q(xP_x-yP_v+P), \tag{16}$$

where P_x, P_v, Q_x, Q_v are the partial derivatives of P and Q with respect to x and y.

When the following derivatives:

$$P_x=y^2 p',\ P_v=xyp',\ Q_x=xyq'+q,\ Q_v=x^2 q',$$

are substituted in (16), each parenthesis is reduced to zero.

CASE 7.—If P and Q satisfy the following equations:

$$P_x = Q_y, \; P_y = -Q_x,$$ (17)

then an integrating factor of (1) is the function:

$$\mu = (P^2 + Q^2)^{-1}.$$ (18)

The proof follows without difficulty by appropriate substitution in equation (4). Of more interest is the observation that (17) are the Cauchy-Riemann conditions which are satisfied by P and Q defined as the components of an analytic function of a complex variable, that is,

$$f(x+iy) = P(x,y) + iQ(x,y).$$ (19)

Both P and Q are solutions of Laplace's equation:

$$\frac{\partial^2 u}{\partial x^2} + \frac{\partial^2 u}{\partial y^2} = 0.$$ (20)

PROBLEMS

1. Solve the following equation:

$$(-3y + 2x^3 y^3) dx + (4x - 3x^4 y^2) dy = 0.$$

2. Show that $\mu = \mu(x^2 y^2)$ is an integrating factor of (1) provided P and Q satisfy the following:

$$xy\left(\frac{P_y - Q_x}{yQ - xP}\right) = \phi(x^2 y^2),$$

where $\phi(z)$ is an arbitrary function.

3. Find the solution of the following equation:

$$(\alpha x^2 + 2\beta xy + \gamma y^2 + \alpha x + \beta y) dx + (\beta x + \gamma y) dy = 0.$$

4. Determine the integrating factor for the equation:

$$\log (x^2 + y^2) dx + 2 \arctan \frac{y}{x} \, dy = 0.$$

Use this factor to obtain its solution.

5. Determine a condition similar to that given in Problem 2 which must be satisfied by P and Q in order that the equation

$$P \, dx + Q \, dy = 0$$

should have as an integrating factor: $\mu = \mu(x/y)$.

6. If P and Q are functions such that

$$P_y - Q_x = P f(y) - Q g(x),$$

show that equation (1) has as an integrating factor $\mu = u(x) \, v(y)$, where u and v are solutions of the following equations:

$$u' + gu = 0, \; v' + fv = 0.$$

7. Some Particular Differential Equations

There exist a number of special cases of differential equations, which are of interest either because they can be integrated by simple devices, or because they throw light upon special aspects of the problem of integration. A few of these will be discussed below.

(a) *Bernoulli's Equation.* This equation, studied by Jacob Bernoulli (1654–1705) in 1695, has the form:

$$y' = f(x)y + g(x) \ y^{\alpha}. \tag{1}$$

If we make the transformation: $y = z^{\beta}$, this equation becomes

$$z' = \frac{1}{\beta} f(x) \ z + \frac{1}{\beta} \ g(x) \ z^{\beta(\alpha-1)+1} \tag{2}$$

Setting $\beta(\alpha-1)+1 = 0$, that is, $\beta = 1/(1-\alpha)$, we obtain the following linear equation:

$$z' = f(x) \ (1-\alpha)z + g(x) \ (1-\alpha), \tag{3}$$

which is integrated by a single quadrature.

(b) *Clairaut's Equation.* This equation, solved by A. C. Clairaut (1713–65), has the form:

$$y = xy' + f(y'), \tag{4}$$

where we shall assume that $f(y')$ has a first derivative.

Special interest attaches to Clairaut's equation, since it has both a *general solution* and a *singular solution*. The latter is not contained in the general solution. To understand the situation, let us take the derivative of (4) from which we obtain:

$$y' = xy'' + y' + f'(y') \ y''. \tag{5}$$

Since this equation can also be written

$$y''(x) \ [x + f'(y')] = 0, \tag{6}$$

it follows that solutions of (4) should be contained in the solutions of one or the other of the equations:

$$(A) \ y''(x) = 0; \ (B) \ x + f'(y') = 0. \tag{7}$$

The solution of equation (A) is clearly: $y = cx + d$, whence $y' = c$. Substituting this value in (4), we obtain what is called the *general solution* of Clairaut's equation, namely,

$$y = cx + f(c). \tag{8}$$

If, now, we substitute $y'=c$ in equation (B) of (7) and eliminate c between it and equation (8), we obtain what is called the *singular solution*.

The reason for this is readily found in the theory of envelopes of a one-parameter system of equations, which we can conveniently represent as follows:

$$F(x,y,c)=0. \tag{9}$$

If an envelope exists, then it is obtained by eliminating c between the equations:

$$F(x, y, c)=0, \quad \frac{\partial}{\partial c} F(x, y, c)=0. \tag{10}$$

The result of this elimination is called the *c-discriminant equation*.

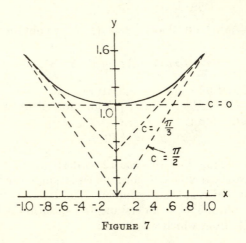

FIGURE 7

It is clear that equation (8) is a one-parameter family of non-parallel lines and that the envelope of the lines is obtained by eliminating c between (8) and the second equation of (10), which in this case reduces to

$$x+f'(c)=0. \tag{11}$$

But this equation is the same as (B) in (7) in which y' has been replaced by c.

The situation is illustrated by the following example:

Example. Discuss the solution of the equation

$$y=xy'+\cos y'. \tag{12}$$

Solution: The general solution is the family of lines:

$$y=cx+\cos c. \tag{13}$$

The singular solution is found to be

$$y = x \arcsin x + \cos (\arcsin x),$$

$$= x \arcsin x + \sqrt{1-x^2}. \tag{14}$$

That the singular solution is the envelope of the family of lines is immediately seen from Figure 7.

An interesting generalization of Clairaut's equation is furnished by the following system:

$$y = xy' + f(y',z'), \; z = xz' + g(y',z'), \tag{15}$$

where f and g are assumed to be functions with first derivatives in both variables.

If we take derivatives of equations (15), we obtain the following system:

$$(x+f_{y'})y'' + f_{z'}z'' = 0,$$
$$g_{z'}y'' + (x+g_{z'})z'' = 0, \tag{16}$$

where $f_{y'}$, $f_{z'}$, $g_{y'}$, and $g_{z'}$ denote partial derivatives of f and g with respect to y' and z'.

It is clear that equations (16) are satisfied if $y'' = z'' = 0$, whence $y' = a$ and $z' = b$. We thus obtain as the general solution of (15) the following two families of lines:

$$y = ax + f(a,b), \; z = bx + g(a,b). \tag{17}$$

But it is also seen that nonzero values of y'' and z'' will satisfy (16) provided the following equation is satisfied identically by y' and z':

$$(x+f_{y'})(x+g_{z'}) - f_{z'}g_{y'} = 0. \tag{18}$$

If y' and z' can be eliminated between (15) and (18) the resulting equation, let us say, $F(x,y,z) = 0$, is that of a surface to which the lines (17) are tangent.

(c) *Chrystal's Equation.* The equation

$$\left(\frac{dy}{dx}\right)^2 + Ax\frac{dy}{dx} + By + Cx^2 = 0, \tag{19}$$

was discussed in some detail by G. Chrystal (1851–1911) in 1896. Like Clairaut's equation, under certain conditions it may have a singular solution.

We first solve (19) for y' and thus obtain:

$$y' = -\frac{1}{2}Ax \pm \frac{1}{2}(A^2x^2 - By - 4Cx^2)^{1/2}. \tag{20}$$

By means of the transformation:

$$4By = (A^2 - 4C - z^2)x^2, \tag{21}$$

equation (20) reduces to the following form:

$$xzz' = A^2 + AB - 4C \pm Bz - z^2, \tag{22}$$

which can be written

$$(A^2 + AB - 4C \pm Bz - z^2)^{-1} z \, dz = dx/x. \tag{23}$$

If a and b are the roots of the equation

$$z^2 \mp Bz + 4C - AB - A^2 = 0, \tag{24}$$

that is, a, $b = \pm \frac{1}{2} B \pm \frac{1}{2} Q$, where $Q^2 = (2A+B)^2 - 16C$, then (23) can be written

$$\frac{z \, dz}{(z-a)(z-b)} = -\frac{dx}{x}. \tag{25}$$

If $a \neq b$, the solution of (25) is found to be

$$x(z-a)^m (z-b)^n = c, \quad m = a/(a-b), \quad n = -b/(a-b), \tag{26}$$

where c is an arbitrary constant.

If $a = b$, that is, if $Q = 0$, equation (26) is replaced by

$$x(z-a) \, \exp[a/(a-z)] = c. \tag{27}$$

Thus the solution is no longer algebraic but transcendental. Hence the analytical character of the solution depends upon Q. If, for example, Q is a rational number, then x and y are connected by a rational function.

A parabolic solution is obtained if one of the roots is zero, which, by (24), is the case when we have

$$A^2 + AB - 4C = 0, \tag{28}$$

or what is the same thing: $Q = B$.

In this case we have

$$x(z \pm B) = c, \tag{29}$$

from which we obtain as the solution of (19) the parabola:

$$4By = -ABx^2 - (c \pm Bx)^2. \tag{30}$$

If we compute the c-discriminant corresponding to this family of parabolas we find that its envelope is the parabola

$$4By = -ABx^2. \tag{31}$$

If we now substitute this function in the original equation (19) to see whether or not it is a singular solution, we find that it is a solution provided the coefficents of the equation satisfy (28).

That this should be the case is not surprising, however, for it will be observed that when the coefficients of (19) satisfy (28), equation (22) reduces to the Clairaut form.

PROBLEMS

1. Show that the singular solution of $y = xy' + 1/y'$ is a parabola.

2. Find a curve such that the product of the distances from two fixed points F and F' to any of its tangents is always equal to a constant b^2. *Hint:* Let the two points be $(c,0)$ and $(-c,0)$ and $P(x,y)$ the point on the curve through which the tangent passes. Show that the conditions of the problem lead to the differential equation:

$$(y - xy')^2 = b^2 + a^2 y'^2, \ a^2 = b^2 + c^2.$$

Hence, show that the solution is an ellipse with semi-axes equal respectively to a and b.

3. Find a curve such that the coordinate axes cut off from any tangent a constant length a. Show that the problem leads to the differential equation

$$(1 + y'^2)(y - xy')^2 = a^2 y'^2,$$

and that the singular solution is the *astroid:*

$$x^{2/3} + y^{2/3} = a^{2/3}.$$

4. The equation

$$y = xf(p) + g(p),$$

where $p = y'$ is called *d'Alembert's equation*. Show that its solution in parametric form is obtained by combining it with the solution of the following equation:

$$\frac{dx}{dp} + \frac{xf'(p) + g'(p)}{f(p) - p} = 0.$$

Hence, solve the equation: $y = (1 + y')x + y'^2$.

8. Singular Solutions

The problem of the existence of singular solutions of differential equations of first order has been the subject of numerous investigations since Clairaut exhibited the first example. Both A. Cayley and J. G. Darboux contributed independent papers on the problem in 1872–73. Extensive memoirs were published by W. P. Workman in 1887 and by M. J. M. Hill in 1888 and 1918. G. Chrystal investigated the problem in 1897 in a paper which contained the equation discussed in the preceding section. A. R. Forsyth, E. Goursat, and E. Picard have all given extensive attention to the problem in their respective treatises on differential equations. An especially lucid treatment of singular solutions will be found in the treatise of E. L. Ince.*

Since there exist a number of convenient sources where the reader may find an adequate discussion of the problem, we shall limit our treatment here to a review of a few salient features of the subject.

Let us consider a differential equation of first order in the form

$$f(x,y,p)=0, \tag{1}$$

where $p=dy/dx$. Let us first designate by $f_p(x,y,p)$ the partial derivative of (1) with respect to p.

Then the equation

$$f_p(x,y,p)=0, \tag{2}$$

together with (1) forms a system from which p, in many cases, can be eliminated. The resulting equation, which we shall denote by

$$g(x,y)=0, \tag{3}$$

defines a curve which is called the *p-discriminant locus.*

This function does not necessarily furnish a solution of (1) since it may contain loci of singular points called *tac-loci, cusp-loci,* and *nodal-loci,* which will not satisfy the equation. Examples of these loci are shown in Figure 8.

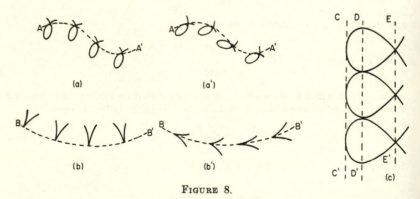

(a)   (a')

(b)   (b')   (c)

FIGURE 8.

If a singular solution, namely, the envelope of the integral curves, exists it will be found in both the p-discriminant and the c-discriminant loci.

If we now differentiate (1) with respect to x, we obtain the equation

$$f_x+f_y\frac{dy}{dx}+f_p\frac{dp}{dx}=0,$$

*For the specific references to these works see the *Bibliography.*

which, by virtue of (2), reduces to

$$f_x(x,y,p) + pf_y(x,y,p) = 0. \tag{4}$$

The fundamental criteria for the existence of a singular solution can now be stated as follows:

(a) *A necessary condition for the existence of a singular solution of equation* (1) *is that the three equations*:

$$f = 0, f_p = 0, f_x + pf_y = 0, \tag{5}$$

shall be simultaneously satisfied by a continuous function of x and y.

(b) *A sufficient condition for the existence of a singular solution is that, in addition to satisfying equations* (5), *the function of x and y must be such that*

$$f_y \neq 0. \tag{6}$$

Thus, referring to Chrystal's equation [(19), Section 7], the singular solution must satisfy the following three equations:

$$p^2 + Apx + By + Cx^2 = 0, \; 2p + Ax = 0, \; Ap + 2Cx + Bp = 0. \tag{7}$$

From the first two equations we get

$$4By = (A^2 - 4C)x^2, \tag{8}$$

and the third equation is satisfied only if $A^2 + AB = 4C$, which reduces (8) to $4By = ABx^2$. The condition, $f_y \neq 0$, introduces the assumption that $B \neq 0$. If this is not the case, then the solution of the original equation degenerates into the parabola:

$$y = -\frac{1}{4}Ax^2 + c,$$

and there is no singular solution.

If we refer to Figure 8, we find the representation of three types of loci. In figure (a) we observe a nodal locus, where the nodes form a series of singular points along the integral curve AA'. Since the nodal loops are not tangent to AA', they do not form an envelope and thus the nodal locus does not provide a singular solution of the equation. But in (a') the situation is different. The nodal loops are now tangent to AA' and thus, since they form an envelope, the nodal locus is a singular solution.

The same situation is shown in (b) and (b'), where in the one case the cusps are not tangent to the integral curve BB', but in the second case are tangent to it. Since the cusps in (b') form an envelope the cusp-locus provides a singular solution for the equation.

A tac-locus DD' is shown in (c) together with an envelope CC' and a nodal locus EE'. The tac-locus, when it exists, may be found in the p-discriminant. A necessary condition for this, but not a sufficient one, is that the following four equations should be simultaneously satisfied:

$$f(x,y,p)=0,\ f_p=0,\ f_x=0,\ f_y=0. \qquad (9)$$

A classical example illustrating this situation is provided by the equation:

$$A^2xp^2=(Bx-a)^2,$$

the general solution of which is readily found to be

$$9A^2(y+c)^2=4x(Bx-3a)^2.$$

The p-discriminant and the c-discriminant are respectively the following:

$$x(Bx-a)^2=0,$$

$$x(Bx-3a)^2=0.$$

Three loci are now observed, namely,

$$x=0,\quad Bx-a=0,\quad Bx-3a=0,$$

the first of which is common to both discriminants. This is the envelope of the general solution and, noting that p is infinite, we see that it is a singular solution of the equation. The second equation is that of the tac-locus and the third that of the nodal-locus.

The obvious intricacies of the problem of finding singular solutions and extraneous loci of singular points led E. B. Wilson to make the following general comment:* "Many authors use a great deal of time and space discussing just what may and what may not occur among the extraneous loci and how many times it may occur. The result is a considerable number of statements which in their details are either grossly incomplete or glaringly false or both. The rules here given for finding singular solutions should not be regarded in any other light than as leading to some expressions which are to be examined, the best way one can, to find out whether or not they are singular solutions."

---

*Advanced Calculus, New York, 1912, p. 233.

Chapter 3

The Riccati Equation

1. The Riccati Equation

IN THE FIRST CHAPTER it was shown that the elimination of the arbitrary constant k from the function

$$y = \frac{a(x) + kb(x)}{c(x) + kd(x)},$$

leads to a nonlinear differential equation of the following form:

$$\frac{dy}{dx} + Q(x)y + R(x)y^2 = P(x). \tag{1}$$

This is called a *Riccati equation*, named after Jacopo Francesco, Count Riccati (1676–1754), who published what is equivalent to the following form in 1724:[*]

$$\frac{dy}{dx} + ay^2 = bx^n. \tag{2}$$

This special equation is frequently referred to as Riccati's equation and (1) is then called the generalized Riccati equation. We shall not recognize this distinction, however, since we wish to refer to a second order equation as the generalization of (1). Riccati does not appear to have contributed essentially to the solution of his equation.

The particular case

$$\frac{dy}{dx} + y^2 + x^2, \tag{3}$$

was first considered by John Bernoulli (1667–1748) as early as 1694, but he confessed his inability to solve it.[†] Some 9 years later in 1703 his brother James Bernoulli (1654–1705) obtained a solution in the form:

$$y = \frac{F(x)}{G(x)},$$

---

[*] *Acta Eruditorum*, Suppl. viii, 1724, pp. 66–73.
[†] In a letter to Leibniz. See Leibniz: *Gesammelte Werke*, 1855, Vol. 3, pp. 50–87.

where $F(x)$ and $G(x)$ are power series in x. Dividing $F(x)$ by $G(x)$ he succeeded in obtaining the following formal solution:*

$$y=\frac{1}{3}x^3+\frac{1}{3^2\cdot7}x^7+\frac{2}{3^3\cdot7\cdot11}x^{11}+\frac{13}{3^4\cdot5\cdot7^2\cdot11}x^{15}+\ \cdots \tag{4}$$

Daniel Bernoulli (1700–1782) made the first essential contribution by publishing in 1725 a solution of (2) for those values of n for which a solution can be obtained in finite terms.†

Stimulated by these early researches, mathematicians turned with much interest to further study of this equation. Its intimate connection with the general linear homogeneous differential equation of second order gave it importance. It was found to have an application in the theory of Bessel functions and it made its appearance in a surprising number of problems. Among the important mathematicians who have studied it we find the names of L. Euler, A. Cayley, C. J. Hargreave, J. Liouville, L. Schafli, J. W. L. Glaisher, A. G. Greenhill, and numerous others. Its bibliography is extensive and reference to a number of these contributions will be found in the *Bibliography* at the end of this volume.

Returning to equation (1) we shall assume that $P(x)$, $Q(x)$ and $R(x)$ are given functions defined, together with their derivatives, within a region D. We shall also assume that $P(x)$ is not identically zero, for in this case we obtain the null equation:

$$\frac{dy}{dx}+Q(x)y+R(x)y^2=0, \tag{5}$$

which is immediately reduced to linear form by means of the transformation:

$$y=\frac{1}{v}\cdot \tag{6}$$

We thus obtain the equation:

$$\frac{dv}{dx}-Q(x)v=R(x), \tag{7}$$

which has the general solution:

$$v=Ce^{\int Qdx}+e^{\int Qdx}\int^x R(t)e^{-\int Qdt}\,dt. \tag{8}$$

---

*See Leibniz: *Gesammelte Werke*, Vol. 3, p. 75.
† *Acta Eruditorum*, 1725, pp. 465-473.

2. Relationship Between the Riccati Equation and the Linear Differential Equation of Second Order

The importance of the Riccati equation in the theory of differential equations is due in part to the following relationship between it and the general linear differential equation of second order.

Thus, given the general Riccati equation,

$$\frac{dy}{dx}+Q(x)y+R(x)y^2=P(x), \tag{1}$$

let us make the transformation:

$$y=\frac{1}{Ru}\frac{du}{dx}=\frac{u'}{Ru}. \tag{2}$$

The resulting equation is the following linear differential equation of second order:

$$R\frac{d^2u}{dx^2}-(R'-QR)\frac{du}{dx}-PR^2u=0. \tag{3}$$

Conversely, there corresponds to the general homogeneous linear differential equation of second order a Riccati equation. Thus, given the equation

$$A(x)\frac{d^2u}{dx^2}+B(x)\frac{du}{dx}+C(x)u=0, \tag{4}$$

we make the transformation:

$$\frac{du}{dx}=(Ry)u, \tag{5}$$

from which we obtain the following Riccati equation:

$$\frac{dy}{dx}+\left(\frac{R'}{R}+\frac{B}{A}\right)y+Ry^2=-\frac{C}{AR}. \tag{6}$$

Comparing this with equation (1) we can then write:

$$Q(x)=\frac{R'(x)}{R(x)}+\frac{B(x)}{A(x)}, \quad P(x)=-\frac{C(x)}{A(x)R(x)}. \tag{7}$$

Since $R(x)$ is an arbitrary function we can determine it so that $Q(x)$ is zero. In this case (6) assumes the simpler form:

$$\frac{dy}{dx}+Ry^2=-\frac{C}{AR}, \tag{8}$$

where $R=\exp\left[-\int (B/A)\,dx\right]$.

3. The Motion of a Body Which Falls in a Medium Where the Resistance Varies as the Square of the Velocity of the Body

A simple example of the application of the Riccati equation is provided by the problem of a body of mass m which falls under gravity in a medium which offers resistance to its motion proportional to the square of the velocity of the body.

If s is the distance passed over by the body in time t, then the motion is defined by the equation

$$m \frac{d^2s}{dt^2} = mg - K \left(\frac{ds}{dt}\right)^2, \tag{1}$$

which can be written

$$m \frac{dv}{dt} = mg - Kv^2, \tag{2}$$

where v is the velocity.

This is a Riccati equation, the solution of which is readily found to be

$$v = \frac{rm}{K} \left(\frac{e^{rt} - ke^{-rt}}{e^{rt} + ke^{-rt}}\right), \tag{3}$$

where we use the abbreviation: $r^2 = Kg/m$.

Let us observe that as t becomes infinite the velocity v approaches the limiting value

$$V = \sqrt{\frac{mg}{K}}, \tag{4}$$

from which we have: $V = g/r = rm/K$. Therefore, we can write (3) in the form

$$v = V \left(\frac{e^{rt} - ke^{-rt}}{e^{rt} + ke^{-rt}}\right). \tag{5}$$

If $v = v_0$ when $t = 0$, and if we use the abbreviation: $u_0 = v_0/V$, then k in (5) has the value: $k = (1 - u_0)/(1 + u_0)$.

Introducing this quantity into (5), we see that v can be written in the form

$$v = V \left(\frac{u_0 + \tanh rt}{1 + u_0 \tanh rt}\right). \tag{6}$$

Assuming finally that $s = 0$ when $t = 0$, we can now determine s as follows:

$$s = \int_0^t v(t)dt = \frac{V}{r} \log(\cosh rt + u_0 \sinh rt),$$

which, when r is replaced by g/V and u_0 by v_0/V, becomes

$$s = \frac{V^2}{g} \log \left(\cosh \frac{gt}{V} + \frac{v_0}{V} \sinh \frac{gt}{V}\right). \tag{7}$$

It is a matter of some interest to test this formula against actual fall in the atmosphere of the earth where the resistance appears to follow rather closely the law assumed in deriving it. The computations given below are based upon the free fall of six men from altitudes varying from 10,600 to 31,400 feet to a terminal altitude of approximately 2,100 feet. The average weight of the men and their equipment was 261.2 pounds.*

In a previous study a terminal velocity as low as 164 feet per second ($=112$ miles per hour) had been reported, but this was for a low altitude fall. Other studies indicated velocities of the order of 202 feet per second ($=138$ m.p.h.) and higher. The long drop from 31,400 to 2,100 feet on which the following analysis is made showed an average velocity of 251.4 feet per second ($=171$ m.p.h.). The statisical analysis of the drop was made by Dr. R. A. Fisher of the Physics Department of Northwestern University. The authors of the study stated that four factors are involved in free fall, namely, (a) the altitude, or air density, (b) the weight of the body, (c) the position of the body, and (d) the amount of spinning and tumbling. Of these probably the air-density factor is the most important, since the average velocity appears to increase with elevations.

The following analysis is based upon the single observed fact that the fall from an altitude of 31,400 to 2,100 feet was accomplished in 116 seconds. The computation is made from formula (7). To find the value of V corresponding to a drop of $s=29,300$ feet in $t=116$ seconds, we set $v_0=0$ and $gt/V=x$ in equation (7) and write the equation as follows:

$$s=\frac{V^2}{g}\log\frac{1}{2}\left(e^x+e^{-x}\right)=\frac{V^2}{g}\left(x-\log 2+e^{-2x}-\frac{1}{2}e^{-4x}+\frac{1}{3}e^{-6x}+\ldots\right). \quad (8)$$

Since x will have a value of the order of 15, it is clear that this equation can be replaced by the approximate one

$$s=Vt-(\log 2)V^2/g. \quad (9)$$

When this is solved for V for $s=29,300$, $t=116$, $g=32.2$, we find the value $V=265.7$ feet per second, which is somewhat larger than the average value of 251.4 given above.

Introducing this value into equation (8), we have

$$s=2192.4\log\cosh(0.12119t)\sim2192.4\ (x-0.69315), \quad (10)$$

where $x=0.12119t$. From this equation the following table of values of $H-s$, $H=31,400$, is computed and compared with those estimated from the barograph carried by the jumper. They are graphically represented in Figure 1.

*A. J. Carlson, A. C. Ivy, L. R. Krasno, and A .H. Andrews: "The Physiology of Free Fall Through the Air: Delayed Parachute Jumps," *Quarterly Bulletin*, Northwestern University Medical School, Vol. 16, 1942, p. 254.

COMPARISON OF OBSERVED FREE FALL WITH ESTIMATES FROM FORMULA

t	$H-s$ (Obs.)	$H-s$ (Comp.)	t	$H-s$ (Obs.)	$H-s$ (Comp.)	t	$H-s$ (Obs.)	$H-s$ (Comp.)
0	31,400	31,400	46.7	20,550	20,511	88.1	9,400	9,512
9.9	30,780	30,099	51.3	19,400	19,289	92.7	8,140	8,289
14.5	30,200	29,003	55.9	18,100	18,067	97.3	7,080	7,067
19.1	28,850	27,823	60.5	16,600	16,845	101.9	5,700	5,845
23.7	27,700	26,616	65.1	15,150	15,623	106.5	4,450	4,623
28.3	26,150	25,397	69.7	14,070	14,401	111.1	3,170	3,401
32.9	24,600	24,146	74.3	12,800	13,178	116.0	2,100	2,100
37.5	23,200	22,955	78.9	11,650	11,956			
42.1	21,750	21,734	83.5	10,400	10,734			

The value of K in equation (2) can be estimated from (4). The average weight of the jumpers with their equipment was 261.9 pounds. Hence, setting $m=261.2$, $g=32.2$, and $V=265.7$, we find $K=0.1191$ in foot-pound units.

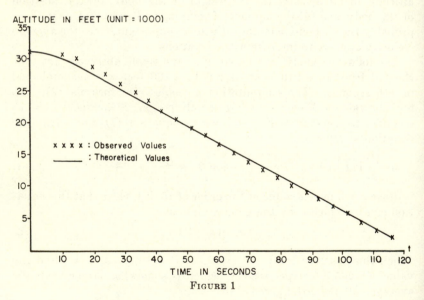

FIGURE 1

4. The Cross-Ratio Theorem for the Riccati Equation

In certain applications, notably in differential geometry, the Riccati equation has been found to have usefulness because of what is called its cross-ratio property.

FIGURE 2.

By the *cross-ratio* of four points on a line (A, B, C, D in Figure 2), the coordinates of which we denote respectively by x_1, x_2, x_3, and x_4, we mean the following ratio:

$$R = \frac{AC}{CB} \bigg/ \frac{AD}{DB} = \frac{(AC)(DB)}{(CB)(AD)} = \frac{(x_1-x_3)(x_2-x_4)}{(x_1-x_4)(x_2-x_3)}. \tag{1}$$

We have seen that the general solution of the Riccati equation can be written

$$y = \frac{g_1 + k\, g_2}{g_3 + k\, g_4}, \tag{2}$$

where the g_i are functions of x, and k is an arbitrary constant. We now consider four particular solutions, let us say, y_1, y_2, y_3, y_4, which are linearly independent of one another, and which are obtained from (2) by letting k assume the four values: k_1, k_2, k_3, k_4. Denoting by G_i the function $g_3 + k_i g_4$, we compute

$$y_i - y_j = \frac{g_1 + k_i\, g_2}{g_3 + k_i\, g_4} - \frac{g_1 + k_j\, g_2}{g_3 + k_j\, g_4},$$

$$= \frac{(k_i - k_j)(g_2\, g_3 - g_1\, g_4)}{G_i\, G_j}. \tag{3}$$

Making use of (3) we now evaluate the cross-ratio R and thus obtain:

$$R = \frac{(y_1 - y_3)(y_2 - y_4)}{(y_1 - y_4)(y_2 - y_3)} = \frac{(k_1 - k_3)(k_2 - k_4)}{(k_1 - k_4)(k_2 - k_3)}, \tag{4}$$

which is thus found to be a constant.

It is from this fact that we derive the cross-ratio theorem: *The cross ratio of any four linearly independent solutions of the Riccati equation is a constant.*

5. Integration of the Riccati Equation

As we have seen in Section 2, the complete solution of a Riccati equation is attained in general only by the integration of a linear differential equation of second order, or by some equivalent algorithm. However, if any particular solution is known, then a great simplification results, for it is then possible to obtain the complete solution by means of quadratures.

To show this, let us consider the equation

$$\frac{dy}{dx} + Q(x)\, y + R(x)\, y^2 = P(x), \tag{1}$$

and let us assume that $y = U$ is a particular integral. We now write

$$y = \frac{1}{u} + U. \tag{2}$$

When this function is substituted in (1), we get

$$\frac{1}{u^2}\left(-\frac{du}{dx} + Qu + R + 2URu\right) + \frac{dU}{dx} + QU + RU^2 = P,$$

which, since U is a solution of (1), reduces to the linear equation

$$\frac{du}{dx} - (2RU + Q)u = R. \tag{3}$$

Since this equation is linear and of first order, it can be solved by two quadratures. Its general solution can be written: $u = ku_0 + u_1$, where k is an arbitrary constant. Therefore, the general solution of (1) can be written

$$y = \frac{1 + u_1U + k\, u_0U}{u_1 + k\, u_0}. \tag{4}$$

Similarly, if two particular solutions are known, the general solution of (1) can be obtained from a single quadrature. To show this, let us assume that the second particular solution of (1) is V, and let us write, as above,

$$y = \frac{1}{v} + V. \tag{5}$$

For the determination of v, we now have the equation

$$\frac{dv}{dx} - (2RV + Q)\, v = R. \tag{6}$$

Multiplying (3) by v and (6) by u, subtracting, and dividing the resulting equation by uv, we get

$$\frac{1}{uv}\left(v\frac{du}{dx} - u\frac{dv}{dx}\right) - 2R\,(U - V) = R\left(\frac{1}{u} - \frac{1}{v}\right). \tag{7}$$

But since we have $1/u = y - U$ and $1/v = y - V$, so that $1/u - 1/v = V - U$, the right-hand member of (7) reduces to $R(V - U)$. Finally, multiplying (7) by u/v, we see that it can be written in the form

$$\frac{d}{dx}\left(\frac{u}{v}\right) - R(U - V)\frac{u}{v} = 0. \tag{8}$$

This equation can be solved by a single integration and we get

$$\frac{u}{v} = k\, e^{\int R(U - V)\, dx} = k\, \phi(x). \tag{9}$$

Since we thus have

$$\frac{u}{v}=\frac{y-V}{y-U}=k\ \phi(x), \tag{10}$$

we solve for y and thus obtain the desired solution in the following explicit form:

$$y=\frac{V-k\ \phi\ U}{1-\ k\ \phi}. \tag{11}$$

If three solutions: $y=U$, $y=V$, $y=W$ of equation (1) are known, then the general solution can be obtained without any quadrature. This follows from the cross-ratio theorem of Section 4. For if in equation (4) of that section, we replace y_1 by y and y_2, y_3, and y_4 respectively by U, V, and W, we get

$$\frac{(y-V)\ (U-W)}{(y-W)\ (U-V)}=k.$$

Solving for y, we then obtain the desired solution in the form

$$y=\frac{V(U-W)-kW(U-V)}{(U-W)-k(U-V)}. \tag{12}$$

6. Solution of the Original Riccati Equation

We shall now return to an examination of the original Riccati equation, which we stated in Section 1 as follows:

$$\frac{dy}{dx}+ay^2=bx^n. \tag{1}$$

This equation has been the subject of many investigations, since it has several unusual properties. Let us first make the transformation: $y=u'/(au)$. Equation (1) is then reduced to the following linear form:

$$\frac{d^2u}{dx^2}-c^2x^n\ u=0,\ \text{where}\ c^2=ab. \tag{2}$$

By means of a second transformation,

$$u=v\ e^{Ax^p} \tag{3}$$

equation (2) becomes

$$\frac{d^2v}{dx^2}+2Apx^{p-1}\frac{dv}{dx}+[Ap(p-1)x^{p-2}+A^2p^2x^{2(p-1)}-c^2x^n]v=0.$$

When A and p assume the following values:

$$A=c/p, \cdot p=1+\frac{1}{2}n, \ p\neq 0, \tag{4}$$

this equation reduces to the following:

$$\frac{d^2v}{dx^2}+2cx^{p-1}\frac{dv}{dx}+c(p-1)x^{p-2}v=0. \tag{5}$$

If we now assume a solution of the following form:

$$v=a_0+a_1\,x^p+a_2\,x^{2p}+a_3\,x^{3p}+\ldots+a_m\,x^{mp}+\ldots, \tag{6}$$

the following relationship between the coefficients a_{m+1} and a_m is readily obtained:

$$a_{m+1}=-\frac{c[(2m+1)p-1]}{(m+1)p\,[(m+1)p-1]}\,a_m. \tag{7}$$

If a_0 is set equal to 1 and m is then given successively the values 0, 1, 2, . . ., the following series results, which provides a formal solution of equation (5):

$$V(x,\,c)=1-\frac{p-1}{p(p-1)}\,cx^p+\frac{(p-1)\,(3p-1)}{p(p-1)2p(2p-1)}\,c^2x^{2p}$$
$$-\frac{(p-1)\,(3p-1)\,(5p-1)}{p(p-1)\,2p(2p-1)\,3p(3p-1)}\,c^3x^{3p}-\ldots \tag{8}$$

Thus, one formal solution of equation (2) can be written:

$$u_1=e^{cx^p/p}\,V(x,\,c). \tag{9}$$

But since it is possible to replace c by $-c$ in (2) without changing the equation, it is clear that a second solution is given by

$$u_2=e^{-cx^p/p}V(x,-c). \tag{10}$$

The solution of equation (1) thus assumes the form

$$y=\frac{u_1'+ku_2'}{a(u_1+ku_2)},$$

where k is an arbitrary constant.

We now observe from (7) that both $V(x,c)$ and $V(x,-c)$ terminate when p is the reciprocal of a positive odd integer, that is, when $p=1/(2m+1)$, $m=0$, 1, 2, etc. From this it follows that equation (1) has a solution expressed in terms of elementary functions, when n has the form:

$$n = \frac{-4m}{2m+1}, \ m = 0, 1, 2, 3, \ldots, \tag{11}$$

that is, for the sequence of values: 0, $-4/3$, $-8/5$, $-12/7$, etc.

But equation (1) also has a solution in terms of elementary functions when n has the form

$$n = \frac{-4m}{2m-1}, \ m = 1, 2, 3, \ldots, \tag{12}$$

that is, for the sequence: -4, $-8/3$, $-12/5$, $-16/7$, etc.

To obtain this second result, let us transform equation (2) by introducing the new variables:

$$u = w/t, \ x = 1/t,$$

from which we get

$$\frac{d^2w}{dt^2} - c^2 t^{-n-4} w = 0.$$

This equation has the same form as (2), and if we let $n = 2q - 2$, then expansion (8) is replaced by the series:

$$W(t, c) = 1 + \frac{q+1}{q(q+1)} \, ct^{-q} + \frac{(q+1)(3q+1)}{q(q+1)2q(2q+1)} \, c^2 t^{-2q} + \ldots.$$

This series terminates when q is the reciprocal of a negative odd integer, that is, when $q = -1/(2m-1)$. Substituting this value in the equation: $n = 2q - 2$, we obtain (12).

The importance of the original Riccati equation defined by (1) is found in its relationship to the theory of the Bessel functions, $J_\nu(x)$ and $Y_\nu(x)$, which are solutions of the equation:

$$L(U) \equiv x^2 \frac{d^2U}{dx^2} + x \frac{dU}{dx} + (x^2 - \nu^2)U = 0. \tag{13}$$

In order to show this connection between the two equations, we first employ the transformation: $t = (x/2)^2$, by means of which (13) becomes:

$$M(U) \equiv t^2 \frac{d^2U}{dt^2} + t \frac{dU}{dt} + \left[t - \left(\frac{1}{2} \nu \right)^2 \right] U = 0. \tag{14}$$

Transforming the dependent variable by writing: $U = t^\beta W$, we obtain

$$t^2 \frac{d^2W}{dt^2} + t(1 + 2\beta) \frac{dW}{dt} + \left\{ \left[\beta^2 - \left(\frac{1}{2} \nu \right)^2 \right] + t \right\} W = 0. \tag{15}$$

If β^2 is now set equal to $(\nu/2)^2$, that is, if $\beta = \pm \frac{1}{2}\nu$, then (15) reduces to the simpler form

$$N(W) \equiv t \frac{d^2W}{dt^2} + (1 \pm \nu) \frac{dW}{dt} + W = 0. \tag{16}$$

We can now write solutions of (13), (14), and (16) as follows:

$$L(U) = 0: \quad U = AJ_\nu(x) + BY_\nu(x); \tag{17}$$

$$M(U) = 0: \quad U = AJ_\nu(2\sqrt{t}) + BY_\nu(2\sqrt{t}); \tag{18}$$

$$N(W) = 0: \quad W = t^{-\beta}[AJ_\nu(2\sqrt{t}) + BY_\nu(2\sqrt{t})], \quad \beta = \pm \frac{1}{2}\nu. \tag{19}$$

Returning now to equation (1), we let $y = u/x$, and thus obtain

$$x \frac{du}{dx} - u + au^2 = bx^h, \, h = n+2. \tag{20}$$

The further transformation: $s = gx^h$, reduces (20) to the form:

$$\frac{du}{ds} - \frac{u}{hs} + \frac{au^2}{hs} = \frac{b}{gh}. \tag{21}$$

If u is now subjected to the transformation:

$$u = \frac{hs}{a} \frac{w'}{w}, \tag{22}$$

equation (21) becomes

$$s \frac{d^2w}{ds^2} + \left(1 - \frac{1}{h}\right) \frac{dw}{ds} - \frac{ab}{gh^2} w = 0.$$

Defining g so that $(-ab/gh^2) = 1$, that is, $g = -ab/h^2$, we then have

$$s \frac{d^2w}{ds^2} + \left(1 - \frac{1}{h}\right) \frac{dw}{ds} + w = 0. \tag{23}$$

If h is set equal to $1/\nu$, then the solution of (23), as shown by (19) above, is the following function:

$$w = s^{\nu/2} [AJ_\nu(2\sqrt{s}) + BY_\nu(2\sqrt{s})]. \tag{24}$$

When ν is not an integer, $J_\nu(x)$ and $J_{-\nu}(x)$ form a fundamental set of solutions of Bessel's equation, so in this case we can replace $Y_\nu(2\sqrt{s})$ by $J_{-\nu}(2\sqrt{s})$ in (24) and thus write the solution of (23) as follows:

$$w = s^{\nu/2} [AJ_\nu(2\sqrt{s}) + BJ_{-\nu}(2\sqrt{s})]. \tag{25}$$

It is well known that when ν is equal to either $\frac{1}{2}(2m-1)$ or $\frac{1}{2}(-2m+1)$, where m is a positive integer, then $J_\nu(x)$ can be expressed in finite form in terms of sines and cosines. A few of these functions are given below as follows:

$$J_{1/2}(x) = P(x)\sin x, \; P(x) = \sqrt{2/\pi x},$$

$$J_{3/2}(x) = P(x)\left[\frac{\sin x}{x} - \cos x\right],$$

$$J_{5/2}(x) = P(x)\left[\left(\frac{3}{x^2}-1\right)\sin x - \frac{3}{x}\sin x\right],$$

$$J_{-1/2}(x) = P(x)\cos x,$$

$$J_{-3/2}(x) = P(x)\left[-\sin x - \frac{\cos x}{x}\right],$$

$$J_{-5/2}(x) = P(x)\left[\frac{3}{x}\sin x + \left(\frac{3}{x^2}-1\right)\cos x\right].$$

Since $h = 1/\nu = n+2$, whence $\nu = 1/(n+2)$, and since $J_\nu(x)$ can be expressed in finite terms when $\nu = \frac{1}{2}(2m-1)$ or $\frac{1}{2}(-2m+1)$, it is clear that the solutions of equation (1) can be expressed in finite form when we have

$$n = \frac{4(1-m)}{2m-1}, \text{ or } n = \frac{4m}{1-2m}, \; m \text{ a positive integer.}$$

For the sequence of values: $m = 1, 2, 3$, etc., we obtain the same values of n given by (11) and (12).

As an example, let us consider the following equation:

$$\frac{dy}{dx} + y^2 = 1.$$

Referring to equation (1), we see that $a = b = 1$, $n = 0$. Hence $h = 2$ and $\nu = \frac{1}{2}$. Thus, equation (23) reduces to

$$s\frac{d^2w}{ds^2} + \left(1 - \frac{1}{2}\right)\frac{dw}{ds} + w = 0,$$

the solution of which is

$$w = As^{1/4}J_{1/2}(2\sqrt{s}) + Bs^{1/4}J_{-1/2}(2\sqrt{s}) = \frac{1}{\sqrt{\pi}}[A\sin(2\sqrt{s}) + B\cos(2\sqrt{s})].$$

From this we get, referring to (22),

$$u = 2s\frac{w'}{w} = \frac{2\sqrt{s}[A\cos(2\sqrt{s}) - B\sin(2\sqrt{s})]}{A\sin(2\sqrt{s}) + B\cos(2\sqrt{s})}.$$

Since s and x are connected by the equation: $4s=-x^2$, whence $2\sqrt{s}=\pm ix$, and since $\pm ix\cos(\pm ix)=\pm ix\cosh x$ and $\pm ix\sin(\pm ix)=-x\sinh x$, we have finally

$$y=\frac{u}{x}=\frac{A'\cosh x+B\sinh x}{A'\sinh x+B\cosh x}.$$

It should be observed, however, that our equation is readily solved without the introduction of Bessel functions, since the associated linear differential equation of second order is merely

$$u''-u=0,$$

which has the solution: $u=A\cosh x+B\sinh x$. But the intricacy of the relationship between the Riccati and the Bessel equations is instructively illustrated by this simple example.

PROBLEMS

Find the solutions of the following equations:

1. $\dfrac{dy}{dx}+y^2=x^{-8/5}$.

2. $\dfrac{dy}{dx}+y^2=x^{-4/3}$.

3. $\dfrac{dy}{dx}+y^2=x^{-4}$.

4. $\dfrac{dy}{dx}+y^2=x^{-8/3}$.

5. Show that equation (1) can be reduced to the following:

$$\frac{dz}{dt}+az^2=bt^{-n-4},$$

by means of the transformation: $y=t/a-zt^2$, $x=1/t$.

6. Show that the transformation: $y=1/z$, $x^{n+1}=(n+1)t$, reduces (1) to the form

$$\frac{dz}{dt}+bz^2=a(n+1)^m t^m, \quad m=-n/(n+1).$$

7. Observing that equation (1) can be solved in terms of elementary functions for $n=0$, make use of the results of Problems 5 and 6 to show that it is also solvable by elementary functions when $n=-4m/(2m\pm1)$, $m=1.\ 2.\ 3.$ etc.

7. Solution of the Riccati Equation by Means of Continued Fractions

An ingenious method of solving the equation

$$\frac{dy}{dx}+ay^2=bx^n, \tag{1}$$

has been devised by the use of the technique of continued fractions.

We first make the transformation

$$y = \frac{u}{x}, \tag{2}$$

which carries (1) into the following form:

$$x\frac{du}{dx} - u + au^2 = bx^p, \tag{3}$$

where $p = n + 2$.

We now make a second transformation as follows:

$$u = \frac{1}{a} + \frac{x^p}{u_1}, \tag{4}$$

by means of which (3) becomes

$$x\frac{du_1}{dx} - (1+p)u_1 + bu_1^2 = ax^p. \tag{5}$$

Following this with a third transformation,

$$u = \frac{1+p}{b} + \frac{x^p}{u_2}, \tag{6}$$

we obtain the equation:

$$x\frac{du_2}{dx} - (1+2p)u_2 + au_2^2 = bx^p. \tag{7}$$

Continuing this through m transformations, we have finally

$$x\frac{du_m}{dx} - (1+mp)u_m + A_m u_m^2 = B_m x^p. \tag{8}$$

where $A_m = a$ and $B_m = b$, when m is even, and where $A_m = b$ and $B_m = a$, when m is odd.

Combining the transformations, we see that the solution of equation (3), and thus, by means of (2), the solution of equation (1) can be written as the following continued fraction:

$$u = xy = \frac{1}{a} + \frac{x^p}{\dfrac{1+p}{b}} + \frac{x^p}{\dfrac{1+2p}{a}} + \frac{x^p}{\dfrac{1+3p}{b}} + \ \cdots \tag{9}$$

If, in equation (8), we set $p = -2/(2m-1)$, where m is an integer, then this equation reduces to the following:

$$x\frac{du_m}{dx} - \frac{1}{2}pu_m + A_m u_m^2 = B_m x^p. \tag{10}$$

But this equation is solved by a quadrature, since its variables are separable. For if we make the transformation: $u_m = x^{p/2}$, then it becomes

$$x^{1-p/2}\frac{dv}{dx} + A_m v^2 = B_m, \tag{11}$$

and the solution of (1) is obtained in finite terms in agreement with what we found in Section 6. The expansion of this solution as a continued fraction is given by (9).

A second solution of equation (1), also expressible as a continued fraction, is obtained by another sequence of transformations. These transformations, beginning with (3), and the equations which result from them are given below as follows:

$$u = \frac{x^p}{u_1}, \qquad x\frac{du_1}{dx} - (p-1)u_1 + bu_1^2 = ax^p,$$

$$u_1 = \frac{p-1}{b} + \frac{x^p}{u_2}, \qquad x\frac{du_2}{dx} - (2p-1)u_2 + au_2^2 = bx^p, \tag{12}$$

$$u_2 = \frac{2p-1}{a} + \frac{x^p}{u_3}, \qquad x\frac{du_3}{dx} - (3p-1)u_3 + bu_3^2 = ax^p,$$

$$\cdot \quad \cdot \quad \cdot \quad \cdot \quad \cdot \quad \cdot \quad \cdot \quad \cdot$$

After k such transformations, the resulting equation is the following:

$$x\frac{du_k}{dx} - (kp-1)u_k + A_k u_k^2 = B_k x^p, \tag{13}$$

where $A_k = a$ and $B_k = b$, when k is even, and where $A_k = b$ and $B_k = a$, when k is odd.

Combining these transformations we obtain the solution of equation (3), and thus, by means of (2), the solution of equation (1), as the following continued fraction:

$$u = xy = \cfrac{x^p}{\cfrac{p-1}{b} + \cfrac{x^p}{\cfrac{2p-1}{a} + \cfrac{x^p}{\cfrac{3p-1}{b} + \cdots}}} \tag{14}$$

As in the previous case, we set $p = 2/(2k-1)$, and equation (13) becomes

$$x\frac{du_k}{dx} - \frac{1}{2}pu_k + A_k u_k^2 = B_k x^p, \tag{15}$$

which can be integrated in finite terms as already explained above.

Since $n = p - 2 = -4(k-1)/(2k-1) = -4m/(2m+1)$, where $m = k-1$, we see that we have the sequence defined by (11) in Section 6.

8. The Character of the Singularities of the Riccati Equation

If one contrasts the solution of the equation: $y'+y^2=0$, with that of the equation: $y'+\frac{1}{2}\,y^3=0$, namely, $y=1/(x+k)$ in the first case with $y=(x+k)^{-\frac{1}{2}}$ in the second, he will observe that the solution of the first equation has a *movable pole*, while that of the second has a *movable branch point*. This difference in the character of the critical points in the two solutions is not an accidental one, but is associated with the fact that y appears as a square in one equation and as a cube in the second.

We shall now show that if an equation has the following form:

$$\frac{dy}{dx}=\frac{P(x,y)}{Q(x,y)}, \tag{1}$$

where $P(x,y)$ and $Q(x,y)$ are polynomials in y, then if its solution is to be free of movable branch points the equation is necessarily a Riccati.

If $P_0=(x_0,y_0)$ is a point such that

$$Q(x_0,y_0)=0,\ P(x_0,y_0)\neq0, \tag{2}$$

then P_0 is a singular point of equation (1).

On the other hand, P_0 is a regular point for the equation

$$\frac{dx}{dy}=\frac{Q(x,y)}{P(x,y)}, \tag{3}$$

in which x has now been advanced to the role of the dependent variable. Hence, in the neighborhood of P_0 the solution of (3) can be expanded as follows:

$$x-x_0=A_1(y-y_0)+A_2(y-y_0)^2+\ \cdot\ \cdot\ \cdot \tag{4}$$

But we see that $A_1=0$, since $Q(x_0,y_0)=0$, and thus the expansion becomes

$$x-x_0=\sum_{r=2}A_r(y-y_0)^r, \tag{5}$$

where some value of A_r, let us say when $r=n\geq2$, will be different from zero. If, then, we invert series (5) to obtain the solution of the original equation (1), we see that $y-y_0$ is expanded in a power series in $(x-x_0)^{1/n}$. We thus reach the conclusion that x_0 is an n-fold branch point. Moreover, since x_0 can be chosen arbitrarily, subject only to the existence of a y_0 which satisfies conditions (2), it can in particular move along an arbitrary curve C and is thus a movable branch point.

From the argument just given it is evident that if the solution of equation (1) is to be free from movable branch points, $Q(x,y)$ must be a function of x alone. That is to say, equation (1) must have the form

$$\frac{dy}{dx}=P_0(x)+P_1(x)\, y+P_2(x)\, y^2+\cdots+P_m(x)\, y^m. \tag{6}$$

That m in (6) cannot exceed 2 is readily shown if we make the transformation: $y=1/z$. We then obtain.

$$\frac{dz}{dx}=-z^2(P_0+P_1/z+P_2/z^2+\cdots+P_m/z^m). \tag{7}$$

The right hand member of this equation must necessarily be a polynomial in z, but this is not possible unless $P_r=0$ for all values of r which exceed 2. We thus reach the conclusion that equation (1) must be a Riccati if its solution is to be free of movable branch points.

9. Abel's Equation

A natural generalization of Riccati's equation is the following:

$$\frac{dy}{dx}=f(x,y), \tag{1}$$

where $f(x,y)$ is a polynomial in y. If, in particular, $f(x,y)$ is a cubic polynomial, that is,

$$f(x,y)=A_0+A_1y+A_2y^2+A_3y^3, \tag{2}$$

where the A_i are functions of x, equation (1) is called *Abel's equation*. Abel's original equation* was written in the form

$$(y+s)\,\frac{dy}{dx}+p+qy+ry^2=0, \tag{3}$$

where p, q, r, and s are functions of x. This equation is converted into (1) by the transformation: $y+s=1/z$, which yields

$$z'=rz+(q-s'-2rs)z^2+(p-qs+rs^2)z^3. \tag{4}$$

If in (2) the A_i are constants, so that $f(x,y)=f(y)$, it is obvious that the roots of the equation $f(y)=0$ are themselves solutions of (1). More generally, the solution has the form:

$$(y-y_1)^a(y-y_2)^b(y-y_3)^c=Ke^{A_3x}, \tag{5}$$

where y_1, y_2, and y_3 are the roots of $f(y)=0$, a, b, c are fixed constants, and K is an arbitrary constant.

Oeuvres, Vol. 2, No. 5.

The case where $A_0=0$ is seen from (4) to be the one actually considered by Abel. If $A_0=0$, $A_1\neq0$, then the transformation

$$y=1/(Bz), \text{ where } B'=-A_1B, \tag{6}$$

reduces (1) to the following form:

$$-B^2z\frac{dz}{dx}=A_2Bz+A_3, \tag{7}$$

which can be written:

$$z\,dz+(P+Qz)dx=0, \tag{8}$$

where $P=A_3/B^2$, $Q=A_2/B$.

It can be shown without difficulty that the function

$$\phi=e^{-k(z+\int Q\,dx)}, \; k \text{ a constant}, \tag{9}$$

is an integrating factor of (8) provided: $Q=kP$, that is, $A_2B=kA_3$. Equation (1) can be put into the canonical form

$$\frac{dz}{dt}=z^3+P(x), \tag{10}$$

by means of the transformation:

$$y=A(x)z(t)+B(x), \; t=\int^x A^2(x)A_3dx, \tag{11}$$

where we write

$$A(x)=\exp Q(x), \; Q(x)=\int^x\left(A_1-\frac{1}{3}\,A_2^2/A_3\right)dx, \; B(x)=-\frac{1}{3}\,(A_2/A_3). \tag{12}$$

Under this transformation $P(x)$ has the following value:

$$P(x)=\frac{1}{A^3(x)A_3}\left[A_0-\frac{1}{3}\left(\frac{A_1A_2}{A_3}\right)+\frac{2}{27}\left(\frac{A_2^3}{A_3^2}\right)+\frac{1}{3}\frac{d}{dx}\left(\frac{A_2}{A_3}\right)\right].$$

PROBLEMS

1. Reduce equation (1) to canonical form when all the A_i are constants.
2. Making use of the results of Problem 1, solve the following equation:

$$y'=ay^3+bx^{-3/2}.$$

3. Solve the following equation:

$$\frac{dy}{dx}=\frac{y}{x}+xy^2-3y^3.$$

10. The Generalized Riccati Equation

Some of the theory of the Riccati equation, which we have given in earlier sections, can be extended to what we shall call the *generalized Riccati equation*. This generalization was introduced by E. Vessiot in 1895 and by G. Wallenberg in 1899.

Such an equation is obtained by the elimination of the parameters in the fraction

$$y = \frac{k_1 v_1 + k_2 v_2 + \cdots + k_n v_n}{k_1 w_1 + k_2 w_2 + \cdots + k_n w_n}, \tag{1}$$

where the v_i and w_i are arbitrary linearly independent functions of x and the k_i are arbitrary constants.

The resulting equation is a nonlinear differential equation of nth order, the solution of which, by a proper transformation, can be expressed in terms of the solutions of a linear equation of order $n+1$. The case $n=2$ obviously reduces to the ordinary Riccati equation, which we have just discussed.

It will be sufficient here for us to consider the generalized Riccati equation of second order. Thus in (1) we set $n=3$. We now multiply y by the denominator of the fraction and then take two derivatives of this equation, thus obtaining the following system of equations:

$$\sum_{n=1}^{3} k_i(w_i y - v_i) = 0, \quad \sum_{n=1}^{3} k_i[(w_i y)' - v_i'] = 0, \quad \sum_{n=1}^{3} k_i[(w_i y)'' - v_i''] = 0. \tag{2}$$

Since this is a homogeneous system in the k_i, it is both necessary and sufficient for the existence of values of the k_i, other than zero, that the determinant of the system shall vanish, that is,

$$\begin{vmatrix} w_1 y - v_1 & w_2 y - v_2 & w_3 y - v_3 \\ (w_1 y)' - v_1' & (w_2 y)' - v_2' & (w_3 y)' - v_3' \\ (w_1 y)'' - v_1'' & (w_2 y)'' - v_2'' & (w_3 y)'' - v_3'' \end{vmatrix} = 0. \tag{3}$$

If we make use of the following abbreviation:

$$(a,b,c) = \begin{vmatrix} a_1 & b_1 & c_1 \\ a_2 & b_2 & c_2 \\ a_3 & b_3 & c_3 \end{vmatrix}, \tag{4}$$

it will be found that equation (3) reduces to the following:

$$(A_0 + A_1 y)y'' + (B_0 + B_1 y)y' - 2A_1(y')^2 + D_0 + D_1 y + D_2 y^2 + D_3 y^3 = 0, \tag{5}$$

where we have

$$A_0=(v,\,v',\,w),\ A_1=(w,\,w',\,v),\ B_0=2(v,\,v',\,w')-(v,\,v'',\,w),$$

$$B_1=2(w,\,w',\,v')=(w,\,w'',\,v),\ D_0=-(v,\,v',\,v''),\ D_3=(w,\,w',\,w''),$$

$$D_1=(v,\,v',\,w'')+(v',\,v'',\,w)-(v,\,v'',\,w'),$$

$$D_2=-(w,\,w',\,v'')-(w',\,w'',\,v)+(w,\,w'',\,v'). \tag{6}$$

By means of a transformation of the form:

$$y=\frac{a-(A_0/A_1)z}{z}, \tag{7}$$

where a is an arbitrary function of x, it is possible to reduce equation (5) to the following form:

$$P_0\frac{d^2z}{dx^2}+(Q_0+Q_1z)\frac{dz}{dx}+R_0+R_1z+R_2z^2+R_3z^3=0, \tag{8}$$

in which, it will be observed, the term in y'^2 has disappeared.

But it is also possible to achieve the same result if A_1 is zero. This can be accomplished without loss of generality if, in equations (6), we replace v_i by rw'_i, where r is an arbitrary function of x.

If we adopt the following abbreviations:

$$W_1=(w,w',w''),\ W_2=(w,w',w^{(3)}),$$

$$W_3=(w,w'',w^{(3)}),\ W_4=(w',w'',w^{(3)}),$$

then the coefficients are seen to reduce as follows:

$$A_0=(rw',\,rw''+r'w',\,w)=r^2(w',\,w'',\,w)=r^2W_1,$$

$$A_1=(w,\,w',\,rw')=0,\ B_0=-r^2W_4-2rr'W_1,$$

$$B_1=3rW_1,\ D_0=-r^3W_4,\ D_1=(2r'^2-rr'')W_1+rr'W_2+r^2W_3,$$

$$D_2=-rW_2-3r'W_1,\ D_3=W_1. \tag{9}$$

Since W_1 is the Wronskian of a set of linearly independent functions, it cannot vanish identically and thus A_0 is not identically zero. Noting this fact, we now substitute the values from (9) into (5), and replace y by z. Equation (5) then reduces to (8), where we have the following explicit values for the coefficients:

$$P_0=1,\quad Q_0=-\frac{2r'}{r}-\frac{W_2}{W_1};\quad Q_1=\frac{3}{r},\quad R_0=-r\frac{W_4}{W_1};$$

$$R_1=\frac{W_3}{W_1}+\frac{2r'^2-rr''}{r^2}+\frac{r'}{r}\frac{W_2}{W_1};\quad R_2=-\frac{W_2}{rW_1}-\frac{3r'}{r^2};\quad R_3=\frac{1}{r^2}. \tag{10}$$

It is to be observed that the following identities exist between these coefficients:

$$9R_3 = Q_1^2, \quad Q_0 Q_1 + Q_1' = 3R_2. \tag{11}$$

Since, moreover, $rQ_1 = 3$, whence

$$r'/r = -Q_1'/Q_1, \quad r''/r = -Q_1''/Q_1 + 2(Q_1'/Q_1)^2,$$

the ratios W_i/W_1 can be explicitly evaluated in terms of the coefficients of (8). We thus obtain the following:

$$\frac{W_2}{W_1} = -Q_0 + 2\frac{Q_1'}{Q_1}; \quad \frac{W_3}{W_1} = R_1 - Q_0\frac{Q_1'}{Q_1} + 2\left(\frac{Q_1'}{Q_1}\right)^2 - \frac{Q_1''}{Q_1}; \quad \frac{W_4}{W_1} = \frac{1}{3}R_0 Q_1. \tag{12}$$

With these ratios we are now in a position to reduce the solution of equation (8) to the solution of a linear equation of third order. For if w_1, w_2, and w_3 are the linearly independent solutions of a linear differential equation of third order, then the equation can be written as follows:

$$\begin{vmatrix} w & w' & w'' & w^{(3)} \\ w_1 & w_1' & w_1'' & w_1^{(3)} \\ w_2 & w_2' & w_2'' & w_2^{(3)} \\ w_3 & w_3' & w_3'' & w_3^{(3)} \end{vmatrix} = 0. \tag{13}$$

But this equation can also be written in the form:

$$\frac{d^3w}{dx^3} - \frac{W_2}{W_1}\frac{d^2w}{dx^2} + \frac{W_3}{W_1}\frac{dw}{dx} + \frac{W_4}{W_1}w - 0, \tag{14}$$

or explicitly in terms of the values defined by (12) as follows:

$$\frac{d^3w}{dx^3} + \left(Q_0 - 2\frac{Q_1'}{Q_1}\right)\frac{d^2w}{dx^2} + \left[R_1 - Q_0\frac{Q_1'}{Q_1} + 2\left(\frac{Q_1'}{Q_1}\right)^2 - \frac{Q_1''}{Q_1}\right]\frac{dw}{dx} + \frac{1}{3}R_0 Q_1 w = 0. \tag{15}$$

Since the solution of equation (8) can be written

$$z = \frac{rw'}{w} = \frac{3}{Q_1}\frac{w'}{w}, \tag{16}$$

it is thus seen that the Riccati equation of second order can be solved by the solution of a linear equation of third order.

A special case of (5) of some interest is obtained if we let $v_i = 1$, which, of course, is a degenerate case since the v_i are no longer linearly independent. We thus obtain the equation:

$$A_1 yy'' + B_1 yy' - 2A_1(y')^2 + D_2 y^2 + D_3 y^3 = 0. \tag{17}$$

By means of the transformation: $y = 1/w$, this equation reduces to the following nonhomogeneous linear equation of second order:

$$A_1 w'' + B_1 w' - D_2 w = D_3. \tag{18}$$

Chapter 4

Existence Theorems

1. Introduction

IN PRECEDING CHAPTERS we have discussed particular devices for the solution of the equation

$$\frac{dy}{dx} = f(x,y), \tag{1}$$

where the function $f(x,y)$ had special properties, which simplified the integration of the equation. We have also investigated certain equations, such as that of Riccati, where the structure of $f(x,y)$ was sufficiently simple so that the question of the existence of an integral was not of major importance, and where the solving algorithms were essentially formal ones.

But it is clear that the definition of a domain within which we may be sure that a solution of equation (1) exists is a matter of great importance in many problems. The definition of such a domain necessarily involves also the definition of an algorithm from which the construction of the solution is at least theoretically possible, however difficult its actual accomplishment may be. With these fundamental matters this chapter will be concerned.

Three essentially different types of existence theorems have been devised, which are usually referred to as (a) *the Calculus of Limits;* (b) *the Method of Successive Approximations;* (c) *the Cauchy-Lipschitz Method.* We shall discuss these in the order named.

2. The Calculus of Limits

The name "the calculus of limits" has been given to an existence theorem originally contributed by A. L. Cauchy (1789–1857), which marked the first systematic and rigorous examination of the problem of the solution of differential equations. "The name (*calcul des limites*)," says Picard, "was not a very fortunate one, but the idea is highly fruitful." The proof as given by Cauchy was quite complicated, but it was modified later by Briot and Bouquet and their ingenious analysis has been the basis of most modern demonstrations.

The theorem may be stated as follows for the case of a differential equation of first order:

Let us assume that for the differential equation

$$\frac{dy}{dx}=f(x,y), \tag{1}$$

the function f(x,y) *is analytic in the neighborhood of the point* $P_0=(x_0,y_0)$. *The differential equation then has a unique solution* y(x), *which is analytic in the neighborhood of* x_0 *and which reduces to* y_0 *when* $x=x_0$. *The solution can be represented explicitly by the series:*

$$y=y_0+y_0'(x-x_0)+\frac{y_0''}{2!}(x-x_0)^2+\frac{y_0^{(3)}}{3!}(x-x_0)^3+\ \cdots, \tag{2}$$

where the derivatives, evaluated at the point $x=x_0$, *are determined from successive differentiations of equation* (1).

The difficulties at once become apparent when we compute the successive coefficients in (2), since these rapidly increase in complexity. Thus, for the first three derivatives, we have

$$\frac{dy}{dx}=f(x,y),\quad \frac{d^2y}{dx^2}=\frac{\partial f}{\partial x}+\frac{\partial f}{\partial y}\frac{dy}{dx}=\frac{\partial f}{\partial x}+\frac{\partial f}{\partial y}f,$$

$$\frac{d^3y}{dx^3}=\frac{\partial^2 f}{\partial x^2}+2\frac{\partial^2 f}{\partial x \partial y}\frac{dy}{dx}+\frac{\partial^2 f}{\partial y^2}\left(\frac{dy}{dx}\right)^2+\frac{\partial f}{\partial y}\frac{d^2y}{dx^2}, \tag{3}$$

$$=\frac{\partial^2 f}{\partial x^2}+2\frac{\partial^2 f}{\partial x \partial y}f+\frac{\partial^2 f}{\partial y^2}f^2+\frac{\partial f}{\partial x}\frac{\partial f}{\partial y}+\left(\frac{\partial f}{\partial y}\right)^2 f.$$

In view of the complexity of these coefficients, it is clear that we cannot assume the convergence of (2) in the neighborhood of P_0 without establishing some measure of the magnitude of the derivatives. For convenience we shall assume that $P_0=(0,0)$.

With this object in view we introduce the equation:

$$\frac{dz}{dx}=F(x,z),\quad \cdot \tag{4}$$

where $F(x,z)$ is a *majorante* (or dominating function) for $f(x,y)$.

To explain this term, let us assume that $f(x,y)$ has been expanded as follows:

$$f(x,y)=\Sigma a_{mn}x^m y^n. \tag{5}$$

Then the function $F(x,y)$, defined as follows:

$$F(x,y)=\sum A_{mn}x^m y^n, \tag{6}$$

is a majorante for $f(x,y)$ provided the coefficients A_{mn} are positive real numbers such that $|a_{mn}| < A_{mn}$ for all values of m and n.

Let us now assume that when the values of x are limited to the interior of a circle C in the complex plane of radius a and that similarly when the values of y lie within a second circle C' of radius b, then series (5) converges. Let us assume further that the maximum value assumed by $f(x,y)$ within the prescribed domain is M. Under these conditions the function

$$F(x,z) = \frac{M}{\left(1 - \dfrac{x}{a}\right)\left(1 - \dfrac{z}{b}\right)}, \tag{7}$$

is a majorante of $f(x,y)$.

If $F(x,z)$ as thus defined is introduced into equation (4), the equation is readily solved and we thus obtain

$$z - z^2/(2b) = -aM \log (1 - x/a),$$

or, explicitly in terms of z:

$$z = b - b\left[1 + \frac{2aM}{b} \log \left(1 - \frac{x}{a}\right)\right]^{\frac{1}{2}}. \tag{8}$$

If the positive sign is chosen for the square root, then $z = 0$ when $x = 0$. When the function under the radical is zero, that is, when $x = x_1$, where

$$x_1 = a(1 - e^{-b/2aM}), \tag{9}$$

then $z = b$.

We thus see that if x lies within a circle C'' of radius x_1, the function

$$\frac{2aM}{b} \log \left(1 - \frac{x}{a}\right)$$

is less in absolute value than 1. Hence, the expansion of the radical in (8) will be a series convergent within C''. It is also readily seen that all the coefficients in the expansion of z are positive. Thus, if x has a value such that $|x| < x_1$, then the absolute value of z will be less than b.

Since $F(x,z)$ is the majorante of $f(x,y)$, it thus follows that when x lies within the circle C'' the absolute value of y will be less than b. Therefore, if y is replaced in $f(x,y)$ by its Taylor's expansion about $P_0 = (0,0)$, the resulting function $G(x)$ will be analytic within C''. From its mode of construction the function $G(x)$ is seen to be identical with dy/dx, an identity which is preserved for their successive derivatives. Thus we have established the convergence of equation (2) for values of x within C''.

As an example of the application of this theorem, let us seek the solution of the equation:

$$\frac{dy}{dx} = xy(y-2), \tag{10}$$

subject to the boundary condition: $y_0 = 1$, $x_0 = 0$.

Writing the equation in the form: $y' = xy^2 - 2xy$, we take successive derivatives. Denoting the nth derivative by D^n and observing by the rule of Leibniz for the nth derivative of a product that

$$D^n(xu) = xD^n u + nD^{n-1} u, \tag{11}$$

we obtain the following sequence of values:

$$y' = xy^2 - 2xy,$$
$$y'' = xDy^2 + y^2 - 2(xy' + y),$$
$$y^{(3)} = xD^2 y^2 + 2Dy^2 - 2(xy'' + 2y'),$$
$$y^{(4)} = xD^3 y^2 + 3D^2 y^2 - 2(xy^{(3)} + 3y''),$$
$$y^{(5)} = xD^4 y^2 + 4D^3 y^2 - 2(xy^{(4)} + 4y^{(3)}),$$
$$y^{(6)} = xD^5 y^2 + 5D^4 y^2 - 2(xy^{(5)} + 5y^{(4)}). \tag{12}$$

Similarly, we next compute

$$Dy^2 = 2yy',$$
$$D^2 y^2 = 2(yy'' + y'y'),$$
$$D^3 y^2 = 2(yy^{(3)} + 2y'y'' + y''y'),$$
$$D^4 y^2 = 2(yy^{(4)} + 3y'y^{(3)} + 3y''y'' + y^{(3)}y'),$$
$$D^5 y^2 = 2(yy^{(5)} + 4y'y^{(4)} + 6y''y^{(3)} + 4y^{(3)}y'' + y^{(4)}y'). \tag{13}$$

Finally, we set $x=0$, $y=1$, and from (12) and (13) compute in succession: $y_0' = 0$, $y_0'' = -1$, $y_0^{(3)} = y_0^{(4)} = y_0^{(5)} = 0$, $y_0^{(6)} = 30$.

When these values are substituted in (2), we obtain the following expansion:

$$y = 1 - \frac{x^2}{2} + \frac{x^6}{24} + \cdots \tag{14}$$

Unfortunately the algorithm which we have used does not provide any ready way to determine the radius of convergence of the series. But we do know from Section 1 of Chapter 2 that this series is the expansion of the function

$$y = \frac{2}{1 + e^{x^2}},$$

which is observed to have poles at $x = \sqrt{\dfrac{\pi}{2}}\,(1 \pm i)$. The radius of convergence of (14) is thus $\sqrt{\pi}$.

Several comments should be made about the calculus of limits. In the first place it can be extended essentially without change to systems of differential equations of first order and to differential equations of second and higher orders. This extension will be described in Section 4 of Chapter 7.

An essential limitation in the method is found in the assumption of the analyticity of the functions involved. For most of the classical equations this is not a significant restriction, but would be for an equation of the form: $y' = \sqrt{y}$ in the neighborhood of the point $(0,0)$. This limitation will not be imposed in the methods which we shall describe in subsequent sections.

The most serious difficulty with the method, however, is found in its solving algorithm. As we have just seen in the relatively simple example given above, the computation of derivatives soon becomes very laborious. We have also observed that there is, in general, no ready method by means of which the radius of convergence of the series can be established. Except for points in the immediate neighborhood of P_0, the reduction of the solution to numerical values rapidly becomes one of great difficulty. However, in Chapter 9, we shall describe a method of continuous analytic continuation which enormously simplifies the whole problem. From this point of view the alogrithm of the calculus of limits provides us with one of the most useful and powerful methods for numerical calculation.

3. The Method of Successive Approximations

Although this method is frequently referred to as the *Method of Picard*, in recognition of the fundamental contribution of this great analyst, its origin can be traced to a much earlier period. Thus the method was applied in 1838 by J. Liouville to the case of linear differential equations of second order and was extended in various directions by J. Caqué in 1864, L. Fuchs in 1870, G. Peano in 1888, and M. Bôcher in 1902. But it was E. Picard (1856–1941), who, in 1890, gave to the theory its most general form* and later made it an essential part of his treatment of differential equations in the second volume of his *Traité d'Analyse.*†

---

*"Mémorie sur la théorie des équations aux dérivées partielles et la méthode des approximations successive," *Journal de Mathématiques*, Vol. 6 (4), 1890, pp. 145–210. In particular, Chap. 5, pp. 197–210.
†Vol. 2, pp. 301–304; 2nd ed., p. 340.

Beginning with the equation

$$\frac{dy}{dx} = f(x,y), \tag{1}$$

we seek a solution which reduces to y_0 when $x = x_0$. For this purpose we write equation (1) in the form

$$y = y_0 + \int_{x_0}^{x} f(x,y)\, dx, \tag{2}$$

which is an *integral equation*, since the unknown function appears under the sign of integration. In the general case this will be an integral equation of nonlinear type.

A first approximation to the solution will be the function y_1 defined as follows:

$$y_1 = y_0 + \int_{x_0}^{x} f(x,y_0)\, dx.$$

Similarly, we get the following sequence of successive approximations:

$$y_2 = y_0 + \int_{x_0}^{x} f(x,y_1)\, dx,$$

$$y_3 = y_0 + \int_{x_0}^{x} f(x,y_2)\, dx, \tag{3}$$

$$* \qquad * \qquad * \qquad * \qquad * \qquad * \qquad *$$

$$y_{n+1} = y_0 + \int_{x_0}^{x} f(x,y_n)\, dx.$$

We shall now consider the following series:

$$y = y_0 + (y_1 - y_0) + (y_2 - y_1) + \ldots + (y_{n+1} - y_n) + \ldots, \tag{4}$$

for which we shall determine conditions under which it will converge and represent the unique solution of the original equation.

We now introduce conditions as follows:

(a) Restricting x and y to the rectangular region R defined by

$$|x - x_0| \leqq a, \qquad |y - y_0| \leqq b, \tag{5}$$

we assume that $f(x,y)$ is continuous in R and has an upper bound, which we shall denote by M. We shall further assume that $a < b/M$. As a matter of convenience, we shall say that x lies in A if it satisfies the first inequality in (5) and that y lies in B if it satisfies the second inequality.

(b) If (x,y) and (x,y') are any two points in R with the same abscissa, then there must exist a positive constant such that

$$|f(x,y)-f(x,y')|<K|y-y'|. \tag{6}$$

Condition (b) is called the *Lipschitz condition* after R. Lipschitz (1832–1903), who introduced it in 1876.[*]

Returning now to a consideration of series (4), we first show that when $|x-x_0|<a$, then $|y_n-y_0|<b$. For this purpose we write

$$|y_1-y_0| \leqq \int_{x_0}^{x} |f(x,y_0)|dx,$$

$$\leqq M|x-x_0| \leqq Ma < b, \tag{7}$$

and thus we see that y_1 lies in B. As a consequence of this $|f(x,y_1)|<M$, which allows us to repeat the argument thus showing that y_2 also lies in B. We now have the elements of an induction by means of which the general proposition is established.

We are now in a position to establish the convergence of (4) for we can write the following inequalities:

$$|y_2-y_1| \leqq \int_{x_0}^{x} |f(x,y_1)-f(x,y_0)|dx,$$

$$\leqq K \int_{x_0}^{x} |y_1-y_0|dx,$$

$$\leqq MK \int_{x_0}^{x} |x-x_0|dx,$$

$$\leqq MK \frac{(x-x_0)^2}{2} \leqq \frac{MKa^2}{2}.$$

By mathematical induction one then establishes the inequality

$$|y_{n+1}-y_n| \leqq MK^n \frac{(x-x_0)^{n+1}}{(n+1)!} \leqq \frac{MK^n a^{n+1}}{(n+1)!}. \tag{8}$$

We thus reach the conclusion that series (4) converges absolutely and uniformly when x is in A. Therefore, the limit function y exists and is continuous in B.

That y, as defined by (4), is indeed a solution of equation (1) is established by the following argument:

We first observe the following limit:

$$y = \lim_{n \to \infty} y_n(x) = y_0 + \lim_{n \to \infty} \int_{x_0}^{x} f[x,y_{n-1}(x)]dx,$$

$$= y_0 + \int_{x_0}^{x} \lim_{n \to \infty} f(x,y_{n-1})dx = y_0 + \int_{x_0}^{x} f(x,y)dx.$$

---

[*]See *Bulletin Sc. Math.*, Vol. 10, 1876, p. 149.

The legitimacy of the inversion of the two limits proceeds from the fact that we have

$$\int_{x_0}^{x} |f(x,y)-f(x,y_{n-1})|dx \leq K \int_{x_0}^{x} |y-y_{n-1}|dx,$$

$$\leq \frac{MK^n a^n}{n!} \left[1 + \frac{Ka}{n+1} + \frac{K^2 a^2}{(n+1)(n+2)} + \cdots \right] |x-x_0|,$$

which approaches zero as $n \to \infty$.

Therefore, since the function $f(x,y)$ is continuous in the interval, the derivative of each y_n exists and is continuous. We can thus differentiate (4) term by term, and thus obtain:

$$\frac{dy}{dx} = \frac{d}{dx} \int_{x_0}^{x} f(x,y)dx = f(x,y).$$

It remains to be proved that y is the unique solution of equation (1). To establish this, we assume that $z(x)$ is any other solution, subject to the restriction

$$|z-y| < b.$$

Hence we have

$$z = y_0 + \int_{x_0}^{x} f(x,z)dx,$$

$$y_{n+1} = y_0 + \int_{x_0}^{x} f(x,y_n)dx,$$

from which it follows that

$$|z-y_{n+1}| = \int_{x_0}^{x} |f(x,z)-f(x,y_n)|dx < K \int_{x_0}^{x} |z-y_n|dx.$$

By the same argument used above it is readily shown that

$$|z-y_n| \leq K^n \frac{b(x-x_0)^n}{n!}.$$

Since the right-hand member approaches zero as $n \to \infty$, we see that

$$z = \lim_{n \to \infty} y_n = y,$$

and the following theorem results:

If f(x,y) *is a function which is subject to the conditions* (a) *and* (b) *stated above, then the differential equation,*

$$\frac{dy}{dx} = f(x,y)$$

has one and only one solution which assumes the value $y=y_0$ *when* $x=x_0$. *This solution is defined by the series (4), which converges within the region* **R**.

4. An Example and Critique of the Method of Successive Approximations

As an example illustrating the application of the theory given in the preceding section, we shall consider the equation

$$\frac{dy}{dx}=xy(y-2), \tag{1}$$

which we shall solve subject to the initial condition: $y_0=1$, $x_0=0$.

In this case we have

$$f(x,y)=xy(y-2),$$

which satisfies conditions (a) and (b) of the theorem throughout any finite region R.

Observing that $f(x,y_0)=-x$, we compute the following sequence of values:

$$y_1=y_0+\int_0^x -x\,dx=1-\frac{1}{2}\,x^2,$$

$$y_2=y_0+\int_0^x \left(-x+\frac{1}{4}\,x^5\right)dx=1-\frac{x^2}{2}+\frac{x^6}{24},$$

$$y_3=y_0+\int_0^x -x\left(1-\frac{x^4}{4}+\frac{x^8}{16}-\frac{x^{12}}{576}\right)dx=1-\frac{x^2}{2}+\frac{x^6}{24}-\frac{x^{10}}{240}+\frac{x^{14}}{8064},$$

$$y_4=y_0+\int_0^x -x\left(1-\frac{x^4}{4}+\frac{x^8}{24}-\frac{17x^{12}}{2880}+\cdots\right)dx,$$

$$=1-\frac{x^2}{2}+\frac{x^6}{24}-\frac{x^{10}}{240}+\frac{17x^{14}}{40320}-\cdots. \tag{2}$$

This expansion is exact to the last term, as one can verify with some effort by expanding the appropriate solution of the differential equation (1), namely,

$$y=\frac{2}{1+e^{x^2}}. \tag{3}$$

It is clear from this, as has already been observed in Section 2, that the radius of convergence of series (2) is $\sqrt{\pi}$, since the solution (3) has poles at $x_1=\sqrt{\frac{\pi}{2}}\,(1+i)$ and $x_2=\sqrt{\frac{\pi}{2}}\,(1-i)$. There is no way to ascertain this fact from the existence theorem, but it serves to indicate the significance of the assumption in condition (a) of Section 3 that

$a \leqq b/M$. For as x approaches x_1, y increases without limit, and, in particular, will equal b however large this has been chosen. Therefore, within R, $M \sim |x_1| b^2$, and $b/M \sim 1/(|x_1| b)$. Since $a \leqq b/M$, x is contained within a safe interval about x_0.

The significance of the Lipschitz condition, namely, condition (b) of Section 3, is readily shown by the following example:

$$\frac{dy}{dx} = \frac{4xy}{x^2+y^2}, \tag{4}$$

the solution of which we shall consider in the neighborhood of the origin, that is to say, the solution for which $y=0$ when $x=0$.

The function

$$f(x,y) = \frac{4xy}{x^2+y^2},$$

is defined to be 0 when $x=y=0$. One can then show without difficulty that it is continuous in a region R, which contains the origin.

We now compute

$$f(x,y') - f(x,y) = \frac{4x(x^2-yy')}{(x^2+y'^2)(x^2+y^2)} \ (y'-y).$$

If in the first factor of the right member we let $y' = \alpha x$, $y = \beta x$, then we have

$$|f(x,y') - f(x,y)| = \frac{4|1-\alpha\beta|}{(1+\alpha^2)(1+\beta^2)} \frac{|y'-y|}{|x|},$$

from which it is evident that the Lipschitz condition is violated in any region R which contains the origin.

The significance of this is at once evident from the solution of equation (4), which can be written:

$$(3x^2-y^2)^2 = cy,$$

where c is an arbitrary constant. This function, for any value of c, satisfies the initial condition: $x=y=0$. Thus, by violating the Lipschitz condition, we have sacrificed the uniqueness of the solution.

5. The Cauchy-Lipschitz Method

The method which bears the name of Cauchy and Lipschitz is an extension to differential equations of the limiting process by which an integral is defined. The first proof was devised by Cauchy somewhere between 1820 and 1830 and appeared in summarized form in 1840 in his *Exercices d'analyse*. A more extended development was given by F. Moigno (1804–84) in his *Leçons de calcul* published in 1844. The introduction to the theory of conditions which attached the name of Lipschitz to the method was made in 1876. A definitive presentation of the proof, which kept in view the principal objective of extending to the general differential equation of first order the existence theorem for the Riemann integral, was made in 1908 by E. Goursat in the second volume of his *Cours d'analyse mathématique*.

In order to understand the basis of the method, let us consider the equation:

$$\frac{dy}{dx} = f(x), \tag{1}$$

the solution of which, subject to the condition that $y = y_0$ when $x = x_0$, is merely the integral:

$$y = y_0 + \int_{x_0}^{x} f(x)dx. \tag{2}$$

But this solution can also be written as the limit of the following sum:

$$Y_n = y_0 + \sum_{i=0}^{n-1} f(x_i)\Delta x_i, \tag{3}$$

where the increments Δx_i cover in some prescribed manner the range from $x = x_0$ to $x = x$, and x_i is any point within the interval Δx_i.

It is the generalization of this idea that is involved in the method of Cauchy-Lipschitz. Thus, let us consider the equation

$$\frac{dy}{dx} = f(x,y), \tag{4}$$

and let us divide the interval (x_0, x) into n parts:

$$\Delta x_0 = x_1 - x_0, \quad \Delta x_i = x_{i+1} - x_i, \quad \Delta x_{n-1} = x - x_{n-1},$$

where $x_i < x_{i+1}$, $x_n = x$.

If we now form the following sequence,

$$y_{i+1} = y_i + f(x_i, y_i)\Delta x_i, \ i = 0, 1, 2, \ldots, n-1, \tag{5}$$

then the series:

$$y_n = y_0 + f(x_0, y_0)\Delta x_0 + f(x_1, y_1)\Delta x_1 + \ldots + f(x_{n-1}, y_{n-1})\Delta x_{n-1},$$

$$= y_0 + \sum_{i=0}^{n-1} f(x_i, y_i)\Delta x_i, \tag{6}$$

is the analogue of (3).

We can now state the following theorem:
Let x *and* y *be restricted to the region* R *defined as follows:*

$$|x-x_0| \leqq a, \ |y-y_0| \leqq b.$$

Let f(x,y) *be a function which satisfies the following conditions when* x *and* y *are in* R:

(1) f(x,y) *is uniformly continuous, that is to say, given an arbitrary value* ε, *there exists a quantity* δ, *independent of* x *and* y, *such that*

$$|f(x,y)-f(x',y)| < \epsilon, \text{ when } |x-x'| < \delta.$$

(2) |f(x,y)| *has a maximum value* M.

(3) f(x,y) *satisfies the Lipschitz condition, that is to say, if* (x,y) *and* (x,y') *are any two points in* R *with the same abscissa, then there exists a constant* K *such that*

$$|f(x,y)-f(x,y')| < K|y-y'|.$$

Under these conditions and with the added restriction that

$$a < b/M,$$

y_n *defined by* (6) *will converge to a limit function* y(x), *which is the unique solution of equation* (4) *satisfying the boundary condition* y=y_0 *when* x=x_0.

The details of the proof of this theorem are considerably more complicated than those which establish the theorems given in Sections 2 and 3 and will be omitted since adequate accounts are given elsewhere.* But an outline of the general argument is readily provided and is instructive.

The proof follows closely that by means of which the convergence of (3) is established. Thus, if M_i is the largest value of $f(x_i)$ in the interval Δx_i and m_i the smallest, then y_n will be some value between the sums

$$S_n = \sum_{i=0}^{n-1} M_i \Delta x_i \text{ and } s_n = \sum_{i=0}^{n-1} m_i \Delta x_i. \tag{7}$$

Under the broad assumption that $f(x)$ is a function of *limited variation*† in the interval $x_0 \leqq x \leqq b$, it can be shown that both S_n and s_n converge uniformly to a common limit S as $\Delta x_i \to 0$. From this fact the existence of a unique integral is thus established.

---

*The original proof of Goursat will be found in Section 30, Vol. 2, of his *Cours d'analyse* (English translation, pp. 68–74). Another proof following the same general argument, but extended in some details, is given by Ince in his *Ordinary Differential Equations*, pp. 75–82.

†The difference $v_i = M_i - m_i$ is called the variation of $f(x)$ in the division x_i. The function $f(x)$ is said to be a function of *limited variation* in (a,b) if the sum Σv_i remains less than some fixed value K for any mode of division of (a,b).

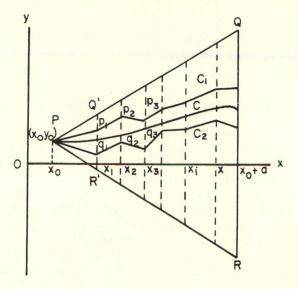

FIGURE 1

In order to obtain the analogue of (3), we first compute the following sequence of values:

$$y_1 = y_0 + f(x_0, y_0)\,\Delta x_0,$$

$$y_2 = y_1 + f(x_1, y_1)\,\Delta x_1,$$

$$* \qquad * \qquad * \qquad * \qquad * \qquad * \qquad *$$

$$y_n = y_{n-1} + f(x_{n-1}, y_{n-1})\,\Delta x_{n-1}. \tag{8}$$

The sum of these quantities then gives series (6), the limit of which is now to be sought for. A triangular region (PQR) is first established enclosed between the lines:

$$x = x_0 + a, \quad y = y_0 + M(x - x_0), \quad y = y_0 - M(x - x_0),$$

as shown in Figure 1.

In the small triangle $PQ'R'$ the function $f(x,y)$ will have an upper bound M_1 and a lower bound m_1 which satisfy the inequality:

$$-M < m_1 \leqq M_1 < M.$$

Lines with slopes equal respectively to M_1 and m_1 are now drawn from P to form the triangle Pp_1q_1. Similarly, lines are next drawn from p_1 and q_1 with slopes equal to M_2 and m_2, the respective upper and lower bounds of $f(x,y)$ in the next segment. In this way continuous arcs C_1 and C_2 are constructed which, with the line $x=x_0+a$, enclose an area that lies entirely within the triangle PQR.

The following sums are now constructed:

$$Y_n=y_0+M_1\Delta x_0+M_2\Delta x_1+\ \ldots\ +M_n\Delta x_{n-1},$$

$$Z_n=y_0+m_1\Delta x_0+m_2\Delta x_1+\ \ldots\ +m_n\Delta x_{n-1}, \qquad (9)$$

which are seen to be the analogues of series (7).

The kernel of the proof is now to show that the conditions imposed by the theorem are sufficient to establish the uniform convergence of Y_n and Z_n to a common limit function $y(x)$ and that this limit function is a unique solution of the differential equation satisfying the given boundary conditions. The graph of this function is a curve C which lies between the bounding arcs C_1 and C_2. The details of the proof are quite intricate and will be omitted.

Since this theorem, as in the case of the other two, contains a solving algorithm it will be instructive to examine its efficacy as a method for solving equations. For this purpose we shall apply it to the equation which we have used previously,

$$\frac{dy}{dx}=xy(y-2), \qquad (10)$$

subject to the boundary condition: $y=1$ when $x=0$.

For this purpose the series of value defined by (8) is now computed over the range $0\leqq x\leqq 1$ for the three cases (a) $\Delta x=0.1$; (b) $\Delta x=0.01$; (c) $\Delta x=0.001$. When each of these series is added to obtain the sum (6), we obtain the following values:

(a) $y_{10}=0.572042$; (b) $y_{100}=0.541086$; (c) $y_{1000}=0.538205$,

which are to be compared with the value correct to six places:

$$y_\infty=0.537883.$$

If we take cognizance of the fact that the estimate of the value of an integral defined by (3) for equal values of the increment can be improved by subtracting from Y_n half the sum of the first and last values, of the series, and if we apply this correction here, we shall obtain for (c) the improved estimate 0.537811. But it is clear from this example that the algorithm provided by the Cauchy-Lipschitz method will not, in general, converge rapidly to a solution. The graphic representations, the curves C_1, C_2, and C, for this example for the case $n=10$ are shown in Figure 2.

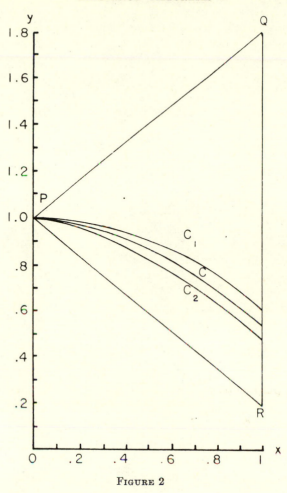

FIGURE 2

Chapter 5

An Introduction to Second Order Equations—The Problems of Conflict and Pursuit

1. Introduction

BEFORE CONSIDERING MORE GENERALLY the problem of the solution of nonlinear differential equations of second order, we shall find it instructive to investigate two problems which, arising in application, exhibit a number of the difficulties of such equations and some of the methods that have been devised to overcome them. These problems are interesting for their own sake and furnish, therefore, a pleasing introduction to a subject which will engage much of our attention in the following pages.

The first of these problems concerns the growth of two populations, which conflict with one another. Although this problem had its origin in earlier studies relating to the growth of collections of individuals, its recent development is due largely to Vito Volterra (1860–1940), one of the founders of the modern theory of integral equations. This mathematician considered the problem of competition between species, the growth and recession of populations one of which preys upon the other, or, in other words, the problem of the *prey and the predator*. In its general form the theory was stated in terms of several nonlinear integro-differential equations of special type. The details are set forth in a work, notable for its originality and depth, published in Paris in 1931 under the title: *Leçons sur la theorie mathematique de la lutte pour la vie*.

The second problem is concerned with the *curve of pursuit*, that is to say, the path generated by a point P, which moves in such a manner that its direction of motion is always toward a second point P', constrained to move along a prescribed path. This problem appears to have originated with Leonardo da Vinci in the 15th century, but its curious difficulties have intrigued the fancy and strained the ingenuity of modern mathematicians. A more detailed history of this problem will be given in a later section.

2. The Logistic Curve

The growth of human populations, as well as that of crystals, plants, animals, and lower organisms, presents a characteristic pattern, which suggested that they might be described by a single equation. The first of the studies of this problem appears to have been made as early as 1844 by P. F. Verhulst, a contemporary and colleague of L. A. J. Quetelet (1796–1874), Belgian statistician and astronomer. Quetelet had made the observation that "when a population is able to develop freely and without obstacles, it grows according to a geometric progression; if the development takes place in the midst of obstacles of all kinds which tend to arrest it, and which operate in a uniform manner, that is to say, if the social state does not change, the population does not increase indefinitely, but tends more and more to become stationary." It was to discover a curve which would meet this requirement that Verhulst initiated his investigations. The curve thus found is called the *logistic*, a term which appears to have been used first by Edward Wright in 1599 to describe an S-shaped curve.

The modern theory and application of the logistic to biological and population studies were initiated by Raymond Pearl and L. J. Reed in 1920 and these subjects are extensively treated in their work on *Studies in Human Biology* published in 1924. More recently the same methods have been applied to the description of certain growth curves observed in economic time series.

As an introduction to the more complex problem of Volterra, we shall discuss the problem of single-population growth. Let us assume that the population has an initial size equal to y_0 and that after the elapse of time t, it has increased to a size which we denote by $y(t)$. The simplest assumption, that of Quetelet's uninhibited growth, is that the rate of increase is proportional to the size of the population, that is to say,

$$\frac{dy}{dt} = Ay. \tag{1}$$

This simple equation yields: $y = y_0 \exp(At)$ as the law of growth, a law which may hold in the initial stages of population increase, but obviously cannot hold over an indefinitely long period.

The assumption made by Verhulst, and later by Pearl and Reed, was that in the normal growth of a population an inhibiting factor appears, which is proportional to $-y^2$. Thus we can replace (1) by the more realistic equation

$$\frac{dy}{dt} = Ay - By^2, \tag{2}$$

which, for convenience, can also be written:

$$\frac{dy}{dt}=ay\left(1-\frac{y}{k}\right). \tag{3}$$

This is a simple null form of the Riccati equation, the solution of which is readily found to be

$$y=\frac{k}{1+Ce^{-at}}, \tag{4}$$

where C is a constant of integration. The S-shaped curve obtained from this equation for positive values of C is called the *logistic curve*.

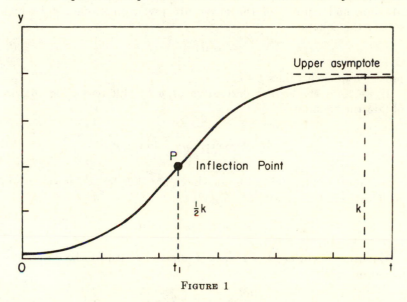

FIGURE 1

It possesses an upper and a lower asymptote, the upper asymptote being the line $y=k$. It has one point of inflection, namely, $P=(t_1,y_1)$ where

$$t_1=\frac{1}{a}\log C, \quad y_1=\frac{1}{2}k. \tag{5}$$

These features are shown in Figure 1.

A natural generalization of equation (3) is found in the following:

$$\frac{dy}{dt}=\phi(t)y(y-k), \tag{6}$$

which can be shown to have the solution:

$$y=\frac{k}{1+Ce^{g(t)}},\tag{7}$$

where we write

$$g(t)=k\int^t\phi(t)dt.\tag{8}$$

If $g(t)$ is a function which varies continuously between $-\infty$ and $+\infty$ as t varies along a segment of the real axis between $t=a$ and $t=b$, then horizontal asymptotes exist, which are the lines $y=0$ and $y=k$. Maxima and minima of the curve are given for values of t which satisfy the equation:

$$\phi'(t)=0,\tag{9}$$

provided at such values $y\neq k/2$.

If we form the second derivative of y by differentiating (6), we obtain the equation:

$$\frac{d^2y}{dt^2}=[\phi'(t)+\phi^2(t)(2y-k)]\,y(y-k).\tag{10}$$

Such points of inflection as y may have are thus found for values of t which satisfy the equation:

$$\phi'(t)+\phi^2(t)(2y-k)=0.\tag{11}$$

Equation (7) belongs to a class of curves defined by the differential equation

$$\frac{dy}{dt}=G(t)F(y/k)y,\tag{12}$$

where $F(z)$ is a function such that $F(1)=0$. The logistic curve and its generalization given above are derived by setting $F(z)=z-1$.

Another specialization of equation (12) is obtained from the choice $F(z)=\log z$, $G(t)=\log b$. The solution is then the Gompertz curve,

$$y=kC^{b^t},$$

where C is an arbitrary constant and b is assumed to be less than 1. The curve is named after Benjamin Gompertz (1779–1865), who used it to graduate the data of the mortality table. The curve resembles the logistic in form.

3. The Problem of Growth in Two Populations Conflicting With One Another

Having now considered the problem of single-population growth, we shall concern ourselves in this and subsequent sections with the problem of two populations conflicting with one another. Although our objective is thus specific, the general problem which is suggested is much broader. In many applied problems one is frequently concerned with the mutual behavior of two variables x and y, both functions of an independent variable t, which are connected by a system of two differential equations:

$$\frac{dx}{dt}=P(x,y), \quad \frac{dy}{dt}=Q(x,y). \tag{1}$$

If it happens that cyclical variations in x cause cyclical variations in y, and if the changes in y lag behind those of x, then it is customary to say that there is *hysteresis* in the relationship between them. This term arose actually in the theory of magnetism, where the magnetism induced in a piece of iron by an imposed field was found to exhibit hysteresis, that is to say, a lag, and the curve which described the phenomenon formed a closed path in the x, y-plane.

This problem can be illustrated in a simple manner by the following system:

$$\frac{dx}{dt}=ax-by, \quad \frac{dy}{dt}=cx-ay, \tag{2}$$

where all the parameters are positive quantities, and where

$$\Delta=bc-a^2>0.$$

These equations state that the growth of both variables is stimulated directly by the magnitude of one of them, but is adversely affected by the magnitude of the second. Although the system is linear and its solution readily obtained, it will serve to illustrate the more complex problem which follows.

In order to find the relationship between x and y, we observe the following equation between the two variables:

$$cxx'-a(xy'+yx')+byy'=0, \tag{3}$$

where x' and y' indicate the derivatives of x and y respectively.

Integrating (3), we obtain the equation

$$cx^2-2axy+by^2=K, \tag{4}$$

where K is an arbitrary constant. From the condition that $\Delta>0$, we see that (4) is an ellipse. This result, of course, we have already obtained earlier in another manner [See (16), Section 5, Chapter 2], since (4) is the solution of the equation:

$$\frac{dy}{dx} = \frac{cx-ay}{ax-by}.\tag{5}$$

But still another derivation of (4) is provided by system (2). If we differentiate the first equation and eliminate x' and y' by substituting their values as defined by the system, we shall obtain

$$\frac{d^2x}{dt^2} + \Delta x = 0.\tag{6}$$

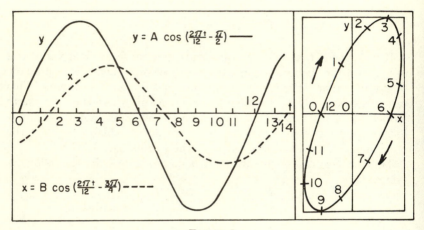

FIGURE 2.

This defines the harmonic

$$x = A \cos(\sqrt{\Delta}\,t + p),\tag{7}$$

where A and p are arbitrary constants. In a similar manner we also obtain

$$y = B \cos(\sqrt{\Delta}\,t + q).\tag{8}$$

Eliminating t between (7) and (8), we obtain

$$B^2x^2 - 2AB \cos(p-q)\, xy + A^2y^2 = A^2B^2 \sin^2(p-q).\tag{9}$$

This equation is observed to be equivalent to (4), also defining an ellipse. If $p=q$, then the ellipse degenerates into two coincident lines. In Figure 2 the two curves defined by (7) and (8) are shown for special values of the parameters. The corresponding ellipse is also graphically represented, the numbers on it agreeing with the numbers on the graphs of the two cosine curves and thus indicating both the position and the direction of motion of the point $P=(x,y)$ as t moves through a complete cycle of y. As we shall see later, the ellipse will be called a *phase trajectory* of the motion, which will itself be described as a *vortex cycle* about the origin as a *singular point*.

The problem of Volterra, namely that of the growth of two conflicting populations, is a generalization of the one which we have just described. The pertinent system of differential equations is derived by the following argument.

We consider two variables N_1 and N_2, which measure respectively the number of individuals in two populations that are assumed to work in opposition to one another. We shall assume that N_2 measures the population of a species (A), which preys upon a second species (B), whose population is measured by N_1. If N_1 is large, then (A), in the presence of so much prey, will flourish and N_2 will increase. But as N_2 increases, the prey will diminish, that is to say, N_1 will decrease, and a period of starvation will set in. Then, as N_1 diminishes, the prey will again begin to increase, and the cycle continues.

The situation which has just been described can be formulated in terms of the following system of equations:

$$\frac{dN_1}{dt} = aN_1 - bN_1N_2,$$

$$\frac{dN_2}{dt} = -cN_2 + dN_1N_2,$$

(10)

where $a, b, c,$ and d are positive numbers.

These differential equations are derived by reasoning as follows. In a bounded environment the number of encounters of the members of the two species will be proportional to N_1N_2, that is, there will be kN_1N_2 encounters per unit of time. If, for every encounter there results an instantaneous diminishing of B_1 members of the first species and a corresponding increase of B_2 members of the second species, then the resulting equations will be

$$\frac{dN_1}{dt} = aN_1 - kB_1\,N_1N_2,$$

$$\frac{dN_2}{dt} = -cN_2 + kB_2\,N_1N_2,$$

where a and c are the growth coefficients that species (A) and (B) would have respectively, if they existed alone. We observe that c is negative, since (B) by assumption depends upon (A) for its source of food. If we let $kB_1 = c$ and $kB_2 = d$, then system (10) is obtained. We also observe that if $N_2 = 0$, then N_1 will increase exponentially, since the predator has disappeared; but if $N_1 = 0$, then N_2 will decrease exponentially, since the source of food of species (B) no longer exists.

This formulation of the "struggle for life" is the work of a number of people, foremost among whom must be mentioned A. J. Lotka and

Volterra. The reader will find an excellent account of Lotka's theory in his useful work: *Elements of Physical Biology*, published in 1925. A comprehensive discussion of the problem, not only from the mathematical point of view, but also from that of the origin and significance of the problem, will be found in Volterra's treatise, to which we have already referred. This work extends Volterra's original investigations, which were first published in the memoirs of the Academia dei Lincei in 1926. Actual application of the mathematical theory to biological material was made by G. F. Gause in *The Struggle for Existence*, published in 1934.

4. Solution of the Problem of Growth of Two Conflicting Populations

We shall now obtain the solution of system (10) of Section 3, which, by means of the transformation: $N_1 = cx/d$, $N_2 = ay/b$, reduces to the following:

$$\frac{dx}{dt} = a(x - xy), \qquad \frac{dy}{dt} = -c(y - xy). \tag{1}$$

If both equations are differentiated and if y and y' are eliminated, the following nonlinear differential equation for the determination of $x(t)$ is obtained:

$$x\frac{d^2x}{dt^2} = \left(\frac{dx}{dt}\right)^2 + acx^2 - cx\frac{dx}{dt} + cx^2\frac{dx}{dt} - acx^3. \tag{2}$$

A similar elimination of x and x' yields the following equation for the determination of $y(t)$:

$$y\frac{d^2y}{dt^2} = \left(\frac{dy}{dt}\right)^2 + acy^2 + ay\frac{dy}{dt} - ay^2\frac{dy}{dt} - acy^3. \tag{3}$$

Unfortunately, neither equation (1) nor equation (2) can be integrated in terms of elementary functions. On the other hand, however, it is possible to obtain the equation of the phase trajectories, that is to say, the solution of the equation:

$$\frac{dy}{dx} = -\frac{c(y - xy)}{a(x - xy)}. \tag{4}$$

To achieve this, we observe that, since x and y satisfy equations (1), we can write

$$cx' + ay' - cx'/x - ay'/y = 0. \tag{5}$$

Integrating this equation, we obtain

$$cx + ay - c\log x - a\log y = K, \tag{6}$$

where K is an arbitrary constant.

This equation can be written in the somewhat more useful form

$$x^{-c}e^{cx} = Cy^{a}e^{-ay}, \quad C = e^{K}. \tag{7}$$

We now have a functional relationship between x and y, from which the graphs of the phase trajectories can be constructed. Since, however, the relationship between the two variables is a complicated transcendental equation, the determination of points on the trajectories can be most easily accomplished by a graphical method devised by Volterra. Thus, let us write

$$\eta = (x^{-1}e^{x})^{c}, \quad \xi = (ye^{-y})^{a}. \tag{8}$$

These two functions are now graphed and the values of x and y are obtained from the linear relationship: $\eta = C\xi$.

As an example of the construction, let us examine Figure 3.* In the second and fourth quadrants of the diagram we have represented the functions:

$$\eta = e^{x}/x, \quad \xi = y^{2}e^{-2y}, \tag{9}$$

which are merely (8) corresponding to $c=1$, $a=2$.

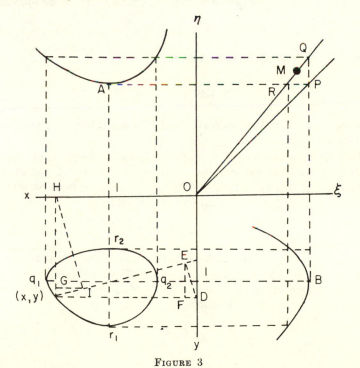

FIGURE 3

*The example and the figure are taken from Volterra's treatise. (*Loc. cit.*)

Now let tangents be drawn from A and B, the minimum and maximum points of η and ξ respectively, and let these tangents intersect in P. The line $\eta = C\xi$ is now drawn and if C exceeds the slope of OP, then (9) will represent a real locus. The points Q and R determine the points q_1, q_2, and r_1, r_2 respectively of the desired locus by the simple construction given in the figure. Other points are similarly determined by means of an identical construction originating from the variable point M in the interval RQ.

If one desires greater accuracy than that possible from the graphical methods just described, a relatively simple numerical approximation is available. To illustrate, let us consider equation (7) in which $a=2$, $c=1$, and let C be determined by the initial condition: $x=1$, $y=3$. We thus have

$$e^x/x = 121.84812 \, y^2 e^{-2y}. \tag{10}$$

Let us now assume that $y=1$, and let us determine the corresponding values of x. We thus wish to solve the equation: $e^x/x = 16.49035$. In order to find the smallest root of this equation, we make use of the following theorem:

If k *is a number less than* 1/e, *then the smallest root of the equation*

$$xe^{-x} = k, \tag{11}$$

is given by the convergent expansion:

$$x = k + k^2 + \frac{3}{2}k^3 + \frac{8}{3}k^4 + \frac{125}{24}k^5 + \ldots + \frac{n^{n-1}}{n!}k^n + \cdots \tag{12}$$

Since $k = 1/16.49035 = 0.06064$, which is less than $1/e$, we readily find that $x = 0.06469$.

To obtain the second value of x corresponding to $y=1$, an approximation is found either graphically or from a table of values of e^x. Let us denote this value by z and write: $x = z + \delta$. When this quantity is substituted in (11), we obtain the equation:

$$(z+\delta)e^{-\delta} = ke^z. \tag{13}$$

Since δ is small, if z has been properly chosen, the left-hand member can be written approximately: $(z+\delta)(1-\delta) \sim z + \delta(1-z)$. Equating this to the right-hand member of (13) and solving for δ, we have for the determination of δ the following:

$$\delta = \frac{z - ke^z}{z-1}. \tag{14}$$

This formula can be used as an iterative device in which z is replaced by successively determined values. If we choose $z=4$ as our first approximation, we get by (14) : $\delta=0.230$ and $x=4.23$. Using this value as a second approximation, we then obtain: $\delta=0.0195$ and $x=4.2495$, which is in error by only one unit in the last place. We thus see that by successive applications of formulas (12) and (14) as many points on (7) as may be desired and to any specified accuracy can be obtained. The complete graph of the phase trajectory is shown in Figure 4.

The next, and considerably more difficult step, is to construct the curves

$$x=x(t), \quad y=y(t). \tag{15}$$

The inherent difficulties of the problem are readily seen, since we are in effect finding specific solutions of equations (2) and (3).

In order to do this we first observe that by means of equations (1) we can write

$$\left[(x-1)\frac{dy}{dt}-(y-1)\frac{dx}{dt}\right]=[c(x-1)^2y+a(y-1)^2x]. \tag{16}$$

Let us now change to the polar coordinates (ρ,ω), referred to the point $(1,1)$, that is, let us write

$$x-1=\rho \cos \omega, \quad y-1=\rho \sin \omega. \tag{17}$$

Equation (16) then assumes the form

$$\frac{d\omega}{dt}=ax\sin^2 \omega+cy \cos^2 \omega. \tag{18}$$

The value of ω is thus defined by the integral

$$\omega=\int_0^t \phi(\omega)dt, \tag{19}$$

where we employ the abbreviation:

$$\phi(\omega)=ax \sin^2 \omega+cy \cos^2 \omega. \tag{20}$$

We now seek values of $\phi(\omega)$. These can be obtained either graphically, if not too much accuracy is required, or by ready computation from (17) if numerical values of x and y are available. If the graphical method is employed, we first observe from Figure 4 that $ED=x \sin \omega$ and hence $FD=x \sin^2 \omega$. Similarly, we have $HI=y \cos \omega$ and $HG=y \cos^2 \omega$. Multiplying these values respectively by a and c and adding them together, we obtain the value of $\phi(\omega)$ for the assumed value of ω. By continuing this graphical process for a sufficient number of values of ω in the interval between 0 and 2, we can construct the graph of $\phi(\omega)$.

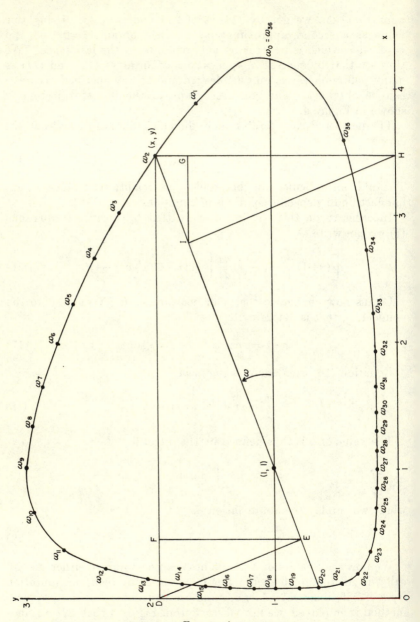

FIGURE 4

TABLE 1. COMPUTATION OF $\phi(\omega)$

ω	x	y	$\sin^2 \omega$	$\cos^2 \omega$	$x \sin^2 \omega$	$y \cos^2 \omega$	$\phi(\omega)$
0°	4. 25	1. 00	0. 0000	1. 0000	0. 00	1. 00	1. 00
12	3. 90	1. 60	0. 0432	0. 9568	0. 17	1. 53	1. 87
21	3. 47	1. 95	0. 1284	0. 8716	0. 45	1. 70	2. 60
31	3. 03	2. 21	0. 2653	0. 7347	0. 80	1. 62	3. 22
40	2. 67	2. 41	0. 4132	0. 5868	1. 10	1. 41	3. 61
50	2. 32	2. 59	0. 5868	0. 4132	1. 36	1. 07	3. 79
61	1. 97	2. 74	0. 7650	0. 2350	1. 51	0. 64	3. 66
70	1. 66	2. 86	0. 8830	0. 1170	1. 47	0. 33	3. 27
80	1. 35	2. 95	0. 9698	0. 0302	1. 31	0. 09	2. 71
90	1. 00	3. 00	1. 0000	0. 0000	1. 00	0. 00	2. 00
100	0. 67	2. 95	0. 9698	0. 0302	0. 65	0. 09	1. 39
119	0. 22	2. 40	0. 7650	0. 2350	0. 17	0. 56	0. 90
129	0. 15	2. 05	0. 6040	0. 3960	0. 09	0. 81	0. 99
140	0. 12	1. 76	0. 4132	0. 5868	0. 05	1. 03	1. 13
149	0. 10	1. 55	0. 2653	0. 7347	0. 03	1. 14	1. 20
159	0. 09	1. 37	0. 1284	0. 8716	0. 01	1. 19	1. 21
169	0. 08	1. 18	0. 0364	0. 9636	0. 003	1. 14	1. 15
180	0. 06	1. 00	0. 0000	1. 0000	0. 00	1. 00	1. 00
192	0. 08	0. 80	0. 0432	0. 9568	0. 003	0. 77	0. 78
201	0. 09	0. 65	0. 1284	0. 8716	0. 01	0. 57	0. 59
211	0. 13	0. 47	0. 2653	0. 7347	0. 03	0. 35	0. 41
220	0. 20	0. 33	0. 4132	0. 5868	0. 08	0. 19	0. 35
230	0. 36	0. 24	0. 5868	0. 4132	0. 21	0. 10	0. 52
250	0. 71	0. 18	0. 8830	0. 1170	0. 63	0. 02	1. 28
270	1. 00	0. 17	1. 0000	0. 0000	1. 00	0. 00	2. 00
300	1. 46	0. 16	0. 7500	0. 2500	1. 10	0. 04	2. 24
319	1. 94	0. 17	0. 4304	0. 5696	0. 83	0. 10	1. 76
327	2. 26	0. 20	0. 2966	0. 7034	0. 67	0. 14	1. 48
337	2. 75	0. 25	0. 1527	0. 8473	0. 42	0. 21	1. 05
347	3. 61	0. 41	0. 0506	0. 9494	0. 18	0. 39	0. 75
360	4. 25	1. 00	0. 0000	1. 0000	0. 00	1. 00	1. 00

The details of the computation of $\phi(\omega)$ are shown in Table 1. Except for a few critical points, the values of x and y were estimated from the graph of the phase trajectory (Fig. 4) drawn to a sufficiently large scale. The graphical representation of the function $\phi(\omega)$ is shown in Figure 5.

The final step in the computation is to find x and y as functions of t. Since these variables are now expressed as functions of ω, as shown in Table 1, the problem is to determine t as a function of ω also. To accomplish this, we note from (19) that $d\omega/dt=\phi(\omega)$, and that $t=0$ when $\omega=0$. Hence, for the determination of t, we have the integral

$$t=\int_0^{\omega} \psi(\omega)d\omega, \tag{21}$$

where $\psi(\omega)=1/\phi(\omega)$.

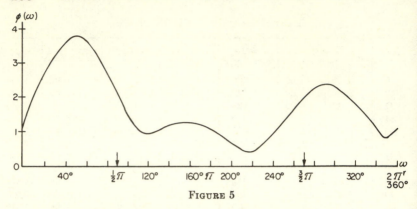

FIGURE 5

TABLE 2. COMPUTATION OF t

n	ω	$\psi(\omega)$	$\Sigma\psi(\omega)$	t	x	y	n	ω	$\psi(\omega)$	$\Sigma\psi(\omega)$	t	x	y
0	0.00	1.00	0.00	0.00	4.25	1.00	18	3.14	1.00	12.59	2.02	0.06	1.00
1	0.17	0.58	1.58	0.14	3.96	1.50	19	3.32	1.22	13.81	2.22	0.08	0.83
2	0.35	0.40	1.98	0.22	3.52	1.91	20	3.49	1.64	15.45	2.47	0.09	0.67
3	0.52	0.32	2.30	0.29	3.07	2.18	21	3.67	2.33	17.78	2.81	0.13	0.49
4	0.70	0.28	2.58	0.34	2.67	2.41	22	3.84	2.86	20.64	3.26	0.20	0.33
5	0.87	0.26	2.84	0.39	2.32	2.59	23	4.01	1.92	22.56	3.68	0.36	0.24
6	1.05	0.27	3.11	0.43	2.03	2.71	24	4.19	1.15	23.71	3.95	0.55	0.20
7	1.22	0.31	3.42	0.48	1.66	2.86	25	4.36	0.78	24.49	4.12	0.71	0.18
8	1.40	0.37	3.79	0.54	1.35	2.95	26	4.54	0.61	25.10	4.24	0.86	0.18
9	1.57	0.50	4.29	0.62	1.00	3.00	27	4.71	0.50	25.60	4.34	1.00	0.17
10	1.74	0.72	5.01	0.72	0.67	2.95	28	4.89	0.46	26.06	4.42	1.15	0.17
11	1.92	1.05	6.06	0.88	0.37	2.70	29	5.06	0.43	26.46	4.49	1.27	0.17
12	2.09	1.11	7.17	1.07	0.21	2.36	30	5.24	0.45	26.94	4.57	1.46	0.17
13	2.27	1.00	8.17	1.25	0.15	2.02	31	5.41	0.50	27.44	4.66	1.65	0.17
14	2.44	0.88	9.05	1.42	0.12	1.76	32	5.59	0.57	28.01	4.75	1.99	0.17
15	2.62	0.83	9.88	1.56	0.10	1.53	33	5.76	0.74	28.75	4.87	2.35	0.20
16	2.79	0.83	10.71	1.71	0.09	1.35	34	5.93	1.04	29.79	5.02	2.93	0.29
17	2.97	0.88	11.59	1.86	0.08	1.16	35	6.11	1.25	31.04	5.22	3.82	0.52
18	3.14	1.00	12.59	2.02	0.06	1.00	36	6.28	1.00	32.04	5.42	4.25	1.00

Expressing ω in radian measure at intervals of $\Delta\omega=2\pi/36=0.1745$, values of t are computed from (21) by numerical integration. In the present case, sufficient accuracy is obtained by means of the trapezoidal formula, that is to say,

$$t_n=\Delta\omega\left\{\sum_{i=0}^{n}\psi(\omega_i)-\frac{1}{2}\left[\psi(\omega_n)+\psi(\omega_0)\right]\right\}. \tag{22}$$

The details of the computation are given in Table 2, and the graphs thus obtained of the functions $x=x(t)$ and $y=y(t)$ are shown through two cycles in Figure 6.

It is evident from the table that the initial values of x and y, namely 1 and 3 respectively, are not attained at $t=0$, but correspond to

$t=0.62$. A linear translation of this value to the origin will correct this discrepancy. It is also observed that the derivative curves for x and y are readily computed from the original system of equations. The four curves, $y=y(t)$, $y'=y'(t)$, $x=x(t)$ and $x'=x'(t)$, are shown in Figure 7. These graphs were obtained from a solution of the system of equations by means of an analogue computer, but of course they can be readily computed numerically.

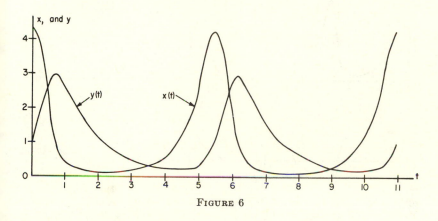

FIGURE 6

5. A Generalization of Volterra's Problem

The problem which we have just studied is a special case of the following more general system:

$$\frac{dx}{dt}=F+Cx+Dy+Gx^2+Hxy+Ky^2,$$

$$\frac{dy}{dt}=E+Ax+By+Lx^2+Mxy+Ny^2. \tag{1}$$

Although the theoretical basis for the study of this equation will be given later in Chapter 11, some anticipation of those results will not be out of place here since they throw light upon the phenomena which we have presented.

In the first place, we have observed that the variations in the phase trajectory were referred to the point (1,1). In the second place, the trajectory is a closed path and as a consequence the integral curves $x=x(t)$ and $y=y(t)$ are periodic. The motion may therefore be characterized as stable and periodic.

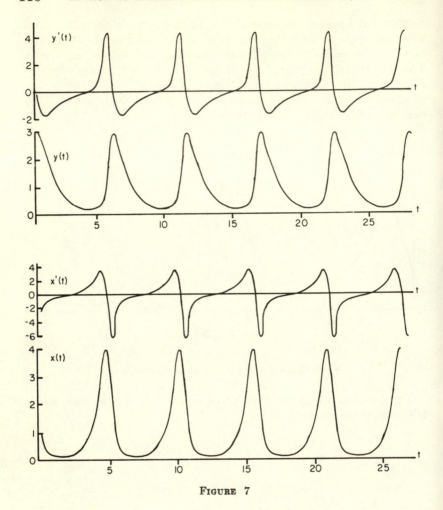

FIGURE 7

We shall first impose upon system (1) the linear transformation:

$$x=w+p, \ y=z+q, \tag{2}$$

from which we obtain the following:

$$\frac{dw}{dt}=F+Cp+Dq+Gp^2+Hpq+Kq^2+(C+2Gp+Hq)w$$
$$+(D+Hp+2Kq)z+Gw^2+Hwz+Kz^2,$$

$$\frac{dz}{dt}=E+Ap+Bq+Lp^2+Mpq+Nq^2+(A+2Lp+Mq)w$$
$$+(B+2Np+Mq)z+Lw^2+Mwz+Nz^2. \tag{3}$$

In general there will exist a set of four points, called the *singular points* of the system, which we shall denote by $P_i=(p_i,q_i)$, $i=1,2,3,4$. These are determined from the intersections of the following conics:

$$Gp^2+Hpq+Kq^2+Cp+Dq+F=0,$$

$$Lp^2+Mpq+Nq^2+Ap+Bq+E=0. \tag{4}$$

As we shall see later in Chapter 11, the character of the solution in the neighborhoods of these singular points is determined by the roots of the equation:

$$\lambda^2-(B'+C')\lambda+B'C'-A'D'=0, \tag{5}$$

where we abbreviate:

$$A'=A+2Lp_i+Mq_i, \qquad B'=B+Mp_i+2Nq_i,$$
$$C'=C+2Gp_i+Hq_i, \qquad D'=D+Hp_i+2Kq_i. \tag{6}$$

If, in particular, the roots of (5) are pure imaginaries, then the corresponding singular point is called a *vortex point*. If certain additional criteria are satisfied by the coefficients, then the phase trajectories are closed paths and the motion is periodic.

In the case of the Volterra system, equations (1) of Section 4, we obtain the following degenerate conics from equations (4):

$$-apq+ap=0, \quad cpq-cq=0,$$

which intersect in the points: $P_1=(0,0)$ and $P_2=(1,1)$.

The point P_2 is a vortex point, since (5) reduces to the equation: $\lambda^2+ac=0$. We have seen by direct analysis that the motion is cyclical about this point.

6. The Hereditary Factor in the Problem of Growth

A natural extension of the problem of single-population growth which we discussed in Section 2 is found in the following integro-differential equation

$$\frac{1}{y}\frac{dy}{dt}=a+by+\int_c^t K(t,s)y(s)ds. \tag{1}$$

This equation was suggested by Volterra as a device to take account of the "hereditary influences" which may exist in the problem of growth.

The introduction of such an inheritance factor in nonbiological phenomena was made by Volterra near the beginning of the present century. Unfortunately little progress has been made since then in the development of this attractive idea, although it has been used to a certain extent by mathematical economists. The name of "hereditary mechanics" was applied to it by E. Picard, who wrote as follows:[*]

"In all this study (of classical mechanics) the laws which express our ideas on motion have been condensed into differential equations, that is to say, relations between variables and their derivatives. We must not forget that we have, in fact, formulated a principle of *nonheredity*, when we suppose that the future of a system depends at a given moment only on its actual state, or in a more general manner, if we regard the forces as depending also on velocities, that the future depends on the actual state and the infinitely neighboring state which precedes. This is a restrictive hypothesis and one which, in appearance at least, is contradicted by the facts. Examples are numerous where the future of a system seems to depend upon former states. Here we have *heredity*. In some complex cases one sees that it is necessary, perhaps, to abandon differential equations and consider functional equations in which there appear integrals taken from a distant time to the present, integrals which will be, in fact, this hereditary part. The proponents of classical mechanics, however, are able to pretend that heredity is only apparent and that it amounts merely to this, that we have fixed our attention upon too small a number of variables. But the situation is just as it was in the simpler one, only under conditions that are more complex."

The following example from Volterra [†] will help to clarify this idea. We know from elementary physics that the relation, to a first approximation, between the couple of torsion, P, and the angle of torsion, W, is given by the linear equation

$$W=kP, \tag{2}$$

where k is a physical constant.

[*]"La mécanique classique et ses approximations successive." *Revista di Scienza.* Vol. 1, 1907, pp. 4–15. In particular, p. 15.

[†] *Leçons sur les équations intégrales et les équations integro-differentielles.* Paris, 1913, pp. 138–139 and 150.

It is reasonable to suppose, however, that W does not depend merely upon the present moment of torsion, but upon all preceding ones as well. The elastic body has experienced fatigue from previous distortions and, in this way, has *inherited*, as it were, characteristics from the past.

To express this analytically, the hereditary part must be represented by an integral which sums the various contributions to the inherited characteristics from some initial time t_0 to the present time t. Thus we replace equation (2) by the integral equation

$$W(t)=kP(t)+\int_{t_0}^{t} K(t,s)P(s)ds, \qquad (3)$$

where $K(t,s)$ is the *coefficient of heredity*.

Assuming that $W(t)$ and $P(t)$ are both periodic functions of the time with the same period, Volterra shows that the coefficient of heredity is then of the form

$$K(t,s)=K(t-s), \qquad (4)$$

a kernel which Volterra characterizes as belonging to the class of the *closed cycle*.

The problem proposed by Volterra's integro-differential equation (1) is one of considerable difficulty, even when the kernel assumes the realtively simple form of the closed cycle. We shall, however, return to it in Chapter 13 and for several cases show how it, and its extension to two variables, give patterns which differ significantly from those which we have discussed in the preceding sections.

7. Curves of Pursuit

As we have said earlier, a second problem which may be formulated in terms of a nonlinear differential equation of second order and which exhibits some of the peculiar characteristics of such equations, is that of curves of pursuit. In Figure 8 we show two curves the first of which (AB) is traced by a point P which moves in such a manner that its direction of motion is always toward a second point P', which moves along the second curve (CD). The velocities of P and P' are usually assumed to be constant, although this is not a necessary assumption. The *curve of pursuit* is the arc AB and the path of the pursued is the arc CD. The problem thus proposed is to construct AB, when CD is given and the velocities of P and P' are known.

The simplest problem of this type is that for which the path of P' is a straight line. A second, and much more difficult one, is that for which CD is a circle. The point P can be taken initially either inside or outside of the circle. This problem has been extensively studied.

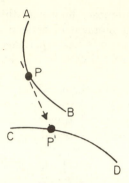

FIGURE 8

Although, as we have said earlier, the problem of pursuit appears to have originated with Leonardo da Vinci, the interest of mathematicians was not awakened until 1732 when Pierre Bouguer (1698–1758), French hydrographer, published a paper in the *Memoires* of the *Histoire de l'Académie Royale des Sciences* under the title: "Sur de nouvelles courbes ausquelles on peut donner le nom de Lignes de Poursuite." In this paper Bouguer proposed and solved the problem: "To find the curve of pursuit, that is to say, the curve by which a vessel moves in pursuing another which flees along a straight line, supposing that the velocities of the two vessels are always in the same ratio."

A history of the problem was given in 1921 by R. C. Archibald and H. P. Manning from which the following notes are taken.* According to these authors the paper of Bouguer was followed in the same work by a shorter solution by P. L. M. de Maupertuis (1698–1759), originator of the principle of least action, who also proposed the problem: "The curve *CE* being given, to find the curve *BM*, such that its tangents *ME* cut upon the curve *CE* arcs proportional to the arcs *BM*."

The first reference to the problem of pursuit, where the curve of the pursued is a circle, appears to have been in an anonymous article published in 1859 in the *Mathematical Monthly* (Vol. 1, p. 249), where the path of the pursuer is discussed for the case where the velocities are equal. No analysis is given. The problem was again proposed by H. Brocard (1845–1922), discoverer of the "Brocard circle", in *Nouvelle Correspondence Mathématique* (Vol. 3, 1877, p. 175). Since no solution was presented, he asked for the differential equation in *Mathesis* (Vol. 3, 1883, p. 232), and this was given by Keelhoff in 1886 in the same journal (Vol. 6, p. 135).

Various formulations of the problem appeared in the literature between 1886 and 1906. In that year L. Dunoyer presented an extensive

---

*American Mathematical Monthly, Vol. 28, 1921, pp. 91–93.

exposition in *Nouvelle Annales de Mathématiques* (See *Bibliography*). Setting up the differential equation in convenient coordinates, he discussed its integration by means of methods based upon the general theory of Poincaré.

A. S. Hathaway in 1920 again proposed the problem in the *American Mathematical Monthly* (Vol. 27) as follows: "A dog at the center of a circular pond makes straight for a duck, which is swimmimg along the edge of a pond. If the rate of swimming of the dog is to the rate of swimming of the duck as $n:1$, determine the equation of the curve of pursuit and the distance the dog swims to catch the duck." An extensive discussion of this problem was made by Hathaway himself the next year in the same journal and also by F. V. Morley. (See *Bibliography*.) The latter used graphical and numerical methods in obtaining the integral curve defined by the differential equation.

The original problem as proposed by Bouguer, namely, where the curve of the pursued is a straight line, is readily solved. It was included by George Boole in his *Differential Equations*, London, 1859, (4th ed., 1877, pp. 252–253).

8. Linear Pursuit

We shall now consider the problem of the curve of pursuit where the path of the pursued is a straight line. Let $P=(x,y)$ be a point on the curve of the pursuer and $P'=(\xi,\eta)$ a point on the path of the pursued. Let the curve traced by P' be represented by the equation

$$f(\xi,\eta)=0. \tag{1}$$

Since the tangent through P passes through P', its equation can be written as follows:

$$\eta-y=\frac{dy}{dx}\,(\xi-x). \tag{2}$$

If we now assume that the ratio of the velocity of P to the velocity of P' is k, then we have $ds/dt=kd\sigma/dt$, that is, $ds=kd\sigma$, where ds and $d\sigma$ are the elements of the arcs of the pursuer and the pursued respectively. We thus obtain the equation

$$dx^2+dy^2=k^2(d\xi^2+d\eta^2); \tag{3}$$

or since y, ξ, and η are functions of x, we can write (3) in the form:

$$1+\left(\frac{dy}{dx}\right)^2=k^2\left[\left(\frac{d\xi}{dx}\right)^2+\left(\frac{d\eta}{dx}\right)^2\right]. \tag{4}$$

Two other equations are obtained by taking the derivatives of (1) and (2) with respect to x. We thus get

$$\frac{\partial f}{\partial \xi}\frac{d\xi}{dx}+\frac{\partial f}{\partial \eta}\frac{d\eta}{dx}=0, \tag{5}$$

$$\frac{d\eta}{dx}=\frac{d^2y}{dx^2}(\xi-x)+\frac{dy}{dx}\frac{d\xi}{dx}. \tag{6}$$

We now have four equations, namely, (1), (2), (5), and (6), for the determination of $d\xi/dx$ and $d\eta/dx$ as functions of x, y, dy/dx, and d^2y/dx^2. When these values are substituted in the right-hand member of equation (4), the differential equation of the curve of pursuit is obtained. It is clearly a nonlinear differential equation of second order.

As an example we shall now apply this theory to determine the curve of pursuit, where the pursued moves along a straight line. For convenience, let us assume that the path is along a line parallel to the y-axis and at a distance a from the origin as shown in Figure 9. Equation (1) then has the simple form

$$\xi=a,$$

and $d\xi/dx=0$.

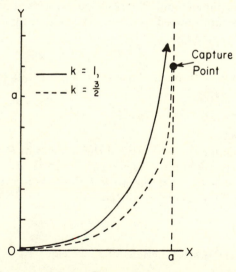

FIGURE 9

From equation (6) we get

$$\frac{d\eta}{dx} = \frac{d^2y}{dx^2}(a-x);$$

and when this is substituted in equation (4), the following differential equation is obtained:

$$1 + \left(\frac{dy}{dx}\right)^2 = k^2(a-x)^2\left(\frac{d^2y}{dx^2}\right)^2.$$

If we now write: $p = dy/dx$, this equation can be written:

$$\frac{k\,dp}{\sqrt{1+p^2}} = \frac{dx}{(a-x)},$$

from which we obtain by integration

$$\frac{dy}{dx} = \frac{1}{2}\left[c(a-x)^{-1/k} - \frac{1}{c}(a-x)^{1/k}\right], \tag{7}$$

where c is an arbitrary constant.

A second integration yields the equation

$$y = \frac{1}{2}\left[\frac{kc}{1-k}(a-x)^{1-1/k} + \frac{k}{c(1+k)}(a-x)^{1+1/k}\right] + c', \tag{8}$$

where c' is arbitrary and $k \neq 1$.

Since both dy/dx and y are 0 when $x=0$, we find from (7) and (8) that

$$c = a^{1/k}, \; c' = ka/(k^2-1).$$

Hence we can write (8) as follows:

$$y = \frac{ka}{k^2-1} + \frac{ka}{2(k^2-1)}\left[(k-1)\left(1-\frac{x}{a}\right)^{1+1/k} - (k+1)\left(1-\frac{x}{a}\right)^{1-1/k}\right]. \tag{9}$$

If $k=1$, then the integration of (7) gives the following solution:

$$y = \frac{a}{4}\left[\left(1-\frac{x}{a}\right)^2 - \log\left(1-\frac{x}{a}\right)^2 - 1\right]. \tag{10}$$

If $k>1$, then point P finally overtakes P', the point of capture being attained when $x=a$. The value of y is thus found to equal $ka/(k^2-1)$. When $k=3/2$, 2, and 3 successively, the corresponding ordinates of the *capture point* are respectively $6a/5$, $2a/3$, and $3a/8$.

In Figure 9 the curves of pursuit are graphed for $k=1$ and $k=3/2$. The values of the ordinates for various values of x/a are given in the following table:

	$k=1$	$k=3/2$		$k=1$	$k=3/2$		$k=1$	$k=3/2$
x/a	y/a	y/a	x/a	y/a	y/a	x/a	y/a	y/a
0	0	0	0. 4	0. 09541	0. 06290	0. 8	0. 56472	0. 34331
0. 1	0. 00518	0. 00345	0. 5	0. 15907	0. 10394	0. 9	0. 90379	0. 51022
0. 2	0. 02157	0. 01435	0. 6	0. 24815	0. 15994	0. 95	1. 24849	0. 64943
0. 3	0. 05084	0. 03370	0. 7	0. 37449	0. 23618	1. 0	∞	1. 20000

We have an interesting variation to the problem just given if the condition that the velocities of the two points are proportional to some constant k is replaced by the condition that the two points remain a fixed distance apart, let us say α. In this case the assumption that

$$(\eta-y)^2+(\zeta-x)^2=\alpha^2, \tag{11}$$

takes the place of the assumption: $ds=k\,d\sigma$.

For the case of linear pursuit, as given above, we have from (2), where $\zeta=\alpha$,

$$(\eta-y)^2=\left(\frac{dy}{dx}\right)^2(\alpha-x)^2.$$

When $(\eta-y)^2$ is eliminated between this equation and (11), we obtain the differential equation:

$$\left(\frac{dy}{dx}\right)^2=\frac{\alpha^2}{(\alpha-x)^2}-1; \tag{12}$$

which, upon integration, yields the solution

$$y=-\sqrt{\alpha^2-(\alpha-x)^2}+\alpha\log\left[\frac{\alpha+\sqrt{\alpha^2-(\alpha-x)^2}}{\alpha-x}\right]+C, \tag{13}$$

where C is an arbitrary constant. Since $y=0$, when $x=0$, we find that $C=0$.

Equation (13) can be reduced to the following form:

$$y=-\alpha\sqrt{1-(1-x/\alpha)^2}+\alpha\operatorname{sech}^{-1}(1-x/\alpha), \tag{14}$$

which is recognized as the equation of the *tractrix*. The tractrix is defined to be the path of a weight which is dragged along a rough horizontal plane by a taut string one end of which is attached to the weight and the other moves along a straight line.

9. Pursuit When the Path of the Pursued Is a Circle

Let us now consider the problem where the pursued, P, is moving with constant speed v along a circle of radius a.   Referring to Figure 10, let Q be the position of the pursuer who moves with velocity kv, where k is a positive constant.   The constant k can be less than, equal to, or greater than unity, and the point Q can lie within, upon, or outside of the circle.

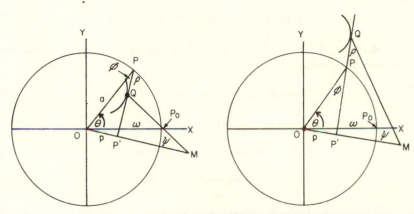

FIGURE 10

Let us assume that pursuit starts at the point P_0 where OX intersects the circle.   Then, when P has moved through an arc θ, Q will have moved through a distance: $s = ka\theta$.   Let PQ be the tangent to the curve of pursuit, and let PQ make an angle ω with the line OX. Let OP' be drawn perpendicular to PQ and denote by p the distance OP'.   The line OP' is extended to M so that $OM = ka$.   If $k > 1$, then M is outside of the circle; if $k = 1$, M is on the circle; and if $k < 1$, then M is inside of the circle.

Let us also denote the distance PQ by ρ and consider ρ as positive when Q is inside of the circle, but negative when Q is outside of the circle.   Finally, we shall denote by ϕ the angle OPP' and by ψ the angle QMP'.

The following relationships between the variables can now be verified from Figure 10:

(a) $\omega=\phi+\theta$; (b) $OP'=a \sin \phi$; (c) $PP'=a \cos \phi$;

(d) $P'M=|ka-a \sin \phi|$; (e) $P'Q=a \cos \phi-\rho$;

(f) $\tan \psi=\dfrac{P'Q}{P'M}=\dfrac{a \cos \phi-\rho}{|ka-a \sin \phi|}.$ (1)

It is clear that if we know θ, ϕ, and ρ the point Q is determined. We shall now derive the differential relationships which exist between these three variables. To determine these let us first observe that the equation of the tangent PQ can be written as follows:

$$x \sin \omega-y \cos \omega=p=a \sin (\omega-\theta).$$ (2)

Similarly, the normal to the tangent through the point Q has the following equation:

$$x \cos \omega+y \sin \omega=a \cos (\omega-\theta)-\rho.$$ (3)

We shall now consider the following differentials:

$$\sin \omega \, dx-\cos \omega \, dy=0,$$ (4)

$$\cos \omega \, dx+\sin \omega \, dy=kad\theta.$$ (5)

The first of these is merely a statement of the fact that dy/dx $=\tan \omega=\sin \omega/\cos \omega$. To establish (5), we observe that

$$ds=\left[1+\left(\frac{dy}{dx}\right)^2\right]^{\frac{1}{2}} dx=(1+\tan^2 \omega)^{\frac{1}{2}}dx=\sec \omega dx=kad\theta,$$

$$ds=\left[1+\left(\frac{dx}{dy}\right)^2\right]^{\frac{1}{2}} dy=(1+\cot^2 \omega)^{\frac{1}{2}}dy=\csc \omega dy=kad\theta.$$

Since $dx=ka \cos \omega d\theta$ and $dy=ka \sin \omega d\theta$, (5) is seen to follow as an immediate consequence.

Observing that ω is a function of x and y, we now form the differentials of equations (2) and (3) and thus obtain:

$$\sin \omega dx-\cos \omega dy+(x \cos \omega+y \sin \omega)d\omega=a \cos (\omega-\theta)(d\omega-d\theta),$$ (6)

$$\cos \omega dx+\sin \omega dy-(x \sin \omega-y \cos \omega)d\omega$$
$$=-a \sin (\omega-\theta)(d\omega-d\theta)-d\rho.$$ (7)

Making use of (2), (3), (4), and (5), we reduce (6) and (7) to the following simpler forms:

$$\rho \, d\omega = a \, \cos (\omega - \theta) d\theta, \tag{8}$$

$$ka \, d\theta = a \, \sin (\omega - \theta) d\theta - d\rho. \tag{9}$$

Introducing the variable $\phi = \omega - \theta$, we get

$$\rho(d\phi + d\theta) = a \, \cos \phi \, d\theta,$$

$$ka \, d\theta = a \, \sin \phi \, d\theta - d\rho,$$

equations which can now be written

$$\rho \, \frac{d\phi}{d\theta} = a \, \cos \phi - \rho, \tag{10}$$

$$\frac{d\rho}{d\theta} = a \, \sin \phi - ka. \tag{11}$$

These equations form a fundamental system, comparable to the system of the Volterra problem defined by (1) in Section 4. As in that problem, differential equations satisfied separately by ϕ and ρ can be obtained without difficulty. If both equations are differentiated once, it will be found that the variables can be separated.

In the first case (the elimination of ρ), we obtain the equation:

$$\cos \phi \, \frac{d^2\phi}{d\theta^2} + (3 \sin \phi - 2k) \frac{d\phi}{d\theta} + (2 \sin \phi - k) \left(\frac{d\phi}{d\theta}\right)^2 + \sin \phi - k = 0. \tag{12}$$

If we make the transformation: $y = \sin \phi$, $\theta = x$, then this equation assumes the form:

$$(1 - y^2) \frac{d^2y}{dx^2} + (3y - 2k) \sqrt{1 - y^2} \, \frac{dy}{dx} + (3y - k) \left(\frac{dy}{dx}\right)^2 + y - k = 0. \tag{13}$$

In the second case (the elimination of ϕ), the second equation is found to be

$$\rho \, \frac{d^2\rho}{d\theta^2} + \rho \, \sqrt{\Delta} - \Delta = 0, \text{ where } \Delta = a^2 - \left[\left(\frac{d\rho}{d\theta}\right)^2 - ka\right]^2. \tag{14}$$

These equations are obviously difficult to solve, but fortunately the interest in this problem is not in attaining values of ϕ and ρ as functions of θ, but rather in the determination of the curve of pursuit itself. The differential equation of this curve, in terms of the variables ρ and ϕ, is obtained by dividing equation (11) by (10), from which we get:

$$\frac{d\rho}{d\phi}=\rho\ \frac{a(\sin\phi-k)}{a\cos\phi-\rho}. \tag{15}$$

We thus see that the curve of pursuit is the phase trajectory of the system defined by equations (10) and (11). But simple as equation (15) appears to be, it presents unusual analytic difficulties. Its solution cannot be obtained in terms of elementary functions. In a later chapter of this book we shall return to it from a more advanced point of view after the proper analysis has been presented.

Fortunately, however, in the problem of pursuit we are able to surmount some of the difficulties in a very simple way, since the nature of the problem itself provides a means for obtaining graphically an approximation of the actual curve. This is illustrated in Figure 11. Equal increments are marked off on the curve of the pursued. A line is drawn from the initial point of the pursuer to the initial point

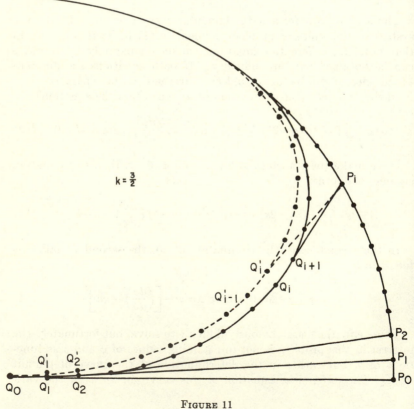

FIGURE 11

of the pursued, that is, from Q_0 to P_0.   On this line a distance $Q_0Q_1 = k(P_0P_1)$ is laid off.   The line formed by all the increments thus obtained by a continuation of this process will form one bound to the desired curve of pursuit.   In order to obtain a second bound the first increment is laid off on the line Q_0P_1, the second increment on the line Q_1P_2, etc.   The curve of pursuit lies between these two bounds and the error in its construction clearly depends upon the size of the initial increments.

The complex nature of the problem of pursuit in a circle is shown in Figure 12, where in (a) the pursuer is initially at the center, and (b) where the pursuer starts outside of the circle at a radius distance from the pursued and on a line which connects both with the center.   In each case $k=\frac{2}{3}$.   We see that the pursuer does not achieve a capture, but in each case the curve of pursuit is asymptotic to a circle of radius ka, which is concentric with the circle of the pursued.   This example thus presents us with a phenomenon, which is often present in nonlinear systems, namely, that of a fixed curve toward which the motion tends asymptotically.   To such a curve Poincaré gave the name of *limit cycle*.   We shall encounter other examples in later chapters.

10. Conditions of Capture

We shall now consider the problem presented by the question: When does the pursuer capture the pursued?   Limiting our discussion to the case of circular pursuit, we shall now prove that the pursuer catches the pursued if $k>1$, but that capture is not achieved if $k \leqq 1$.*

For this proof, we make in equation (11) of Section 9 the transformation: $ds=ka\,d\theta$ and thus obtain:

$$ds = \frac{k}{\sin \phi - k}\,d\rho.$$

Integrating from the initial arc length s_0, we then have

$$s - s_0 = \int_{\rho_0}^{\rho} \frac{k}{\sin \phi - k}\,d\rho. \tag{1}$$

If k is greater than 1, the integrand is continuous.   By the theorem of mean value we can then write

$$s - s_0 = \frac{k(\rho_0 - \rho)}{k - \sin \phi'},$$

where ϕ' is some value between 0 and $\frac{1}{2}\pi$.

---

*In this proof we follow arguments given by A. S. Hathaway.   (See *Bibliography*.)

From this equation we derive the inequality

$$s - s_0 < \frac{k\rho_0}{k-1}.$$

Therefore, since the distance s that the pursuer can go from s_0 is less than a fixed value, the pursued must be captured.

If k is less than 1, the integrand of (1) becomes infinite when $\sin \phi = k$. Since ϕ varies between 0 and $\frac{1}{2}\pi$, there will exist one value for which the infinite value is attained. That the order of the infinity of the integrand is at least as great as 1, is shown from the derivative

$$\frac{d^2\rho}{ds^2} = \frac{\cos \phi}{k} \frac{d\phi}{ds} = \frac{\cos \phi}{k} \frac{d\phi}{d\rho} \frac{d\rho}{ds} = \frac{\cos \phi \, (\cos \phi - \rho)}{k^2},$$

which remains finite when $\sin \phi = k$.

It follows from this that the integral (1) is infinite. Hence, when $k < 1$, $s - s_0$ is infinite and there is no capture.

If $k = 1$, we consider the inequality

$$s - s_0 > \frac{\rho_0 - \rho}{1 - \sin \phi_0},$$

where ϕ_0 is an initial value corresponding to ρ_0. The inequality is seen to hold for all values of s except s_0, since the integrand of (1) has been replaced by its smallest value.

Let us now make the assumption that capture takes place in a finite distance. As s approaches the capture point, $\rho \to 0$ and the inequality becomes

$$s - s_0 \geq \frac{\rho_0}{1 - \sin \phi_0}.$$

But we have

$$\frac{\rho_0}{1 - \sin \phi_0} = \frac{a \cos \phi_0}{1 - \sin \phi_0} = \frac{a \cos \phi_0 - \rho_0}{1 - \sin \phi_0} = \frac{a(1 + \sin \phi_0)}{\cos \phi_0} = a \tan \psi_0, \quad (2)$$

where $\tan \psi$ is the angle QMP' of Figure 10. In this case, when $k = 1$, M lies on the circle and moves toward P as its limiting position.

If we assume that $s - s_0$ is finite, we can choose the initial point (ρ_0, ψ_0) as near to the pursued as we wish. But under these conditions $\psi \to 0$ and $\phi \to \frac{1}{2}\pi$, and the assumption that $s - s_0$ remains finite is seen to be contradicted from (2). Hence, it follows that there is no capture.

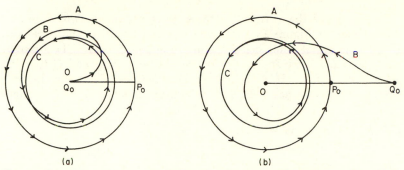

FIGURE 12.—A is path of pursued; B path of pursuer; C the limit cycle.   Ratio
of speeds (k) is 2/3.

11. General Pursuit Curve

It is evident from the discussion given above of the two special cases
of pursuit, namely, where the path of the pursued is either a straight
line or a circle, that the general problem is one of great analytical
difficulty.   The differential equation which determines the path of
the pursuer, if it can be explicitly determined, is a nonlinear equation of second order.   That these analytical difficulties are inherent
in the problem is readily seen from the existence of double points
and nodal points in pursuit paths derived from relatively simple
paths taken by the pursued.   It is also readily observed that these
singularities are functionally connected with the initial points from
which pursued and pursuer start.

Under these circumstances, we see that the problem for general
pursuit is most readily solved by the graphical method described
above in Section 9.   We first trace the curve of the pursued.   This
curve can be defined in any convenient set of coordinates, usually
Cartesian or polar, denoted, let us say, by ξ, η.   The curve is then
represented by the equation:

$$f(\xi, \eta) = 0, \tag{1}$$

as in Section 8.

Upon this curve an initial point, P_0, is chosen from which the pursuit starts.   Another point, Q_0, is similarly designated for the initial
position of the pursuer.   Two conditions are now imposed: (a) that
the pursuer always moves toward the pursued, and (b) that the distance from P_0 to the pursued measured along the curve defined by
(1) is a fixed ratio of the distance of the pursuer from Q_0 as measured
along the curve of pursuit.   Now, using the graphical technique
described in Section 9, the curve of pursuit is traced.

To illustrate the curious patterns thus obtained two examples are given. In Figure 13 the curve of the pursued is the parabola: $y^2 = 4x$, the initial point being the origin of coordinates. Two pursuers A and B start respectively at the positions $(0,4)$ and $(1,0)$. All speeds are equal. It is seen that both curves of pursuit become asymptotic to the parabola and no capture is achieved. The effect

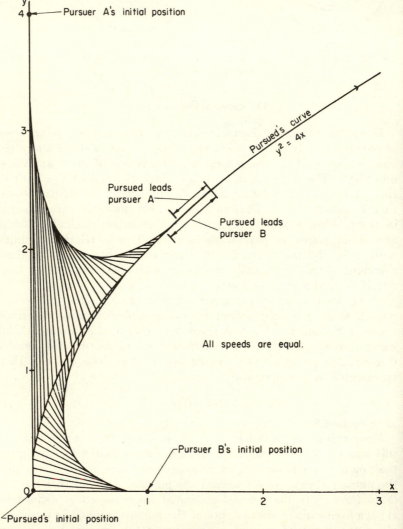

FIGURE 13.—Pursuit curve in which the path of the pursued is the parabola: $y^2 = 4x$. The ratio of the speeds (k) is 1.

of the choice of different origins of pursuit is shown in the lead which one pursuer gains over the other.

In the second example, Figure 14, the curve of the pursued is the spiral: $\rho = 10\,\theta$ and the origin is at the point $(0,0)$. Even though the pursuer starts at the advantageous point where the curve of the pursued crosses the Y-axis, and thus need not move at all to achieve capture, the curve of pursuit never again touches that of the pursued, but is asymptotic to it, when the ratio of velocities (k) is 1. If k is greater than 1, the pursued is captured, but if k is less than 1, the curve of pursuit finally becomes asymptotic to the spiral $\rho = 10k\,\theta$.

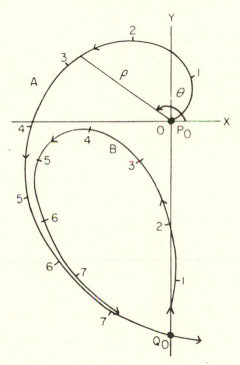

FIGURE 14.—Pursuit curve in which the path of the pursued (A) is the spiral: $\rho = 10\,\theta$. The initial point of (A) is P_0. The path of the pursuer begins at Q_0. The ratio of the speeds (k) is 1.

Chapter 6

Elliptic Integrals, Elliptic Functions, and Theta Functions

1. Introduction

ONE CANNOT PROCEED very far in the theory of nonlinear equations before he encounters solutions that are expressed in terms of elliptic integrals or elliptic functions. Elliptic integrals are the natural generalizations of the inverse circular functions and elliptic functions are similar generalizations of the direct circular functions. In order to understand this relationship, let us consider the following nonlinear equation:

$$\left(\frac{dy}{dx}\right)^2 = a + by + cy^2. \tag{1}$$

If the discriminant: $\Delta = b^2 - 4ac$, is a positive number, then the solution of equation (1) is expressed in terms of the circular sine, and if Δ is a negative number, the solution is expressed in terms of the hyperbolic sine. Explicitly we have for the two cases the following solutions:

$$y = -\frac{\sqrt{\Delta}}{2c} \sin\left(\sqrt{-c}\,x + p\right) - \frac{b}{2c}, \qquad c < 0,$$

$$y = \frac{\sqrt{-\Delta}}{2c} \sinh\left(\sqrt{c}\,x + p\right) - \frac{b}{2c}, \qquad c > 0. \tag{2}$$

The natural generalization of equation (1) is the following nonlinear equation:

$$\left(\frac{dy}{dx}\right)^2 = a + by + cy^2 + dy^3 + ey^4, \tag{3}$$

which introduces the elliptic functions in its solution.

If the roots of the quartic equation

$$a + by + cy^2 + dy^3 + ey^4 = 0, \tag{4}$$

are denoted by α, β, γ, δ, then equation (3) can be written in the form:

$$\left(\frac{dy}{dx}\right)^2 = e(y - \alpha)(y - \beta)(y - \gamma)(y - \delta). \tag{5}$$

If, in particular, $e=k^2$ and if the roots are respectively 1, -1, $1/k$, and $-1/k$, then we have the canonical form

$$\left(\frac{dy}{dx}\right)^2 = (1-y^2)(1-k^2y^2). \tag{6}$$

In this case, if $y(0)=0$ and $y'(0)=1$, the particular solution of the equation is the function $y=\mathrm{sn}(x,k)$, where $\mathrm{sn}(x,k)$ is the elliptic sine of Jacobi, which we shall examine in detail later.

The following differential equation of second order:

$$\frac{d^2y}{dx^2} = A + By + Cy^2 + Dy^3, \tag{7}$$

is also solved by means of elliptic functions. To show this we differentiate equation (3) and thus obtain

$$2y'y'' = by' + 2cyy' + 3dy'y^2 + 4ey'y^3,$$

which reduces to (7) when the factor y' is removed and the constants are properly equated.

It will appear later that the following differential equations are also solved in terms of elliptic integrals:

$$\left(\frac{dy}{dx}\right)^2 = \frac{1-y^2}{1-k^2y^2}, \tag{8}$$

$$\left(\frac{dy}{dx}\right)^2 = (1+ny^2)^2(1-y^2)(1-k^2y^2). \tag{9}$$

If we take derivatives of both sides of equation (8) and reduce the resulting equation, we find that it is equivalent to the following:

$$(1-k^2y^2)\frac{d^2y}{dx^2} - k^2y\left(\frac{dy}{dx}\right)^2 + y = 0. \tag{10}$$

Finally, if we make the transformation: $y=\sin\theta$, in equations (6) and (8), we obtain the following equivalent equations respectively:

$$\left(\frac{d\theta}{dx}\right)^2 = 1 - k^2\sin^2\theta, \tag{11}$$

$$\left(\frac{d\theta}{dx}\right)^2 = \frac{1}{1-k^2\sin^2\theta}. \tag{12}$$

2. Elliptic Integrals

By an *elliptic integral* of the first kind we shall mean the function

$$F(x,k) = \int_0^x \frac{dx}{\sqrt{(1-x^2)(1-k^2x^2)}}, \qquad k^2 < 1, \tag{1}$$

or its equivalent,

$$F(\phi,k) = \int_0^\phi \frac{d\theta}{\sqrt{1-k^2 \sin^2 \theta}}, \qquad k^2 < 1, \tag{2}$$

where x in the first form is connected with ϕ in the second form by means of the equation

$$x = \sin \phi. \tag{3}$$

The parameter k is called the *modulus* of the elliptic integral and the quantity k', defined by the equation: $k'^2 = 1 - k^2$, is called the *complementary modulus* with respect to k. It is often convenient to write

$$k = \sin \alpha, \tag{4}$$

in which case we have: $k' = \cos \alpha$.

The quantity,

$$K = F(1,k) = F\left(\frac{1}{2}\pi,k\right), \tag{5}$$

is called the *complete elliptic integral of first kind*, and the quantity,

$$K' = F(1,k') = F\left(\frac{1}{2}\pi,k'\right), \tag{6}$$

is called the *complementary complete integral of first kind*.

By an *elliptic integral of second kind* we shall mean the function

$$E(x,k) = \int_0^x \sqrt{\frac{1-k^2x^2}{1-x^2}}\,dx, \qquad k^2 < 1, \tag{7}$$

or its equivalent

$$E(\phi,k) = \int_0^\phi \sqrt{1-k^2 \sin^2 \theta}\,d\theta, \qquad k^2 < 1, \tag{8}$$

where, as in the first case, x and ϕ are connected by means of equation (3).

The functions,

$$E = E(1,k) = E\left(\frac{1}{2}\pi, k\right), \tag{9}$$

$$E' = E(1,k') = E\left(\frac{1}{2}\pi, k'\right), \tag{10}$$

where $k'^2 = 1 - k^2$, are called respectively the *complete* and the *complementary complete elliptic integrals of second kind*.

By an *elliptic integral of third kind* we mean the function

$$\Pi(x,n,k) = \int_0^x \frac{dx}{(1+nx^2)\sqrt{(1-x^2)(1-k^2x^2)}}, \quad k^2 < 1, \tag{11}$$

or its equivalent form

$$\Pi(\phi,n,k) = \int_0^\phi \frac{d\theta}{(1+n\sin^2\theta)\sqrt{1-k^2\sin^2\theta}}, \quad k^2 < 1, \tag{12}$$

where, as before, x and ϕ are connected by $x = \sin\phi$.

In some problems, such, for example, as the rectification of the hyperbola, one requires the value of the integral

$$s = \int_0^x \sqrt{\frac{1+Ax^2}{1+Bx^2}} \, dx, \text{ where } A > B > 0. \tag{13}$$

This integral can be expressed in terms of both F and E as follows:

$$s = \frac{1}{\sqrt{A}} F[\phi, \sqrt{(A-B)/A}] - \frac{\sqrt{A}}{B} E[\phi, \sqrt{(A-B)/A}] + \frac{\sqrt{A}}{B} \tan\phi \, \Delta\phi, \tag{14}$$

where we abbreviate:

$$\tan\phi = \sqrt{A}x \text{ and } \Delta\phi = \sqrt{1-[(A-B)/A]\sin^2\phi}.$$

PROBLEMS

1. Show that the length of an arc of the ellipse: $b^2x^2 + a^2y^2 = a^2b^2$, measured from the end of the minor axis, is $aE(\phi,e)$, where e is the eccentricity.

2. Show that the length of the arc of the hyperbola: $b^2x^2 - a^2y^2 = a^2b^2$, measured from any point (x,y), is given by

$$s = \frac{b^2}{ae} F(\phi, 1/e) - aeE(\phi, 1/e) + ae \tan\phi\Delta\phi, \tag{15}$$

where $\tan\phi = aey/b^2$, $\Delta\phi = \sqrt{1-\sin^2\phi/e^2}$ and e is the eccentricity.

3. Show that the length of the arc of the lemniscate: $\rho^2 = a^2\cos 2\theta$, measured from its vertex to any point $(\rho,0)$, is given by

$$s = \frac{1}{2}\sqrt{2}aF\left(\frac{\sqrt{2}}{2}, \phi\right), \tag{16}$$

where $\sin^2\phi = 2\sin^2\theta$.

4. Show that when $y = \sin \frac{1}{2} z$, the equation: $z'' + k^2 \sin z = 0$, is transformed into the following:

$$y'' = ay + by^3.$$

5. Given the transformation:

$$z^2 = \left(\frac{\beta - \delta}{\alpha - \delta}\right)\left(\frac{x - \alpha}{x - \beta}\right), \quad k^2 = \left(\frac{\beta - \gamma}{\alpha - \gamma}\right)\left(\frac{\alpha - \delta}{\beta - \delta}\right),$$

establish the following identity:

$$\frac{dz}{\sqrt{(1 - z^2)(1 - k^2 z^2)}} = \frac{1}{2}\sqrt{(\beta - \delta)(\alpha - \gamma)}\left[\frac{dx}{\sqrt{(x - \alpha)(x - \beta)(x - \gamma)(x - \delta)}}\right].$$

6. Given the transformation:

$$z^2 = \frac{\alpha - \gamma}{y - \gamma}, \quad k^2 = \frac{\beta - \gamma}{\alpha - \gamma},$$

prove the following:

$$\frac{dz}{\sqrt{(1 - z^2)(1 - k^2 z^2)}} = -\frac{1}{2}\sqrt{(\alpha - \gamma)}\left[\frac{dy}{\sqrt{(y - \alpha)(y - \beta)(y - \gamma)}}\right].$$

3. Expansions of the Complete Elliptic Integrals of First and Second Kinds

For convenience we introduce the notation

$$\Delta\theta = \sqrt{1 - k^2 \sin^2 \theta}. \tag{1}$$

Since the reciprocal of $\Delta\theta$ is the integrand of the integral which defines $F(\phi, k)$, we first obtain the following expansion:

$$\frac{1}{\Delta(\theta)} = 1 + \frac{1}{2} k^2 \sin^2 \theta + \frac{1\cdot 3}{2\cdot 4} k^4 \sin^4 \theta + \frac{1\cdot 3\cdot 5}{2\cdot 4\cdot 6} k^6 \sin^6 \theta + \cdots \tag{2}$$

Integrating this series term by term and observing the integral

$$\int_0^{\frac{1}{2}\pi} \sin^{2n} \theta \, d\theta = \frac{1\cdot 3\cdot 5 \cdots (2n - 1)}{2\cdot 4\cdot 6 \cdots 2n} \frac{\pi}{2}, \tag{3}$$

we obtain the following expansion for $K = F\left(\frac{1}{2}\pi, k\right)$:

$$K = \frac{1}{2}\pi\left[1 + \left(\frac{1}{2}\right)^2 k^2 + \left(\frac{1\cdot 3}{2\cdot 4}\right)^2 k^4 + \left(\frac{1\cdot 3\cdot 5}{2\cdot 4\cdot 6}\right)^2 k^6\right.$$

$$\left. + \cdots + \left(\frac{(2n)!}{2^{2n}(n!)^2}\right)^2 k^{2n} + \cdots \right] \tag{4}$$

Similarly, if we expand $\Delta(\theta)$ and integrate the resulting series term by term, we obtain the following expansion for $E=E\left(\frac{1}{2}\pi,k\right)$:

$$E=\frac{1}{2}\pi\left[1-\left(\frac{1}{2}\right)^2 k^2-\left(\frac{1\cdot3}{2\cdot4}\right)^2\frac{k^4}{3}-\left(\frac{1\cdot3\cdot5}{2\cdot4\cdot6}\right)^2\frac{k^6}{5}\right.$$
$$\left.-\cdots-\left(\frac{(2n)!}{2^{2n}(n!)^2}\right)^2\frac{k^{2n}}{(2n-1)}-\cdots\right]. \qquad (5)$$

In both expansions (4) and (5), $k^2<1$.

If we introduce the customary notation of the hypergeometric function,

$$F(\alpha,\beta,\gamma;\,x)=1+\frac{\alpha\cdot\beta}{1\cdot\gamma}x+\frac{\alpha(\alpha+1)\beta(\beta+1)}{1\cdot2\cdot\gamma(\gamma+1)}x^2+\cdots, \qquad (6)$$

which satisfies the differential equation,

$$x(1-x)\frac{d^2y}{dx^2}+[\gamma-(\alpha+\beta+1)x]\frac{dy}{dx}-\alpha\cdot\beta y=0, \qquad (7)$$

then K and E can be expressed as follows:

$$K=\frac{1}{2}\pi F\left(\frac{1}{2},\frac{1}{2},1;k^2\right),\quad E=\frac{1}{2}\pi F\left(\frac{1}{2},-\frac{1}{2},1;k^2\right). \qquad (8)$$

The expansions of the complementary integrals K' and E' are given as follows:

$$K'=\log\left(\frac{4}{k}\right)+\left(\frac{1}{2}\right)^2 k^2\left[\log\left(\frac{4}{k}\right)-1\right]+\left(\frac{1\cdot3}{2\cdot4}\right)^2 k^4\left[\log\left(\frac{4}{k}\right)-1-\frac{2}{3\cdot4}\right]$$
$$+\left(\frac{1\cdot3\cdot5}{2\cdot4\cdot6}\right)^2 k^6\left[\log\left(\frac{4}{k}\right)-1-\frac{2}{3\cdot4}-\frac{2}{5\cdot6}\right]+\cdots, \qquad (9)$$

$$E'=1+\left(\frac{1}{2}\right)k^2\left[\log\left(\frac{4}{k}\right)-\frac{1}{1\cdot2}\right]+\left(\frac{1}{2}\right)^2\frac{3}{4}k^4\left[\log\left(\frac{4}{k}\right)-\frac{2}{1\cdot2}-\frac{1}{3\cdot4}\right]$$
$$+\left(\frac{1\cdot3}{2\cdot4}\right)^2\frac{5}{6}k^6\left[\log\left(\frac{4}{k}\right)-\frac{2}{1\cdot2}-\frac{2}{3\cdot4}-\frac{1}{5\cdot6}\right]+\cdots. \qquad (10)$$

The quantities K and E, as functions of k, are connected by means of the following equations:

$$\frac{dE}{dk}=\frac{1}{k}(E-K),\qquad \frac{dK}{dk}=\frac{1}{k\cdot k'^2}(E-k'^2K). \qquad (11)$$

The four functions, K, K', E and E', satisfy the following remarkable identity originally due to Legendre:

$$K\,E'+K'\,E-K\,K'=\frac{1}{2}\pi. \qquad (12)$$

PROBLEMS

1. Derive the first equation in (11) by differentiating: $E = \int_0^{\pi/2} (1 - k^2 \sin^2 \theta)^{1/2} d\theta$
with respect to k.

2. Derive the following formulas:

$$\frac{dE'}{dk} = \frac{k}{k'^2} (K' - E'), \qquad \frac{dK'}{dk} = \frac{1}{k \cdot k'^2} (k^2 K' - E'). \tag{13}$$

3. Show that $K E' + K' E - K K'$ is a constant by differentiating with respect to k.

4. Prove that K and K' are solutions of the following equation:

$$k(1-k^2) \frac{d^2u}{dk^2} + (1 - 3k^2) \frac{du}{dk} - ku = 0. \tag{14}$$

5. Prove that E and $E' - K'$ are solutions of the equation:

$$(1-k^2) \frac{d}{dk} \left(k \frac{du}{dk} \right) + ku = 0. \tag{15}$$

6. Show that the general solution of the Riccati

$$\frac{dy}{dk} + y^2[k(1-k^2)] = k, \tag{16}$$

is the function:

$$y = [c(E - k'^2 K) + c'(k^2 K' - E')]/[cK + c' K'], \tag{17}$$

where c and c' are arbitrary constants.

7. Show that the function

$$y = [c(E - K) + c' E']/[cE + c'(E' - K')], \tag{18}$$

where c and c' are arbitrary constants, is the general solution of the Riccati

$$\frac{dy}{dk} + y^2/k + k/(1-k^2) = 0. \tag{19}$$

4. Expansions of the Elliptic Integrals of First and Second Kinds

The functions $F(\phi, k)$ and $E(\phi, k)$ can be expanded in terms of K and E and functions of $\sin \phi$ and $\cos \phi$. To obtain such expansions we define $S_{2n}(\phi)$ as follows:

$$S_{2n}(\phi) = \int_0^\phi \sin^{2n} \theta d\theta, \tag{1}$$

and observe the values:

$$S_0(\phi) = \phi,$$

$$S_2(\phi) = \frac{1}{2} \phi - \frac{1}{2} \sin \phi \cos \phi,$$

$$S_4(\phi) = \frac{3}{8} \phi - \frac{1}{8} \sin \phi \cos \phi (3 + 2 \sin^2 \phi),$$

$$S_6(\phi) = \frac{5}{16} \phi - \frac{1}{48} \sin \phi \cos \phi (15 + 10 \sin^2 \phi + 8 \sin^4 \phi). \tag{2}$$

Functions of higher order can be computed by means of the iterative formula:

$$S_{2n}(\phi) = -\frac{1}{2n}\sin^{2n-1}\phi\cos\phi + \frac{2n-1}{2n}S_{2n-2}(\phi). \tag{3}$$

When the series expansion of $1/\Delta(\theta)$ given by (2) of Section 3 is integrated term by term between $\theta=0$ and $\theta=\phi$, we obtain:

$$F(\phi,k) = S_0(\phi) + \frac{1}{2}k^2 S_2(\phi) + \frac{1\cdot3}{2\cdot4}k^4 S_4(\phi) + \frac{1\cdot3\cdot5}{2\cdot4\cdot6}k^6 S_6(\phi) + \ldots \tag{4}$$

When the explicit values of $S_{2n}(\phi)$ are substituted, the following series results:

$$F(\phi,k) = \frac{2K}{\pi}\phi - \sin\phi\cos\phi\left(\frac{1\cdot1}{2\cdot2}k^2 + \frac{1\cdot3}{2\cdot4}A_4 k^4 + \frac{1\cdot3\cdot5}{2\cdot4\cdot6}A_6 k^6\right.$$
$$\left. + \frac{1\cdot3\cdot5\cdot7}{2\cdot4\cdot6\cdot8}A_8 k^8 + \ldots + \frac{(2n)!}{2^{2n}(n!)^2}A_{2n}k^{2n} + \ldots\right), \tag{5}$$

where the coefficients A_{2n} have the following values:

$$A_4 = \frac{1}{4}\sin^2\phi + \frac{3}{4\cdot2},$$

$$A_6 = \frac{1}{6}\sin^4\phi + \frac{5}{6\cdot4}\sin^2\phi + \frac{5\cdot3}{6\cdot4\cdot2},$$

$$A_8 = \frac{1}{8}\sin^6\phi + \frac{7}{8\cdot6}\sin^4\phi + \frac{7\cdot5}{8\cdot6\cdot4}\sin^2\phi + \frac{7\cdot5\cdot3}{8\cdot6\cdot4\cdot2},$$

$$A_{10} = \frac{1}{10}\sin^8\phi + \frac{9}{10\cdot8}\sin^6\phi + \frac{9\cdot7}{10\cdot8\cdot6}\sin^4\phi + \frac{9\cdot7\cdot5}{10\cdot8\cdot6\cdot4}\sin^2\phi + \frac{9\cdot7\cdot5\cdot3}{10\cdot8\cdot6\cdot4\cdot2},$$

$$\tag{6}$$

* * * * * * *

The expansion of $E(\phi,k)$ is similarly accomplished and we obtain

$$E(\phi,k) = \frac{2E}{\pi}\phi + \sin\phi\cos\phi\left(\frac{1}{2\cdot2}k^2 + \frac{1}{2\cdot4}A_4 k^4 + \frac{1\cdot3}{2\cdot4\cdot6}A_6 k^6\right.$$
$$\left. + \frac{1\cdot3\cdot5}{2\cdot4\cdot6\cdot8}A_8 k^8 + \ldots + \frac{(2n)!}{2^{2n}(n!)^2(2n-1)}A_{2n}k^{2n} + \ldots\right), \tag{7}$$

where A_4, A_6, etc., are defined as above.

5. Differential Equations Satisfied by the Complete Elliptic Integrals

The complete elliptic integrals K, K' and E, E', regarded as functions of the parameter k, can be shown to furnish solutions of certain linear differential equations. The derivation of these equations follows readily from equations (11) and (13) of Section 3 and proofs will be omitted.

Thus the differential equation

$$L(Y) \equiv (1-k^2) \frac{d^2Y}{dk^2} + \frac{1-3k^2}{k} \frac{dY}{dk} - Y = 0, \tag{1}$$

has for solutions the functions K and K'.

Similarly, the differential equation

$$M(Z) \equiv (1-k^2) \frac{d^2Z}{dk^2} + \frac{1-k^2}{k} \frac{dZ}{dk} + Z = 0, \tag{2}$$

has for solutions E and E'.

Nonhomogeneous linear differential equations of second order are also readily constructed for which $F(\phi,k)$ and $E(\phi,k)$, regarded as functions of k, provide particular integrals. Thus, let us define the function

$$G(\phi,k) = \frac{\sin \phi \cos \phi}{\Delta^3(\phi)}, \tag{3}$$

and write the differential equation:

$$L(Y) = -G(\phi,k). \tag{4}$$

This equation has the solution: $Y = F(\phi,k) + AK + BK'$, where A and B are arbitrary constants.

Similarly, the equation

$$M(Z) = G(\phi,k), \tag{5}$$

has the solution: $Z = E(\phi,k) + AE + BE'$.

6. The Computation and Tables of Elliptic Integrals

The computation of elliptic integrals is in general most easily accomplished by means of what are called *Landen's transformations* after their originator J. Landen (1719–90).[*]

[*]"An Investigation of a General Theorem for Finding the Length of any Arc of any Conic Hyperbola by Means of Two Elliptical Arcs, etc." *Philosophical Trans.*, Vol. 65, 1775, p. 285; also, *Math. Mem.*, Vol. 1, London, 1780, p. 33.

The transformation for the elliptic integral of first kind is given in the following form:

$$F(\phi,k)=k_n\sqrt{(k_1 \cdot k_2 \cdot k_3 \ldots k_{n-1})/k}\, F(\phi_n,k_n),\tag{1}$$

where we use the abbreviations:

$$k_0=k,\quad k_p=\frac{2\sqrt{k_{p-1}}}{1+k_{p-1}},\quad \phi_0=\phi,\quad \sin(2\phi_p-\phi_{p-1})=k_{p-1}\sin \phi_{p-1}.\tag{2}$$

If k is less than 1, then $k_n \to 1$ as $n \to \infty$. This may be proved from the following equality:

$$1-k_{n+1}=\rho_n(1-k_n),\tag{3}$$

where we have

$$\rho_n=\frac{1-\sqrt{k_n}}{1+\sqrt{k_n}}\,\frac{1}{1+k_n}.\tag{4}$$

It is evident that ρ_n is always less than 1. Therefore, if $k_0=k$ is less than 1, we shall have: $1-k_1=\rho_0(1-k)<1$, $1-k_2=\rho_1(1-k)=\rho_0\rho_1(1-k)<1$, and thus

$$1-k_n=\rho_0\rho_1\rho_2 \ldots \rho_{n-1}(1-k),\tag{5}$$

which approaches zero as $n \to \infty$.

As k_n diminishes, one can show from (2) that ϕ_n also diminishes, but approaches a limit different from zero, let us say Φ. Thus, for the limiting value of $F(\phi_n,k_n)$ in (1) we shall have

$$\lim_{n\to\infty} F(\phi_n,k_n)=F(\Phi,1)=\int_0^\Phi \frac{d\phi}{\sqrt{1-\sin^2 \phi}}=\log \tan\left(\frac{1}{4}\pi+\frac{1}{2}\Phi\right).\tag{6}$$

Formula (1) thus assumes the following useful form:

$$F(\phi,k)=\sqrt{(k_1 \cdot k_2 \cdot k_3 \ldots)/k}\,\log \tan\left(\frac{1}{4}\pi+\frac{1}{2}\Phi\right).\tag{7}$$

The above form of the transformation can be used effectively in computing the value of $F(\phi,k)$ provided k is sufficiently large, that is to say, when k is close to 1. In the contrary case, when k is small, the following form of the transformation is more useful since the convergence is more rapid:

$$F(\phi,k)=(1+K_1)(1+K_2) \ldots (1+K_n)F(\Phi_n,K_n)/2^n,\tag{8}$$

where we employ the abbreviations:

$$K_0=k, \ K_p=\frac{1-\sqrt{1-K_{p-1}^2}}{1+\sqrt{1-K_{p-1}^2}}, \ \Phi_0=\phi, \ \tan(\Phi_p-\Phi_{p-1})=\sqrt{1-K_{p-1}^2}\ \tan \Phi_{p-1}. \tag{9}$$

Since in the limit, $K_n \to 0$, we have

$$\lim_{n \to \infty} F(\Phi_n,K_n)=F(\Phi,0)=\int_0^{\Phi} d\phi=\Phi; \tag{10}$$

and thus formula (8) becomes

$$F(\phi,k)=\lim_{n \to \infty}(1+K_1)(1+K_2)\ldots(1+K_n)\Phi/2^n. \tag{11}$$

Similar formulas hold for the elliptic integral of second kind:

$$E(\phi,k)=F(\phi,k)\left[1+k\left(1+\frac{2}{k_1}+\frac{2^2}{k_1k_2}+\ldots+\frac{2^{n-1}}{k_1k_2\ldots k_{n-1}}\right.\right.$$
$$\left.\left.-\frac{2^n}{k_1k_2\ldots k_{n-1}}\right)\right]-k\left[\sin\phi+\frac{2\sin\phi_1}{\sqrt{k}}+\frac{2^2\sin\phi_2}{\sqrt{k\,k_1}}+\ldots\right.$$
$$\left.+\frac{2^{n-1}\sin\phi_{n-1}}{\sqrt{k\,k_1\ldots k_{n-1}}}-\frac{2^n\sin\phi_n}{\sqrt{k\,k_1\ldots k_{n-1}}}\right], \tag{12}$$

where we employ the abbreviations given in (2).

Formula (12) is useful for computation when k is close to 1; in the contrary case, when k is small, the following series converges more rapidly:

$$E(\phi,k)=F(\phi,k)\left[1-\frac{1}{2}k^2\left(1+\frac{1}{2}K_1+\frac{1}{2^2}K_1K_2+\frac{1}{2^3}K_1K_2K_3+\ldots\right)\right]$$
$$+k\left(\frac{1}{2}\sqrt{K_1}\sin\Phi_1+\frac{1}{2^2}\sqrt{K_1K_2}\sin\Phi_2+\frac{1}{2^3}\sqrt{K_1K_2K_3}\sin\Phi_3+\ldots\right), \tag{13}$$

where we use the same abbreviations as in (9).

The complete integrals K and E can be computed from the following series, derived respectively from (8) and (13):

$$K=F\left(\frac{1}{2}\pi,k\right)=\frac{\pi}{2}(1+K_1)(1+K_2)(1+K_3)\ldots, \tag{14}$$

$$E=E\left(\frac{1}{2}\pi,k\right)=F\left(\frac{1}{2}\pi,k\right)\left[1-\frac{1}{2}k^2\left(1+\frac{1}{2}K_1\right.\right.$$
$$\left.\left.+\frac{1}{2^2}K_1K_2+\frac{1}{2^3}K_1K_2K_3+\ldots\right)\right]. \tag{15}$$

Extensive computations of the elliptic integrals are found in the classical tables of A. M. Legendre (1752–1833), which appeared in Volume 2 of his *Traité des fonctions elliptiques et des intégrales Euleriennes avec tables pour en faciliter le calcul numériques*, published in Paris in three volumes between 1825 and 1828. These tables include 9 or 10 place values of $F(\phi,\alpha)$ and $E(\phi,\alpha), k=\sin \alpha$, at intervals of $1°$ for both values. In addition, tables are provided for $F\left(\phi,\frac{1}{4}\pi\right)$ and $E\left(\phi,\frac{1}{4}\pi\right)$ to 12 decimal places for ϕ between $0°$ and $90°$ at intervals of $\frac{1°}{2}$ and for $\log_{10} F\left(\phi,\frac{1}{4}\pi\right)$ and $\log E\left(\phi,\frac{1}{4}\pi\right)$ to 12 decimal places for ϕ between $0°$ and $90°$ at intervals of $0.1°$.

Numerous other tables of elliptic integrals have been computed, descriptions of which will be found in *An Index of Mathematical Tables* by A. Fletcher, J. C. P. Miller, and L. Rosenhead, London, 1946. A short table of both $F(\phi,k)$ and $E(\phi,k)$, based on Legendre's tables, is given below.

In order to find values of the functions beyond $\phi=\frac{1}{2}\pi$, for values of $k^2 \leqq 1$, we make use of the following formulas:

$$F(n\pi+\theta,k)=2nK+F(\theta,k), \; E(n\pi+\theta,k)=2nE+E(\theta,k), \tag{16}$$

$$F(n\pi-\theta,k)=2nK-F(\theta,k), \; E(n\pi-\theta,k)=2nE-E(\theta,k). \tag{17}$$

It is possible to find real values of the integrals for values of $k^2>1$, provided k is less than csc ϕ. The region for which real values of the elliptic integrals exist is shown in Figure 1.

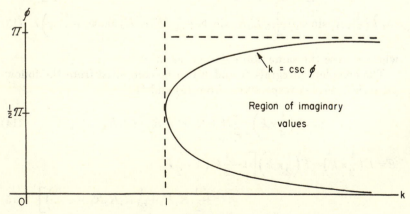

FIGURE 1

In this case we make use of the following transformations:

$$F(\phi,k)=\frac{1}{k}\int_0^{\phi'}\frac{dt}{\sqrt{1-(1/k^2)\sin^2 t}}=\frac{1}{k}\,F(\phi',1/k), \qquad (18)$$

$$E(\phi,k)=(1-k^2)F(\phi,k)+k\int_0^{\phi'}\sqrt{1-(1/k^2)\sin^2 t}\,dt,$$

$$=\left(\frac{1-k^2}{k}\right)F(\phi',1/k)+kE(\phi',1/k), \qquad (19)$$

where we abbreviate: $\phi'=\arcsin\,(k\,\sin\,\phi)$.

PROBLEMS

1. Use formula (11) to compute $F\left(\frac{1}{6}\pi,\frac{1}{2}\right)$.

2. Given $F\left(\frac{1}{6}\pi,\frac{1}{2}\sqrt{3}\right)=0.5422$, find the value of $E\left(\frac{1}{6},\frac{1}{2}\sqrt{3}\right)$ by means of (12).

3. Evaluate the integral: $\int_0^{\pi/6}\frac{d\phi}{\sqrt{1-2\sin^2\phi}}$.   *Answer:* 05841.

4. Show that $\frac{1}{2}\int_0^x\frac{dx}{[x(1-x)(1-k^2x^2)]^{1/2}}=F(\sin^{-1}\sqrt{x},k)$, $0<x<1$.

5. Given that $F[g(x),k]=\int_0^x\frac{dx}{[(1+x^2)(1+k'^2x^2)]^{1/2}}$, find $g(x)$.

6. If $K=\frac{2\sqrt{k}}{1+k}$ and $\tan\,\phi=\frac{\sin 2\psi}{k+\cos 2\psi}$, prove that $F(\phi,k)=\frac{2}{1+k}\,F(\psi,K)$.

7. Prove the identity:
$$(1-k^2)\Pi(\phi,-1,k)=(1-k^2)F(\phi,k)-E(\phi,k)+\tan\,\phi\Delta\phi.$$

8. If $k=(\sqrt{2}-1)/(\sqrt{2}+1)$, prove that $K'\left(\frac{\pi}{2},k\right)=2K\left(\frac{\pi}{2},k\right)$.

9. Given the following integrals:
$$A=\int_0^{\pi/2}\frac{\sin^2\phi}{\Delta\phi}\,d\phi,\qquad B=\int_0^{\pi/2}\frac{\cos^2\phi}{\Delta\phi}\,d\phi,\qquad C=\int_0^{\pi/2}\frac{(\sin\phi\cos\phi)^2}{(\Delta\phi)^3}\,d\phi,$$
show that: $A=(K-E)/k^2$, $B=K-A$, $C=(A-B)/k^2$.

10. Given the identity:
$$F(x,k)+F(y,k)=F\left[\frac{x\sqrt{(1-y^2)(1-k^2y^2)}+y\sqrt{(1-x^2)(1-k^2x^2)}}{1-k^2x^2y^2},k\right],$$

show that for $k=0$, this reduces to the formula for the addition of $\sin^{-1} x$ and $\sin^{-1} y$, and that for $k=1$, to the formula:

$$\tan^{-1} x+\tan^{-1} y=\tan^{-1}\left(\frac{x+y}{1-xy}\right).$$

$$\text{TABLE OF } F(\phi, k)=\int_0^\phi \frac{d\theta}{\sqrt{1-k^2\sin^2\theta}}, \quad k=\sin\alpha.$$

ϕ	$\alpha = \text{arc sin } k$								
	0°	10°	15°	30°	45°	60°	70°	80°	90°
1°	0.0175	0.0175	0.0175	0.0175	0.0175	0.0175	0.0175	0.0175	0.0175
2°	0.0349	0.0349	0.0349	0.0349	0.0349	0.0349	0.0349	0.0349	0.0349
3°	0.0524	0.0524	0.0524	0.0524	0.0524	0.0524	0.0524	0.0524	0.0524
4°	0.0698	0.0698	0.0698	0.0698	0.0698	0.0699	0.0699	0.0699	0.0699
5°	0.0873	0.0873	0.0873	0.0873	0.0873	0.0873	0.0874	0.0874	0.0874
6°	0.1047	0.1047	0.1047	0.1048	0.1048	0.1049	0.1049	0.1049	0.1049
7°	0.1222	0.1222	0.1222	0.1222	0.1223	0.1224	0.1224	0.1225	0.1225
8°	0.1396	0.1396	0.1397	0.1397	0.1399	0.1400	0.1400	0.1401	0.1401
9°	0.1571	0.1571	0.1571	0.1572	0.1574	0.1576	0.1577	0.1577	0.1577
10°	0.1745	0.1746	0.1746	0.1748	0.1750	0.1752	0.1753	0.1754	0.1754
15°	0.2618	0.2619	0.2620	0.2625	0.2633	0.2641	0.2645	0.2647	0.2648
20°	0.3491	0.3493	0.3495	0.3508	0.3526	0.3545	0.3555	0.3561	0.3564
25°	0.4363	0.4367	0.4372	0.4397	0.4433	0.4470	0.4490	0.4504	0.4509
30°	0.5236	0.5243	0.5251	0.5294	0.5356	0.5422	0.5459	0.5484	0.5493
35°	0.6109	0.6119	0.6133	0.6200	0.6300	0.6408	0.6471	0.6513	0.6528
40°	0.6981	0.6997	0.7016	0.7116	0.7267	0.7436	0.7535	0.7604	0.7629
45°	0.7854	0.7876	0.7903	0.8044	0.8260	0.8512	0.8665	0.8774	0.8814
50°	0.8727	0.8756	0.8792	0.8982	0.9283	0.9647	0.9876	1.0044	1.0107
55°	0.9599	0.9637	0.9683	0.9683	1.0337	1.0848	1.1186	1.1444	1.1542
60°	1.0472	1.0519	1.0577	1.0896	1.1424	1.2125	1.2619	1.3014	1.3170
65°	1.1345	1.1402	1.1474	1.1869	1.2545	1.3489	1.4199	1.4810	1.5065
70°	1.2217	1.2286	1.2373	1.2853	1.3697	1.4944	1.5959	1.6918	1.7354
75°	1.3090	1.3171	1.3273	1.3846	1.4879	1.6492	1.7927	1.9468	2.0276
80°	1.3963	1.4056	1.4175	1.4846	1.6085	1.8125	2.0119	2.2653	2.4362
81°	1.4137	1.4234	1.4356	1.5046	1.6328	1.8461	2.0584	2.3387	2.5421
82°	1.4312	1.4411	1.4536	1.5247	1.6572	1.8799	2.1057	2.4157	2.6603
83°	1.4486	1.4588	1.4717	1.5448	1.6817	1.9140	2.1537	2.4965	2.7942
84°	1.4661	1.4765	1.4897	1.5649	1.7062	1.9482	2.2024	2.5811	2.9487
85°	1.4835	1.4942	1.5078	1.5850	1.7308	1.9826	2.2518	2.6694	3.1313
86°	1.5010	1.5120	1.5259	1.6052	1.7554	2.0172	2.3017	2.7612	3.3547
87°	1.5184	1.5297	1.5439	1.6253	1.7801	2.0519	2.3520	2.8561	3.6425
88°	1.5359	1.5474	1.5620	1.6454	1.8047	2.0867	2.4026	2.9537	4.0481
89°	1.5533	1.5651	1.5801	1.6656	1.8294	2.1216	2.4535	3.0530	4.7413
90°	1.5708	1.5828	1.5981	1.6858	1.8541	2.1565	2.5046	3.1534	∞

$$\text{TABLE OF } E(\phi, k) = \int_0^{\phi} \sqrt{1 - k^2 \sin^2 \theta} \, d\theta, \ k = \sin \alpha$$

ϕ	$\alpha = \text{arc sin } k$								
	0°	10°	15°	30°	45°	60°	70°	80°	90°
1°	0.0175	0.0175	0.0175	0.0175	0.0175	0.0175	0.0175	0.0175	0.0175
2°	0.0349	0.0349	0.0349	0.0349	0.0349	0.0349	0.0349	0.0349	0.0349
3°	0.0524	0.0524	0.0524	0.0524	0.0523	0.0523	0.0523	0.0523	0.0523
4°	0.0698	0.0698	0.0698	0.0698	0.0698	0.0698	0.0698	0.0698	0.0698
5°	0.0873	0.0873	0.0873	0.0872	0.0872	0.0872	0.0872	0.0872	0.0872
6°	0.1047	0.1047	0.1047	0.1047	0.1046	0.1046	0.1046	0.1045	0.1045
7°	0.1222	0.1222	0.1222	0.1221	0.1220	0.1220	0.1219	0.1219	0.1219
8°	0.1396	0.1396	0.1396	0.1395	0.1394	0.1393	0.1392	0.1392	0.1392
9°	0.1571	0.1571	0.1570	0.1569	0.1568	0.1566	0.1565	0.1565	0.1564
10°	0.1745	0.1745	0.1745	0.1743	0.1741	0.1739	0.1738	0.1737	0.1736
15°	0.2618	0.2617	0.2616	0.2611	0.2603	0.2596	0.2592	0.2589	0.2588
20°	0.3491	0.3489	0.3486	0.3473	0.3456	0.3438	0.3429	0.3422	0.3420
25°	0.4363	0.4359	0.4354	0.4330	0.4296	0.4261	0.4243	0.4230	0.4226
30°	0.5236	0.5229	0.5221	0.5179	0.5120	0.5061	0.5029	0.5007	0.5000
35°	0.6109	0.6098	0.6085	0.6019	0.5928	0.5833	0.5782	0.5748	0.5736
40°	0.6981	0.6966	0.6947	0.6851	0.6715	0.6575	0.6497	0.6446	0.6428
45°	0.7854	0.7832	0.7806	0.7672	0.7482	0.7282	0.7171	0.7097	0.7071
50°	0.8727	0.8698	0.8663	0.8483	0.8227	0.7954	0.7801	0.7697	0.7660
55°	0.9599	0.9562	0.9517	0.9284	0.8949	0.8588	0.8382	0.8242	0.8192
60°	1.0472	1.0426	1.0368	1.0076	0.9650	0.9184	0.8914	0.8728	0.8660
65°	1.1345	1.1288	1.1218	1.0858	1.0329	0.9743	0.9397	0.9152	0.9063
70°	1.2217	1.2149	1.2065	1.1632	1.0990	1.0266	0.9830	0.9514	0.9397
75°	1.3090	1.3010	1.2911	1.2399	1.1635	1.0759	1.0217	0.9814	0.9659
80°	1.3963	1.3870	1.3755	1.3161	1.2266	1.1225	1.0565	1.0054	0.9848
81°	1.4137	1.4042	1.3924	1.3312	1.2391	1.1316	1.0630	1.0096	0.9877
82°	1.4312	1.4214	1.4093	1.3464	1.2516	1.1406	1.0695	1.0135	0.9903
83°	1.4486	1.4386	1.4261	1.3616	1.2640	1.1495	1.0758	1.0173	0.9925
84°	1.4661	1.4558	1.4430	1.3767	1.2765	1.1584	1.0821	1.0209	0.9945
85°	1.4835	1.4729	1.4598	1.3919	1.2889	1.1673	1.0882	1.0244	0.9962
86°	1.5010	1.4901	1.4767	1.4070	1.3012	1.1761	1.0944	1.0277	0.9976
87°	1.5184	1.5073	1.4936	1.4221	1.3136	1.1848	1.1004	1.0309	0.9986
88°	1.5359	1.5245	1.5104	1.4372	1.3260	1.1936	1.1064	1.0340	0.9994
89°	1.5533	1.5417	1.5273	1.4523	1.3383	1.2023	1.1124	1.0371	0.9998
90°	1.5708	1.5589	1.5442	1.4675	1.3506	1.2111	1.1184	1.0401	1.0000

7. Gauss's Limit

Closely related to the methods of Landen is a limit due to K. F. Gauss,* which we describe as follows:

Given $k'=\sqrt{1-k^2}$, $a_1=\frac{1}{2}(1+k')$, $b_1=\sqrt{1\cdot k'}$, $a_n=\frac{1}{2}(a_{n-1}+b_{n-1})$,

$b_n=\sqrt{a_{n-1}\cdot b_{n-1}}$, then we have as limits:

$$\lim_{n=\infty} a_n=\lim_{n=\infty} b_n=M(1,k')=\pi/2K, \quad M(1,k)=\pi/2K'.$$

As an example, let us set $k'=\frac{1}{2}$. We then obtain

$$\begin{array}{ll}
a_1=0.750000\,, & b_1=0.707107, \\
a_2=0.728553\,, & b_2=0.728238, \\
a_3=0.7283955, & b_3=0.7283955.
\end{array}$$

From the tables we find, corresponding to $k=\frac{1}{2}\sqrt{3}$, that is, $\alpha=60°$, the value $K=2.15651\ 5648$. Dividing $\pi=3.14159\ 2654$ by $2K$ we obtain

$$\frac{\pi}{2K}=0.72839\ 5515\cdot$$

This remarkable limit of Gauss has been further investigated by H. Geppert,† who has shown that similar limits exist for the *arithmetic-harmonic* means of 1 and k', and for the *geometric-harmonic* means of 1 and k'.

Thus, in the first case, let us write:

$$a_1=\frac{1}{2}(1+k'), \quad b_1=\frac{2k'}{1+k'}, \text{ (the harmonic mean of 1 and } k'),$$

$$a_n=\frac{1}{2}(a_{n-1}+b_{n-1}), \quad b_n=\frac{2a_{n-1}b_{n-1}}{a_{n-1}+b_{n-1}}.$$

It can then be shown that

$$\lim_{n=\infty} a_n=\lim_{n=\infty} b_n=\sqrt{k'}.$$

In the second case, let us write

$$a_1=\sqrt{1\cdot k}, \qquad b_1=\frac{2k'}{1+k'},$$

$$a_n=\sqrt{a_{n-1}\cdot b_{n-1}}, \quad b_n=\frac{2a_{n-1}\cdot b_{n-1}}{a_{n-1}+b_{n-1}}.$$

We then have the limits

$$\lim_{n=\infty} a_n=\lim_{n=\infty} b_n=2k'K/\pi.$$

---

* *Werke*, Vol. 3, pp. 361–387, in particular, p. 370.
†"Über iterative Algorithmen," *Mathematische Annalen*, Vol. 107, 1933, pp. 387–399.

ELLIPTIC FUNCTIONS

8. The Elliptic Functions of Jacobi

The elliptic functions described in this section were first defined simultaneously, but independently, by C. G. J. Jacobi (1804–51) and N. H. Abel (1802–29) in 1827, although K. F. Gauss (1777–1855) had developed many of their properties as early as 1809. The theory of these functions, as well as that of Theta functions, was exhaustively set forth in Jacobi's great treatise *Fundamenta nova theoriae functionum ellipticarum*, published in 1829 in Königsberg.

The elliptic functions of Jacobi are defined as inverses of the elliptic integral of first kind. Thus, if we write

$$u = \int_0^\phi \frac{d\phi}{\sqrt{1 - k^2 \sin^2 \phi}}, \tag{1}$$

we then define the following functions:

$$\text{sn}\,(u, k) = \sin\,\phi, \ \text{cn}\,(u, k) = \cos\phi, \ \text{dn}\,(u, k) = \sqrt{1 - k^2 \sin^2 \phi} = \Delta(\phi),$$

$$\text{am}\,(u, k) = \phi, \ \text{tn}\,(u, k) = \frac{\text{sn}\,(u, k)}{\text{cn}\,(u, k)} = \tan\,\phi. \tag{2}$$

In some applications, where k is a fixed value, or where the discussion concerns only the variable u, it is convenient to write sn u for sn (u, k), cn u for cn (u, k), etc.

As in the case of the circular functions, it is often important to have a notation for the inverses of the Jacobi elliptic functions. It is clear that we can write,

$$u = \text{sn}^{-1}(\sin\,\phi, k) = \text{cn}^{-1}(\cos\,\phi, k) = \text{dn}^{-1}(\Delta\phi, k),$$

$$= \text{am}^{-1}(\phi, k) = \text{tn}^{-1}(\tan\,\phi, k). \tag{3}$$

Since $u = F(\phi, k)$, these values are readily computed from tables of the elliptic integral. Thus we have

$$\text{sn}^{-1}\left(\frac{1}{2}, \frac{1}{2}\sqrt{3}\right) = F(30°, 60°) = 0.54223.$$

9. Properties of the Elliptic Functions of Jacobi

The Jacobi elliptic functions are rich in special values and identical relationships. A few of these are given below as follows:

$$\text{sn}(0) = 0, \quad \text{cn}(0) = 1, \quad \text{dn}(0) = 1, \quad \text{am}(0) = 0; \tag{1}$$

$$\text{sn}^2 u + \text{cn}^2 u = 1, \tag{2}$$

$$dn^2u - k^2cn^2u = 1 - k^2 = k'^2, \tag{3}$$

$$k^2sn^2u + dn^2u = 1; \tag{4}$$

$$sn(-u) = -sn\ u, \quad cn(-u) = cn\ u, \quad dn(-u) = dn\ u,$$
$$am(-u) = -am\ u; \tag{5}$$

$$sn(u, 0) = \sin u, \quad cn(u,0) = \cos u, \quad dn(u, 0) = 1; \tag{6}$$

$$sn(u,1) = \tanh u = \frac{e^u - e^{-u}}{e^u + e^{-u}}, \quad cn(u,1) = dn(u,1) = \operatorname{sech} u = \frac{1}{e^u + e^{-u}}. \tag{7}$$

Introducing the imaginary argument ui, we obtain the following identities:

$$sn(ui,k) = i\ \frac{sn(u,k')}{cn(u,k')}, \quad cn(ui,k) = \frac{1}{cn(u,k')}, \quad dn(ui,k) = \frac{dn(u,k')}{cn(u,k')};$$

$$sn(u,k) = -i\ \frac{sn(ui,k')}{cn(ui,k')}, \quad cn(u,k) = \frac{1}{cn(ui,k')}, \quad dn(u,k) = \frac{dn(ui,k')}{cn(ui,k')}. \tag{8}$$

Elliptic functions belong to the class of doubly periodic functions in which $2K$ and $K'i$ play roles similar to π in the theory of the circular functions and πi in the theory of the hyperbolic functions. The nature of this relationship is shown by the following identities:

$$sn(u \pm K) = \pm \frac{cn\ u}{dn\ u}, \qquad sn(u \pm 2K) = -sn\ u; \tag{9}$$

$$sn(u \pm 3K) = \mp \frac{cn\ u}{dn\ u}, \quad sn(u \pm 4K) = sn\ u, \quad sn(u + K'i) = \frac{1}{k\ sn\ u} \tag{10}$$

$$sn(u + 2mK + 2nK'i) = (-1)^m sn\ u, \text{ where } m \text{ and } n \text{ are integers;} \tag{11}$$

$$cn(u \pm K) = \mp k'\ \frac{cn\ u}{dn\ u}, \quad cn(u \pm 2K) = -cn\ u; \tag{12}$$

$$cn(u \pm 3K) = \pm k'\ \frac{sn\ u}{dn\ u}, \quad sn(u \pm 4K) = cn\ u, \quad cn(u + K'i) = -i\ \frac{dn\ u}{k\ sn\ u}; \tag{13}$$

$$cn(u + 2mK + 2nK'i) = (-1)^{m+n} cn\ u, \text{where } m \text{ and } n \text{ are integers;} \tag{14}$$

$$dn(u \pm K) = \frac{k'}{dn\ u}, \quad dn(u \pm 2K) = dn\ u; \tag{15}$$

$$dn\ (u + K'i) = -i\ \frac{cn\ u}{sn\ u}; \tag{16}$$

$$dn(u + 2mK + 2nK'i) = (-1)^n\ du, \text{ where } m \text{ and } n \text{ are integers.} \tag{17}$$

Most of the identities just given follow readily from the definitions. Thus, to prove (2), we merely observe: $\text{sn}^2 u + \text{cn}^2 u = \sin^2\phi + \cos^2\phi = 1$.

To establish such an identity as $\text{sn}(u+2K) = -\text{sn } u$, we consider the integral

$$v = \int_0^{\phi+\pi} \frac{d\phi}{\Delta(\phi)} = \int_0^\pi \frac{d\phi}{\Delta(\phi)} + \int_\pi^{\phi+\pi} \frac{d\phi}{\Delta(\phi)},$$

$$= 2K + \int_\pi^{\phi+\pi} \frac{d\phi}{\sqrt{1-k^2\sin^2\phi}}. \tag{18}$$

We now make the transformation: $\phi = \pi + \theta$, and thus obtain

$$v = 2K + \int_0^\phi \frac{d\theta}{\sqrt{1-k^2 \sin^2 \theta}} = 2K + u. \tag{19}$$

Therefore, since by (18) we have

$$\text{sn } v = \sin (\phi+\pi) = -\sin \phi,$$

we now establish by (19) the identity

$$\text{sn}(u+2K) = -\text{sn } u.$$

Such an identity as $\text{sn}(u+K) = \text{cn } u/\text{dn } u$ is obtained from the addition formulas, which are given in Section 11.

10. Derivatives and Integrals of the Elliptic Functions

The derivatives of the elliptic functions are easily found if we make use of the relationship

$$\frac{du}{d\phi} = \frac{1}{\sqrt{1-k^2 \sin^2 \phi}} = \frac{1}{\text{dn } u}. \tag{1}$$

Thus, since $\text{sn } u = \sin \phi$, we get

$$\frac{d}{du} \text{ sn } u = \cos \phi \frac{d\phi}{du} = \text{cn } u \text{ dn } u. \tag{2}$$

The following derivatives are similarly obtained:

$$\frac{d^2}{du^2} \text{ sn } u = 2k^2 \text{ sn}^3 u - (1+k^2) \text{ sn } u, \tag{3}$$

$$\frac{d}{du} \text{ cn } u = -\text{sn } u \text{ dn } u, \tag{4}$$

$$\frac{d^2}{du^2} \text{ cn } u = (2k^2-1) \text{ cn } u - 2k^2 \text{ cn}^3 u, \tag{5}$$

$$\frac{d}{du} \operatorname{dn} u = -k^2 \operatorname{sn} u \operatorname{cn} u, \tag{6}$$

$$\frac{d^2}{du^2} \operatorname{dn} u = (2-k^2)\, 2 \operatorname{dn} u - 2 \operatorname{dn}^3 u, \tag{7}$$

$$\frac{d}{du} \operatorname{am} u = \operatorname{dn} u. \tag{8}$$

We also record the following table of integrals, which is readily verified by differentiation:

$$\int \operatorname{sn} u\, du = \frac{1}{k} \log (\operatorname{dn} u - k \operatorname{cn} u), \tag{9}$$

$$\int \operatorname{cn} u\, du = \frac{1}{k} \operatorname{arc\ cos} (\operatorname{dn} u), \tag{10}$$

$$\int \operatorname{dn} u\, du = \operatorname{arc\ sin} (\operatorname{sn} u), \tag{11}$$

$$\int \frac{1}{\operatorname{sn} u}\, du = \log \left(\frac{\operatorname{dn} u - \operatorname{cn} u}{\operatorname{sn} u} \right), \tag{12}$$

$$\int \frac{1}{\operatorname{cn} u}\, du = \frac{1}{k'} \log \left(\frac{\operatorname{dn} u + k' \operatorname{sn} u}{\operatorname{cn} u} \right), \tag{13}$$

$$\int \frac{1}{\operatorname{dn} u}\, du = \frac{1}{k'} \operatorname{arc\ cos} \left(\frac{\operatorname{cn} u}{\operatorname{dn} u} \right), \tag{14}$$

$$\int \frac{\operatorname{cn} u}{\operatorname{sn} u}\, du = \log \left(\frac{1 - \operatorname{dn} u}{\operatorname{sn} u} \right), \tag{15}$$

$$\int \frac{\operatorname{sn} u}{\operatorname{cn} u}\, du = \frac{1}{k'} \log \left(\frac{\operatorname{dn} u + k'}{\operatorname{cn} u} \right), \tag{16}$$

$$\int \frac{\operatorname{dn} u}{\operatorname{cn} u}\, du = \log \left(\frac{1 + \operatorname{sn} u}{\operatorname{cn} u} \right). \tag{17}$$

11. Addition Theorems

The following identities exhibit the addition properties of the Jacobian elliptic functions:

$$\operatorname{sn} (u \pm v) = \frac{\operatorname{sn} u \operatorname{cn} v \operatorname{dn} v \pm \operatorname{cn} u \operatorname{sn} v \operatorname{dn} u}{1 - k^2 \operatorname{sn}^2 u \operatorname{sn}^2 v}; \tag{1}$$

$$\operatorname{cn} (u \pm v) = \frac{\operatorname{cn} u \operatorname{cn} v \mp \operatorname{sn} u \operatorname{sn} v \operatorname{dn} u \operatorname{dn} v}{1 - k^2 \operatorname{sn}^2 u \operatorname{sn}^2 v}, \tag{2}$$

$$\text{cn}(u \pm v) = \text{cn}\ u\ \text{cn}\ v \mp \text{sn}\ u\ \text{sn}\ v\ \text{dn}(u \pm v);\ ^*$$

$$\text{dn}(u \pm v) = \frac{\text{dn}\ u\ \text{dn}\ v \mp k^2\ \text{sn}\ u\ \text{sn}\ v\ \text{cn}\ u\ \text{cn}\ v}{1 - k^2\ \text{sn}^2 u\ \text{sn}^2 v},$$

$$= \text{dn}\ u\ \text{dn}\ v \mp k^2\ \text{sn}\ u\ \text{sn}\ v\ \text{cn}(u \pm v); \tag{3}$$

$$\text{tn}(u \pm v) = \frac{\text{tn}\ u\ \text{dn}\ v \pm \text{tn}\ v\ \text{dn}\ u}{1 \mp \text{tn}\ u\ \text{tn}\ v\ \text{dn}\ u\ \text{dn}\ v}; \tag{4}$$

$$\text{sn}(u+v) + \text{sn}(u-v) = \frac{2\ \text{sn}\ u\ \text{cn}\ v\ \text{dn}\ v.}{1 - k^2\ \text{sn}^2 u\ \text{sn}^2 v}; \tag{5}$$

$$\text{sn}(u+v) - \text{sn}(u-v) = \frac{2\ \text{sn}\ v\ \text{cn}\ u\ \text{dn}\ u.}{1 - k^2\ \text{sn}^2 u\ \text{sn}^2 v}; \tag{6}$$

$$\text{cn}(u+v) + \text{cn}(u-v) = \frac{2\ \text{cn}\ u\ \text{cn}\ v}{1 - k^2 \text{sn}^2 u\ \text{sn}^2 v}; \tag{7}$$

$$\text{cn}(u+v) - \text{cn}(u-v) = -\frac{2\ \text{sn}\ u\ \text{sn}\ v\ \text{dn}\ u\ \text{dn}\ v.}{1 - k^2\ \text{sn}^2 u\ \text{sn}^2 v}; \tag{8}$$

$$\text{dn}(u+v) + \text{dn}(u-v) = \frac{2\text{dn}\ u\ \text{dn}\ v}{1 - k^2\ \text{sn}^2 u\ \text{sn}^2 v}; \tag{9}$$

$$\text{dn}(u+v) - \text{dn}(u-v) = -\frac{2k^2\ \text{sn}\ u\ \text{sn}\ v\ \text{cn}\ u\ \text{cn}\ v}{1 - k^2\ \text{sn}^2 u\ \text{sn}^2 v} \tag{10}$$

These formulas have been established in a number of ways, but considerable computation is usually involved in verifying them. One method which is quite effective is illustrated as follows. To establish the formula for $\text{sn}(u+v)$, let us represent the right hand member of (1) by $F(u, v)$. This function is also a function of $z = u+v$, and consequently the Jacobian of F and z must vanish, that is

$$\begin{vmatrix} \dfrac{\partial F}{\partial u} & \dfrac{\partial F}{\partial v} \\[2mm] \dfrac{\partial z}{\partial u} & \dfrac{\partial z}{\partial v} \end{vmatrix} = \frac{\partial F}{\partial u} - \frac{\partial F}{\partial v} = 0. \tag{11}$$

The proof of formula (1) is thus given by first establishing (11). The computation is tedious, but straightforward, and can be facilitated by first taking the logarithm of F. Since we have now shown that $F(u, v) = F(u+v)$, identification of the function with the elliptic sine is made by setting $v=0$, from which we have $F(u) = \text{sn}\ u$.

The products of the functions of sums and differences can be derived from the preceding identities.   Thus, we have

$$\operatorname{sn}(u+v)\,\operatorname{sn}(u-v)=\frac{\operatorname{sn}^2u-\operatorname{sn}^2v}{1-k^2\operatorname{sn}^2u\,\operatorname{sn}^2v}=\frac{\operatorname{cn}^2v+\operatorname{sn}^2u\,\operatorname{dn}^2v}{1-k^2\operatorname{sn}^2u\,\operatorname{sn}^2v}-1,$$

$$=\frac{1}{k^2}\frac{\operatorname{dn}^2v+k^2\operatorname{sn}^2u\,\operatorname{cn}^2v}{1-k^2\operatorname{sn}^2u\,\operatorname{sn}^2v}-1; \tag{12}$$

$$\operatorname{cn}(u+v)\operatorname{cn}(u-v)=\frac{\operatorname{cn}^2u-\operatorname{sn}^2v+k^2\,\operatorname{sn}^2u\,\operatorname{sn}^2v}{1-k^2\operatorname{sn}^2u\,\operatorname{sn}^2v},$$

$$=\frac{\operatorname{cn}^2u+\operatorname{cn}^2v}{1-k^2\operatorname{sn}^2u\,\operatorname{sn}^2v}-1=1-\frac{\operatorname{sn}^2u\,\operatorname{dn}^2v+\operatorname{sn}^2v\,\operatorname{dn}^2u}{1-k^2\operatorname{sn}^2u\,\operatorname{sn}^2v}; \tag{13}$$

$$\operatorname{dn}(u+v)\,\operatorname{dn}(u-v)=\frac{1-k^2\operatorname{sn}^2u-k^2\operatorname{sn}^2v+k^2\operatorname{sn}^2u\,\operatorname{sn}^2v}{1-k^2\operatorname{sn}^2u\,\operatorname{sn}^2v},$$

$$=\frac{\operatorname{dn}^2u+\operatorname{dn}^2v}{1-k^2\operatorname{sn}^2u\,\operatorname{sn}^2v}-1; \tag{14}$$

$$\operatorname{sn}(u\pm v)\operatorname{cn}(u\mp v)=\frac{\operatorname{sn}u\operatorname{cn}u\operatorname{dn}v\pm\operatorname{sn}v\operatorname{cn}v\operatorname{dn}u}{1-k^2\operatorname{sn}^2u\operatorname{sn}^2v}; \tag{15}$$

$$\operatorname{sn}(u\pm v)\operatorname{dn}(u\mp v)=\frac{\operatorname{sn}u\operatorname{dn}u\operatorname{cn}v\pm\operatorname{sn}v\operatorname{dn}v\operatorname{cn}u}{1-k^2\operatorname{sn}^2u\operatorname{sn}^2v}; \tag{16}$$

$$\operatorname{cn}(u\pm v)\operatorname{dn}(u\mp v)=\frac{\operatorname{cn}u\operatorname{dn}u\operatorname{cn}v\operatorname{dn}v\mp k'^2\operatorname{sn}u\operatorname{sn}v}{1-k^2\operatorname{sn}^2u\operatorname{sn}^2v}; \tag{17}$$

$$[1\pm\operatorname{sn}(u+v)]\,[1\pm\operatorname{sn}(u-v)]=\frac{(\operatorname{cn}v\pm\operatorname{sn}u\operatorname{dn}v)^2}{1-k^2\operatorname{sn}^2u\operatorname{sn}^2v}. \tag{18}$$

12. Double-Angle and Half-Angle Formulas

From the identities of the preceding section we can derive the following formulas involving double-angles and half-angles:

$$\operatorname{sn}2u=\frac{2\operatorname{sn}u\operatorname{cn}u\operatorname{dn}u}{1-k^2\operatorname{sn}^4u}; \tag{1}$$

$$\operatorname{cn}2u=\frac{\operatorname{cn}^2u-\operatorname{sn}^2u\operatorname{dn}^2u}{1-k^2\operatorname{sn}^4u}=\frac{1-2\operatorname{sn}^2u+k^2\operatorname{sn}^4u}{1-k^2\operatorname{sn}^4u}; \tag{2}$$

$$\operatorname{dn}2u=\frac{\operatorname{dn}^2u-k^2\operatorname{sn}^2u\operatorname{cn}^2u}{1-k^2\operatorname{sn}^4u}=\frac{1-2k^2\operatorname{sn}^2u+k^2\operatorname{sn}^4u}{1-k^2\operatorname{sn}^4u}; \tag{3}$$

$$\operatorname{tn}2u=\frac{2\operatorname{tn}u\operatorname{dn}u}{1-\operatorname{tn}^2u\operatorname{dn}^2u}. \tag{4}$$

Using formulas (2) and (3) we compute

$$\frac{1-\operatorname{cn} 2u}{1+\operatorname{dn} 2n}=\frac{2\operatorname{sn}^2 u-2k^2\operatorname{sn} u}{2-2k^2\operatorname{sn}^2 u}=\frac{2\operatorname{sn}^2 u(1-k^2\operatorname{sn}^2 u)}{2(1-k^2\operatorname{sn}^2 u)}=\operatorname{sn}^2 u. \tag{5}$$

Replacing u by $\frac{1}{2}u$, we then obtain the formula

$$\operatorname{sn}^2 \frac{1}{2}u=\frac{1-\operatorname{cn} u}{1+\operatorname{dn} u}. \tag{6}$$

If we multiply the numerator and denominator of this fraction by $(1-\operatorname{dn} u)$ and make use of the identities of Section 9, we obtain

$$\frac{(1-\operatorname{cn} u)(1-\operatorname{dn} u)}{(1+\operatorname{dn} u)(1-\operatorname{dn} u)}=\frac{(1-\operatorname{cn} u)(1-\operatorname{dn} u)}{1-\operatorname{dn}^2 u}=\frac{(1-\operatorname{cn} u)(1-\operatorname{dn} u)}{k^2\operatorname{sn}^2 u}$$

$$=\frac{(1-\operatorname{cn} u)(1-\operatorname{dn} u)}{k^2(1-\operatorname{cn}^2 u)}=\frac{1-\operatorname{cn} u}{k^2(1+\operatorname{cn} u)}. \tag{7}$$

Similarly, if we multiply the numerator and denominator of the fraction in (6) by $(k'^2-k^2\operatorname{cn} u+\operatorname{dn} u)$ and simplify, we obtain

$$\operatorname{sn}^2 \frac{1}{2}u=\frac{\operatorname{dn} u-\operatorname{cn} u}{k'^2+\operatorname{dn} u-k^2\operatorname{cn} u}. \tag{8}$$

Combining these results, we get the following identities:

$$\sin^2 \frac{1}{2}u=\frac{1-\operatorname{cn} u}{1+\operatorname{dn} u}=\frac{1-\operatorname{dn} u}{k^2(1+\operatorname{cn} u)}=\frac{\operatorname{dn} u-\operatorname{cn} u}{k'^2+\operatorname{dn} u-k^2\operatorname{cn} u}. \tag{9}$$

In a similar manner the following half-angle formulas are established:

$$\operatorname{cn}^2\frac{1}{2}u=\frac{\operatorname{dn} u+\operatorname{cn} u}{1+\operatorname{dn} u}=\frac{k^2\operatorname{cn} u-k'^2+\operatorname{dn} u}{k^2(1+\operatorname{cn} u)}=\frac{k'^2(1+\operatorname{cn} u)}{k'^2+\operatorname{dn} u-k^2\operatorname{cn} u}; \tag{10}$$

$$\operatorname{dn}^2 \frac{1}{2}u=\frac{k'^2+\operatorname{dn} u+k^2\operatorname{cn} u}{1+\operatorname{dn} u}=\frac{\operatorname{cn} u+\operatorname{dn} u}{1+\operatorname{cn} u},=\frac{k'^2(1+\operatorname{dn} u)}{k'^2+\operatorname{dn} u-k^2\operatorname{cn} u}; \tag{11}$$

$$\operatorname{tn}^2 \frac{1}{2}u=\frac{1-\operatorname{cn} u}{\operatorname{dn} u+\operatorname{cn} u}=\frac{1-\operatorname{dn} u}{k^2\operatorname{cn} u-k'^2+\operatorname{dn} u}=\frac{\operatorname{dn} u-\operatorname{cn} u}{k'^2(1+\operatorname{cn} u)}. \tag{12}$$

PROBLEMS

Establish the following values:

1. $\operatorname{sn} K=1$, $\quad \operatorname{sn} iK'=\infty$, $\quad \operatorname{sn}(K+iK')=1/k$;
   $\operatorname{cn} K=0$, $\quad \operatorname{cn} iK'=\infty$, $\quad \operatorname{cn}(K+iK')=-ik'/k$;
   $\operatorname{dn} K=k'$, $\quad \operatorname{dn} iK'=\infty$, $\quad \operatorname{dn}(K+iK')=0$. $\tag{13}$

2.

$$\operatorname{sn}\tfrac{1}{2}K=(1+k')^{-1/2},\ \operatorname{sn}\tfrac{1}{2}iK'=ik^{-1/2},\ \operatorname{sn}\tfrac{1}{2}(K+iK')=(2k)^{-1/2}[(1+k)^{1/2}+i(1-k)^{1/2}];$$

$$\operatorname{cn}\tfrac{1}{2}K=[k'/(1+k')]^{1/2},\ \operatorname{cn}\tfrac{1}{2}iK'=[(1+k)/k]^{1/2},\ \operatorname{cn}\tfrac{1}{2}(K+iK')=(1-i)(k'/2k)^{1/2};$$

$$\operatorname{dn}\tfrac{1}{2}K=k'^{1/2},\ \operatorname{dn}\tfrac{1}{2}iK'=(1+k)^{1/2},\ \operatorname{dn}\tfrac{1}{2}(K+iK')=\left(\tfrac{1}{2}k'\right)^{1/2}[(1+k')^{1/2}-i(1-k')^{1/2}].$$

(14)

13. Expansions of the Elliptic Functions in Powers of u

By evaluating successive derivatives of the Jacobi elliptic functions at $u=0$, the following expansions of the functions have been obtained:*

$$\operatorname{sn}(u,k)=u-(1+k^2)\frac{u^3}{3!}+(1+14k^2+k^4)\frac{u^5}{5!}-(1+135k^2+135k^4+k^6)\frac{u^7}{7!}$$

$$+(1+1228k^2+5478k^4+1228k^6+k^8)\frac{u^9}{9!}$$

$$-(1+11069k^2+165826k^4+165826k^6$$

$$+11069k^8+k^{10})\frac{u^{11}}{11!}+\ \ldots\ ;\qquad (1)$$

$$\operatorname{cn}(u,k)=1-\frac{u^2}{2!}+(1+4k^2)\frac{u^4}{4!}-(1+44k^2+16k^4)\frac{u^6}{6!}$$

$$+(1+408k^2+912k^4+64k^6)\frac{u^8}{8!}-(1+3688k^2+30768k^4$$

$$+15808k^6+256k^8)\frac{u^{10}}{10!}+(1+33212k^2+8070640k^4+1538560k^6$$

$$+259328k^8+1024k^{10})\frac{u^{12}}{12!}+\ \ldots\ ;\qquad (2)$$

$$\operatorname{dn}(u,k)=1-k^2\frac{u^2}{2!}+(4k^2+k^4)\frac{u^4}{4!}-(16k^2+44k^4+k^6)\frac{u^6}{6!}$$

$$+(64k^2+912k^4+408k^6+k^8)\frac{u^8}{8!}-(256k^2+15808k^4$$

$$+30768k^6+3688k^8+k^{10})\frac{u^{10}}{10!}+(1024k^2+259328k^4$$

$$+1538560k^6+870640k^8+33212k^{10}+k^{12})\frac{u^{12}}{12!}+\ \ldots\ ;\qquad (3)$$

*These coefficients are taken from C. Gudermann: "Theorie der Modular-Functionen und der Modular-Integrale, "*Journal für Math.*, Vol. 19, 1839, pp. 45–83; in particular, pp. 79–81.

$$\text{am}(u,k) = u - k^2 \frac{u^3}{3!} + (4k^2 + k^4)\frac{u^5}{5!} - (16k^2 + 44k^4 + k^6)\frac{u^7}{7!}$$

$$+ (64k^2 + 912k^4 + 408k^6 + k^8)\frac{u^9}{9!}$$

$$- (256k^2 + 15808k^4 + 30768k^6 + 3688k^8 + k^{10})\frac{u^{11}}{11!}$$

$$+ (1024k^2 + 259328k^4 + 1538560k^6 + 870640k^8 + 33212k^{10}$$

$$+ k^{12})\frac{u^{13}}{13!} + \ldots \quad (4)$$

14. The Poles of the Elliptic Functions

Both the circular and the hyperbolic functions have zeros in the finite plane and both are periodic, but neither sin x, cos x, nor sinh x, cosh x have poles. The corresponding elliptic functions, however, are doubly periodic and have polar singularities in the finite plane. This difference is readily illustrated by the properties of the elliptic sine.

The function sn$(x.k)$ is doubly periodic with the periods: $\Omega_1 = 4K$ and $\Omega_2 = 2K'i$. If, in the complex plane, one forms a set of rectangles with corners at the points: $4mK + 2nK'i$, as shown in Figure 2, then the behavior of sn(x, k) in each of the rectangles is identical by virtue of the periodic properties given by equation (11) of Section 9.

The function sn(x,k) is analytic except at the points: $4mK + (2n+1)K'i$, where it has simple poles of residue $1/k$, and at the points: $(4m+2)K + (2n+1)K'i$, where it has simple poles of residue $-1/k$.

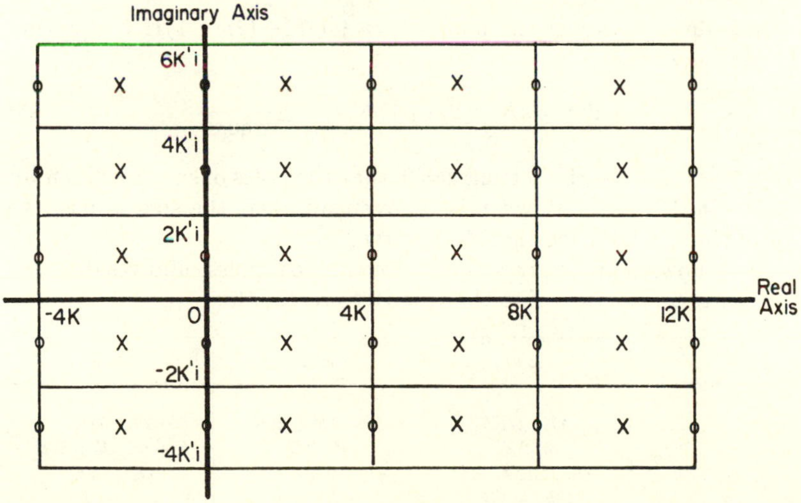

FIGURE 2

The position of these poles is indicated in Figure 2 by circles (poles of residue $1/k$) and crosses (poles of residue $-1/k$). The zeros of $\operatorname{sn}(x,k)$ are found at the points: $2mK+2nK'i$.

From this it is clear that $\operatorname{sn}(x,k)$ may be expected to have a property similar to that of $\tan \theta$, which can be expressed in terms of its reciprocal by a linear transformation of θ, that is

$$\tan (\theta+\frac{1}{2}\pi)=-\frac{1}{\tan \theta}.$$

That this is, indeed, the case can be shown as follows:
From formula (1) of Section 11 we have

$$\operatorname{sn}(u+v)=\frac{\operatorname{sn} u \operatorname{cn} u \operatorname{dn} v+\operatorname{cn} u \operatorname{sn} v \operatorname{dn} u}{1-k^2 \operatorname{sn}^2 u \operatorname{sn}^2 v}, \tag{1}$$

where $\operatorname{sn} u$, $\operatorname{cn} u$, and $\operatorname{dn} u$ are connected by the equations:

$$\operatorname{sn}^2u+\operatorname{cn}^2u=1, \quad k^2\operatorname{sn}^2u+\operatorname{dn}^2u=1. \tag{2}$$

Let us now choose v so that $k^2 \operatorname{sn}^2v=1$, from which it follows that $\operatorname{dn}^2v=0$. Equation (1) then reduces to

$$\operatorname{sn}(u+v)=\operatorname{sn} v \frac{\operatorname{dn} u}{\operatorname{cn} u}=\pm\frac{1}{k\operatorname{sn}(u+K)}. \tag{3}$$

Hence, if we write: $u+K=w$ and $u+v=w'$, equation (3) can be written:

$$\operatorname{sn} w'=\pm\frac{1}{k \operatorname{sn} w}, \tag{4}$$

where $w'=w+v-K$.

Since $\operatorname{dn} v$ is zero at the points: $(2n+1)K+(2m+1)K'i$, equation (4) can be written:

$$\operatorname{sn}[w+2nK+(2m+1)K'i]=\pm\frac{1}{k \operatorname{sn} w}. \tag{5}$$

From this it follows, by setting $w=0$, that the poles of $\operatorname{sn} u$ are given by $2nK+(2m+1)K'i$. When n is an even number, the sign is positive and when n is odd, the sign is negative.

The following table gives the periods, zeros, poles, and residues for $\operatorname{sn} u$, $\operatorname{cn} u$, and $\operatorname{dn} u$ within the primitive rectangle:

	$\operatorname{sn} u$	$\operatorname{cn} u$	$\operatorname{dn} u$
Periods	$4K, 2iK'$	$4K, 2K+2iK'$	$2K, 4iK'$
Zeros	$0, 2K$	$K, 3K$	$K+iK, K+3iK'$
Poles	$iK', 2K+iK'$	$iK', 2K+iK'$	$iK', 3iK'$
Residues	$1/k, -1/k$	$-i/k, i/k$	$-i, i$

15. The Zeta Elliptic Function of Jacobi

Let us now write

$$E(\phi,k)=\int_0^\phi \Delta(\phi,k)d\phi=\int_0^u \mathrm{dn}^2\,u\,du=E(u),\tag{1}$$

where, as previously, $u=F(\phi,k)$. The function $E(u)$, thus associated with the elliptic integral of second kind, is an elliptic function. Since $\phi=\mathrm{am}\,u$, we see that we can write (1) in the form:

$$E(u)=E(\mathrm{am}\,u,k).\tag{2}$$

Unfortunately $E(u)$ is not periodic in either $2K$ or $2K'i$, so Jacobi, who first studied it, found it more convenient to introduce the following new function:

$$Z(u)=E(u)-uE/K,\tag{3}$$

where K and E are respectively the complete elliptic functions of first and second kind. The function $Z(u)$, called the Zeta elliptic function of Jacobi, is singly periodic of period $2K$.

By methods which are straightforward, but somewhat complicated, one can show that both $E(u)$ and $Z(u)$ have the following addition formulas:

$$E(u)+E(v)-E(u+v)=k^2\,\mathrm{sn}\,u\,\mathrm{sn}\,v\,\mathrm{sn}(u+v),$$
$$Z(u)+Z(v)-Z(u+v)=k^2\,\mathrm{sn}\,u\,\mathrm{sn}\,v\,\mathrm{sn}(u+v).\tag{4}$$

From the second of these we have

$$Z(u+2K)=Z(u),\tag{5}$$

which follows from the equations:

$$E(2K)=E(\pi,k^2)=2E,\quad Z(2K)=E(2K)-2E=0,$$
$$Z(u)+Z(2K)-Z(u+2K)=0.$$

PROBLEMS

1. Show that $Z(u+2Ki)=Z(u)-\pi i/K$.
2. Prove that

$$\frac{dZ}{du}=\mathrm{dn}^2 u-E/K,\quad \frac{d^2Z}{du^2}=-2k^2\,\mathrm{sn}\,u\,\mathrm{cn}\,u\,\mathrm{dn}\,u.$$

3. Verify the following integral:

$$\int\frac{du}{\mathrm{sn}^2 u}=u\,Z'(0)-Z(u)-\frac{\mathrm{cn}\,u\,\mathrm{dn}\,u}{\mathrm{sn}\,u}+C.$$

4. Show by differentiation that

$$\int\mathrm{dn}^2 u\,du=u+\mathrm{dn}\,u\,\mathrm{tn}\,u-\int\left(\frac{\mathrm{dn}\,u}{\mathrm{cn}\,u}\right)^2 du.$$

5. Establish the identity

$$E(iu,k) = i \int \left[\frac{\text{dn}\,(u,k')}{\text{cn}\,(u,k')} \right]^2 du,$$

and then, by the formula in Problem 4, prove that

$$E(iu,k) = i[u + \text{dn}\,(u,k')\text{tn}\,(u,k') - E(u,k')].$$

6. Use the result of Problem 5 to show that

$$E(2iK,k) = 2i(K' - E').$$

7. Show that

$$E(u + 2K + 2\imath K') = E(u) + 2E + 2i(K' - E').$$

8. Establish the identity

$$\text{sn}^2(iu,k) + \text{sn}^2(u,k') = \text{sn}^2(iu,k)\text{sn}^2(u,k').$$

16. The Elliptic Functions of Weierstrass

Directed in his approach by the expansion

$$\frac{1}{\sin^2 x} = \sum_{n=-\infty}^{\infty} \frac{1}{(x - n\pi)^2}, \tag{1}$$

Karl Weierstrass (1815–97) defined a new function, denoted by the symbol, $\wp(x)$, by means of the following series:

$$\wp(x) = \frac{1}{x^2} + \sum_{m,n}' \frac{1}{(x - 2m\omega - 2n\omega')^2} - \frac{1}{(2m\omega + 2n\omega')^2}. \tag{2}$$

The summation is taken over all positive and negative integral values of m and n, including zero, except when m and n are simultaneously zero. The quantities ω and ω' are two numbers the ratio of which is not real.

In order to relate this function to the Jacobi elliptic functions which we have just described, let us write

$$u = \wp(x). \tag{3}$$

It can then be shown that the relationship between x and u can be expressed as the elliptic integral:

$$x = \int_u^\infty \frac{ds}{\sqrt{R}} = \wp^{-1}(u), \tag{4}$$

where we write in customary notation

$$R = 4s^3 - g_2 s - g_3 = 4(s - e_1)(s - e_2)(s - e_3). \tag{5}$$

The relationship between the parameters g_2, g_3 and the quantities e_1, e_2, e_3 is readily expressed by the following equations:

$$e_1+e_2+e_3=0, \quad e_1e_2+e_1e_3+e_2e_3=-\frac{1}{4}\,g_2, \quad e_1e_2e_3=\frac{1}{4}\,g_3. \tag{6}$$

The differential equation satisfied by $\wp(x)$ is readily obtained by differentiating (4), from which we get

$$\left(\frac{du}{dx}\right)^2=4u^3-g_2u-g_3. \tag{7}$$

When it is necessary to exhibit the values of the parameters explicitly, $\wp(x)$ is customarily written: $\wp(x, g_2, g_3)$.

In order to connect the Weierstrass function with the Jacobi elliptic functions, we shall now establish the following identity:

$$\wp(x)=e_3+\frac{e_1-e_3}{\mathrm{sn}^2(\lambda x, k)}, \tag{8}$$

where we write

$$\lambda^2=e_1-e_3, \qquad k^2=\frac{e_2-e_3}{e_1-e_3}. \tag{9}$$

Writing equation (8) in the simpler form: $u=e_3+(e_1-e_3)\,\mathrm{sn}^{-2}\lambda x$, and making use of the formulas in Sections 9 and 10, we compute and simplify $(du/dx)^2$ as follows:

$$\left(\frac{du}{dx}\right)^2=\frac{4(e_1-e_3)^2\lambda^2}{\mathrm{sn}^6\lambda x}\,\mathrm{cn}^2\lambda x\;\mathrm{dn}^2\lambda x,$$

$$=\frac{4(e_1-e_3)^2\lambda^2}{\mathrm{sn}^6\lambda x}\,(1-\mathrm{sn}^2\lambda x)\,(1-k^2\mathrm{sn}^2\lambda x),$$

$$=\frac{4(e_1-e_3)^2\lambda^2}{\mathrm{sn}^2\lambda x}\left(\frac{1}{\mathrm{sn}^2\lambda x}-1\right)\left(\frac{1}{\mathrm{sn}^2\lambda x}-k^2\right),$$

$$=4(e_1-e_3)^{-1}\lambda^2(u-e_3)\left(\frac{e_1-e_3}{\mathrm{sn}^2\lambda x}-e_1+e_3\right)\left[\frac{e_1-e_3}{\mathrm{sn}^2\lambda x}-k^2(e_1-e_3)\right],$$

$$=4\,(e_1-e_3)^{-1}\lambda^2\,(u-e_3)\,(u-e_1)\,[u-e_3-k^2(e_1-e_3)].$$

If λ and k are now defined by (9), then we obtain the equation:

$$\left(\frac{du}{dx}\right)^2=4(u-e_1)\,(u-e_2)\,(u-e_3)=4u^3-g_2u-g_3. \tag{10}$$

The general solution of this equation is $u=\wp(x+c)$, where c is an arbitrary constant. This fact leads to the interesting and often useful conclusion that $\wp(x)$ can be written as follows:

$$\wp_e(x)=e_3+(e_2-e_3)\,\mathrm{sn}^2(\lambda x, k). \tag{11}$$

This results from equation (4) of Section 14, where it was shown that a linear transformation of the variable w in $1/\text{sn } w$ changes this function into $k \text{ sn } w'$. It is obvious that the variable x in (11) is different from that in (8), but both λ and k have the values given in (9).

From equation (8) we are immediately able to deduce the existence of the half-periods ω and ω', such that

$$\wp(x+2\omega)=\wp(x), \quad \wp(x+2\omega')=\wp(x). \tag{12}$$

Since the periods of $\text{sn}^2 z$ are $2K$ and $2iK'$, the values of ω and ω' in terms of them are found from the equations: $\lambda\omega=K$ and $\lambda\omega'=iK'$, that is,

$$\omega=\frac{K}{\sqrt{e_1-e_3}} \text{ and } \omega'=\frac{iK'}{\sqrt{e_1-e_3}}. \tag{13}$$

The corresponding values of $\wp(\omega)$, $\wp(\omega+\omega')$, and $\wp(\omega')$ are obtained by use of the special values of $\text{sn } z$ given in (13), Section 12. We thus find

$$\wp(\omega)=e_1, \quad \wp(\omega+\omega')=e_2, \quad \wp(\omega')=e_3. \tag{14}$$

It is also useful to be able to express the Jacobi elliptic functions in terms of $u=\wp(\dot{x})$. This is accomplished by solving equation (8) for $\text{sn}(\lambda x,k)$. We thus obtain

$$\text{sn}(\lambda x,k)=\frac{\sqrt{e_1-e_3}}{\sqrt{u-e_3}}, \tag{15}$$

and similarly for $\text{cn}(\lambda x,k)$ and $\text{dn}(\lambda x,k)$:

$$\text{cn}(\lambda x,k)=\frac{\sqrt{u-e_1}}{\sqrt{u-e_3}}, \quad \text{dn}(\lambda x,k)=\frac{\sqrt{u-e_2}}{\sqrt{u-e_3}}. \tag{16}$$

Two other matters are of interest in connection with the analysis just given. The first of these relates to the equation which determines values of k^2 in terms of g_2 and g_3. To obtain this equation we must evaluate e_1, e_2, and e_3 as functions of λ and k. This is accomplished by means of the following equations:

$$e_1+e_2+e_3=0, \quad e_1-e_3=\lambda^2, \quad k^2e_1-e_2+(1-k^2)e_3=0, \tag{17}$$

where the first is from (6) and the other two from (9).

We thus obtain explicitly,

$$e_1=\frac{1}{3}\lambda^2(2-k^2), \quad e_2=\frac{1}{3}\lambda^2(2k^2-1), \quad e_3=-\frac{1}{3}\lambda^2(k^2+1). \tag{18}$$

When these values are substituted in the equations

$$e_1e_2+e_1e_3+e_2e_3=-\frac{1}{4}g_2, \quad e_1e_2e_3=\frac{1}{4}g_3, \tag{19}$$

there results

$$\frac{1}{9}\lambda^4(1-k^2+k^4)=\frac{1}{12}g_2,$$

$$\frac{1}{27}\lambda^6(1+k^2)(2-k^2)(1-2k^2)=\frac{1}{4}g_3. \tag{20}$$

Eliminating λ between these equations, we obtain

$$108\,g_3^2(1-k^2+k^4)^3=g_2^3(1+k^2)^2(2-k^2)^2(1-2k^2)^2, \tag{21}$$

where g_2 and g_3 are assumed $\neq 0$.

If $g_2=0$, $g_3\neq0$, then k^2 satisfies the equation:

$$1-k^2+k^4=0, \tag{22}$$

and if $g_3=0$, $g_2\neq0$, then k^2 has the values, -1, 2, and $1/2$.

The second matter refers to the periods 2ω and $2\omega'$ defined by (13). Since these are numbers whose ratio is not real, we can form from them what is called the *period parallelogram*. Thus, representing 2ω and $2\omega'$ graphically as shown in Figure 3, we see that $2(\omega+\omega')$ forms with them and the origin a parallelogram. By adjoining the sum of multiples of 2ω and $2\omega'$, namely, $2m\omega+2n\omega'$, to the plane, we can construct a net of congruent parallelograms. Within each of

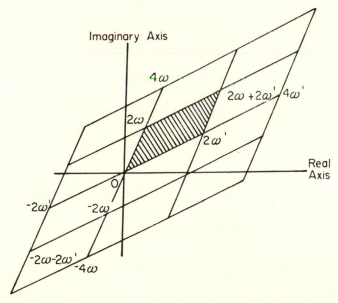

FIGURE 3

these the properties of $\wp(x)$ are identical by virtue of the periodic properties given in equations (12).

Although we have already given explicit values for the half-periods in (13), there is some interest in evaluating them directly from the fundamental integral (4). Referring to (14), we see that ω, $\omega+\omega'$, and ω' are defined respectively by the following integrals:

$$\omega=\int_{e_1}^{\infty} \frac{ds}{\sqrt{R}}, \quad \omega+\omega'=\int_{e_2}^{\infty} \frac{ds}{\sqrt{R}}, \quad \omega'=\int_{e_3}^{\infty} \frac{ds}{\sqrt{R}}. \tag{23}$$

For simplicity in the argument, we shall assume that e_1, e_2, and e_3 are real, and that $e_1 > e_2 > e_3$. We now make the transformation:

$$s=e_3+(e_1-e_3)/x^2, \tag{24}$$

from which it follows, observing (5), that we have

$$\omega=\int_{e_1}^{\infty} \frac{ds}{\sqrt{R}}=\frac{1}{\lambda} \int_0^1 \frac{dx}{\sqrt{(1-x^2)(1-k^2x^2)}}=K/\lambda, \tag{25}$$

where λ and k^2 are defined by (9). A similar analysis applies to the evaluation of ω'.

We shall consider more explicitly one case of special interest, namely, when g_2 and g_3 are both real constants.* Since one of the roots is real, we shall assume that this is e_2 and write R in the following form:

$$R=4(s-e_2)\{(s-m)^2+n^2\}, \tag{26}$$

where we abbreviate: $m=-\dfrac{1}{2} e_2$, $g_2=3e_2^2-4n^2$, $g_3=e_2^3+4n^2e_2$.

If we now define a constant H as follows:

$$H^2=(e_2-m)^2+n^2=\frac{9}{4} e_2^2+n^2, \tag{27}$$

then u, as given by (4), can be written in the form:

$$u=\int_s^{\infty} \frac{ds}{\sqrt{R}}=\frac{1}{2\sqrt{H}} \operatorname{cn}^{-1}\left\{\frac{s-e_2-H}{s-e_2+H}, k\right\}, \tag{28}$$

where $kk'=n/(2H)$.

Setting $s=e_2$, we then obtain the real period: $\Omega=2\Omega_2$, where we have

$$\Omega_2=\omega+\omega'=\frac{1}{2\sqrt{H}} \operatorname{cn}^{-1}(-1,k)=\frac{1}{2\sqrt{H}} F(\pi,k)=\frac{1}{\sqrt{H}}K, \tag{29}$$

where K is the complete elliptic integral corresponding to k.

---

*See A. G. Greenhill: *The Applications of Elliptic Functions*. London, 1892, Section 61, p. 50.

As an example, let us consider what has been called the *equianharmonic* case,* where $g_2=0$. We then have $e_2=a\sqrt[3]{g_3}$, where $a=1/\sqrt[3]{4}$. Since $n^2=\dfrac{3}{4}\,e_2^2$ and $H^2=3e_2^2$, we find $k^2=\dfrac{1}{2}-\dfrac{1}{4}\sqrt{3}$ and $k=(\sqrt{3}-1)/(2\sqrt{2})$. Since $\sin\,\alpha=k$, this gives $\alpha=15°$.

Hence the desired half-period has the following value:

$$\Omega_2=\sqrt[3]{2}K(15°)/(\sqrt[6]{g_3}\,\sqrt[4]{3})=(1/\sqrt[6]{g_3})\,\,1.52995\,4037. \tag{30}$$

The Functions $\zeta(x)$ and $\sigma(x)$.

In addition to $\wp(x)$, Weierstrass introduced two additional functions denoted respectively by $\zeta(x)$ and $\sigma(x)$. The first of these is defined by the equation

$$\frac{d}{dx}\,\zeta(x)=-\wp(x), \tag{31}$$

together with the condition,

$$\lim_{x\to 0}\left[\zeta(x)-\frac{1}{x}\right]=0. \tag{32}$$

This function has a quasi-periodicity given by the equations:

$$\zeta(x+2\omega)=\zeta(x)+2\zeta(\omega),$$

$$\zeta(x+2\omega')=\zeta(x)+2\zeta(\omega'). \tag{33}$$

The second function was similarly defined by the equation:

$$\frac{d}{dx}\log\,\sigma(x)=\zeta(x), \tag{34}$$

together with the condition:

$$\lim_{x\to 0}\left[\frac{\sigma(x)}{x}\right]=1. \tag{35}$$

This function also has a quasi-periodicity as follows:

$$\sigma(x+2\omega)=-e^{2\eta\,(x+\omega)}\,\sigma(x),$$

$$\sigma(x+2\omega')=-e^{2\eta'\,(x+\omega')}\,\sigma(x), \tag{36}$$

where we write: $\eta=\zeta(\omega),\,\eta'=\zeta(\omega')$.

---

*H. Burkhardt: *Elliptische Funktionen*. Berlin and Leipzig, 3d ed., 1920, p. 75.

THETA FUNCTIONS

17. Theta Functions

Any study of elliptic functions would be incomplete which did not contain some account of the theory of Theta functions. These functions were first developed systematically by C. G. J. Jacobi in his treatise *Fundamenta nova theoriae functionum ellipticarum*, which we have mentioned earlier in Section 8. Such functions had appeared previously in connection with the partition function of Euler,[*] namely, $\prod_{n=1}^{\infty} (1-x^n z)^{-1}$, and in Fourier's analytical theory of heat (1822). But it was reserved to Jacobi to give an almost complete account of their properties.

Four principal Theta functions have been recognized and these are defined by means of the following series:

$$\theta_1(x,q)=2q^{\frac{1}{4}} (\sin x-q^2 \sin 3x+q^6 \sin 5x-q^{12} \sin 7x+ \ldots),$$

$$=2 \sum_{n=0}^{\infty} (-1)^n q^{(n+\frac{1}{2})^2} \sin (2n+1)x; \tag{1}$$

$$\theta_2(x,q)=2q^{\frac{1}{4}} (\cos x+q^2 \cos 3x+q^6 \cos 5x+q^{12} \cos 7x+ \ldots),$$

$$=2 \sum_{n=0}^{\infty} q^{(n+\frac{1}{2})^2} \cos (2n+1)x; \tag{2}$$

$$\theta_3(x,q)=1+2q \cos 2x+2q^4 \cos 4x+2q^9 \cos 6x+ \ldots,$$

$$=1+2 \sum_{n=1}^{\infty} q^{n^2} \cos 2nx; \tag{3}$$

$$\theta_4(x,q)=1-2q \cos 2x+2q^4 \cos 4x-2q^9 \cos 6x+ \ldots,$$

$$=1+2 \sum_{n=1}^{\infty} (-1)^n q^{n^2} \cos 2nx. \tag{4}$$

In these functions q is assumed to be a number lying within the unit circle so that $|q|<1$. It can be represented conveniently in the form

$$q=e^{\pi i \tau}, \tag{5}$$

where τ is a complex number of the form: $\tau=r+si$, in which $s>0$.

It is sometimes convenient to represent these functions in terms of a single variable, in which case we shall use the notation:

$$\theta_i(x)=\theta_i(x,q). \tag{6}$$

---

[*]*Introductio in Analysin Infinitorum*, Lausanne, 1748, I, 304.

With this understanding, we can define the functions at $x=0$ as follows:

$$\theta_2(0)=2q^{\frac{1}{4}}(1+q^2+q^6+q^{12}+q^{20}+\ldots)=2q^{\frac{1}{4}}\sum_{n=0}^{\infty}q^{n^2+n};$$

$$\theta_3(0)=1+2q+2q^4+2q^9+2q^{16}+\ldots)=1+2\sum_{n=1}^{\infty}q^{n^2};$$

$$\theta_4(0)=1-2q+2q^4-2q^9+2q^{16}-\ldots)=1+2\sum_{n=1}^{\infty}(-1)^nq^{n^2}.\quad(7)$$

When it is desirable to express the Theta functions as functions of τ instead of q, one can use the notation: $\theta_1(x|\tau)$, $\theta_2(x|\tau)$, etc.

Certain properties of the functions are readily deduced from the series. Thus we see that $\theta_1(x)$ is an odd function, but that $\theta_2(x)$, $\theta_3(x)$, and $\theta_4(x)$ are even functions. One verifies also that

$$\theta_m(x+\pi)=A_m\theta_m(x),\quad \theta_m(x+\pi\tau)=B_m\frac{1}{q}e^{-2ix}\theta_m(x),\quad(8)$$

where: $A_1=A_2=-1$, $A_3=A_4=1$; $B_1=B_4=-1$, $B_2=B_3=1$.

Introducing the notation $Q=q^{1/4}e^{ix}$, one can show that the following relationships exist between the four Theta functions:

$$\theta_1(x)=-\theta_2\left(x+\frac{1}{2}\pi\right)=-iQ\ \theta_3\left(x+\frac{1}{2}\pi+\frac{1}{2}\pi\tau\right)=-iQ\ \theta_4\left(x+\frac{1}{2}\pi\tau\right),$$

$$\theta_2(x)=Q\ \theta_3\left(x+\frac{1}{2}\pi\tau\right)=Q\ \theta_4\left(x+\frac{1}{2}\pi+\frac{1}{2}\pi\tau\right)=\theta_1\left(x+\frac{1}{2}\pi\right),$$

$$\theta_3(x)=\theta_4\left(x+\frac{1}{2}\pi\right)=Q\ \theta_1\left(x+\frac{1}{2}\pi+\frac{1}{2}\pi\tau\right)=Q\ \theta_2\left(x+\frac{1}{2}\pi\tau\right),$$

$$\theta_4(x)=-iQ\ \theta_1\left(x+\frac{1}{2}\pi\tau\right)=iQ\ \theta_2\left(x+\frac{1}{2}\pi+\frac{1}{2}\pi\tau\right)=\theta_3\left(x+\frac{1}{2}\pi\right).\quad(9)$$

These identities, although not quite obvious, can be established readily. Thus, observing that we can write

$$\theta_4(x)=1+\sum_{n=1}^{\infty}(-1)^nq^{n^2}(e^{2nix}+e^{-2nix}),$$

we have

$$-iQ\ \theta_4(x+\tfrac{1}{2}\pi\tau)=-iQ\left[1+\sum_{n=1}^{\infty}(-1)^n(q^{n^2+n}e^{2nix}+q^{n^2-n}e^{-2nix})\right],$$

$$=-iq^{\frac{1}{4}}\left[(e^{ix}-e^{-ix})+\sum_{n=1}^{\infty}(-1)^nq^{n^2+n}e^{2nix+ix}\right.$$

$$\left.+\sum_{n=1}^{\infty}(-1)^{n+1}q^{(n+1)^2-(n+1)}e^{-2(n+1)ix+ix}\right],$$

$$=2q^{\frac{1}{4}}\sin\ x-iq^{\frac{1}{4}}\sum_{n=1}^{\infty}(-1)^nq^{n^2+n}[e^{(2n+1)ix}-e^{-(2n+1)ix}],$$

$$=2\sum_{n=0}^{\infty}(-1)^nq^{(n+\frac{1}{2})^2}\sin\ (2n+1)x.\quad(10)$$

We have thus shown that $\theta_1(x) = -iQ \; \theta_4\left(x+\frac{1}{2}\pi\tau\right)$, and the other identities can be similarly established.

A large number of relationships have been discovered between the four functions, among which the following identities will be of special use to us:

$$\theta_1^2(x)\theta_4^2(0) = \theta_3^2(x)\theta_2^2(0) - \theta_2^2(x)\theta_3^2(0),$$

$$\theta_2^2(x)\theta_4^2(0) = \theta_4^2(x)\theta_2^2(0) - \theta_1^2(x)\theta_3^2(0),$$

$$\theta_3^2(x)\theta_4^2(0) = \theta_4^2(x)\theta_3^2(0) - \theta_1^2(x)\theta_2^2(0),$$

$$\theta_4^2(x)\theta_4^2(0) = \theta_3^2(x)\theta_3^2(0) - \theta_2^2(x)\theta_2^2(0). \tag{11}$$

For the derivation of these and other relationships the reader is referred to treatises on the subject. Modern methods of proof depend usually upon the application of theorems in the theory of functions of a complex variable, but the identities were derived originally by Jacobi from purely algebraic arguments.

If we denote by G the infinite product

$$G = (1-q^2)(1-q^4)(1-q^6) \; \cdot \; \cdot \; \cdot, \tag{12}$$

then the Theta functions can be expressed as the following infinite products:

$$\theta_1(x) = 2Gq^{1/4} \sin x \prod_{n=1}^{\infty} (1-2q^{2n} \cos 2x + q^{4n}),$$

$$\theta_2(x) = 2Gq^{1/4} \cos x \prod_{n=1}^{\infty} (1+2q^{2n} \cos 2x + q^{4n}),$$

$$\theta_3(x) = G \prod_{n=1}^{\infty} (1+2q^{2n-1} \cos 2x + q^{4n-2}),$$

$$\theta_4(x) = G \prod_{n=1}^{\infty} (1-2q^{2n-1} \cos 2x + q^{4n-2}). \tag{13}$$

The zeros of the Theta functions are obtained from equations (9). For example, from these we have the following identity:

$$\theta_1(x) = -iQ^2\theta_1(x+\pi+\pi\tau), \tag{14}$$

and thus, if x_0 is any zero of $\theta_1(x)$, so also is $x_0+\pi+\pi\tau$.

More generally, by the same argument, it can be shown that if x_0 is any zero of any one of the Theta functions, then

$$z = x_0 + m\pi + n\pi\tau, \tag{15}$$

is also a zero for any integral values of m and n.

Since $x_0 = 0$ is a zero of $\theta_1(x)$, it follows from the first relations in (9) that $\frac{1}{2}\pi$, $\frac{1}{2}\pi + \frac{1}{2}\pi\tau$, and $\frac{1}{2}\pi\tau$ are primitive zeros respectively of $\theta_2(x)$, $\theta_3(x)$, and $\theta_4(x)$.

18. The Differential Equation of the Theta Functions

The relationships which exist between both the elliptic integrals and the elliptic functions, which we have described in earlier sections, and the Theta functions are obtained most readily by means of the following identity:

$$\frac{d}{dx}\left(\frac{\theta_1(x)}{\theta_4(x)}\right) = \theta_4^2(0)\,\frac{\theta_2(x)}{\theta_4(x)} \cdot \frac{\theta_3(x)}{\theta_4(x)}. \tag{1}$$

The derivation of this equation is difficult and the reader is referred for it to treatises on the Theta function.

If we now make use of the abbreviation:

$$y = \frac{\theta_1(x)}{\theta_4(x)}, \tag{2}$$

square both members of (1), and replace $\theta_2^2(x)\,\theta_4^2(0)$ and $\theta_3^2(x)\,\theta_4^2(0)$ respectively by $\theta_4^2(x)\,\theta_2^2(0) - \theta_1^2(x)\,\theta_3^2(0)$ and $\theta_4^2(x)\,\theta_3^2(0) - \theta_1^2(x)\,\theta_2^2(0)$, obtained from equations (11) of Section 17, we derive the following differential equations:

$$\left(\frac{dy}{dx}\right)^2 = [\theta_2^2(0) - y^2\,\theta_3^2][\theta_3^2(0) - y^2\,\theta_2^2(0)]. \tag{3}$$

Making use of the transformation:

$$z = \frac{\theta_3(0)}{\theta_2(0)}\,y, \qquad u = \theta_3^2(0)\,x, \tag{4}$$

and employing the abbreviation:

$$k = \frac{\theta_2^2(0)}{\theta_3^2(0)}, \tag{5}$$

we obtain equation (3) in the canonical form:

$$\left(\frac{dz}{du}\right)^2 = (1 - z^2)\,(1 - k^2 z^2). \tag{6}$$

Since this equation has the solution: $z = \operatorname{sn}(u,k)$, and also the solution defined by (2) and (4), we achieve the desired relationship between the Jacobi elliptic sine and the Theta functions as follows:

$$\operatorname{sn}(u,k) = \frac{\theta_3(0)}{\theta_2(0)} \frac{\theta_1[u/\theta_3^2(0)]}{\theta_4[u/\theta_3^2(0)]}. \tag{7}$$

That the arbitrary constant which enters linearly with u in the solution of (6) has been properly determined is seen from the fact that both sides of (7) equal 0 when $u = 0$.

Observing the equations: $\operatorname{cn}^2 u = 1 - z^2$ and $\operatorname{dn}^2 u = 1 - k^2 z^2$, and making use of the relationships between the Theta functions given in (11) of Section 17, one readily obtains the following formulas:

$$\operatorname{cn}(u,k) = \frac{\theta_4(0)}{\theta_2(0)} \cdot \frac{\theta_2(v)}{\theta_4(v)}, \quad \operatorname{dn}(u,k) = \frac{\theta_4(0)}{\theta_3(0)} \cdot \frac{\theta_3(v)}{\theta_4(v)}, \tag{8}$$

where $v = u/\theta_3^2(0)$.

It is often useful to express k' in terms of Theta functions, the desired formula being the following:

$$k' = \frac{\theta_4^2(0)}{\theta_3^2(0)}. \tag{9}$$

This is readily derived by observing that

$$k'^2 = 1 - k^2 = \frac{\theta_3^4(0) - \theta_2^4(0)}{\theta_3^4(0)}. \tag{10}$$

If we now set $x = 0$ in the last formula of (11), Section 17, we obtain the identity:

$$\theta_4^4(0) = \theta_3^4(0) - \theta_2^4(0), \tag{11}$$

from which (9) follows as an immediate consequence.

The values of K and K' have the following equivalent forms in terms of Theta functions:

$$K = \frac{1}{2} \pi \, \theta_3^2(0), \quad K' = -\frac{1}{2} \pi i \tau \theta_3^2(0), \tag{12}$$

from which it follows that

$$K'/K = -i\tau, \quad q = e^{-\pi K'/K}. \tag{13}$$

Formulas (12) following immediately from (7) by observing that the periods of $\operatorname{sn}(u,k)$ are $4K$ and $2K'i$ and that the corresponding periods of $\theta_1(v)/\theta_4(v)$, given by (8) of Section 17, are 2π and $\pi\tau$. Hence we equate: $2K/\theta_3^2(0) = \pi$, and $2K'i/\theta_3^2(0) = \pi\tau$.

From these formulas one now has the following series, which are useful for the computation of the periods as well as k and k' when q is given:

$$\left(\frac{2K}{\pi}\right)^{\frac{1}{2}}=\theta_3(0)=1+2q+2q^4+2q^9+\ \cdot\ \cdot\ \cdot,$$

$$\left(\frac{2kK}{\pi}\right)^{\frac{1}{2}}=\theta_2(0)=2q^{\frac{1}{4}}(1+q^2+q^6+q^{12}+q^{20}+\ \cdot\ \cdot\ \cdot\ ),\qquad(14)$$

$$\left(\frac{2k'K}{\pi}\right)^{\frac{1}{2}}=\theta_4(0)=1-2q+2q^4-2q^9+\ \cdot\ \cdot\ \cdot\ \cdot$$

$$K'=-\frac{K}{\pi}\log q.$$

PROBLEMS

1. Given $q=\frac{1}{2}$, find the values of k, K, and K'.

2. If $K=3$, invert the first equation in (14) to estimate q. Use this value to find k and K'.

3. Defining τ' by the equation: $\tau\tau'=-1$, establish the following:

$$\theta_1(x\,|\,\tau)=-iF(x)\theta_1(x\tau'\,|\,\tau'),\quad \theta_2(x\,|\,\tau)=F(x)\theta_4(x\tau'\,|\,\tau'),$$

$$\theta_3(x\,|\,\tau)=F(x)\theta_3(x\tau'\,|\,\tau'),\quad \theta_4(x\,|\,\tau)=F(x)\theta_2(x\tau'\,|\,\tau'),$$

where we write: $F(x)=(-i\tau)^{-1/2}\exp\ (i\tau'x^2/\pi)$.

4. Use the results of Problem 3 to prove that

$$\mathrm{sn}\,(iu,k)=i\ \mathrm{tn}(u,k').$$

19. Representation of the Jacobi Elliptic Functions as Fourier Series

From the results of the preceding section we are now able to give a representation of the Jacobi elliptic functions as Fourier series. Since $\theta_3^2(0)=2K/\pi$, from which it follows that $u/\theta_3^2(0)=\pi u/2K$, we shall adopt the following notation:

$$v=\frac{\pi}{2K}\,u.\qquad(1)$$

In terms of the variable v, we can then write the following expansions:

$$\mathrm{sn}(u,k)=\frac{\theta_3(0)}{\theta_2(0)}\cdot\frac{\theta_1(v)}{\theta_4(v)},$$

$$=\left[\frac{1+2q+2q^4+2q^9+\ \cdot\ \cdot\ \cdot}{\cdot 1+q^2+q^6+q^{12}+\ \cdot\ \cdot\ \cdot}\right]\left[\frac{\sin v-q^2\sin 3v+q^6\sin 5v-\ \cdot\ \cdot\ \cdot}{1-2q\ \cos\ 2v+2q^4\cos\ 4v-\ \cdot\ \cdot\ \cdot}\right];$$

$$(2)$$

$$cn(u,k) = \frac{\theta_4(0)}{\theta_2(0)} \cdot \frac{\theta_2(v)}{\theta_4(v)},$$

$$= \left[\frac{1 - 2q + 2q^4 - 2q^9 + \ldots}{1 + q^2 + q^6 + q^{12} + \ldots}\right] \left[\frac{\cos v + q^2 \cos 3v + q^6 \cos 5v + \ldots}{1 - 2q \cos 2v + 2q^4 \cos 4v - \ldots}\right];$$

$$(3)$$

$$dn(u,k) = \frac{\theta_4(0)}{\theta_3(0)} \cdot \frac{\theta_3(v)}{\theta_4(v)},$$

$$= \left[\frac{1 - 2q + 2q^4 - 2q^9 + \ldots}{1 + 2q + 2q^4 + 2q^9 + \ldots}\right] \left[\frac{1 + 2q \cos 2v + 2q^4 \cos 4v + \ldots}{1 - 2q \cos 2v + 2q^4 \cos 4v - \ldots}\right].$$

$$(4)$$

This representation of the elliptic functions as the ratios of Fourier series suggests the possibility of representing them as single Fourier series. That this is, indeed, the case is readily seen from the second ratio of (2). If we write the denominator in the form: $1 - 2z$, where we use the abbreviation:

$$z = \cos 2v - q^3 \cos 4v + q^8 \cos 6v - \ldots, \quad (5)$$

then the ratio itself can be written as the following product:

$$(\sin v - q^2 \sin 3v + q^6 \sin 5v - \ldots)(1 + 2qz + 4q^2z^2 + 8q^3z^3 + \ldots). \quad (6)$$

Since both powers of $\cos nv$ and such products as $\cos {}^r nv \times \cos {}^s mv$ can be reduced to the linear sums of cosines of the form $\cos (av + bv)$, where a and b are integers (or zero), it is seen that, at least formally, the second term in (6) can be reduced to a Fourier series in cosines. Since, furthermore, the product $\sin mv \cos nv$ is reducible to the sum of two sine terms, the product given by (6) can be reduced formally to the sum of terms of the form $\sin pv$, that is to say, to a Fourier series in sines. The same argument applies to (3) and (4), both of which reduce formally to Fourier series in cosine terms.

The explicit expansions are as follows:*

$$sn\,(u,k) = \frac{2\pi}{Kk} \sum_{n=0}^{\infty} \frac{q^{n+1/2}}{1 - q^{2n+1}} \sin (2n+1)v, \quad (7)$$

$$cn\,(u,k) = \frac{2\pi}{Kk} \sum_{n=0}^{\infty} \frac{q^{n+1/2}}{1 + q^{2n+1}} \cos (2n+1)v, \quad (8)$$

$$dn\,(u,k) = \frac{\pi}{2K} + \frac{2\pi}{K} \sum_{n=1}^{\infty} \frac{q^n}{1 + q^{2n}} \cos 2nv. \quad (9)$$

---

*The explicit derivation of these series by different methods will be found in Whittaker and Watson: *Modern Analysis*, p. 510, and in Greenhill: *Elliptic Functions*, pp. 285–286. The origin of the series is found in Jacobi's *Fundamenta Nova*, p. 101.

If we write: $v=x+iy$ and $\tau=r+si$, $s>0$, then these series converge for all values of v within the strip

$$|y|<\frac{1}{2}\pi s, \qquad (10)$$

and represent the functions there. This follows from the fact that the functions, which form the left members, are analytic except at their poles and that the series converge provided $\exp(\pm nv/2+n\pi i\tau)<1$.

A useful form can be given to the coefficients of the harmonic terms in the three series by observing that $q=\exp(-\pi K'/K)$. Thus, in (7), we can write the coefficient of $\sin(2n+1)v$ as follows:

$$\frac{2\pi}{Kk}\frac{q^{n+1/2}}{1-q^{2n+1}}=\frac{\pi}{Kk}\frac{1}{\frac{1}{2}(q^{-n-1/2}-q^{n+1/2})}=\frac{\pi}{Kk}\frac{1}{\sinh\left[\left(n+\frac{1}{2}\right)\pi K'/K\right]}. \qquad (11)$$

Similarly, the coefficients of the harmonic terms in (8) and (9) can be written respectively as follows:

$$\frac{2\pi}{Kk}\frac{q^{n+1/2}}{1+q^{2n+1}}=\frac{\pi}{Kk}\frac{1}{\cosh\left[\left(n+\frac{1}{2}\right)\pi K'/K\right]}, \qquad (12)$$

$$\frac{2\pi}{K}\frac{q^n}{1+q^{2n}}=\frac{\pi}{K}\frac{1}{\cosh(n\pi K'/K)}. \qquad (13)$$

20. The Elliptic Modular Functions

Because of their importance in the practical application of the theory of elliptic functions, three quantities have been defined which are called *elliptic modular functions*. These are the following:

$$f(\tau)=\frac{\theta_2^4(0|\tau)}{\theta_3^4(0|\tau)}, \qquad g(\tau)=\frac{\theta_4^4(0|\tau)}{\theta_3^4(0|\tau)}, \qquad h(\tau)=-\frac{f(\tau)}{g(\tau)}, \qquad (1)$$

where the variable τ, as previously defined, is connected with q, K, and K' by means of the equations:

$$\tau=-\frac{i}{\pi}\log q=\frac{iK'}{K}. \qquad (2)$$

From equations (5) and (9) of Section 18, we see that

$$f(\tau)=k^2, \quad g(\tau)=k'^2, \quad h(\tau)=-k^2/k'^2=-k^2/(1-k^2). \qquad (3)$$

Observing that $k^2+k'^2=1$, and that $q=e^{\pi i \tau}=e^{\pi i (\tau+2)}=-e^{\pi i(\tau+1)}$, we obtain the following relationships:

$$f(\tau+2)=f(\tau), \quad g(\tau+2)=g(\tau), \quad f(\tau)+g(\tau)=1, \quad f(\tau+1)=h(\tau). \quad (4)$$

If τ' is defined by the equation: $\tau\tau'=-1$, then these equations can be supplemented by the following:

$$f(\tau')=g(\tau), \quad g(\tau')=f(\tau). \quad (5)$$

By means of equations (13), Section 17, the modular functions are readily expressed as the following infinite products in terms of the variable q:

$$f(\tau)=16q\left[\frac{(1+q^2)(1+q^4)(1+q^6)\ldots}{(1+q)(1+q^3)(1+q^5)\ldots}\right]^8; \quad (6)$$

$$g(\tau)=\left[\frac{(1-q)(1-q^3)(1-q^5)(1-q^7)\ldots}{(1+q)(1+q^3)(1+q^5)(1+q^7)\ldots}\right]^8. \quad (7)$$

Since $f(\tau)=k^2$, it is clear that we can write

$$k^{1/4}=\sqrt{2}\;\sqrt[8]{q}F(q), \quad (8)$$

where we employ the abbreviation:

$$F(q)=\left[\frac{(1+q^2)(1+q^4)(1+q^6)\ldots}{(1+q)(1+q^3)(1+q^5)\ldots}\right]. \quad (9)$$

The expansion of $F(q)$ as a power series in q has been given by L. A. Schoncke (1807–53) to the term q^{26} as follows:[*]

$$F(q)=1-q+2q^2-3q^3+4q^4-6q^5+9q^6-12q^7+16q^8$$

$$-22q^9+29q^{10}-38q^{11}+50q^{12}-64q^{13}+82q^{14}-105q^{15}+132q^{16}$$

$$-166q^{17}+208q^{18}-258q^{19}+320q^{20}-395q^{21}+484q^{22}-592q^{23}$$

$$+722q^{24}-876q^{25}+1{,}060q^{26}+\ldots. \quad (10)$$

By means of formulas (8) and (10) it is possible to compute k when q is given. The inverse problem, to compute q when k is given, is more difficult, and extensive analysis has been devoted to the problem associated with this inversion.

The practical method makes use of the following quantity:

$$\epsilon=\frac{1}{2}\left(\frac{1-\sqrt{k'}}{1+\sqrt{k'}}\right), \quad (11)$$

---

[*] "Aequationes modulares pro transformationis Functionum Ellipticarum," *Journal für Math.*, Vol. 16. 1837, pp. 97–130.

which we see lies between 0 and $\frac{1}{2}$ when k or k' is a value between 0 and 1.

From formula (9) of Section 18 we see that $\sqrt{k'}=\theta_4(0)/\theta_3(0)$, and thus, referring to the definitions of the functions given in Section 17, we can write (11) as follows:

$$\epsilon=\frac{1}{2}\left[\frac{\theta_3(0)-\theta_4(0)}{\theta_3(0)+\theta_4(0)}\right]=\frac{1}{2}\left[\frac{\theta_2(0,q^4)}{\theta_3(0,q^4)}\right], \tag{12}$$

$$=\frac{q+q^9+q^{25}+q^{49}+q^{81}+\cdots}{1+2q^4+2q^{16}+2q^{36}+2q^{64}+\cdots}. \tag{13}$$

This series is now inverted and q obtained as the following series in ϵ:

$$q=\epsilon+2\epsilon^5+15\epsilon^9+150\epsilon^{13}+1{,}707\epsilon^{17}+20{,}910\epsilon^{21}+268{,}616\epsilon^{25}+\ldots, \tag{14}$$

a series which clearly converges rapidly for values of ϵ less than $\frac{1}{2}$.

If k' is small, so that ϵ is close to $\frac{1}{2}$, then the convergence of (14) can be improved by using k instead of k' in formula (11) and computing q' instead of q. The value of q is then found from the equation:

$$\log q \cdot \log q'=\pi^2. \tag{15}$$

It is also possible to compute q directly in terms of k^2 from a formula provided by C. Hermite (1822–1901), although usually (14), because of its rapid convergence, is to be preferred. Hermite's expansion is as follows:[*]

$$q=a_1k^2+a_2k^4+a_3k^6+a_4k^8+a_5k^{10}+\ldots, \tag{16}$$

where we have

$$2^4a_1=1,\ 2^5a_2=1,\ 2^{10}a_3=21,\ 2^{11}a_4=31,\ 2^{19}a_5=6{,}257,\ 2^{20}a_6=10{,}293,$$

$$2^{25}a_7=279{,}025,\ 2^{26}a_8=483{,}127,\ 2^{36}a_9=435{,}506{,}703,$$

$$2^{37}a_{10}=776{,}957{,}575,\ 2^{42}a_{11}=22{,}417{,}045{,}555,\ 2^{43}a_{12}=40{,}784{,}671{,}953.$$

The numerical values of these coefficients to ten significant figures are given in the following table:

$a_1=0.062500\ 00000,$	$a_5=0.0119342\ 80396,$	$a_9=0.0063374\ 56623,$
$a_2=0.031250\ 00000,$	$a_6=0.0098161\ 69739,$	$a_{10}=0.0056531\ 10384,$
$a_3=0.020507\ 81250,$	$a_7=0.0083155\ 93004,$	$a_{11}=0.0050970\ 46040,$
$a_4=0.015136\ 71875,$	$a_8=0.0071991\ 53304,$	$a_{12}=0.0046366\ 80382.$

[*] *Oeuvres*, Vol. 4, pp. 470–487.

21. Solution of the Quintic Equation By Modular Functions

An instructive application of the modular functions is found in the solution of the quintic equation:

$$z^5 + a_1 z^4 + a_2 z^3 + a_3 z^2 + a_4 z + a_5 = 0. \tag{1}$$

By means of a Tschirnhausen transformation* it is possible to reduce (1) to the canonical form:

$$x^5 - x - a = 0. \tag{2}$$

How this is actually done will not concern us here, since the transformation involves algebraic processes of considerable complexity. Fundamentally, however, only square and cube roots are used in the reduction.

For this equation Hermite found a very elegant solution in terms of the modular functions as follows:†

Introducing the functions

$$\phi(\tau) = \sqrt[4]{k} = \sqrt[8]{f(\tau)}, \quad \psi(\tau) = \sqrt[4]{k'} = \sqrt[8]{g(\tau)}, \tag{3}$$

Hermite defined

$$\Phi(\tau) = \left[\phi(5\tau) + \phi\left(\frac{\tau}{5}\right)\right]\left[\phi\left(\frac{\tau+16}{5}\right)\right.$$
$$\left. -\phi\left(\frac{\tau+4\cdot16}{5}\right)\right]\left[\phi\left(\frac{\tau+2\cdot16}{5}\right) - \phi\left(\frac{\tau+3\cdot16}{5}\right)\right], \tag{4}$$

and showed that the quantities

$$\Phi(\tau), \quad \Phi(\tau+16), \quad \Phi(\tau+2\cdot16), \quad \Phi(\tau+3\cdot16), \quad \Phi(\tau+4\cdot16) \tag{5}$$

are roots of the following quintic equation:

$$\Phi^5 - 2^4 5^3 \phi^4(\tau)\psi^{16}(\tau)\Phi - 2^6\sqrt{5^5}\phi^3(\tau)\psi^{16}(\tau)[1+\phi^8(\tau)] = 0. \tag{6}$$

If we make the transformation:

$$\Phi = 2\sqrt[4]{5^5}\phi(\tau)\psi^4(\tau)x, \tag{7}$$

then equation (6) reduces to the canonical form defined by (2), where a has the following value:

$$a = \frac{2}{\sqrt[4]{5^5}}\frac{1+\phi^8(\tau)}{\phi^2(\tau)\psi^4(\tau)} = \frac{2(1+k^2)}{\sqrt[4]{5^5}k^{1/2}k'}. \tag{8}$$

---

*For a history of this problem see F. Cajori's *History of Mathematics*, New York, 1919, pp. 349–350. For the transformation itself consult: M. Serret: *Cours d'algèbra supérieure*, Vol. 1, art. 192.

†"Sur la resolution de l'équation du cinquième degré," *Comptes Rendus*, Vol. 48, 1858 (I), p. 508; see also Hermite's *Oeuvres*, Vol. 2, 1908, pp. 5–12.

It is clear from (8) that, given a, k can be found as the root of the quartic equation:

$$4(1+k^2)^2 - \sqrt{5^5}a^2k(1-k^2) = 0. \tag{9}$$

To solve this equation, we write $A = \frac{1}{2}\sqrt[4]{5^5}\,a$ and determine α from the equation:

$$\sin \alpha = 4/A^2. \tag{10}$$

The modulus k is then one of the following values:

$$k = \tan\frac{\alpha}{4}, \quad \tan\frac{\alpha+2\pi}{4}, \quad \tan\frac{\pi-\alpha}{4}, \quad \tan\frac{3\pi-\alpha}{4}. \tag{11}$$

Choosing any one of these quantities for the modulus, the desired roots of equation (2) are the following:

$$B\Phi(\tau), \quad B\Phi(\tau+16), \quad B\Phi(\tau+2\cdot16), \quad B\Phi(\tau+3\cdot16), \quad B\Phi(\tau+4\cdot16), \tag{12}$$

where we write:

$$B = \frac{1}{2\sqrt[4]{5^3}\phi(\tau)\psi^4(\tau)} = \frac{1}{2\sqrt[4]{5^3k}\,k'}. \tag{13}$$

The actual numerical application of this theory to the solution of equation (2) would be quite difficult except for the fortunate circumstance that $\Phi(\tau)$ has the following expansion:

$$\Phi(\tau) = \sqrt{2^3 5}\,\sqrt[8]{Q^3}(1+Q-Q^2+Q^3-8Q^5-9Q^6+8Q^7-9Q^8+\ldots), \tag{14}$$

where $Q = \sqrt[5]{q}$.

The details of the numerical solution of the equation

$$x^5 - x - 2 = 0,$$

are given below as follows:

Since $A = 5\sqrt[4]{5}$, we compute:

$$\sin \alpha = 0.07155\ 41753, \quad \tan\frac{\alpha}{4} = 0.01790\ 57586 = k, \quad k' = 0.99983\ 9679.$$

Hence, by (16), of Section 20, we have:

$$q = 0.00002\ 0041725, \quad Q = \sqrt[5]{q} = 0.11491\ 7725, \quad B = 0.40884\ 9953.$$

We also compute

$$C=\sqrt{2^3 5}\ \sqrt[8]{Q^3}=2.80979\ 8187,\ BC=1.14878\ 5857,$$

and the following powers of Q:

$$Q=0.11491\ 7725,\qquad Q^5=0.00002\ 0042,$$

$$Q^2=0.01320\ 6084,\qquad Q^6=0.00000\ 2303,$$

$$Q^3=0.00151\ 7613,\qquad Q^7=0.00000\ 0265,$$

$$Q^4=0.00017\ 4401,\qquad Q^8=0.00000\ 0030.$$

Substituting in (14), we get: $\Phi(\tau)=3.09934\ 7991$, and hence

$$x_1=B\Phi(\tau)=1.26716\ 8280,$$

which is correct to six places.

In order to obtain the pair of conjugate complex roots x_2 and x_3, we first consider

$$Q(\tau+16)=q\left[\frac{(\tau+16)\pi i}{5}\right]=Qe^{16\pi i/5},$$

$$=Q\left(\cos\frac{16}{5}+i\sin\frac{16}{5}\right),$$

$$=Q(-0.80901\ 6994-i\ 0.58778\ 5252),$$

from which we obtain

$$Q^n(\tau+16)=Q^n e^{16n\pi i/5},\quad Q^{3/8}(\tau+16)=Q^{3/8}e^{6\pi i/5}.$$

When these values are substituted in (14), we obtain

$$\Phi(\tau+16)=C(e^{6\pi i/5}+Qe^{22\pi i/5}-Q^2 e^{38\pi i/5}+\ \ldots),$$

$$=C(-0.77868\ 9960-0.46496\ 7885i),$$

$$=-2.18796\ 1638-1.30646\ 5921i.$$

Multiplying this value by B, we then obtain the second root.

$$x_2 = -0.89454\ 8013 - 0.53414\ 8530i,$$

and from its conjugate, the third root x_3.

The other two roots are similarly obtained from the evaluation of

$$B\Phi\left(\tau + \frac{32}{5}\right).$$

We thus find: $x_4 = 0.26096\ 4068 + 1.1772\ 2613\ i$, and from its conjugate, the fifth root x_5. These roots are correct to the sixth place.

22. Tables of the Elliptic Functions

In the numerical solution of differential equations of the type described in Section 1, tables of the elliptic functions are required. Tables of sn u, cn u, and dn u have been provided to different arguments by L. M. Milne-Thomson in *Die elliptischen Funktionen von Jacobi*, Berlin, 1931, and by G. W. and R. M. Spenceley in *Smithsonian Elliptic Function Tables*, Washington, D.C., 1947.

The table of Milne-Thomson gives the values of the three functions to five decimal places for values of u at intervals of 0.01 corresponding to the values of k^2 at intervals of 0.1 from $k^2 = 0$ to $k^2 = 0.9$. The range of u was from 0 to 2.00 for k^2 between 0 and 0.5, from 0 to 0.25 for k^2 from 0.6 to 0.8, and from 0 to 3.00 for $k^2 = 0.9$. For $k^2 = 1$, we have sn $u = \tanh u$ and cn $u = $ dn $u = $ sech u.

The table of G. W. and R. M. Spenceley was computed to 12 decimal places for values of $k = \sin \alpha$ at each degree from $0°$ to $89°$. The values were given as functions of u for values of r between 0 and 90 at unit intervals, where u was defined as follows:

$$u = \left(\frac{r}{90}\right)K,$$

K being the complete elliptic integral corresponding to α.

The following tables are four-decimal approximations of sn u, cn u, and dn u, in which the variables are those used in the Spenceley table.

$$\text{TABLE OF } su(u,k), \quad u=\left(\frac{r}{90}\right)K, \quad k=\sin \alpha.$$

					$\alpha=\text{arc sin } k$				
r	0°	10°	15°	30°	45°	60°	70°	80°	89°
0	0.0000	0.0000	0.0000	0.0000	0.0000	0.0000	0.0000	0.0000	0 0000
1	0.0175	0.0176	0.0178	0.0187	0.0206	0.0240	0.0278	0.0350	0.0603
2	0.0349	0.0352	0.0355	0.0375	0.0412	0.0479	0.0556	0.0700	0.1202
3	0.0523	0.0527	0.0532	0.0562	0.0617	0.0718	0.0833	0.1047	0.1792
4	0.0698	0.0703	0.0710	0.0748	0.0823	0.0956	0.1109	0.1393	0.2370
5	0.0872	0.0878	0.0887	0.0935	0.1027	0.1193	0.1383	0.1734	0.2931
6	0.1045	0.1053	0.1063	0.1121	0.1231	0.1429	0.1655	0.2072	0.3473
7	0.1219	0.1228	0.1240	0.1306	0.1435	0.1664	0.1925	0.2405	0.3992
8	0.1392	0.1402	0.1415	0.1491	0.1637	0.1897	0.2192	0.2733	0.4487
9	0.1564	0.1576	0.1591	0.1676	0.1838	0.2128	0.2456	0.3054	0.4956
10	0.1736	0.1749	0.1766	0.1859	0.2038	0.2357	0.2717	0.3369	0.5398
15	0.2588	0.2607	0.2630	0.2764	0.3018	0.3465	0.3960	0.4826	0.7191
20	0.3420	0.3443	0.3473	0.3639	0.3953	0.4495	0.5081	0.6061	0.8360
25	0.4226	0.4253	0.4286	0.4477	0.4832	0.5432	0.6062	0.7064	0.9069
30	0.5000	0.5029	0.5065	0.5269	0.5646	0.6268	0.6900	0.7851	0.9481
35	0.5736	0.5765	0.5802	0.6011	0.6389	0.7000	0.7599	0.8451	0.9713
40	0.6428	0.6457	0.6493	0.6696	0.7059	0.7630	0.8171	0.8900	0.9842
45	0.7071	0.7098	0.7132	0.7321	0.7654	0.8165	0.8632	0.9231	0.9914
50	0.7660	0.7685	0.7715	0.7882	0.8174	0.8612	0.8998	0.9471	0.9953
55	0.8192	0.8212	0.8238	0.8379	0.8623	0.8979	0.9285	0.9643	0.9975
60	0.8660	0.8677	0.8697	0.8810	0.9002	0.9278	0.9507	0.9766	0.9987
65	0.9063	0.9075	0.9091	0.9175	0.9316	0.9515	0.9677	0.9853	0.9993
70	0.9397	0.9405	0.9416	0.9472	0.9567	0.9698	0.9802	0.9914	0.9996
75	0.9659	0.9664	0.9670	0.9704	0.9758	0.9834	0.9893	0.9955	0.9998
80	0.9848	0.9850	0.9853	0.9868	0.9893	0.9927	0.9954	0.9981	0.9999
81	0.9877	0.9879	0.9881	0.9893	0.9914	0.9941	0.9963	0.9985	1.0000
82	0.9903	0.9904	0.9906	0.9916	0.9932	0.9954	0.9971	0.9988	1.0000
83	0.9925	0.9927	0.9928	0.9936	0.9948	0.9965	0.9978	0.9991	1.0000
84	0.9945	0.9946	0.9947	0.9953	0.9962	0.9974	0.9984	0.9993	1.0000
85	0.9962	0.9963	0.9963	0.9967	0.9973	0.9982	0.9989	0.9995	1.0000
86	0.9976	0.9976	0.9976	0.9979	0.9983	0.9988	0.9993	0.9997	1.0000
87	0.9986	0.9987	0.9987	0.9988	0.9990	0.9994	0.9996	0.9998	1.0000
88	0.9994	0.9994	0.9994	0.9995	0.9996	0.9997	0.9998	0.9999	1.0000
89	0.9998	0.9999	0.9999	0.9999	0.9999	0.9999	1.0000	1.0000	1.0000
90	1.0000	1.0000	1.0000	1.0000	1.0000	1.0000	1.0000	1.0000	1.0000
K •	1.5708	1.5828	1.5981	1.6858	1.8541	2.1565	2.5046	3.1534	5.4349

$$\text{TABLE OF } cn(u,k), \quad u=\left(\frac{r}{90}\right) K, \quad k=\sin \alpha.$$

					$\alpha=$arc sin k				
r	0°	10°	15°	30°	45°	60°	70°	80°	89°
0	1.0000	1.0000	1.0000	1.0000	1.0000	1.0000	1.0000	1.0000	1.0000
1	0.9998	0.9998	0.9998	0.9998	0.9998	0.9997	0.9996	0.9994	0.9982
2	0.9994	0.9994	0.9994	0.9993	0.9992	0.9989	0.9985	0.9975	0.9928
3	0.9986	0.9986	0.9986	0.9984	0.9981	0.9974	0.9965	0.9945	0.9838
4	0.9976	0.9975	0.9975	0.9972	0.9966	0.9954	0.9938	0.9903	0.9715
5	0.9962	0.9961	0.9961	0.9956	0.9947	0.9929	0.9904	0.9848	0.9561
6	0.9945	0.9944	0.9943	0.9937	0.9924	0.9897	0.9862	0.9783	0.9378
7	0.9925	0.9924	0.9923	0.9914	0.9897	0.9861	0.9813	0.9706	0.9169
8	0.9903	0.9901	0.9899	0.9888	0.9865	0.9818	0.9757	0.9619	0.8937
9	0.9877	0.9875	0.9873	0.9859	0.9830	0.9771	0.9694	0.9522	0.8685
10	0.9848	0.9846	0.9843	0.9826	0.9790	0.9718	0.9624	0.9415	0.8418
15	0.9659	0.9654	0.9648	0.9610	0.9534	0.9381	0.9183	0.8759	0.6949
20	0.9397	0.9388	0.9378	0.9314	0.9185	0.8933	0.8613	0.7954	0.5487
25	0.9063	0.9051	0.9035	0.8942	0.8755	0.8396	0.7953	0.7079	0.4213
30	0.8660	0.8644	0.8622	0.8499	0.8254	0.7792	0.7238	0.6194	0.3181
35	0.8192	0.8171	0.8144	0.7992	0.7693	0.7141	0.6500	0.5346	0.2379
40	0.7660	0.7636	0.7605	0.7428	0.7083	0.6463	0.5765	0.4559	0.1768
45	0.7071	0.7044	0.7010	0.6813	0.6436	0.5774	0.5048	0.3847	0.1310
50	0.6428	0.6399	0.6362	0.6154	0.5760	0.5083	0.4362	0.3211	0.0967
55	0.5736	0.5706	0.5669	0.5458	0.5064	0.4401	0.3713	0.2647	0.0711
60	0.5000	0.4971	0.4935	0.4731	0.4354	0.3732	0.3100	0.2149	0.0519
65	0.4226	0.4200	0.4166	0.3978	0.3635	0.3078	0.2523	0.1707	0.0375
70	0.3420	0.3397	0.3368	0.3205	0.2911	0.2440	0.1978	0.1312	0.0266
75	0.2588	0.2570	0.2546	0.2417	0.2185	0.1816	0.1459	0.0952	0.0181
80	0.1736	0.1724	0.1707	0.1617	0.1457	0.1204	0.0961	0.0620	0.0112
81	0.1564	0.1553	0.1538	0.1456	0.1311	0.1082	0.0863	0.0556	0.0100
82	0.1392	0.1381	0.1368	0.1295	0.1165	0.0961	0.0766	0.0493	0.0088
83	0.1219	0.1210	0.1198	0.1134	0.1020	0.0841	0.0669	0.0430	0.0076
84	0.1045	0.1037	0.1027	0.0972	0.0874	0.0720	0.0573	0.0368	0.0065
85	0.0872	0.0865	0.0857	0.0810	0.0728	0.0600	0.0477	0.0306	0.0053
86	0.0698	0.0692	0.0686	0.0649	0.0583	0.0480	0.0381	0.0244	0.0043
87	0.0523	0.0519	0.0514	0.0487	0.0437	0.0360	0.0286	0.0183	0.0032
88	0.0349	0.0346	0.0343	0.0324	0.0291	0.0240	0.0190	0.0122	0.0021
89	0.0175	0.0173	0.0172	0.0162	0.0146	0.0120	0.0095	0.0061	0.0011
90	0.0000	0.0000	0.0000	0.0000	0.0000	0.0000	0.0000	0.0000	0.0000
E	1.5708	1.5589	1.5442	1.4675	1.3506	1.2111	1.1184	1.0401	1.0008

$$\text{TABLE OF } dn(u,k), \quad u=\left(\frac{r}{90}\right)K, \quad k=\sin\alpha.$$

				$\alpha=\text{arc sin } k$					
r	5°	10°	15°	30°	45°	60°	70°	80°	89°
0	1.0000	1.0000	1.0000	1.0000	1.0000	1.0000	1.0000	1.0000	1.0000
1	1.0000	1.0000	1.0000	1.0000	0.9999	0.9998	0.9997	0.9994	0.9982
2	1.0000	1.0000	1.0000	0.9998	0.9996	0.9991	0.9986	0.9976	0.9928
3	1.0000	1.0000	0.9999	0.9996	0.9990	0.9981	0.9969	0.9947	0.9838
4	1.0000	0.9999	0.9998	0.9993	0.9983	0.9966	0.9946	0.9906	0.9715
5	1.0000	0.9999	0.9997	0.9989	0.9974	0.9946	0.9915	0.9853	0.9561
6	1.0000	0.9998	0.9996	0.9984	0.9962	0.9923	0.9878	0.9790	0.9378
7	0.9999	0.9998	0.9995	0.9979	0.9948	0.9896	0.9835	0.9715	0.9169
8	0.9999	0.9997	0.9993	0.9972	0.9933	0.9864	0.9786	0.9631	0.8937
9	0.9999	0.9996	0.9992	0.9965	0.9915	0.9829	0.9730	0.9537	0.8686
10	0.9999	0.9995	0.9990	0.9957	0.9896	0.9789	0.9669	0.9434	0.8418
15	0.9997	0.9990	0.9977	0.9904	0.9770	0.9539	0.9282	0.8799	0.6950
20	0.9996	0.9982	0.9960	0.9833	0.9601	0.9211	0.8787	0.8024	0.5489
25	0.9993	0.9973	0.9938	0.9746	0.9398	0.8824	0.8219	0.7184	0.4216
30	0.9990	0.9962	0.9914	0.9647	0.9169	0.8398	0.7613	0.6342	0.3185
35	0.9987	0.9950	0.9887	0.9538	0.8921	0.7953	0.7001	0.5543	0.2385
40	0.9984	0.9937	0.9858	0.9423	0.8665	0.7505	0.6406	0.4814	0.1777
45	0.9981	0.9924	0.9828	0.9306	0.8409	0.7071	0.5848	0.4167	0.1321
50	0.9978	0.9911	0.9799	0.9191	0.8160	0.6662	0.5339	0.3607	0.0982
55	0.9974	0.9898	0.9770	0.9080	0.7926	0.6287	0.4886	0.3133	0.0732
60	0.9971	0.9886	0.9743	0.8977	0.7712	0.5954	0.4493	0.2738	0.0548
65	0.9969	0.9875	0.9719	0.8886	0.7524	0.5666	0.4162	0.2417	0.0414
70	0.9966	0.9866	0.9699	0.8807	0.7365	0.5428	0.3893	0.2164	0.0318
75	0.9964	0.9858	0.9682	0.8744	0.7238	0.5241	0.3685	0.1974	0.0251
80	0.9963	0.9853	0.9669	0.8698	0.7146	0.5108	0.3537	0.1841	0.0207
81	0.9963	0.9852	0.9667	0.8691	0.7132	0.5087	0.3515	0.1821	0.0201
82	0.9963	0.9851	0.9666	0.8684	0.7119	0.5069	0.3495	0.1803	0.0195
83	0.9963	0.9850	0.9664	0.8679	0.7108	0.5053	0.3478	0.1787	0.0190
84	0.9962	0.9850	0.9663	0.8674	0.7098	0.5039	0.3462	0.1774	0.0186
85	0.9962	0.9849	0.9662	0.8670	0.7090	0.5027	0.3449	0.1762	0.0183
86	0.9962	0.9849	0.9661	0.8666	0.7083	0.5017	0.3439	0.1753	0.0180
87	0.9962	0.9848	0.9660	0.8664	0.7078	0.5010	0.3431	0.1746	0.0177
88	0.9962	0.9848	0.9660	0.8662	0.7074	0.5004	0.3425	0.1741	0.0176
89	0.9962	0.9848	0.9659	0.8661	0.7072	0.5001	0.3421	0.1738	0.0175
90	0.9962	0.9848	0.9659	0.8660	0.7071	0.5000	0.3420	0.1736	0.0175
q	0.0005	0.0019	0.0043	0.0180	0.0432	0.0858	0.1311	0.2066	0.4033

Chapter 7

Differential Equations of Second Order

1. Introduction

IN PRECEDING CHAPTERS we have studied a few problems which led in their solution to certain special differential equations of second order. In this chapter we shall consider the general problem of such equations, which, for convenience, we can write in the form

$$F(x,y,y',y'')=0. \tag{1}$$

Let us assume that $F(x,y,y',y'')$, regarded as a function of the four variables x,y,y',y'', is continuous in the neighborhood of the point: $P_0=(x_0,y_0,y_0',y_0'')$ and possesses continuous first derivatives there. If, furthermore, the first derivative of F with respect to y'' does not vanish at P_0, then, by the theory of implicit functions, there exists a unique continuous function y'' of x,y,y', let us say, $y''=f(x,y,y')$, which satisfies equation (1) and which assumes the value y_0'' when $x=x_0$, $y=y_0$, $y'=y_0'$. Therefore, in the neighborhood of P_0, we can write equation (1) in the explicit form:

$$\frac{d^2y}{dx^2}=f(x,y,y'). \tag{2}$$

2. The Origin of Differential Equations of Second Order

Nonlinear differential equations of second order occur frequently in connection with applied problems, a circumstance which has led to considerable interest in them in recent years.

The prototype of some of them is found in one of the earliest examples, namely, the equation which describes the oscillation of the simple pendulum. This equation, derived in Section 4 of Chapter 1, was found to have the form:

$$\frac{d^2z}{dx^2}+k^2 \sin z=0. \tag{1}$$

By means of the transformation: $y = \sin \frac{1}{2} z$, the problem can be reduced to the solution of the equation,

$$\frac{d^2 y}{dx^2} = ay + by^3,$$

which is integrated by elliptic functions.

A generalization of (2) is found in the equation:

$$\frac{d^2 y}{dx^2} + k \frac{dy}{dx} + ay + by^3 = f \cos mx, \tag{3}$$

which introduces a damping term and an impressed harmonic force. The first systematic study of this equation was made by G. Duffing in 1918 in an extensive investigation of forced vibrations and for this reason it is frequently referred to as Duffing's equation.

In his investigation of the orbital motion of planets under the assumptions of general relativity, that is, the problem of the perihelion shift, Albert Einstein was led to the solution of the following equation:

$$\frac{d^2 y}{dx^2} + y = a + by^2. \tag{4}$$

It is a matter of some historical interest to note that P. Gerber in 1898, in an investigation of the velocity of gravitation, was led to the same problem and derived the correct perihelion shift for Mercury by solving the following equation:

$$(1 + \gamma y) \frac{d^2 y}{dx^2} + y = \alpha - \beta \left(\frac{dy}{dx} \right)^2. \tag{5}$$

Both equations are solved by means of elliptic functions.

An example of special interest is the one already examined in Chapter 4 in connection with Volterra's problem of the prey and the predator. This problem led to a nonlinear differential equation of the following form:

$$y \frac{d^2 y}{dx^2} = \left(\frac{dy}{dx} \right)^2 + acy^2 + (ay - ay^2) \frac{dy}{dx} - acy^3. \tag{6}$$

The significant thing about this equation is the existence of periodic solutions for positive values of a and c.

The significance of a term containing a power, other than one, of the first derivative was pointed out as early as 1883 by Lord Rayleigh in a discussion of the damping of a vibratory system under a viscosity factor. Lord Rayleigh showed how a steady state might be maintained if a term proportional to the cube of y' were introduced. This interesting equation can be written:

$$a \frac{d^2y}{dx^2} + \left[-b + k \left(\frac{dy}{dx} \right)^2 \right] \frac{dy}{dx} + cy = 0. \tag{7}$$

Years later these ideas became important in studying electrical circuits associated with triode oscillators. The well-known equation of B. van der Pol, which can be written in the form:

$$a \frac{d^2y}{dx^2} - \epsilon (1 - y^2) \frac{dy}{dx} + y = 0, \tag{8}$$

appeared in connection with this phenomenon. This equation, however, can be derived by a relatively simple transformation from that of Lord Rayleigh. [See (D), Section 3.]

Another similar equation found in the theory of currents limited by a space charge between coaxial cables, and called the equation of Langmuir, is the following:

$$3y \frac{d^2y}{dx^2} + 4y \frac{dy}{dx} + \left(\frac{dy}{dx} \right)^2 - 1 + y^2 = 0. \tag{9}$$

One of the early theories about the behavior of a spherical cloud of gas acting under the mutual attraction of its molecules and subject to the thermodynamics of gases led R. Emden to a consideration of the equation:

$$\frac{d^2y}{dx^2} + \frac{2}{x} \frac{dy}{dx} + y^n = 0. \tag{10}$$

The solution of this equation subject to the conditions: $y=1$, $y'=0$, when $x=0$, is a classic chapter in astrophysics.

These examples are perhaps sufficient to illustrate the variety of applied problems which have contributed to interest in the theory of nonlinear differential equations of second order.

It is also possible, of course, to obtain equations of this kind by the elimination of two parameters from some given function, as we have shown earlier in Section 4 of Chapter 1. For example, if A and B are eliminated from the equation

$$y = \log \sin (Ax + B), \tag{11}$$

we obtain the equation:

$$2 \frac{d^2y}{dx^2} = (\coth y - 1) \left(\frac{dy}{dx} \right)^2, \tag{12}$$

for which y, as given by (11), is the general solution.

The derivation of the generalized Riccati equation of second order by the elimination of the arbitrary parameters in the ratio of two linear forms, as described in Section 10 of Chapter 3, is another example of the origin of such nonlinear equations.

3. Classification of Nonlinear Differential Equations of Second Order

From the examples which we have given in the preceding section it will be seen that most of the equations are special cases of the following second order nonlinear differential equation:

$$A(y)\frac{d^2y}{dx^2}+B(y)\frac{dy}{dx}+C(y)\left(\frac{dy}{dx}\right)^2+D(y)=0, \tag{1}$$

where the coefficients are the polynomials:

$$A(y)=A_0+A_1y+\ldots+A_my^m, \qquad B(y)=B_0+B_1y+\ldots+B_ny^n,$$

$$C(y)=C_0+C_1y+\ldots+C_py^p, \qquad D(y)=D_0+D_1y+\ldots+D_qy^q. \tag{2}$$

The quantities A_i, B_i, C_i, and D_i are assumed to be functions of x and the exponents m, n, p, and q are integers.

That this equation does not include all cases of interest is evident, however, from equation (10) of Section 2, where n may have non-integral values, and from equation (12), where the coefficient of y'^2 includes coth y. But an examination of 249 examples of nonlinear differential equations of second order given by E. Kamke in the extensive list of such equations in the first volume of his *Differentialgleichungen* (1943) shows that 132, or somewhat more than half, are subsumed under equation (1).

As a practical matter, therefore, we shall classify all equations which are included under (1) as equations of *polynomial class* and all others either as equations of *transcendental class*, where transcendental functions of the dependent variable occur, or as equations of *algebraic class*, where no transcendental functions are involved.

Equation (12) of Section 2 belongs to the transcendental class and equation (1) appears to belong to this class, but is transformed in a simple manner into an equation of polynomial class. Another example of an equation of transcendental class is furnished by the following:

$$\frac{d^2y}{dx^2}+a\left(\frac{dy}{dx}\right)^2+b\sin y=0, \tag{3}$$

which defines the motion of the simple pendulum with a damping factor proportional to the square of the velocity.

Such equations as

$$\frac{d^2y}{dx^2}+a\frac{dy}{dx}+by-y^{3/2}=0, \tag{4}$$

$$\frac{d^2y}{dx^2}=(ay+bx+c)\left[\left(\frac{dy}{dx}\right)^2+1\right]^{3/2}, \tag{5}$$

are examples of equations belonging to the algebraic class.

Progress in the understanding of nonlinear differential equations has been made largely through the study of examples included under a few special classes of equations. Some of these classes we shall now describe.

(A) *Equations Solved by Elliptic Functions*

In this class we find those equations the solutions of which can be reduced to functions which satisfy the equation:

$$\frac{d^2y}{dx^2}=A+By+Cy^2+Dy^3, \tag{6}$$

where A, B, C, and D are constants.

It will be shown later that there exists a solution for this equation, which assumes the specific values: $P_0=(x_0,y_0,y_0')$ and which is analytic in the neighborhood of P_0. Since, moreover, the equation is invariant with respect to the linear transformation: $x=x'-x_0$, one of the arbitrary parameters is x_0. From the theory of elliptic functions it is clear that the only singularities are movable poles.

An interesting example of an equation the solution of which can be reduced to the solution of (6) is the following:

$$4(y-y^2)\frac{d^2y}{dx^2}=3(1-2y)\left(\frac{dy}{dx}\right)^2+4q(x)(y-y^2)\frac{dy}{dx}. \tag{7}$$

This equation, which is due to B. Gambier, has the solution:

$$y=1/[1-\wp^2(u)], \tag{8}$$

where u is a solution of the equation: $u''-q(x)u'=0$, and $\wp(u)$ is the elliptic function of Weierstrass corresponding to $g_2=4$, $g_3=0$, that is, $\wp(u)=\wp(u,4,0)$.

Since the verification of Gambier's example is not entirely a trivial matter, it may be of interest to show it here. Let us first consider the equation:

$$\frac{d^2y}{du^2}=\frac{3}{4}\left(\frac{1}{y}+\frac{1}{y-1}\right)\left(\frac{dy}{du}\right)^2. \tag{9}$$

If we now transform the independent variable from u to x by means of the relationship: $u=u(x)$, we obtain the equation:

$$\frac{d^2y}{dx^2}-\left(\frac{u''}{u'}\right)\frac{dy}{dx}=\frac{3}{4}\left(\frac{1}{y}+\frac{1}{y-1}\right)\left(\frac{dy}{dx}\right)^2,$$

which, when u''/u' is set equal to $q(x)$, reduces to (7).

In order to solve (9) we now write: $(1-\wp^2)y=1$ and differentiate twice. We thus obtain:

$$(1-\wp^2)y'-2\wp\wp'y=0, \quad (1-\wp^2)y''-4\wp\wp'y'-(2\wp\wp')'y=0.$$

Dividing both of these equations by $(1-\wp^2)$ and noting (8), we get

$$y'=2\wp\wp'y^2,$$

from the first, and the following from the second:

$$y''=2y^2[4\wp^2\wp'^2y+(\wp')^2+\wp\wp''],$$
$$=2y^3[4\wp^2\wp'^2+(\wp')^2(1-\wp^2)+\wp\wp''(1-\wp^2)],$$
$$=2y^3[3\wp^2\wp'^2+\wp\wp''-\wp^3\wp'']. \tag{10}$$

Referring now to equation (7), Section 16, Chapter 6, we obtain the following derivatives:

$$\wp'^2=4\wp^3-g_2\wp-g_3, \quad \wp''=6\wp^2-\frac{1}{2}g_2.$$

When these are substituted in (10), the following equation results:

$$y''=y^3[12\wp^5+(20-5g_2)\wp^3-6g_3\wp^2-3g_2\wp-2g_3]. \tag{11}$$

A similar reduction of the right hand member of (9) gives us the following:

$$\frac{3}{4}\left(\frac{1}{y}-\frac{1}{y-1}\right)\left(\frac{dy}{du}\right)^2=y^3[12\wp^5+(12-3g_2)\wp^3-3g_3\wp^2-3g_3\wp-3g_3]. \tag{12}$$

Comparing (11) with (12), we see that they are equivalent only if $g_2=4$, $g_3=0$. The solution of (9) is thus: $y=1/[1-\wp^2(u)]$, where $\wp(u)=\wp(u,4,0)$.

(B) *Equations in Which Critical Points Are Fixed Points*

The class of equations solved by elliptic functions suggested to E. Picard, P. Painlevé, B. Gambier, and their associates, the problem of classifying the general nonlinear differential equation of second order by special categories with respect to the character of the singular points of the solutions.

The form of the equation adopted for the investigation was (1), that is to say, the equations studied belonged to the polynomial class.

The problem proposed was to establish conditions under which the *critical points* of a solution, that is to say, *branch points* and *essential singularities*, would be *fixed* points instead of *movable* points. Thus any function which was the solution of an equation in this class would have only poles as movable singularities. Clearly the equations described in (A) would be included in this category.

The investigation resulted in the discovery of 50 canonical types of equations with the desired property. Of these all but 6 were found to be integrable in terms of elementary or classical functions, or transcendents defined by linear equations. But the remaining 6 equations required the introduction of new transcendental functions for their solution. These functions are called *Painlevé transcendents*.

The Painlevé equations are given explicitly as follows:

(I) $\quad \dfrac{d^2y}{dx^2} = 6y^2 + \lambda x.$

(II) $\quad \dfrac{d^2y}{dx^2} = 2y^3 + xy + \mu.$

(III) $\quad xy\dfrac{d^2y}{dx^2} = x\left(\dfrac{dy}{dx}\right)^2 - y\dfrac{dy}{dx} + ax + by + cy^3 + dxy^4.$

(IV) $\quad y\dfrac{d^2y}{dx^2} = \dfrac{1}{2}\left(\dfrac{dy}{dx}\right)^2 - \dfrac{1}{2}a^2 + 2(x^2-b)y^2 + 4xy^3 + \dfrac{3}{2}y^4.$

(V) $\quad x^2(y-y^2)\dfrac{d^2y}{dx^2} = \dfrac{1}{2}x^2(1-3y)\left(\dfrac{dy}{dx}\right)^2 - xy(1-y)\dfrac{dy}{dx} + ay^2(1-y)^3$

$$+ b(1-y)^3 + cxy(1-y) + dx^2y^2(1+y).$$

(VI) $\quad y(1-y)(x-y)\dfrac{d^2y}{dx^2} = \dfrac{1}{2}\left[x - 2(x+1)y + 3y^2\right]\left(\dfrac{dy}{dx}\right)^2$

$$+ \dfrac{y(1-y)}{x(1-x)}\left[x^2 + (1-2x)y\right]\dfrac{dy}{dx} + \dfrac{1}{2x^2(1-x)^2}\left[ay^2(1-y)^2(x-y)^2\right.$$

$$- bx(1-y)^2(x-y)^2 - c(1-x)y^2(x-y)^2 - dx(1-x)y^2(1-y)^2\big].$$

The first three equations were originally given by Painlevé. Their solutions have no singularities other than movable poles. The last three equations are due to Gambier.

The 50 canonical types discovered in this investigation are recorded in Appendix 1.

(C) The Generalized Riccati Equation of Second Order

The essential features of this equation have already been introduced in Section 10 of Chapter 3. Its significance is found in two facts. First, the arbitrary constants appear in the ratio of two linear terms. In the second place, the solution of the general equation can be reduced to the solution of a linear differential equation of third order.

The generalized Riccati equation of second order belongs to the polynomial class of nonlinear equations. It is characterized by the following conditions: $A(y)$ and $B(y)$ of equations (2) are linear functions of y; $C(y)$ is equal to $-2A_1$, where A_1 is the coefficient of y in $A(y)$; $D(y)$ is a cubic polynomial in y.

(D) Equations Having Periodic Solutions

Extensive investigations have been made of a class of nonlinear equations the solutions of which are periodic. Obviously this class includes the equations solved by elliptic functions, but there are many other categories which, under special conditions, introduce functions which are periodic. Initiated by the researches of H. Poincaré in 1882, by M. A. Liapounoff in 1892, and by I. Bendixson in 1901, there has been an unusual activity in the study of such equations in recent years. Much of the literature of the subject is associated with the development of what has been called *nonlinear mechanics*, which we mentioned in Chapter 1.

Equations of this class are illustrated by the equations of Volterra, which we have already discussed in Chapter 4, and by Van der Pol's equation (8) in Section 1. A celebrated example of this class of equations was due to Lord Rayleigh, who first discussed it in a paper in 1883 and included it in the second edition (1894) of his *Theory of Sound*. (See *Bibliography*.)

Rayleigh argued as follows: The solution of the equation

$$u'' + ku' + n^2 u = 0, \tag{13}$$

defines a steady vibration if $k=0$; but if k is positive, the vibrations will die down, and if k is negative they will increase without limit.

Let us now add to (13) a term proportional to the cube of u', that is,

$$u'' + ku' + k'u'^3 + n^2 u = 0. \tag{14}$$

If k and k' are both positive the resulting motion will again die out, and if both are negative, the motion will increase without limit. But if k and k' have different signs, then the two terms which contain them can be written

$$ku'(1 - au'^2), \quad a > 0, \tag{15}$$

and the motion is no longer unidirectional.

If k is initially negative and the initial value of u' is sufficiently small so that the term in parentheses is positive, the motion will expand until $(1-au'^2)$ becomes negative. Thereupon the motion will begin to damp, u' will diminish until the term (15) is again negative, and the motion once more increases. For small values of k and k' Rayleigh gave the following approximate solution of (14):

$$u = A \sin nt + \frac{k'nA^3}{32}\cos 3nt, \tag{16}$$

where A is defined by

$$k + \frac{3}{4}k'n^2A^2 = 0. \tag{17}$$

Commenting on the situation, Rayleigh said: "If k be negative and k' positive, the vibration becomes steady and assumes the amplitude determined by (17). A smaller vibration increases up to this point, and a larger vibration falls down to it. If on the other hand k be positive, while k' is negative, the steady vibration abstractly possible is unstable, a departure in either direction from the amplitude given by (17) tending always to increase."

It is of considerable interest to see how Rayleigh's equation (14) can be transformed into that of Van der Pol, equation (8) of Section 2. For this purpose let us write (14) in the following form:

$$\frac{d^2u}{dt^2} + \left[-b + c\left(\frac{du}{dt}\right)^2\right]\frac{du}{dt} + n^2u = 0. \tag{18}$$

We now introduce the following change of variables:

$$pt = x, \qquad q\frac{du}{dt} = y, \tag{19}$$

where p and q are constants.

Equation (18) now becomes

$$\frac{p}{q}\frac{dy}{dx} + \left(-b + \frac{c}{q^2}y^2\right)\frac{y}{c} + n^2u = 0.$$

Differentiating this equation and simplifying, we obtain

$$\frac{d^2y}{dx^2} - \frac{b}{p}\left(1 - \frac{3c}{b}y^2\right)\frac{dy}{dx} + \frac{n^2}{p^2}y = 0. \tag{20}$$

If we now let $p = n$, $q = \sqrt{3c/b}$, and introduce the abbreviation: $\epsilon = b/p$, then equation (20) assumes the usual form of the Van der Pol equation:

$$\frac{d^2y}{dx^2} - \epsilon\,(1-y^2)\,\frac{dy}{dx} + y = 0. \tag{21}$$

(E) Miscellaneous Equations Solved by Special Devices

The literature of the subject contains a number of special examples of nonlinear differential equations of second order, which can be integrated in terms of elementary functions, or which can be reduced to the solution of equations of simpler type. But each such equation is a special case and general rules will not apply.

An interesting example is provided by the following equation:

$$\frac{d^2y}{dx^2} - 2a^2y^3 + 2abxy - b = 0, \tag{22}$$

which has as an integral any solution of the Riccati equation:

$$\frac{dy}{dx} + ay^2 - bx = 0. \tag{23}$$

This is readily proved by taking the derivative of (23), from which we have:

$$\frac{d^2y}{dx^2} + 2ay\frac{dy}{dx} - b = 0.$$

If dy/dx is now replaced by its value from (23), equation (22) is obtained.

But the fact that any solution of (23) is also a solution of (22) does not help in obtaining the complete solution of the second order equation. This situation is quite different from that which pertains in the case of linear differential equations of second order where the knowledge of one solution makes it possible to obtain the general solution by a single integration.

The difficulties of the problem are readily seen if we set $b=0$. In this case the solution of (23) is the simple algebraic function:

$$y = \frac{a}{a^2x + k}$$

where k is an arbitrary constant, but the general solution of (22) is given by the elliptic integral:

$$x = \int^y (a^2y^4 + k')^{-\frac{1}{2}}\, dy.$$

Only when $k'=0$ and the proper sign is taken for the radical is the solution of (22) also a solution of (23).

If $b \neq 0$, the relationship between (22) and (23) is even more interesting. Referring to Chapter 3, we see that the solution of (23) is $y = u'/(au)$, where u is a solution of the linear equation:

$$u'' - axu = 0.$$

The solution of this equation is explicitly

$$u= \sqrt{x}\,[C_1\,J_{\frac{1}{3}}(s)+C_2\,J_{-\frac{1}{3}}(s)], \quad s=\frac{2}{3}\,i\,\sqrt{a}x^{3/2}.$$

From this we see that the solution of (23) is expressed in terms of known transcendental functions.

But if we now apply to (22) the transformation:

$$y=\lambda z, \quad x=t/\mu, \quad \text{where } \lambda=-\left(\frac{2b}{a^2}\right)^{1/3}, \quad \mu=(2ab)^{1/3},$$

then the equation reduces to the following:

$$\frac{d^2z}{dt^2}=2z^3+tz-\frac{1}{2},$$

which is a special case of the second Painlevé transcendent.

Another example which illustrates the difficulties just mentioned is furnished by Emden's equation for the case where $n=5$, that is,

$$\frac{d^2y}{dx^2}+\frac{2}{x}\frac{dy}{dx}+y^5=0.$$

A particular solution containing one arbitrary constant is found to be

$$y=\left(\frac{3a}{x^2+3a^2}\right)^{1/2};$$

but this fact does not help in obtaining the complete solution, which is not known.

4. Existence Theorems

The theorems which were given in Chapter 4 defining conditions for the existence of a solution of the general differential equation of first order can be extended without essential change to differential equations of higher order. The three types of existence theorem given there are adaptable to the more general problem, although that of Cauchy-Lipschitz involves some complexities in the formulas involved.

In order not to repeat arguments which differ little from those already given, we shall state, without entering into the details of the proof, the existence theorem from the calculus of limits for a system of two equations of first order. It will be seen that this theorem also includes as a special case the existence theorem for a differential equation of second order.

In a somewhat abbreviated form, we shall extend the proof of the method of successive approximations to these cases. The generalization to systems of equations in n dependent variables and to differential equations of higher order should then be easily understood from these arguments.

We shall be concerned with the following system of two equations of first order:

$$\frac{dy}{dx}=f(x,y,z), \qquad \frac{dz}{dx}=g(x,y,z), \tag{1}$$

where the functions $f(x,y,z)$ and $g(x,y,z)$ are subject to limitations imposed by the theorems.

The following equation of second order:

$$\frac{d^2y}{dx^2}=F(x,y,y'), \tag{2}$$

is readily converted into a special case of system (1) by writing:

$$\frac{dy}{dx}=z, \qquad \frac{dz}{dx}=F(x,y,z). \tag{3}$$

A solution of (1) is now sought within some domain R of the variables involved, which reduces to given initial values: $y=y_0$ and $z=z_0$, when $x=x_0$.

In the calculus of limits we assume that the functions $f(x, y, z)$ and $g(x, y, z)$ are analytic in the neighborhood of the initial values.

Under this assumption, *the equations of system* (1) *have a unique solution, given by the functions* $y=y(x)$ *and* $z=z(x)$, *which are analytic in the neighborhood of* $x=x_0$ *and which reduce respectively to* y_0 *and* z_0 *when* $x=x_0$. *These solutions can be represented explicitly by the following series:*

$$y=y_0+y_0'(x-x_0)+\frac{y_0''}{2!}(x-x_0)^2+\frac{y_0^{(3)}}{3!}(x-x_0)^3+\ldots,$$

$$z=z_0+z'(x-x_0)+\frac{z_0''}{2!}(x-x_0)^2+\frac{z_0^{(3)}}{3!}(x-x_0)^3+\ldots, \tag{4}$$

where the derivatives, evaluated at the point $x=x_0$, *are obtained from successive differentiations of the equations* (1).

As in the simpler case of an equation of first order, the proof of the theorem depends upon the determination of a majorante for system (1). Such a majorante is provided by the system:

$$\frac{dY}{dx}=\frac{dZ}{dx}=F(x,Y,Z), \tag{5}$$

where we have

$$F(x,Y,Z)=\frac{M}{\left(1-\dfrac{x}{a}\right)\left(1-\dfrac{Y}{b}\right)\left(1-\dfrac{Z}{c}\right)}. \tag{6}$$

The use of this majorante in establishing the convergence of the series (4) differs in no essential manner from the use of a similar majorante in the case of one equation of first order. The reader is referred to Section 2 of Chapter 4 for the details of the argument.

In establishing conditions for the existence of a solution of system (1) by the method of successive approximations, we proceed as follows:

Since $y=y_0$ and $z=z_0$ when $x=x_0$ we first write the solution in the following form:

$$y=y_0+\int_{x_0}^{x}f(x,y,z)dx, \qquad z=z_0+\int_{x_0}^{x}g(x,y,z)\,dx. \tag{7}$$

The variations of the variables x, y, and z are now restricted to the interior of a region R defined as follows:

$$|x-x_0|\leqq a, \quad |y-y_0|\leqq b,\ |z-z_0|\leqq c, \tag{8}$$

and we assume that within R both $f(x, y, z)$ and $g(x, y, z)$ are continuous functions and have upper bounds less in absolute value than a positive constant M. We shall assume further that a is the smaller of the two values b/M and c/M.

We now introduce the following Lipschitz condition:

If (x, y, z) and (x, y', z') are any two points in R which have the same x-coordinate, then there exist two positive numbers K and L such that

$$|f(x, y, z)-f(x, y', z')|\leqq K|y-y'|+L|z-z'|,$$
$$|g(x, y, z)-g(x, y', z')|\leqq K|y-y'|+L|z-z'|. \tag{9}$$

As in the case of a single equation of first order, a series of successive approximations are obtained, which assume the following general form:

$$y_n(x)=y_0+\int_{x_0}^{x}f[x, y_{n-1}(x), z_{n-1}(x)]\,dx,$$
$$z_n(x)=z_0+\int_{x_0}^{x}g[x, y_{n-1}(x), z_{n-1}(x)]\,dx. \tag{10}$$

By an argument which differs in no essential detail from that given in Section 3 of Chapter 4, the conditions imposed above are sufficient to guarantee, first, the uniform convergence of the following series:

$$y(x)=y_0+(y_1-y_0)+(y_2-y_1)+\ \ldots\ +(y_n-y_{n-1})+\ \ldots,$$
$$z(x)=z_0+(z_1-z_0)+(z_2-z_1)+\ \ldots\ +(z_n-z_{n-1})+\ \ldots, \tag{11}$$

and, second, that there exists no other system of integrals, which assume the prescribed values y_0 and z_0.

5. The Problem of the Pendulum

As we have already seen in Section 4 of Chapter 1, the mathematical description of the vibration of the simple pendulum is formulated in terms of the equation:

$$\frac{d^2\theta}{dt^2} + \frac{g}{L}\sin\theta = 0, \tag{1}$$

where L is the length of the pendulum, g the acceleration of gravity, and θ the angular displacement of the pendulum from its position of equilibrium.

As we have seen earlier, the solution of this equation can be achieved by means of elliptic integrals and expressed in terms of elliptic functions. To obtain the solution in convenient form, we proceed as follows:

The first integral is readily seen to be the following:

$$\frac{1}{2}\left(\frac{d\theta}{dt}\right)^2 - \frac{g}{L}\cos\theta = C, \tag{2}$$

where C is an arbitrary constant.

If $\theta = \omega$ is the maximum displacement of the pendulum from its equilibrium position, then $\theta' = 0$ for this value, and we thus can evaluate C, for which we find $C = -(g\cos\omega)/L$.

We now solve (2) for $d\theta/dt$, and thus obtain the equation

$$\frac{d\theta}{dt} = \sqrt{\frac{2g}{L}}\sqrt{\cos\theta - \cos\omega}, \tag{3}$$

which can also be written in the form

$$dt = \sqrt{\frac{L}{2g}}\frac{d\theta}{\sqrt{\cos\theta - \cos\omega}}. \tag{4}$$

In order to reduce the right-hand member of this equation to standard form, we introduce the transformation:

$$\cos\theta = 1 - 2k^2\sin^2\phi, \qquad k = \sin\tfrac{1}{2}\omega, \tag{5}$$

and observe the following relationships:

$$\cos\theta - \cos\omega = 2k^2\cos^2\phi,$$

$$\sin\theta = 2k\sin\phi\sqrt{1 - k^2\sin^2\phi},$$

$$\sin\theta\,d\theta = 4k^2\sin\phi\cos\phi\,d\phi. \tag{6}$$

When these values are substituted in (4), we obtain the following standard form for dt:

$$dt = \sqrt{\frac{L}{g}} \frac{d\phi}{\sqrt{1 - k^2 \sin^2 \phi}}. \tag{7}$$

Therefore the time T required for the pendulum to swing from its position of equilibrium at $\theta = 0$ to a displacement of $\theta = \theta_0$ is given by the integral

$$T = \sqrt{\frac{L}{g}} \int_0^{\phi_0} \frac{d\phi}{\sqrt{1 - k^2 \sin^2 \phi}}, \tag{8}$$

where ϕ_0 is obtained from the equation:

$$\sin^2 \phi_0 = \frac{1 - \cos \theta_0}{2k^2} = \frac{\sin^2 \frac{1}{2}\theta_0}{k^2}, \tag{9}$$

that is to say,

$$\phi_0 = \arc \sin \left(\frac{\sin \frac{1}{2}\theta_0}{k} \right). \tag{10}$$

In terms of elliptic integrals, we can write (8) as follows:

$$T = \sqrt{\frac{L}{g}} \, F(\phi_0, k). \tag{11}$$

The period of the simple pendulum is defined to be the time required to make a complete oscillation between positions of maximum displacement. To determine this position of maximum displacement, we combine (6) with (3) and thus write

$$\frac{d\theta}{dt} = 2k \sqrt{\frac{g}{L}} \cos \phi. \tag{12}$$

Since the desired value of θ is that for which $d\theta/dt = 0$, we see that this corresponds to $\phi = \frac{1}{2}\pi$. Therefore, if we let $P(k)$ be the period of the pendulum, we get

$$P(k) = 4\sqrt{\frac{L}{g}} \int_0^{\pi/2} \frac{d\phi}{\sqrt{1 - k^2 \sin^2 \phi}},$$

$$= 4\sqrt{\frac{L}{g}} \, K(k), \tag{13}$$

where $K(k)$ is the complete elliptic integral of first kind.

When $k = 0$, this reduces to

$$P = 2\pi \sqrt{\frac{L}{g}}. \tag{14}$$

In order to find the actual motion of the pendulum, that is, the displacement as a function of t, we integrate (7) and thus obtain

$$t = \sqrt{\frac{L}{g}} \int_0^\phi \frac{d\phi}{\sqrt{1 - k^2 \sin^2 \phi}}. \tag{15}$$

From the definitions of Section 8, Chapter 6, this equation can be written

$$\operatorname{sn}\left(t\sqrt{\frac{g}{L}}, k\right) = \sin \phi = \frac{1}{k} \sin \frac{1}{2}\theta, \tag{16}$$

from which we obtain finally

$$\theta = 2 \arcsin\left[k \operatorname{sn}\left(t\sqrt{\frac{g}{L}}, k\right)\right], \quad k = \sin \frac{1}{2}\omega. \tag{17}$$

PROBLEMS

1. A pendulum is displaced through an angle of 45°. Compute its period. *Answer* 6.53 $\sqrt{L/g}$ seconds.

2. A *seconds pendulum* is one that makes a full swing in one second. Using the standard value $g = 32.174$ ft./sec.², show that the length of the seconds pendulum is 39.11 inches.

3. If a seconds pendulum is displaced through an angle of 90°, determine the time required for it to make one complete oscillation. *Answer* 2.36 seconds.

4. Answer Problem 3 if the initial displacement is 10°. *Answer* 2.004 seconds.

5. Find the ratio of the periods of a pendulum for which $k = \sin 30°$ and the pendulum for which $k = 0$. *Answer* 1.0732.

6. Show that $P(k)$ has the expansion

$$P(k) = 2\pi\sqrt{\frac{L}{g}}\left[1 + \left(\frac{1}{2}\right)^2 k^2 + \left(\frac{1\cdot3}{2\cdot4}\right)^2 k^4 + \left(\frac{1\cdot3\cdot5}{2\cdot4\cdot6}\right)^2 k^6 + \dots\right].$$

7. Use the expansion of Problem 6 to compute $P(\frac{1}{2})$. *Answer* 6.743 $\sqrt{L/g}$.

8. If $L = g/16$ and if $k = \frac{1}{2}$, graph the displacement of the pendulum as a function of time through one complete oscillation.

6. The Equation: $y'' = 6y^2$

In order to illustrate the manner in which arbitrary constants enter into the solution of nonlinear equations of second order, we shall consider from various points of view the relatively simple equation

$$\frac{d^2y}{dx^2} = 6y^2. \tag{1}$$

Referring to Section 1, Chapter 6, we see that the solution of this equation can be expressed in terms of elliptic functions, and, as we shall observe shortly, the solution is actually $y = \wp(x)$. But the problem in which we shall be interested in this and in subsequent sections is the manner in which the arbitrary constants enter the general solution and their relationship to various types of expansions. In this way, more, perhaps, than in any other, one can observe the complexities which arise when the property of linearity has been abandoned.

We shall first show that the solution of (1) can be written in the form

$$y(x) = C^2 \left[\frac{-k^2}{1+k^2} + \frac{1}{\operatorname{sn}^2\{C(x-x_1),k\}} \right], \tag{2}$$

where C and x_1 are arbitrary constants and k^2 is a root of the equation:

$$1 - k^2 + k^4 = 0 \tag{3}$$

In order to prove this, let us assume the solution in the form

$$y = A + \frac{B}{\operatorname{sn}^2(Cv)}, \quad v = x - x_1. \tag{4}$$

Taking two derivatives of y, we get

$$\frac{d^2y}{dx^2} = \frac{-2BC^2}{\operatorname{sn}^6(Cv)} \left[\operatorname{sn}^3(Cv) \frac{d^2}{dx^2} \operatorname{sn}(Cv) - 3 \operatorname{sn}^2(Cv) \left\{ \frac{d}{dx} \operatorname{sn}(Cv) \right\}^2 \right]. \tag{5}$$

Making use of the identities

$$\left\{ \frac{d}{dx} \operatorname{sn} x \right\}^2 = \operatorname{cn}^2 x \, dn^2 x = (1 - \operatorname{sn}^2 x)(1 - k^2 \operatorname{sn}^2 x),$$

$$\frac{d^2}{dx^2} \operatorname{sn} x = 2k^2 \operatorname{sn}^3 x - (1 + k^2) \operatorname{sn} x,$$

we reduce (5) as follows:

$$\frac{d^2y}{dx^2} = \frac{-2BC^2}{\operatorname{sn}^4(Cv)} [-k^2 \operatorname{sn}^4(Cv) + 2 \operatorname{sn}^2(Cv) + 2k^2 \operatorname{sn}^2(Cv) - 3],$$

$$= \frac{6BC^2}{\operatorname{sn}^4(Cv)} + 2k^2BC^2 - \frac{4BC^2(1+k^2)}{\operatorname{sn}^2(Cv)}. \tag{6}$$

Since $y'' = 6y^2$, by equating six times the square of (4) to (6), we get the following identity between the functions:

$$\frac{6BC^2}{\operatorname{sn}^4(Cv)} + 2k^2BC^2 - \frac{4BC^2(1+k^2)}{\operatorname{sn}^2(Cv)} = 6A^2 + \frac{6B^2}{\operatorname{sn}^4(Cv)} + \frac{12AB}{\operatorname{sn}^2(Cv)}.$$

Identifying coefficients, we thus obtain

$$6A^2=2k^2BC^2, \qquad 6B^2=6BC^2, \qquad -4BC^2(1+k^2)=12AB,$$

from which we get the following values for A and B:

$$B=C^2, \qquad A=-\frac{1}{3}\,C^2(1+k^2)=\frac{\sqrt{3}}{3}\,kC^2. \qquad (7)$$

From the second equation of (7), after squaring and deleting the common factor, we obtain:

$$1-k^2+k^4=0. \qquad (8)$$

That is to say, k^2 is either of the complex roots of the equation: $z^3+1=0$, namely, $k^2=1+\omega$, or $1+\omega^2$, where ω is a complex cube root of unity.

We further observe that $A^2=k^2C^4/3$, $1/A=-3/[C^2(1+k^2)]$, and hence we get

$$A=-\frac{C^2k^2}{1+k^2}. \qquad (9)$$

The solution of equation (1) can also be written in the form

$$y=\wp(x), \qquad (10)$$

where $\wp(x)$ is the elliptic function of Weierstrass described in Section 16, Chapter 6. If we take the derivative of equation (7) of that section, we have

$$\frac{d^2u}{dx^2}=6u^2-\frac{1}{2}\,g_2. \qquad (11)$$

When u is replaced by y, and g_2 set equal to zero, equation (1) results.

That $\wp(x)$ is a solution of (1) can also be proved directly. Let us write

$$\wp(x)=e_3+\frac{e_1-e_3}{\mathrm{sn}^2(Cx,k)}, \qquad (12)$$

where, by virtue of the fact that $g_2=0$, we have

$$C^2=e_1-e_3, \qquad k^2=\frac{e_2-e_3}{e_1-e_3}, \qquad e_1+e_2+e_3=0, \qquad e_1e_2+e_1e_3+e_2e_3=0. \qquad (13)$$

Referring to equation (2), we see that we must prove that $-C^2k^2/(1+k^2)$ is equal to e_3. But this follows from equations (13), since we have

$$\frac{-C^2k^2}{1+k^2} = \frac{-(e_1-e_3)(e_2-e_3)}{e_1+e_2-2e_3} = \frac{e_1e_2-e_1e_3-e_2e_3+e_3^2}{3e_3},$$

$$= \frac{(-2e_1e_3-2e_2e_3+e_3^2)}{3e_3} = \frac{3e_2^3}{3e_3} = e_3.$$

Although the form in which we have taken the solution, namely, that given in (2), will be useful in the next section, it should be observed that the solution can also be written equally well as follows:

$$y = C^2\left[\frac{-k^2}{1+k^2} + k^2\text{sn}^2(Cv', k)\right], \qquad v' = x-x_0, \tag{14}$$

where x_0 is an arbitrary constant. This second form of the solution is obtained by means of (4) in Section 14 of Chapter 6.

Although we have now achieved the complete solution of equation (1) in terms of Jacobi elliptic functions in either of the two forms (2) and (14), it is clear that something yet remains. Since k^2 is a complex number, the solution is not real. Even though one separated y into real and imaginary components, let us say, $y = U+iV$, neither U nor V separately is a solution as in the case of linear differential equations. For if we substitute y into (1) we get

$$U'' + iV'' = 6(U^2-V^2) + 12\ UVi,$$

from which it follows that: $U'' = 6(U^2-V^2)$ and $V^2 = 12\ UV$.

The problem of achieving a real solution will be discussed in Section 9.

7. The Solution of $y''=6y^2$ as a Laurent Series

We proceed next to express the solution of the equation

$$\frac{d^2y}{dx^2} = 6y^2, \tag{1}$$

as a Laurent series in the variable v, where $v = x-x_1$. From equation (2) of Section 6, we see that the solution has a pole of second order for $v=0$, and thus we assume that y can be written as the following series:

$$y = \frac{a_{-2}}{v^2} + \frac{a_{-1}}{v} + a_0 + a_1v + a_2v^2 + a_3v^3 + \ \dots \ . \tag{2}$$

When this series is substituted in equation (1) and the coefficients of equal powers are equated, the following values of a_n are determined:

$$a_{-2}=1, \quad a_{-1}=a_0=a_1=a_2=a_3=0, \quad a_4=h, \quad a_5=a_6= \ldots =a_9=0,$$

$$a_{10}=\frac{h^2}{13}a_{11}=a_{12}= \ldots =a_{15}=0, \quad a_{16}=\frac{h^3}{247}= \ldots , \quad (3)$$

where h is an arbitrary constant.

Substituting these values in (2), we thus obtain

$$y=\frac{1}{v^2}+hv^4+\frac{h^2}{13}\,v^{10}+\frac{h^3}{247}\,v^{16}+ \cdots . \quad (4)$$

To relate this expansion to the solution given in the preceding section, namely, equation (2) of that section, we first write sn z as follows:

$$\text{sn } z=z+A_1z^3+A_2z^5+A_3z^7+A_4z^9+ \ldots ,$$

where the A_i are obtained explicitly from the expansions given in Section 13 of Chapter 6. We now write

$$\text{sn}^2z=z^2+2A_1z^4+(A_1^2+2A_2)z^6+(2A_3+2A_1A_2)z^8$$

$$+(A_2^2+2A_4+2A_1A_3)z^{10}+(2A_5+2A_1A_4+2A_2A_3)z^{12}+ \ldots ,$$

$$=z^2+B_1z^4+B_2z^6+B_3z^8+B_4z^{10}+B_5z^{12}+ \ldots . \quad (5)$$

From the explicit values of A_i given in Section 13 of Chapter 6, we find the following expressions for B_i:

$$B_1=-\frac{1}{3}\,(1+k^2), \quad B_2=\frac{1}{45}\,(2+13k^2+2k^4), \quad B_3=-\frac{1}{315}\,(1+30k^2+30k^4+k^6),$$

$$B_4=\frac{1}{28350}\,(4+502k^2+1752k^4+502k^6+4k^8), \quad (6)$$

$$B_5=-\frac{2}{467775}\,(1+509k^2+4951k^4+495k^6+509k^8+k^{10}).$$

Assuming next that

$$\frac{z^2}{\text{sn}^2z}=1+C_1z^2+C_2z^4+C_3z^6+C_4z^8+C_5z^{10}+ \ldots , \quad (7)$$

---

*Additional values of B_i in terms of A_i are as follows:

$B_6=A_3^2+2A_6+2A_1A_5+2A_2A_4;$

$B_7=2A_7+2A_1A_6+2A_2A_5+2A_3A_4;$

$B_8=A_4^2+2A_8+2A_1A_7+2A_2A_6+2A_3A_5;$

$B_9=2A_9+2A_1A_8+2A_2A_7+2A_3A_6+2A_4A_5.$

we compute the following product:

$$(1+C_1z^2+C_2z^4+ \ldots)(1+B_1z^2+B_2z^4+ \ldots)$$

$$=1+(B_1+C_1)z^2+(C_2+B_1C_1+B_2)z^4+(C_3+B_1C_2$$

$$+B_2C_1+B_3)z^6+(C_4+B_1C_3+B_2C_2+B_3C_1+B_4)z^8$$

$$+(C_5+B_1C_4+B_2C_3+B_3C_2+B_4C_1+B_5)z^{10}+ \ldots \quad (8)$$

Since this product is identically equal to 1, we set the coefficients of z^2, z^4, etc. equal to zero and thus compute successively the values of the C_i from the values of the B_i. These are found explicitly to be the following:

$$C_1=-B_1=\frac{1}{3}(1+k^2), \quad C_2=-B_2-B_1C_1=\frac{1}{15}(1-k^2+k^4),$$

$$C_3=\frac{1}{189}(2-3k^2-3k^4+2k^6), \quad C_4=\frac{1}{675}(1-k^2+k^4)^2=\frac{1}{3}C_2^2,$$

$$C_5=\frac{1}{10395}(1-k^2+k^4)(2-3k^2-3k^4+2k^6)=\frac{3}{11}C_2C_3. \quad (9)$$

But we have shown in Section 6 that $1-k^2+k^4=0$. Therefore we have $C_2=C_4=C_5=0$. If we now replace z by Cv and substitute the expansion

$$\frac{1}{\operatorname{sn}^2(Cv)}=\frac{1}{C^2v^2}+C_1+C_2C^2v^2+C_3C^4v^4+ \ldots ,$$

into the right hand member of equation (2) of Section 6, we get

$$y(x)=C^2\left[\frac{-k^2}{1+k^2}+\frac{1}{C^2v^2}+C_1+ C_2C^2v^2+C_3C^4v^4+ \ldots \right].$$

We observe finally, that when $1-k^2+k^4=0$, we have

$$C_1-\frac{k^2}{1+k^2}=0, \text{ and } C_2=C_4=C_5=0.$$

Therefore, if we write

$$h=C^4C_3=\frac{C^4}{189}(2-2k^2-3k^4+2k^6)=\frac{C^4}{63}(1-2k^2), \quad (10)$$

we obtain as the Laurant expansion of the solution of equation (1) the following:

$$y=\frac{1}{v^2}+hv^4+0 \cdot v^6+0 \cdot v^8+ \cdots .$$

This, we see, agrees with the expansion (4) above.

8. The Solution of $y'' = 6y^2$ as a Taylor's Series

In the preceding section the solution of the equation

$$\frac{d^2y}{dx^2} = 6y^2, \tag{1}$$

was obtained in the form of a Laurent series. One of the arbitrary constants in the solution was exhibited as the value x_1, where the function had a pole of second order. In other words, a solution of (1) can be found which has a pole of second order at any specified point in the plane. The equation thus provides us with an example of a solution which has a movable pole.

But in the solution of differential equations, one usually seeks a solving function which is analytic in the neighborhood of a point $x = x_0$ and which assumes at that point prescribed values of y and y', let us say, y_0 and y_0'.

That such a solution exists for equation (1) at every point in the plane is assured by the existence theorems of Section 4. We thus have the peculiar situation that every point in the plane can be a polar singularity of the solution and yet every point can also be a regular point. This apparent paradox is readily dispelled in the present case by observing that the solution of (1), which was shown in Section 6 first to have the form

$$y = C^2 \left[\frac{-k^2}{1+k^2} + \frac{1}{\mathrm{sn}^2(Cv,k)} \right], \quad v = x - x_1, \tag{2}$$

was later exhibited as the function

$$y = C^2 \left[\frac{-k^2}{1+k^2} + k^2 \, \mathrm{sn}^2(Cv',k) \right], \quad v' = x - x_0. \tag{3}$$

In both solutions x_0 and x_1 are arbitrary constants.

Since $\mathrm{sn}(x,k)$ is analytic in the neighborhood of zero, it is clear that the function defined in (3) is analytic in the neighborhood of x_0. The Taylor's series is readily found to be

$$y = y_0 + y_0' (x - x_0) + \frac{y_0''}{2!} (x - x_0)^2 + \frac{y_0^{(3)}}{3!} (x - x_0)^3 + \ldots, \tag{4}$$

where y_0 and y_0' are specified arbitrarily, y_0'' is obtained from the differential equation, and $y_0^{(3)}$ and higher derivatives are evaluated from successive derivatives of (1).

The first few of these derivatives are the following:

$$y^{(3)} = 12yy', \quad y^{(4)} = 12(yy'' + y'^2) - 12(6y^3 + y'^2),$$

$$y^{(5)} = 12(yy^{(3)} + 3y'y'') = 360y^2y', \quad y^{(6)} = 720y(3y^3 + y'^2), \tag{5}$$

$$y^{(7)} = 720y'(24y^3 + y'^2), \quad y^{(8)} = 12960y^2(8y^3 + 5y'^2).$$

The relationship between the two expansions, equation (4) of Section 7 and equation (4) of this section, is readily understood if x in the former is replaced by x_0 and y by y_0. Equation (4) of Section 7 is now differentiated and x and y' replaced respectively by x_0 and y_0'. Denoting $x_0 - x_1$ by v_0, we thus have explicitly

$$y_0 = \frac{1}{v_0^2} + h\, v_0^4 + \frac{h^2}{13} v_0^{10} + \frac{h^3}{247} v_0^{16} + \ldots,$$

$$y_0' = -\frac{2}{v_0^3} + 4h\, v_0^3 + \frac{10}{13} h^2\, v_0^9 + \frac{16}{247} h^3\, v_0^{15} + \ldots \tag{6}$$

These two equations form a system for the determination of the unknown parameters h and v in terms of the given values x_0, y_0, and y_0'. From the value of v_0 thus determined, it is then possible to obtain x_1, which is a polar point for y. This value determines the region of convergence of series (4), which is thus limited to the interior of the circle with center at x_0 and radius equal to $|x_1 - x_0|$.

System (6) is generally difficult to solve unless $|v_0|$ is small, some value less than 1. By a method which will be described in a later chapter, it is possible to begin with an initial set of values (X_0, Y_0, Y_0') and determine new values (x_0, y_0, y'_0) for which $|v_0|$ is less than 1. Under these conditions terms in (6) which contain powers of h higher than 1 can usually be neglected for a first approximation and equations (6) reduce to the following:

$$y_0 = \frac{1}{v_0^2} + h\, v_0^4,$$

$$y_0' = -\frac{2}{v_0^3} + 4h\, v_0^3. \tag{7}$$

Eliminating h from these equations, we obtain the following cubic:

$$y_0'\, v_0^3 - 4y_0 v_0^2 + 6 = 0, \tag{8}$$

from which an approximation for v_0 can be made. The corresponding value of h is then found from the equation:

$$h = (y_0 v_0^2 - 1)/v_0^6. \tag{9}$$

Even though $|v_0|$ is not less than 1, the approximation obtained from (8) may be a good one. Thus, for the initial values: $x_0=0$, $y_0=1$, $y_0'=0$, we find that $v_0=x_0=\frac{1}{2}\sqrt{6}=1.2247$ and that $h=4/27$. As we shall show in the next section, these parameters have the exact values: 1.21432 (to five decimal places) and 1/7 respectively.

From what has been said we see that equation (4) provides a method for obtaining values of y in the neighborhood of x_0 and the first of equations (6) for computing values in the neighborhood of the pole x_1. But in general, the rate of convergence of both series is such that neither usually provides a convenient algorithm for obtaining values midway between x_0 and x_1. Another method for computing these values will be described in a later chapter.

9. The Equation: $y''=6y^2-\frac{1}{2}g_2$

Although the equation

$$\frac{d^2y}{dx^2}=6y^2-\frac{1}{2}g_2, \tag{1}$$

appears to have a restricted form, the more general equation

$$\frac{d^2u}{dz^2}=Au^2+B, \tag{2}$$

is readily reduced to it by means of the transformation:

$$u=py, \quad z=qx, \quad p^2=-12B/(Ag_2), \quad q^4=-3g_2/(AB). \tag{3}$$

The solution of equation (1) is given by

$$y=\wp(x),$$

and the formal expansions of both y and y' about an arbitrary polar singularity x_1 are the following series, where g_3 is an arbitrary constant:

$$y=\wp(x)=\frac{1}{v^2}+\frac{1}{20}g_2v^2+\frac{1}{28}g_3v^4+\frac{1}{1200}g_2^2v^6+\frac{3}{6160}g_2g_3v^8+\cdots,$$

$$y'=\wp'(x)=-\frac{2}{v^3}+\frac{1}{10}g_2v+\frac{1}{7}g_3v^3+\frac{1}{200}g_2^2v^5+\frac{3}{770}g_2g_3v^7+\cdots. \tag{4}$$

In these expansions $v=x-x_1$.

The half periods, Ω and Ω', are defined by the equations:

$$\wp(\Omega)=e_1, \quad \wp(\Omega+\Omega')=e_2, \quad \wp(\Omega')=e_3, \tag{5}$$

where e_1, e_2, and e_3 are the roots of the equation:

$$4s^3-g_2s-g_3=0. \tag{6}$$

For the equianharmonic case (See Section 16, Chapter 6) we have $g_2=0$, which reduces (1) to the equation already discussed in the preceding three sections. We can now complete that case by obtaining the real solution. For this purpose we consider the roots of (6), which we now write: $e_1=a\omega$, $e_2=a$, $e_3=a\omega^2$, where ω and ω^2 are the complex cube roots of 1, and $a=(g_3/4)^{1/3}=0.6299605\ (g_3)^{1/3}$. We shall assume that g_3 is a positive real number.

The real half-period Ω_2, defined by $\wp(\Omega_2)=e_2$, has been computed in Section 16, Chapter 6, and shown to be $1.52995\ 4037/(g_3)^{1/6}$.

A table of the values of the solution of equation (1) was first published by A. G. Greenhill and A. G. Hadcock in 1889 and this table appears in all the editions of the *Funktionentafeln* of E. Jahnke and F. Emde. The values of y, given in the Greenhill-Hadcock table, correspond to an argument r defined in terms of v by the equation: $r=108v/\Omega_2$. Since $v=0$ is a pole, it follows that $v=2\Omega_2$ is also a pole. Thus r ranges between 0 and 360. The midpoint of the range, which we shall denote by v_0, corresponds to $r=180$. Its value is thus Ω_2. The initial values used by the computers were: $y_0=a=0.62996$, which corresponds to a choice of $g_3=1$; $y_0'=0$; $v_0=\Omega_2$.

The table just described can be used for the evaluation of a solution of the equianharmonic case of (1), that is, where $g_2=0$, for the initial values: $y=u_0$, $y'=0$, $x=z_0$, in which both u_0 and z_0 are arbitrary.

For this purpose let us write equation (1) and the corresponding equation defining the first derivative as follows:

$$\frac{d^2y}{dx^2}=6y^2, \quad \left(\frac{dy}{dx}\right)^2=4y^3-g_3. \tag{7}$$

A particular solution is given in terms of the initial conditions: $y=y_0$, $y'=y_0'=0$, $x=x_0$, and a second solution is now sought corresponding to the initial values $(u_0,0,z_0)$.

Observing first that the arbitrary constant g_3 has the value: $g_3=4y_0^3$, we next make the transformation:

$$y=Pu, \quad x-x_0=Q(z-z_0), \tag{8}$$

by means of which equations (7) become

$$\frac{d^2u}{dz^2}=6PQ^2u^2, \quad \left(\frac{du}{dx}\right)^2=Q^2(4Pu^3-g_3/P^2). \tag{9}$$

We now write: $PQ^2=1$, and since $u=u_0$ when $y=y_0$, we have $P=y_0/u_0$ and $Q^2=u_0/y_0$. Since $u_0'=0$, we obtain from the second equation of (9), $g_3=4u_0^3P^3$, which is seen to be consistent with the value of g_3 already obtained above.

In particular, if we now let $y_0=a=1/\sqrt[3]{4}, x_0=\Omega_2$, as in the table mentioned above, and if $u_0=1$, $z_0=0$, then the values corresponding to the new boundary conditions are obtained from the tabulated values by the following equations:

$$u=y/P=\sqrt[3]{4}\,y=1.5874011\,y,$$

$$z=(x-x_0)/Q=(x-\Omega_2)=(x-1.5299540)/\sqrt[3]{2}, \qquad (10)$$

$$=0.7937006\,(x-1.5299540).$$

In terms of r, for which values of y are tabulated, $x=\Omega_2(r/180)$. Thus, when $r=240$, we have $x=\frac{4}{3}\Omega_2$, $y=1.0000$. Hence, $u=1.5874$, when $z=0.7937\left(\frac{4}{3}-1\right)\Omega_2=0.4048$.

The values for u' are computed from (9), which now becomes

$$\left(\frac{du}{dz}\right)^2=4u^3-4. \qquad (11)$$

The half-period of the function $u(z)$ is equal to $1.5299540/\sqrt[3]{2}=1.2143254$. This value we shall denote by z_1. The expansions of $u(z)$ and $u'(z)$ in the neighborhood of z_1 are obtained from equations (4) in which appropriate substitutions have been made with special reference to the higher terms given in (3) of Section 7. We thus have

$$u=\frac{1}{v^2}+\frac{1}{7}v^4+\frac{1}{637}v^{10}+\frac{1}{84721}v^{16}+\cdots,$$

$$u'=-\frac{2}{v^3}+\frac{4}{7}v^3+\frac{10}{637}v^9+\frac{16}{84721}v^{15}+\cdots, \qquad (12)$$

where $v=z-z_1$.

The graphical representations of $u(z)$ and $u'(z)$ are shown in Figure 1. It will be observed that $u(z)$ is an even function and $u'(z)$ an odd function. Both functions are periodic with real period equal to $2.4286508=0.7730636\pi$. Values of u and u' are given in the

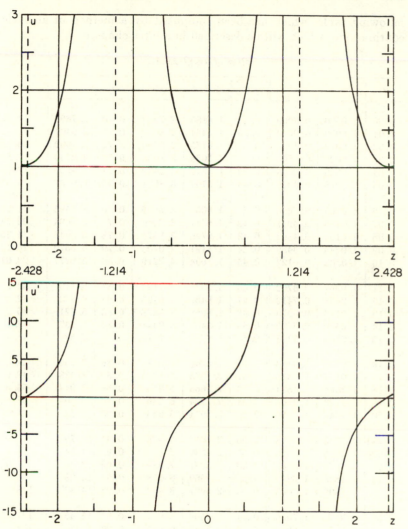

FIGURE 1.—Graphical representation of $u(z)$ and $u'(z)$.

following table, which has been computed by a method of analytic continuation, which will be described in a later chapter.

Values of $u = \wp(z,0,4)$.

z	u	u'	z	u	u'	z	u	u'
0. 00	1. 0000	0. 0000	0. 35	1. 4185	2. 7238	0. 70	3. 7895	14. 62
0. 01	1. 0003	0. 0600	0. 36	1. 4464	2. 8469	0. 71	3. 9401	15. 51
0. 02	1. 0012	0. 1201	0. 37	1. 4755	2. 9750	0. 72	4. 1000	16. 48
0. 03	1. 0027	0. 1803	0. 38	1. 5059	3. 1083	0. 73	4. 2700	17. 53
0. 04	1. 0048	0. 2408	0. 39	1. 5377	3. 2472	0. 74	4. 4509	18. 67
0. 05	1. 0075	0. 3015	0. 40	1. 5709	3. 3921	0. 75	4. 6438	19. 91
0. 06	1. 0108	0. 3626	0. 41	1. 6055	3. 5435	0. 76	4. 8495	21. 27
0. 07	1. 0148	0. 4241	0. 42	1. 6417	3. 7016	0. 77	5. 0694	22. 74
0. 08	1. 0193	0. 4862	0. 43	1. 6796	3. 8671	0. 78	5. 3048	24. 35
0. 09	1. 0245	0. 5489	0. 44	1. 7191	4. 0403	0. 79	5. 5570	26. 12
0. 10	1. 0303	0. 6122	0. 45	1. 7604	4. 2219	0. 80	5. 8277	28. 07
0. 11	1. 0367	0. 6762	0. 46	1. 8036	4. 4124	0. 81	6. 119	30. 2
0. 12	1. 0438	0. 7412	0. 47	1. 8487	4. 6125	0. 82	6. 433	32. 6
0. 13	1. 0515	0. 8070	0. 48	1. 8958	4. 8228	0. 83	6. 771	35. 2
0. 14	1. 0599	0. 8739	0. 49	1. 9452	5. 0441	0. 84	7. 137	38. 1
0. 15	1. 0690	0. 9418	0. 50	1. 9968	5. 2772	0. 85	7. 534	41. 4
0. 16	1. 0788	1. 0110	0. 51	2. 0508	5. 523	0. 86	7. 964	44. 9
0. 17	1. 0892	1. 0815	0. 52	2. 1073	5. 782	0. 87	8. 433	48. 9
0. 18	1. 1004	1. 1534	0. 53	2. 1665	6. 056	0. 88	8. 944	53. 5
0. 19	1. 1123	1. 2269	0. 54	2. 2284	6. 346	0. 89	9. 504	58. 6
0. 20	1. 1249	1. 3019	0. 55	2. 2934	6. 653	0. 90	10. 12	64. 3
0. 21	1. 1383	1. 3788	0. 56	2. 3615	6. 978	0. 91	10. 79	70. 9
0. 22	1. 1525	1. 4575	0. 57	2. 4330	7. 322	0. 92	11. 54	78. 4
0. 23	1. 1675	1. 5382	0. 58	2. 5081	7. 689	0. 93	12. 36	86. 9
0. 24	1. 1833	1. 6211	0. 59	2. 5869	8. 078	0. 94	13. 28	96. 8
0. 25	1. 1999	1. 7063	0. 60	2. 6697	8. 492	0. 95	14. 30	108
0. 26	1. 2174	1. 7939	0. 61	2. 7568	8. 934	0. 96	15. 45	121
0. 27	1. 2358	1. 8842	0. 62	2. 8485	9. 405	0. 97	16. 74	137
0. 28	1. 2551	1. 9772	0. 63	2. 9450	9. 909	0. 98	18. 20	155
0. 29	1. 2753	2. 0733	0. 64	3. 0467	10. 45	0. 99	19. 85	177
0. 30	1. 2965	2. 1725	0. 65	3. 1540	11. 02	1. 00	21. 74	203
0. 31	1. 3188	2. 2751	0. 66	3. 2673	11. 64	1. 21	∞	∞
0. 32	1. 3420	2. 3813	0. 67	3. 3870	12. 31			
0. 33	1. 3664	2. 4913	0. 68	3. 5135	13. 02			
0. 34	1. 3919	2. 6054	0. 69	3. 6475	13. 79			
0. 35	1. 4185	2. 7238	0. 70	3. 7895	14. 62			

10. The Equation: $y'' = Ay + By^3$

The equation

$$\frac{d^2y}{dx^2} = Ay + By^3, \tag{1}$$

is readily solved by observing the following identity taken from Section 10 of Chapter 6:

$$\frac{d^2}{du^2} \operatorname{sn} u = 2k^2 \operatorname{sn}^3 u - (1 + k^2) \operatorname{sn} u. \tag{2}$$

Replacing u by λx and making an identification of constants, we readily obtain the general solution of (1) in the form:

$$y = C \operatorname{sn}(\lambda v, k), \quad v = x - x_0, \tag{3}$$

where λ and x_0 are arbitrary constants, and where k and C are determined as follows:

$$k^2 = -\frac{(\lambda^2 + A)}{\lambda^2}, \quad C^2 = -\frac{2(\lambda^2 + A)}{B}. \tag{4}$$

Since k and C are thus functions of λ, it is clear that either of them, but not both, may be chosen arbitrarily.

Making use of the addition formulas for sn u given in Section 9 of Chapter 6, we can give other forms to the general solution. Three of these are listed below as follows:

$$y = C \operatorname{cn}(\lambda v, k), \quad v = x - x_0,$$

where
$$k^2 = (\lambda^2 + A)/(2\lambda^2), \quad C^2 = -(\lambda^2 + A)/B; \tag{5}$$

$$y = C \operatorname{dn}(\lambda v, k), \quad v = x - x_0,$$

where
$$k^2 = (2\lambda^2 - A)/\lambda^2, \quad C^2 = -2\lambda^2/B; \tag{6}$$

$$y = C/\operatorname{sn}(\lambda v, k), \quad v = x - x_0,$$

where
$$k^2 = -(\lambda^2 - A)/\lambda^2, \quad C^2 = 2\lambda^2/B. \tag{7}$$

In Section 5 of this chapter the equation

$$\frac{d^2y}{dx^2} + n^2 \sin y = 0, \tag{8}$$

was solved and the solution given in the form:

$$y = 2 \arcsin [k \operatorname{sn}(t, k)], \quad k = \sin \frac{1}{2}\omega, \quad t = nx, \tag{9}$$

where $y' = 0$, when $y = \omega$.

As has been stated earlier, it is possible to convert (8) into an equation of type (1) by means of the transformation: $y = \sin \frac{1}{2} z$, but it will be useful to us later to study the form of the equation obtained by replacing $\sin y$ by the first two terms of its Taylor's expansion. Thus we have

$$\frac{d^2y}{dx^2} + n^2y - \frac{n^2}{6} y^3 = 0. \tag{10}$$

The solution of this equation, comparable to (9), can be written:

$$y = \omega \, \text{sn}(\mu t, k), \quad t = nx, \tag{11}$$

where we have

$$k^2 = \omega^2 / (12 - \omega^2), \quad \mu^2 = 1 - \omega^2 / 12. \tag{12}$$

The classical approximation of equation (8) is, of course, the linear equation:

$$\frac{d^2y}{dx^2} + n^2y = 0, \tag{13}$$

the solution of which, comparable to (9), that is, $y = \omega$, when $y' = 0$, is merely

$$y = \omega \sin t, \quad t = nx. \tag{14}$$

This function gives a close approximation to the solution of (8) when ω is small, but when ω is of the order of $\pi/3$ the departure is large. It will be of interest to compare the values of the three functions, defined respectively by (9), (11), and (14), when $\omega = \pi/3$.

That (9) and (11) give values very close to one another is seen if we compute the value of t for which the functions have attained their maximum values, namely, when $y = \omega$. In the first case [equation (9)], when $\omega = \pi/3$, $k = \sin(\pi/6) = \frac{1}{2}$. Hence, when $\text{sn}\left(t, \frac{1}{2}\right) = 1$, $t = 1.68575$.

In the second case [equation (11)], we compute

$$k^2 = \frac{\pi^2}{108 - \pi^2} = 0.10058, \quad k = 0.31714,$$

$$\mu^2 = \left(1 - \frac{\pi^2}{108}\right) = 0.90861, \quad \mu = 0.95321.$$

Setting $k = \sin \alpha$, we find $\alpha = 18°.48992$. Corresponding to this value, we have $K(k) = 1.61270$, from which we get

$$t = K/\mu = 1.69185.$$

This exceeds the value obtained in the first case by only 0.36 of one percent.

The corresponding value of t for equation (14) is $\frac{1}{2}\pi = 1.57080$, which is nearly seven percent less than the exact value. This difference is graphically illustrated in Figure 2, which shows the two functions, defined respectively by (9) and (14), over a complete cycle.

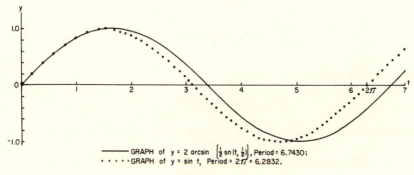

———— GRAPH of y = 2 arcsin $\left[\frac{1}{2}\,\text{sn}\,(t,\,\frac{1}{2})\right]$, Period = 6.7430;
· · · · ·GRAPH of y = sin t, Period = 2π = 6.2832.

FIGURE 2

11. Solution of the General Elliptic Equation

The solution of the general elliptic equation,

$$y'' = A + By + Cy^2 + Dy^3, \tag{1}$$

is achieved by means of a theory of transformations, which reduces the integral to a standard form.

If we multiply (1) by y', integrate, and simplify the resulting expression, we obtain the following equation:

$$(y')^2 = a + 2Ay + By^2 + \frac{2}{3}Cy^3 + \frac{1}{2}Dy^4, \tag{2}$$

where a is an arbitrary constant, which is to be determined from the initial conditions.

We now write (2) in the more convenient form:

$$y'^2 = a + by + cy^2 + dy^3 + ey^4, \tag{3}$$

and seek a transformation:

$$z = z(y), \tag{4}$$

by means of which (3) is reduced to the standard form:

$$\left(\frac{dz}{dx}\right)^2 = (1-z^2)\,(1-k^2z^2) = \Delta^2(z). \tag{5}$$

Since $z=\mathrm{sn}(x,k)$, we can express the desired solution y in terms of the Jacobi elliptic functions by inverting equation (4). It will be convenient to consider two cases: (I) When the right-hand member of (3) is a cubic; (II) when the right-hand member is a quartic.

Case I. Let us write (3) as follows:

$$\left(\frac{dy}{dx}\right)^2 = h^2(y-\alpha)\,(y-\beta)\,(y-\gamma), \tag{6}$$

from which we have

$$\frac{1}{h\Delta_1(y)}\frac{dy}{dx} = 1, \tag{7}$$

where we abbreviate: $\Delta_1^2(y) = (y-\alpha)(y-\beta)(y-\gamma)$, and where h may have either sign.

If we now write

$$z^2 = \frac{\alpha-\gamma}{y-\gamma}, \quad k^2 = \frac{\beta-\gamma}{\alpha-\gamma}, \quad M^2 = \frac{\alpha-\gamma}{4}, \tag{8}$$

we obtain

$$\frac{1}{\Delta(z)}\frac{dz}{dx} = -\frac{M}{\Delta_1(y)}\frac{dy}{dx}, \tag{9}$$

and hence

$$\frac{1}{h\Delta_1(y)}\frac{dy}{dx}\cdot\frac{1}{\Delta(z)}\frac{dz}{d(-hMx)} = 1. \tag{10}$$

From this it follows that

$$z = \mathrm{sn}(-hMx,\,k), \quad v = x-x_0, \tag{11}$$

from which y, the solution of (6), is obtained by the inversion of the first equation in (8).

As an example, let us assume that $\alpha=e_1$, $\beta=e_2$, $\gamma=e_3$ and $h^2=4$, which identifies y with $\wp(v)$. We readily find from equations (8):

$$y = e_3 + \frac{e_1-e_3}{z^2}, \quad z^2 = \mathrm{sn}^2(\lambda v,k),$$

$\lambda^2 = e_1-e_3$, $k^2 = (e_2-e_3)/(e_1-e_3)$, which is the expression given for $\wp(v)$ in Section 16, of Chapter 6.

Other transformations, similar to (8), are also available, such, for example, as the following for which k^2 and M^2 are the same as in (8):

$$\text{(a) } z^2=\frac{y-\alpha}{y-\beta}; \quad \text{(b) } z^2=\left(\frac{\alpha-\gamma}{\beta-\gamma}\right)\left(\frac{\beta-y}{\gamma-y}\right); \quad \text{(c) } z^2=\frac{y-\gamma}{\beta-\gamma}. \tag{12}$$

Case II. We now replace (6) and (7) by the following:

$$\left(\frac{dy}{dx}\right)^2=h^2\Delta_2(y), \quad \frac{1}{h\Delta_2(y)}\frac{dy}{dx}=1, \tag{13}$$

where we abbreviate:

$$\Delta_2^2(y)=(y-\alpha)(y-\beta)(y-\gamma)(y-\delta). \tag{14}$$

If we now write

$$z^2=\frac{(\beta-\delta)}{(\alpha-\delta)}\frac{(y-\alpha)}{(y-\beta)}, \quad k^2=\frac{(\beta-\gamma)(\alpha-\delta)}{(\alpha-\gamma)(\beta-\delta)}, \quad M^2=\frac{(\beta-\delta)(\alpha-\gamma)}{4}, \tag{15}$$

we obtain

$$\frac{1}{\Delta(z)}\frac{dz}{dx}=\frac{M}{\Delta_2(y)}\frac{dy}{dx}, \tag{16}$$

and thus the equation:

$$z=\operatorname{sn}(hMv, k), \quad v=x-x_0. \tag{17}$$

Other transformations similar to (15) are available, such, for example, as the following:

$$\text{(a) } z^2=\frac{(\alpha-\gamma)(\beta-x)}{(\beta-\gamma)(\alpha-x)}; \text{ (b) } z^2=\frac{(\beta-\delta)(x-\gamma)}{(\beta-\gamma)(x-\delta)}; \text{ (c) } z^2=\frac{(\alpha-\gamma)(\delta-x)}{(\alpha-\delta)(\gamma-x)}, \tag{18}$$

where k^2 and M^2 are the same as in (15).

Chapter 8

Second Order Differential Equations of Polynomial Class

1. Introduction

WE HAVE ALREADY DEFINED in Chapter 7 and have indicated the importance of second order differential equations of *polynomial class*. This class of nonlinear equations has been defined by the equation:

$$A(y)\frac{d^2y}{dx^2}+B(y)\frac{dy}{dx}+C(y)\left(\frac{dy}{dx}\right)^2+D(y)=0, \tag{1}$$

where the coefficients are the polynomials:

$$A(y)=A_0+A_1y+ \ldots +A_my^m,\ B(y)=B_0+B_1y+ \ldots +B_ny^n,$$

$$C(y)=C_0+C_1y+ \ldots +C_py^p,\ D(y)=D_0+D_1y+ \ldots +D_qy^q. \tag{2}$$

The quantities A_i, B_i, C_i, and D_i are assumed to be functions of x and the exponents m, n, p, and q are integers.

The simplest case of this equation, where $A(y)=1$, $B(y)=C(y)=0$, and $D(y)$ is a polynomial of third degree with constant coefficients, has already been studied in the preceding chapter. The general solution can be expressed in terms of elliptic functions.

It is the purpose of this chapter to extend our knowledge of equation (1) by certain special devices and by the consideration of particular equations whose solutions are reasonably tractable to elementary analysis.

2. The Linear Fractional Transformation

Since the generalized Riccati equation described in Section 10 of Chapter 3 belongs to the polynomial class and since its general solution is expressible as the quotient of two linear forms, the importance of investigating equation (1) of Section 1 by means of a *linear fractional transformation* is suggested.

This transformation, denoted by T, we shall write in the form:

$$T)\qquad y=\frac{a+bz}{c+dz},\quad \Delta=ad-bc\neq0, \tag{1}$$

where a, b, c, d are functions of x.

213

The linear fractional transformation has the following properties:*

(1) The inverse transformation, T^{-1}, that is to say, the transformation in which z in (1) is expressed in terms of y, is also a linear fractional transformation.

(2) The successive application of two transformations S and T, that is, the product transformation, $P = ST$, is a linear transformation the determinant of which is the product of the determinants of the two transformations. In general, however, ST is different from TS.

(3) If four points are transformed by T into four other points, the cross-ratios of both sets of points are equal. That is to say, the linear fractional transformation leaves invariant the cross-ratio of four points.

We shall apply the transformation T to the following special form of equation (1) of Section 1:

$$L(y) = (A_0 + A_1 y) \frac{d^2 y}{dx^2} + (B_0 + B_1 y) \frac{dy}{dx} + C_0 \left(\frac{dy}{dx} \right)^2$$

$$+ D_0 + D_1 y + D_2 y^2 + D_3 y^3 = 0. \quad (2)$$

The transformed equation, $TL(y) = M(z) = 0$, assumes the following form:

$$f_1(x,z) \frac{d^2 z}{dx^2} + f_2(x,z) \left(\frac{dz}{dx} \right)^2 + f_3(x,z) z \frac{dz}{dx} + f_4(x,z) \frac{dz}{dx} + f_5(x,z) = 0, \quad (3)$$

where we employ the abbreviations:

$$f_i(x,z) = \psi_1^i(x) + \psi_2^i(x) z + \psi_3^i(x) z^2 + \psi_4^i(x) z^3 + \psi_5^i(x) z^4. \quad (4)$$

The values of $\psi_j^i(x)$ are given explicitly in *Appendix* 2. Since these functions are seen to be identically zero when $i = 1, 2, 3, j = 4, 5$; $i = 4, j = 5$, and since, moreover, we have

$$\psi_4^4(x) + \psi_3^3(x) = 0, \quad (5)$$

the transformed equation, $M(z) = 0$, reduces to the following:

$$A(z) \frac{d^2 z}{dx^2} + B(z) \frac{dz}{dx} + C(z) \left(\frac{dz}{dx} \right)^2 + D(z) = 0, \quad (6)$$

where the coefficients have the following explicit forms:

$$A(z) = \psi_1^1 + \psi_2^1 z + \psi_3^1 z^2,$$
$$B(z) = \psi_1^4 + (\psi_2^4 + \psi_1^3) z + (\psi_3^4 + \psi_2^3) z^2,$$
$$C(z) = \psi_1^2 + \psi_2^2 z,$$
$$D(z) = \psi_1^5 + \psi_2^5 z + \psi_3^5 z^2 + \psi_4^5 z^3 + \psi_5^5 z^4. \quad (7)$$

---

*For an extensive account of the properties of the linear fractional transformation see L. R. Ford: *Automorphic Functions*. New York, 2d ed., 1951, Chapter 1.

From this representation of the transformed equation, we obtain the following theorem:

THEOREM 1. *The effect of the transformation* T *on* $L(y)=0$ *is to produce an equation of similar form in which, in general, the exponents of the polynomial coefficients have been increased by unity.*

But an example is readily given which shows that the equation $TL(y)=M(z)=0$ is not always one in which the exponents have been increased. For example, the generalized Riccati equation, discussed in Section 10 of Chapter 3, is a special case of equation (2). But the generalized Riccati equation is obtained by the elimination of the arbitrary constant parameters in the function,

$$y=\frac{k_1v_1+k_2v_2+k_3v_3}{k_1w_1+k_2w_2+k_3w_3},\tag{8}$$

and the same equation would clearly be obtained if the elimination were made on the reciprocal of y. Thus the transformation: $y=1/z$ would not increase the exponents of the generalized Riccati equation.

It is thus a matter of interest to formulate general conditions under which the transformation T does not increase the exponents of (2). We shall speak of such a transformation as one which preserves exponents.

It is clear that exponents are preserved if the coefficients $A(z)$, $B(z)$, $C(z)$, and $D(z)$ have a common linear factor in z, let us say, $p+qz$, where p and q are functions of x. But since a linear transformation preserves exponents, as we shall soon show, such a transformation can remove p and it is thus sufficient to consider the case where z is a factor of the coefficients.

An examination of (7) shows that a transformation which preserves exponents must satisfy one or the other of the following sets of conditions:

(I): (a) $\psi_3^1=0$; (b) $\psi_3^4+\psi_2^3=0$; (c) $\psi_2^2=0$; (d) $\psi_5^5=0$;

(II): (a') $\psi_1^1=0$; (b') $\psi_1^4=0$; (c') $\psi_1^2=0$; (d') $\psi_1^5=0$. (9)

Referring to the explicit values of the functions as given in *Appendix 2*, we see that conditions (I) are equivalent to the following:

Case (I)

(a) and (c): $d(A_0d+A_1b)=0$;

(b): $2(d'\Delta-d\Delta')(A_0d+A_1b)+2C_0\Delta(bd'-b'd)-d\Delta(B_0d+B_1b)=0$;

(d): $(A_0d+A_1b)[d(b''d-bd'')+2d'(bd'-b'd)]+C_0(b'd-bd')^2+$

$\qquad d(B_0d+B_1b)(b'd-bd')+d(d^3D_0+bd^2D_1+b^2dD_2+b^3D_3)=0.$ (10)

If $d=0$, we see that all the conditions are fulfilled. Hence we have the theorem:

THEOREM 2. *If* T *is a linear transformation, exponents in* $M(z)=0$ *are preserved.*

If, however, $d\neq0$, the conditions given above reduce to the following:

$$\text{(a) and (c): } \rho=\frac{b}{d}=-\frac{A_0}{A_1}; \quad \text{(b): } 2C_0\rho'=B_0+B_1\rho;$$

$$\text{(d): } 3C_0\rho'^2+D_0+D_1\rho+D_2\rho^2+D_3\rho^3=0. \tag{11}$$

The following theorem is a consequence of these conditions:

THEOREM 3. *If the coefficients of equation* $L(y)=0$ *satisfy the conditions* (b) *and* (d) *of* (11), *then the transformation*

$$T) \qquad y=\frac{a-A_0z}{c+A_1z}, \tag{12}$$

preserves the exponents of $TL(y)=M(z)=0$, *provided* c *and* d *are any functions such that* $A_1a+A_0c\neq0$.

We can also derive the following theorem as a consequence of conditions (11):

THEOREM 4. *If in equation* $L(y)=0$, *the functions* A_0 *and* A_1 *are linearly dependent, and if the determinant*

$$\begin{vmatrix} A_0 & A_1 \\ B_0 & B_1 \end{vmatrix} \tag{13}$$

vanishes identically, then the transformation T *defined by* (12) *preserves the exponents of* $M(z)=0$, *provided* a *and* c *are functions such that* $A_1a+A_0c\neq0$, *and the functions* D_1 *satisfy the condition:*

$$D_0+D_1\rho+D_2\rho^2+D_3\rho^3=0, \tag{14}$$
where $\rho=-A_0/A_1$.

Proof: Since A_0 and A_1 are linearly dependent, their ratio is a constant. Therefore, $\rho'=0$, from which it follows that $\rho=-B_0/B_1$. Thus determinant (13) must vanish. But since $\Delta=d(a-\rho c)\neq0$, we see that a and c must be functions such that $A_1a+A_0c\neq0$.

Conditions (II), which also preserve the exponents of equation (6), are found to have the following explicit form:

Case (II)

(a'): $c(A_0c+A_1a)=0$;

(b'): $(4c'\Delta-2c\Delta')(A_0c+A_1a)+2C_0\Delta(ac'-a'c)-c\Delta(B_0c+B_1a)=0$;

(c'): $2d(A_0c+A_1a)+\Delta C_0=0$;

(d'): $(A_0c+A_1a)c(a'c-ac')-2c'(a'c-ac')+C_0(a'c-ac')^2$
$\qquad +(B_0c+B_1a)c(a'c-ac')+c^4D_0+ac^3D_1+a^2c^2D_2+a^3cD_3=0.$ (15)

These conditions are all satisfied provided

$$c=0, \ C_0=-2A_1, \ ad\neq0. \tag{16}$$

We thus have the theorem:

THEOREM 5.   *The exponents of* $M(z)=0$ *are preserved under the transformation:*

$$T) \qquad y=\frac{a+\delta z}{z}, \tag{17}$$

where a *and* δ *are arbitrary, provided* $C_0=-2A_1$.

The conditions (II) are also satisfied when we have

$$r=\frac{a}{c}=-\frac{A_0}{A_1}=-\frac{B_0}{B_1}, \ C_0=0, \ D_0+D_1r+D_2r^2+D_3r^3=0, \ A_0d+A_1b\neq0. \tag{18}$$

The following theorem may thus be stated:

THEOREM 6.   *The exponents of* $M(z)=0$ *are preserved under the transformation:*

$$T) \qquad y=\frac{-A_0+bz}{A_1+dz}, \tag{19}$$

provided:

$$1) \ C_0=0; \quad 2) A_0B_1-A_1B_0=0; \quad 3) \ A_0d+A_1b\neq0;$$
$$4) \ D_0+D_1r+D_2r^2+D_3r^3=0,$$

where $r=-A_0/A_1$.

A useful corollary of Theorem 5 is obtained if we observe that

$$\psi_2^2=-2\psi_3^1=2\Delta d(A_0d+A_1b). \tag{20}$$

For if $\delta=b/d$ is taken equal to $-A_0/A_1$, then both ψ_2^2 and ψ_3^1 are zero. Therefore, the function $A(z)$, defined in (7), reduces to $z\psi_2^1=-az\Delta A_1$, and $C(z)$, also defined in (7), vanishes identically.

We thus have the theorem:

THEOREM 7.   *Under the transformation:*

$$T) \qquad y=\frac{a-(A_0/A_1)z}{z}, \tag{21}$$

the exponents of $M(z)=0$ *are preserved provided* $C_0=-2A_1$. *Moreover, the term in* z'^2 *disappears and in the coefficient of* z'' *the multiplier of* z *is zero.*

Under the indicated transformation (21), equation (6) reduces to the following:

$$\psi_2^1\frac{d^2z}{dx^2}+[(\psi_2^4+\psi_1^3)+(\psi_3^4+\psi_2^3)z]\frac{dz}{dx}+(\psi_2^5+\ldots+\psi_5^5z^3)=0. \qquad (22)$$

Since in (21) the parameter a is still available for definition, it is possible to make one further reduction in equation (22).

The only coefficients which can be reduced to zero without assuming that $a=0$, are the following three:

$$P=\psi_2^4+\psi_1^3, \quad Q=\psi_3^5, \quad R=\psi_4^5.$$

The three conditions,

$$P=0, \quad Q=0, \quad R=0,$$

reduce respectively to the following equations in a:

$$2A_1a'-B_1a=0;$$
$$A_1aa''-2A_1a'^2+B_1aa'+(D_2+3bD_3)a^2=0, \quad b=-A_0/A_1;$$
$$(-4A_1b'+B_0+bB_1)a'+(b''A_1+b'B_1+D_1+2bD_2+3b^2D_3)a=0. \qquad (23)$$

Since the first and third of these equations are linear equations of first order in a, it is clear that a can be determined by a single integration.

But the second equation can be reduced to a linear equation of second order by means of the transformation: $a=1/w$. We thus obtain

$$A_1w''+B_1w'-(D_2+3bD_3)w=0, \quad b=-A_0/A_1. \qquad (24)$$

These results are formulated in the following theorem:

THEOREM 8. *If the equation reduced by the transformation (21) is written in the following form:*

$$P_0\frac{d^2z}{dx^2}+(Q_0+Q_1z)\frac{dz}{dx}+R_0+R_1z+R_2z^2+R_3z^3=0, \qquad (25)$$

then the function a *in the transformation can be selected as the solution of a linear differential equation of first order so that either* Q_0 *or* R_2 *is zero. The coefficient* R_1 *can be reduced to zero if* a *is the reciprocal of any solution of the differential equation of second order given by (24).*

3. Applications of the Linear Fractional Transformation

In Section 10 of Chapter 3 it was shown that the Riccati differential equation of second order had the following form:

$$(A_0+A_1y)y''+(B_0+B_1y)y'-2A_1(y')^2+D_0+D_1y+D_2y^2+D_3y^3=0. \qquad (1)$$

This equation is characterized by two facts. In the first place, it is derived by the elimination of the parameters in the fraction:

$$y = \frac{k_1 v_1 + k_2 v_2 + k_3 v_3}{k_1 w_1 + k_2 w_2 + k_3 w_3}, \tag{2}$$

where the v_i and the w_i are arbitrary linearly independent functions of x. In the second place, the quantity A_1 appears both as a multiplier of $(y')^2$ and in the coefficient of y''. Both of these characteristic features are related to linear fractional transformations.

If we denote the numerator of (2) by U and the denominator by V, that is, $y = U/V$, it is clear that the fraction

$$z = \frac{aU + bV}{cU + dV}, \tag{3}$$

where a, b, c, and d are arbitrary functions of x, subject to the condition that $ad - bc \neq 0$, will have the same form as (2). Hence the differential equation in z obtained by the elimination of the constant parameters, k_i, will be a Riccati equation of second order.

But since $U = yV$, the fraction (3) is the linear fractional transformation

$$T') \qquad z = \frac{ay + b}{cy + d}. \tag{4}$$

Hence, observing that the inverse of T' is also a linear fractional transformation, we reach the conclusion that *the form of the Riccati equation of second order is unchanged by a linear fractional transformation*.

Let us now consider the second characteristic feature of the Riccati equation, namely, that the coefficient of the term $(y')^2$ in (1) is equal to $-2A_1$, where A_1 is the multiplier of y in the coefficient of y''. Since the equation is unchanged in form by a linear fractional transformation, the possibility exists that the term $(y')^2$ might be removed by a proper choice of the elements of the transformation. This is, indeed, the case. By means of the transformation:

$$T) \qquad y = \frac{a - (A_0/A_1)z}{z}, \tag{5}$$

where a is an arbitrary function of x, equation (1) is reduced to the following form (see Theorem 7, Section 2):

$$P_0 \frac{d^2 z}{dx^2} + (Q_0 + Q_1 z)\frac{dz}{dx} + R_0 + R_1 z + R_2 z^2 + R_3 z^3 = 0. \tag{6}$$

We now observe that equation (1) by proper specialization includes the general nonhomogeneous linear differential equation of second order, which, for convenience, we shall write in the form:

$$Py'' + Qy' + Ry = F. \tag{7}$$

Applying to this equation the transformation:

$$S) \qquad y = \frac{a+bz}{c+dz}, \quad \Delta = ad - bc \neq 0, \tag{8}$$

where a, b, c, and d are constants, we obtain the following Riccati equation:

$$P(c+dz)z'' + Q(c+dz)z' - 2Pd(z')^2 - (R/\Delta)(c+dz)^2(a+bz)$$
$$+ (F/\Delta)(c+dz)^3 = 0. \tag{9}$$

Since this equation can be restored to its original form (7) by applying to it the inverse transformation S^{-1}, we see that there exist special Riccati equations of second order, which can be solved by reduction to a linear equation of second order.

If we further simplify (9) by setting $F=0$, then (7) reduces to its homogeneous form. But we have seen in Chapter 3 that the solution of the Riccati differential equation of first order can be reduced to the solution of a homogeneous linear equation of second order. We thus reach the conclusion that *the special Riccati equation of second order given by (9), where* F$=0$, *is equivalent to the general Riccati equation of first order.*

PROBLEMS

1. Given the transformations:

$$S) \quad y = \frac{a+bz}{c+dz}, \quad \Delta = ad - bc; \qquad T) \quad z = \frac{a'+b'w}{c'+d'w}, \quad \Delta' = a'd' - b'c',$$

show that $P = ST$ is a linear fractional transformation in w with determinant equal to $\Delta \cdot \Delta'$.

2. Given the points: $x_1 = 3$, $x_2 = 11$, $x_3 = 8$, $x_4 = 23$, show that their cross-ratio R (see Section 4, Chapter 3) equals -1. Now compute four other points z_i by the transformation:

$$z_i = \frac{2+3x_i}{-4+x_i},$$

and show that the value of R is unchanged.

3. Prove explicitly that if four points are transformed by a linear fractional transformation, the cross-ratio is unchanged.

4. If the transformation S in Problem 1 leaves two points unchanged, show that they are roots of the following equation:

$$cz^2 + (d-a)z - b = 0.$$

5. Prove that there is only one linear fractional transformation that transforms three distinct points: x_1, x_2, x_3, into three other given distinct points: x_1', x_2', x_3'.

6. Solve the following equation:

$$2(3+4y)y'' + 5(3+4y)y' - 8y'^2 + (3+4y)^2(1+2y) = 0.$$

4. Transformations of the Independent Variable

We now consider the transformation of the independent variable, where we write: $x = x(t)$.

The equation,

$$A(y)\frac{d^2y}{dx^2} + B(y)\frac{dy}{dx} + C(y)\left(\frac{dy}{dx}\right)^2 + D(y) = 0, \tag{1}$$

then assumes the form:

$$\dot{x}A^*(y)\frac{d^2y}{dt^2} + [B^*(y)\dot{x}^2 - \ddot{x}A^*(y)]\frac{dy}{dt} + \dot{x}\,C^*(y)\left(\frac{dy}{dt}\right)^2 + \dot{x}^3D^*(y) = 0, \tag{2}$$

where $A^*(y)$ indicates that the functions of x and y in $A(y)$ have been transformed, and where $\dot{x}$ and $\ddot{x}$ denote respectively the first and second derivatives of $x(t)$.

Specifically, if the transformation, denoted by S, is the linear fractional one:

$$S) \qquad x = \frac{\alpha + \beta t}{\gamma + \delta t}, \quad \Delta = \alpha\delta - \beta\gamma \neq 0, \tag{3}$$

then equation (2) becomes:

$$(\gamma + \delta t)^4 A^*(y)\frac{d^2y}{dt^2} + (\gamma + \delta t)^2[2\delta(\gamma + \delta t)A^*(y) - \Delta B^*(y)]\frac{dy}{dt}$$
$$+ (\gamma + \delta t)^4 C^*(y)\left(\frac{dy}{dt}\right)^2 + \Delta^2 D^*(y) = 0. \tag{4}$$

If, in this equation, we set $\delta = 0$, and use the abbreviation: $\mu = \beta/\gamma$, then we have

$$A^*(y)\frac{d^2y}{dt^2} + \mu B^*(y)\frac{dy}{dt} + C^*(y)\left(\frac{dy}{dt}\right)^2 + \mu^2 D^*(y) = 0. \tag{5}$$

From this we derive the following theorem:

THEOREM 9.   *If the coefficients of equation (1) do not contain the variable* x *then the solutions are invariant with respect to the linear transformation:* x = t + p, *where* p *is an arbitrary constant.*

This follows from setting $\mu = 1$ in equation (5).

Although the contents of Theorem 9 may at first appear to be trivial, its implications are quite otherwise. For it suggests the possibility of examining equation (1) for solutions which are *automorphic functions*. While it is beyond the scope of this work to develop the theory of such functions, certain of their properties will be of interest through their connection with elliptic functions.

Automorphic functions are associated with groups of linear fractional transformations.* By a *group* (G) of such transformations, we shall mean a set: $T_1, T_2, \cdots, T_N$, where N may be either finite or infinite, which has the following properties:

(1) An inverse of any transformation of the set is itself a member of the set, that is, $T_p^{-1} = T_q$, for some q.

(2) The succession of any two transformations is a transformation of the set, that is, $T_i T_j = T_q$ for some q.

Examples of such groups are the following:

(a) The anharmonic, or cross ratio, group:

$$x = t, \quad \frac{1}{t}, \quad 1-t, \quad \frac{1}{1-t}, \quad \frac{t-1}{t}, \quad \frac{t}{t-1}. \tag{6}$$

(b) The group of simply periodic functions:

$$x = t + m\omega, \tag{7}$$

where ω is a constant and m assumes any integral value, including zero.

(c) The group of the doubly period functions:

$$x = t + m\omega + n\omega', \tag{8}$$

where ω and ω' are constants and m and n any pair of integers, including zero.

An automorphic function is one that remains unchanged with respect to the elements of a group of linear fractional transformations. More precisely, we shall say that $f(x)$ is automorphic with respect to a group (G) of such transformations provided:

(1) $f(x)$ is a single valued function analytic within a domain D.

(2) If x lies within D, then every element T_n of the group is also in D.

(3) $f[T_n(x)] = f(x)$.

Examples are readily given, since it is clear that any rational function of $e^{2\pi i x}$ is simply automorphic with respect to group (b) above and that any rational function of $\wp(x)$ is automorphic with respect to group (c).

---

* For an exhaustive treatment of this subject the reader is referred to L. E. Ford (*loc. cit.*).

Thus, returning to Theorem 9, we see that equation (1) will include as special cases equations which have as solutions functions that are automorphic with respect to groups (b) and (c). An interesting example is furnished by Gambier's equation [(9), Section 3, Chapter 7], where the solution belongs to group (c).

But the fact that an equation may be invariant with respect to the transformation S does not carry with it any implication that its solution is automorphic with respect to any group of S. This is illustrated by the following two equations:

$$yy''+y'^2-1=0, \quad yy''-y'^2-1=0, \tag{9}$$

both of which are invariant with respect to the linear transformation: $x=t+p$.

But the solution of the first equation is $y=[(x+p)^2+q]^{1/2}$, where p and q are arbitrary constants, which is not an automorphic function; while the solution of the second is $y=(1/q)\cosh q(x+p)$, which belongs to the group (b) of simply periodic functions, and is thus automorphic with respect to this group.

An instructive example is furnished by the following equation:

$$\frac{dy}{dx}\frac{d^3y}{dx^3}-\frac{3}{2}\left(\frac{d^2y}{dx^2}\right)^2+k\left(\frac{dy}{dx}\right)^4=0, \tag{10}$$

which will be found to be invariant with respect to the transformation S, given by (3) above, if $\Delta=1$.

Replacing y' by z in (10), we obtain the equation:

$$z\frac{d^2z}{dx^2}-\frac{3}{2}\left(\frac{dz}{dx}\right)^2+kz^4=0, \tag{11}$$

which is a particular case of (1) and has a solution invariant with respect to a linear transformation, but not invariant with respect to S.

The general solution of (11) is readily found to be

$$z=\frac{aq}{(x+p)^2+q^2}, \quad a^2=2/k, \tag{12}$$

where p and q are arbitrary constants. When z is integrated, we obtain the following general solution of (10):

$$y=a\arctan\left(\frac{x+p}{q}\right)+r, \tag{13}$$

where r is the third arbitrary constant.

But since equation (10) is invariant with respect to the transformation S, it is clear that the solution (13) can also be written

$$y=a \ \arctan\left(\frac{t+P}{Q}\right)+R, \qquad (14)$$

where P, Q, R are arbitrary constants, and where

$$t=\frac{\alpha-\gamma x}{-\beta+\delta x}, \qquad (15)$$

s S^{-1}, the inverse of S defined by (3).

This does not mean, however, that $y(x)$ is itself an invariant of the transformation, but that for any choice of the parameters of the transformation the two forms of y can be equated by proper adjustment of the arbitrary constants.

PROBLEMS

1. Prove by explicit substitution that equation (10) is invariant with respect to a linear fractional transformation.

2. Given $k=2$, determine the arbitrary constants so that y as given by (13) is equal to y as given by (14).

3. The following expression is called the *Schwarzian derivative* of f with respect to x:

$$D(f)_x=\frac{2f'(x)f^{(3)}(x)-3[f''(x)]^2}{2[f''(x)]^2}.$$

Prove that

$$D\left(\frac{af+b}{cf+d}\right)_x=D(f)_x.$$

4. Referring to Problem 3, prove that for the transformation: $x=x(t)$, we have:

$$D(f)_x=D(f)_t\left(\frac{dt}{dx}\right)^2+D(t)_x.$$

5. If, in equation (10) $k=0$, show that the solution is: $y=(ax+b)/(cx+d)$.

6. Show that $y=(ax+b)/(cx-a)$ is a solution of the equation:

$$(y-x)y''=2y(1+y').$$

7. Given the equation:

$$yy''-\frac{3}{2}\ y'^2-2g(\dot x)y^2=0,$$

show that its solution is

$$y=\frac{d}{dx}\left(\frac{u}{v}\right),$$

where u and v are any two solutions of the linear equation:

$$\frac{d^2z}{dx^2}+g(x)z=0.$$

5. Equations With Fixed Critical Points and Movable Poles

In discussing the solution of the equation

$$\frac{d^2y}{dx^2}=6y^2, \tag{1}$$

it was shown (Section 6, Chapter 7) that the general solution is $y=\wp(x)$ and that this function can be written in the form

$$\wp(x)=\frac{1}{v^2}+P(v), \tag{2}$$

where $v=x-a$, a arbitrary, and $P(v)$ is a convergent power series in v. It was thus shown explicitly that any point in the plane can be made a polar singularity. Furthermore, this movable pole was the only singularity of the solution.

This example suggests that one obvious generalization of elliptic functions would be to define a class of functions, solutions of second order differential equations, which would share the fundamental property of elliptic functions that their only movable singularities would be poles. In other words, if we use the term *critical* points to denote branch points and essential singularities, then the members of the class mentioned above would possess only critical points that are non-movable, that is to say, *fixed*.

A brief description of this problem has been given in Chapter 7. Its investigation was undertaken by E. Picard, P. Painlevé, B. Gambier and their associates around the beginning of the present century. The problem was not an easy one and required the examination of a large number of equations. In this work we shall not attempt to describe fully the methods employed by these investigators, since they produced many memoirs on the subject, most of which have been listed in the *Bibliography*. The problem has been extensively presented by E. L. Ince in his treatise on *Ordinary Differential Equations* (1927) with an abundance of detail. It will be sufficient for our purpose to indicate the general method of approach.

The first problem was to find the form of the differential equations of second order which would have solutions with the desired property. It was found that this equation could be written as follows:

$$\frac{d^2y}{dx^2}=P(x,y)\left(\frac{dy}{dx}\right)^2+Q(x,y)\frac{dy}{dx}+R(x,y), \tag{3}$$

where P, Q, and R are rational functions of y. Thus the equation is a member of the general class of polynomial equations.

For the further limitation of the functions P, Q, and R two necessary conditions were discovered as follows:

I. $P(x,y)$ must be either identically zero, or must have one of the five following forms:

$$\text{(A)} \ \frac{m+1}{m(y-a_1)}+\frac{m-1}{m(y-a_2)}, \ m \geqq 1.$$

$$\text{(B)} \ \frac{1}{2}\sum_{n=1}^{4}\frac{1}{y-a_n}.$$

$$\text{(C)} \ \frac{2}{3}\sum_{n=1}^{3}\frac{1}{y-a_n}.$$

$$\text{(D)} \ \frac{3}{4}\left(\frac{1}{y-a_1}+\frac{1}{y-a_2}\right)+\frac{1}{2}\left(\frac{1}{y-a_3}\right).$$

$$\text{(E)} \ \frac{1}{6}\sum_{n=1}^{3}\frac{n+2}{y-a_n}.$$

The quantities a_n are arbitrary functions of x. They are not necessarily different and one of them may be infinite.

II. The coefficients Q and R must have the following form:

$$Q(x,y)=\frac{m(x,y)}{p(x,y)}, \quad R(x,y)=\frac{n(x,y)}{p(x,y)}, \tag{4}$$

where $p(x,y)$, of degree p in y, is the least common denominator of the partial fractions in $P(x,y)$ and $m(x,y)$ and $n(x,y)$ are polynomials in y of degrees not exceeding $p+1$ and $p+3$ respectively.

But these conditions are very far from being sufficient and a long and arduous investigation was initiated by the French analysts to separate equations with the desired property from the total class of equations which satisfied the criteria given in (I) and (II). The magnitude of the task can be inferred from the first case, where $P(x,y)$ is zero, which meant that all equations of the form

$$\frac{d^2y}{dx^2}=Q(x,y)\frac{dy}{dx}+R(x,y), \tag{5}$$

where $Q(x,y)$ is linear in y and $R(x,y)$ is a cubic function of y, had to be separately studied.

This investigation led to the discovery of 10 equations, which we have listed explicitly in *Appendix* 1. The solutions of these equations, with two exceptions, are expressed in terms of the classical transcendents or in terms of functions satisfying a linear equation. For example, the sixth equation in the list, namely,

$$\frac{d^2y}{dx^2}=-[3y+q(x)]\frac{dy}{dx}-q(x)y^2-y^3, \tag{6}$$

has for its solution,

$$y=-\frac{u'}{u}, \text{ where } u^{(3)}+q(x)u''=0. \tag{7}$$

This is observed to be a special case of the generalized Riccati equation of second order.

The two exceptions mentioned above are the first and second Painlevé transcendents, which we have already described in Section 3 of Chapter 7. They are the fourth and ninth equations listed in *Appendix* 1.

In their study of the remaining five cases listed in (I), the French analysts found it desirable to express $P(x,y)$ in a convenient canonical form. This was accomplished by means of a linear fractional transformation of the dependent variable. By proper specialization, it was found that $P(x,y)$, when it was not identically zero, must have one of the following seven forms:

(a) $\dfrac{1}{y}$;

(b) $\dfrac{m-1}{my}$, m an integer greater than 1;

(c) $\dfrac{3y-1}{2y(y-1)}$;

(d) $\dfrac{2(2y-1)}{3y(y-1)}$;

(e) $\dfrac{3(2y-1)}{4y(y-1)}$;

(f) $\dfrac{7y-4}{6y(y-1)}$;

(g) $\dfrac{3y^2-2y(\alpha+1)+\alpha}{2y(y-1)(y-\alpha)}$,   where α is a function of x.

This long study finally resulted in the discovery of a total of 50 special cases, the last 40 of which are included under the 7 forms listed above. These equations are reproduced in *Appendix* 1, from which we get the following count for each of the categories: (a) 6; (b) 20, in three of which m is unspecified, but in the remainder $m=2$, 3, 4, 5; (c) 4; (d) 2; (e) 5; (f) 1; (g) 2. In one of these two cases in the last category α is a constant and in the other $\alpha=x$.

For all these 40 cases, with the exception of four, the solution can be expressed in terms of classical transcendentals, or reduced to the solution of a linear equation. As an example, consider equation (43) in the list in the *Appendix*, namely,

$$(y-y^2)\frac{d^2y}{dx^2}=\frac{3}{4}(1-2y)\left(\frac{dy}{dx}\right)^2,\tag{8}$$

which we have already discussed in Chapter 7 [Equation (9), Section 3]. There it was shown that the solution can be expressed in terms of the elliptic function $\wp(x)$, the singularities of which are only movable poles.

The four exceptional equations are (13), (31), (39), and (50), which, together with the two already mentioned, form the six transcendents of Painlevé. They cannot be solved in terms of the classical function, nor can their solutions be expressed in terms of the solutions of a linear equation. These six equations have been listed explicitly in Section 3 of Chapter 7. An extensive analysis of the first two will be given in the following pages.

In the use of the list of equations in *Appendix* 1 it is important to keep in mind that a much larger set of equations, with solutions satisfying the fundamental criterion, can be generated from it by applying to each member the following transformation:

$$T)\qquad y=\frac{a+bz}{c+dz},\quad \Delta=ad-bc\neq 0,$$

$$x=x(t),$$

where a, b, c, d are analytic functions of x and $x(t)$ is an analytic function of t.

Thus, in the investigation of an equation not included in the canonical list, one first applies to it the transformation T to see whether it can be reduced to some member of the list, since the inverse of T is a transformation of the same kind.

For example, the equation

$$4(y-y^2)\frac{d^2y}{dx^2}=3(1-2y)\left(\frac{dy}{dx}\right)^2+4q(x)(y-y^2)\frac{dy}{dx},\tag{9}$$

which we discussed in Section 3 of Chapter 7, is not included in the standard list. But if we make the transformation:

$$y=z,\ u=u(x)$$

where u is a solution of the equation: $u''-q(x)u'=0$, then (9) reduces to (8) expressed in terms of the variables z and u.

6. The First Painlevé Transcendent

It will be convenient to consider the equation which defines the first Painlevé transcendent in the following form:

$$\frac{d^2y}{dx^2} = 6y^2 + \lambda x, \tag{1}$$

where λ is an arbitrary parameter. That the parameter can be set equal to 1 without essentially changing the generality of the equation is seen from the fact that the transformation: $x = \lambda^{-1/5}t$, $y = \lambda^{2/5}w$, reduces (1) to the form

$$\frac{d^2w}{dt^2} = 6w^2 + t. \tag{2}$$

However, the transformation does not admit the important limiting case where $\lambda = 0$. For this reason, since it will be useful to compare the solution of equation (1) with the solution of the equation

$$\frac{d^2y}{dx^2} = 6y^2, \tag{3}$$

the parametric form will be kept.

Since the general solution of (1) is characterized by the existence of a movable pole, we shall expand this solution in the following series:

$$y = \frac{a_{-2}}{v^2} + \frac{a_{-1}}{v} + a_0 + a_1 v + a_2 v^2 + a_3 v^3 + \ldots, \quad v = x - x_1. \tag{4}$$

The first eight coefficients have the following values:

$$a_{-2} = 1, \, a_{-1} = a_0 = a_1 = 0, \, a_2 = -\lambda x_1/10, \, a_3 = -\lambda/6, \, a_4 = h, \, a_5 = 0, \tag{5}$$

where both x_1 and h are arbitrary constants.

The explicit values of a_n through $n = 16$ are given in *Appendix 3*. Others can be computed from the following recursion formula:

$$a_n = \frac{6}{n^2 - n - 12} \sum_{k=-1}^{n-3} a_k a_{n-k-2}, \quad n > 4. \tag{6}$$

Series (4) thus gives the expansion of the solution in the neighborhood of the singular point x_1 and can be used effectively in computing values of y near the pole provided both h and x_1 are known. Unfortunately, however, it is customary to specify as initial conditions values of y and y', let us say, y_0 and y_0', at some regular point: $x = x_0$. This specification thus defines both h and x, but does not suggest how these constants are to be determined.

The existence of a solution of (1) determined by the initial conditions: $y = y_0$, $y' = y_0'$, is provided by the theorems of Section 4, Chapter 7. Its analytical expansion is given by the series:

$$y = y_0 + y_0'(x - x_0) + \frac{y_0''}{2!}(x - x_0)^2 + \frac{y_0^{(3)}}{3!}(x - x_0)^3 + \ldots, \qquad (7)$$

where y_0'' is evaluated in terms of y_0 and y_0' by means of the differential equation, and $y_0^{(3)}$ and higher derivatives are found from the successive derivatives of (1).

The first few of these derivatives are the following:

$$y^{(3)} = 12yy' + \lambda,$$

$$y^{(4)} = 12(yy'' + y'^2) = 12(6y^3 + \lambda xy + y'^2),$$

$$y^{(5)} = 12(yy^{(3)} + 3y'y'') = 360y^2y' + 12\lambda y + 36\lambda xy'. \qquad (8)$$

Other values of $y^{(n)}$ through $n = 15$ are given in *Appendix 3*.

Connection between the two expansions (4) and (7) is readily established if x in (4) is replaced by x_0 and y by y_0. Equation (4) is differentiated and x and y' replaced respectively by x_0 and y_0'. Denoting $x_0 - x_1$ by v_0, we have explicitly

$$y_0 = \frac{1}{v_0^2} - \frac{\lambda x_1}{10}v_0^2 - \frac{\lambda}{6}v_0^3 + hv_0^4 + \frac{\lambda^2 x_1^2}{300}v_0^6 + \frac{\lambda^2 x_1}{150}v_0^7 + \ldots,$$

$$y_0' = -\frac{2}{v_0^3} - \frac{\lambda x_1}{5}v_0 - \frac{1}{2}\lambda v_0^2 + 4hv_0^3 + \frac{\lambda^2 x_1^2}{50}v_0^5 + \frac{7\lambda^2 x_1^2}{150}v_0^6 + \ldots \qquad (9)$$

These equations can now be put into better form for our purpose if x_1 is replaced by $x_0 - v_0$. We thus obtain

$$y_0 = \frac{1}{v_0^2} - \frac{\lambda x_0}{10}v_0^2 - \frac{\lambda}{15}v_0^3 + hv_0^4 + \frac{\lambda^2 x_0^2}{300}v_0^6 + \ldots,$$

$$y_0' = -\frac{2}{v_0^3} - \frac{\lambda x_0}{5}v_0 - \frac{\lambda}{5}v_0^2 + 4hv_0^3 + \frac{\lambda^2 x_0^2}{50}v_0^5 + \frac{\lambda^2 x_0^2}{150}v_0^6 + \ldots \qquad (10)$$

The two equations in (10) form a system for the determination of the unknown parameters h and v_0 in terms of the given initial values x_0, y_0, and y_0'. From the value of v_0 it is possible to obtain x_1, which is the polar point for y. This value defines the region of convergence of (7), which is thus limited to the interior of the circle with center at x_0 and radius equal to $|x_1 - x_0|$.

We have already applied this analysis to equation (3) in Section 8 of Chapter 7. We shall now extend it to equation (1). For this purpose it will be convenient to use only the first four terms in each of the series in (10).   Since h appears only to the first degree, it can be eliminated and the following quintic is obtained for the approximate determination of v_0:

$$\lambda v_0^5 + 3\lambda x_0 v_0^4 - 15 y_0' v_0^3 + 60 y_0 v_0^2 - 90 = 0. \tag{11}$$

In general, however, the value of $x_0 - x_1$ will be too large to allow a good approximation of v_0 by equating it to the proper root of (11). This difficulty would be overcome, however, if equation (7) and its derivative could be used to obtain new values of y and y' which correspond to a value of x in the neighborhood of x_1.   Unfortunately, the rapidity of the convergence of (7) decreases as x approaches x_1 and one is thus limited in the use of this device.

However, a method of continuous analytic continuation, which is described in Chapter 9, has been devised to overcome the difficulties in the convergence of (7).   By means of it, tables have been computed for y and y' for various values of λ corresponding to the initial conditions: $x_0 = 0$, $y_0 = 1$, $y_0' = 0$.

These values are recorded in Table I in the *Appendix* from which the graph shown in Figure 1 has been constructed.   The graphs of y for $y_0 = 0$, $y_0' = 1$, $\lambda = 0$, 1, and 5, which are shown in Figure 2, was obtained by means of the differential analyzer of the Radiation Laboratory at the University of California and were made by John Killeen. It will be observed that the slopes of the functions corresponding to various values of the parameter are less for the second choice of initial conditions than for the first.   That this should be the case is readily seen for the parameter $\lambda = 0$, if we compare the values of the polar

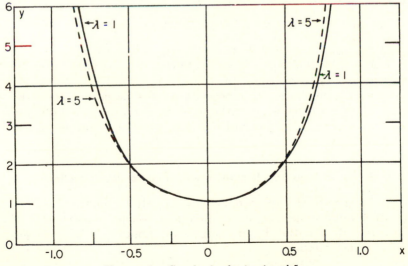

FIGURE 1.—Graph of y for $\lambda = 1$ and 5.

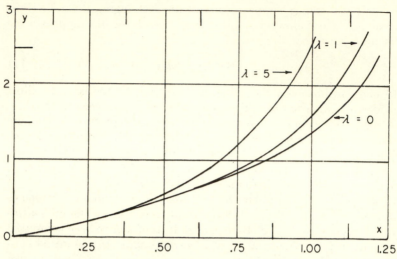

FIGURE 2.—Graph of y for $y_0=0$, $y_0'=1$; $\lambda=0$, 1, and 5.

points. For the first set of conditions we found that $x_1=1.21432$. A similar computation for the second set of conditions yields the value: $x_1=1.52995$.

With the values available in Table I it is now possible to use equation (11) effectively in determining x_1 corresponding to the initial conditions: $x_0=0$, $y_0=1$, $y_0'=0$. Thus, when $\lambda=5$, we have for $x_0=0.90$ the values $y_0=12.78$, $y_0'=91.9$. When these quantities are substituted in (11) and proper simplifications made, we obtain the following equation for the determination of v_0:

$$v_0^5+2.70\,v_0^4-275.7\,v_0^3-153.36\,v_0^2-18=0.$$

From the root, $v_0=-0.279$, we thus determine: $x_1=x_0-v_0=1.179$. This same value is obtained if we use the much larger values of y and y' corresponding to $x=0.96$.

A similar computation to determine the first negative pole corresponding to $\lambda=5$ was made with the initial values: $x_0=-1.00$, $y_0=15.31$, $y_0'=-119$. The value of this pole was thus found to be $x_2=-1.256$.

7. The Boutroux Transformation of the First Painlevé Equation

In a long memoir published in 1913–14, M. P. Boutroux studied the asymptotic properties of the first and second transcendents of Painlevé. In initiating his investigations, Boutroux made the following transformation:

$$y=x^{1/2}w, \quad t=\frac{4}{5}\,x^{5/4}, \tag{1}$$

upon the variables of the equation:

$$\frac{d^2y}{dx^2}=6y^2+\lambda x,\tag{2}$$

and thus obtained:

$$\frac{d^2w}{dt^2}=6w^2+\lambda-\frac{1}{t}\frac{dw}{dt}+\frac{4}{25t^2}\,w.\tag{3}$$

For certain regions of t, when the absolute value of t is sufficiently large, Boutroux showed that w is asymptotic to the solution of the equation:

$$\frac{d^2W}{dt^2}=6W^2+\lambda.\tag{4}$$

The solution of this equation has already been shown (Section 9, Chapter 7) to be

$$W=\wp(Ct,k),\tag{5}$$

where, in this case, the constants C and k are given in terms of the roots of the equation

$$4s^3+2\lambda s+g=0,\tag{6}$$

in which g is arbitrary.

The value of k is determined from the equation of twelfth degree defined by equation (21) in Section 16 of Chapter 6 and is seen to be a function of both λ and g, and thus of both λ and C. The value of C is determined from the boundary conditions imposed initially upon the solution of (4).

The period of $\wp(z)$ is the period of $\mathrm{sn}^2(z)$ and is thus one-half the period of $\mathrm{sn}(z)$. Let us denote this by Ω. The period of W is thus $\Omega_1=\Omega/C$, and since Ω_1 is a function of both C and λ, we can denote this by writing: $\Omega_1=\Omega_1(C,\lambda)$.

For given boundary conditions, let us say at the origin, C is a constant, and hence, for a specified λ, Ω_1 is a constant. If the boundary conditions are real, then Ω_1 is real and, as we have seen, there will exist a series of values: $t_1+m\Omega_1$ for which $W(t)$ is infinite.

Thus, for sufficiently large values of t, the poles of w will be asymptotic to the poles of W. Hence, by means of the transformation (1), we can obtain the asymptotic behavior of the poles of y for large values of x.

Let T_1 and T_2 be successive poles of W, and let their asymptotic images for y be respectively X_1 and X_2. By virtue of (1), we can then write:

$$T_2-T_1=\frac{4}{5}\,(X_2^{5/4}-X_1^{5/4}).\tag{7}$$

If we now denote the distance between X_1 and X_2 by δ, then from (7) we get

$$(X_1+\delta)^{5/4}-X_1^{5/4}=X_1^{5/4}\left(1+\frac{5}{4}\frac{\delta}{X_1}+\frac{5}{32}\frac{\delta^2}{X_1^2}+\ldots\right)-X_1^{5/4}\sim\frac{5}{4}\Omega_1.$$

Since X_1 is large, we thus have

$$\delta\sim\Omega_1 X_1^{-1/4}. \tag{8}$$

The conclusion is thus reached that *the distance between the real poles of the solution of (2) approaches zero asymptotically as the fourth root of the distance of the poles from the origin.*

8. Definition of a New Transcendental Function

Since the solution of the equation

$$\frac{d^2y}{dx^2}=6y^2, \tag{1}$$

can be written in the form (Section 6, Chapter 7)

$$y=a+\frac{b}{\mathrm{sn}^2(Cv)}, \quad v=x-x_1, \tag{2}$$

the possibility suggests itself of defining a new transcendental function, $S(z)$, which will be defined by the equation

$$S^2(Cv)=\frac{B}{y(v)-A}, \tag{3}$$

where $y(v)$ is a solution of the equation

$$\frac{d^2y}{dx^2}=6y^2+\lambda x. \tag{4}$$

It is now our object to determine the parameters in such a way that $S(z)$ reduces to $\mathrm{sn}(z)$ when $\lambda=0$. Fortunately the parameter λ will enter into the coefficients in such a way that $S(z)$ can be written as the sum

$$S(z)=\mathrm{sn}(z)+\lambda\phi(z,\lambda), \tag{5}$$

where $\phi(z)$ is the function which we wish to determine explicitly.

Let us assume that we can write $S^2(Cv)$ as follows:

$$S^2(Cv)=B(v^2+b_4v^4+b_6v^6+b_7v^7+\ldots). \tag{6}$$

Introducing into (3) the expansion of y given by (4) of Section 6, we have

$$S^2(Cv) = \frac{Bv^2}{1 - Av^2 + a_2v^4 + a_3v^5 + a_4v^6 + \ldots}. \tag{7}$$

Equating (6) and (7), we then obtain the following equation for the determination of the coefficients b_i:

$$Bv^2 = B(v^2 + b_4v^4 + b_6v^6 + b_7v^7 + \ldots)(1 - Av^2 + a_2v^4 + a_3v^5 + \ldots). \tag{8}$$

Equating the coefficients of like powers, we obtain the following values for the b_i:

$$b_4 = A; \ b_6 = A^2 - a_2; \ b_7 = a_3 -; \ b_8 = A^3 - 2Aa_2 - a_4; \ b_9 = -2Aa_3;$$
$$b_{10} = A^4 - 3a_2A^2 - 2a_4A + a_2^2 - a_6; \ b_{11} = -3a_3A^2 + 2a_2a_3 - a_7;$$
$$b_{12} = A^5 - 4a_2A^3 - 3a_4A^2 + (3a_2^2 - 2a_6)A + 2a_2a_4 + a_3^2 - a_8;$$
$$b_{13} = -4a_3A^3 + (6a_2a_3 - 2a_7)A + 2a_3a_4 - a_9;$$
$$b_{14} = A^6 - 5a_2A^4 - 4a_4A^3 + (6a_2^2 - 3a_6)A^2 + (6a_2a_4 + 3a_3^2 - 2a_8)A$$
$$- a_2^3 + 2a_2a_6 + a_4^2 - a_{10}. \tag{9}$$

In order to obtain the expansion of $S(Cv)$ itself, it is now necessary to extract the square root of the series given in (6). In order to have an algorithm for the computation of the coefficients, let us consider the following expansions:

$$f^2(v) = F_0 + F_1v + F_2v^2 + F_3v^3 + \ldots + F_nv^n + \ldots, \tag{10}$$

and its square-root:

$$f(v) = f_0 + f_1v + f_2v^2 + f_3v^3 + \ldots + f_nv^n + \ldots. \tag{11}$$

Squaring (11), we obtain the series:

$$f^2(v) = f_0^2 + 2f_0f_1v + (f_1^2 + 2f_0f_2)v^2 + \ldots + \left(\sum_{j=0}^{n} f_jf_{n-j}\right)v^n + \ldots. \tag{12}$$

Equating the coefficients of (12) to (10), we get

$$f_0^2 = F_0, \quad 2f_0f_1 = F_1, \quad f_1^2 + 2f_0f_2 = F_2, \ldots, \sum_{j=0}^{n} f_jf_{n-j} = F_n. \tag{13}$$

From these equations we can compute successively the values of f_0, f_1, f_2, etc., in terms of the F_i.

We now apply this algorithm to determine from (6) and (9) the coefficients of the expansion of $S(Cv)$, which we write as follows:

$$S(Cv) = \sqrt{B}(S_1v + S_2v^2 + S_3v^3 + \ldots + S_nv^n + \ldots). \tag{14}$$

We thus obtain the following values for S_i:

$$S_1=1, \quad S_2=0, \quad S_3=\frac{1}{2}b_4=\frac{1}{2}A, \quad S_4=0, \quad S_5=\frac{1}{2}b_4^2-\frac{1}{8}b_4^2=\frac{3}{8}A^2-\frac{1}{2}a_2,$$

$$S_6=\frac{1}{2}b_7=-\frac{1}{2}a_3=\frac{\lambda}{12}, \quad S_7=\frac{1}{2}b_8-\frac{1}{4}b_4b_6+\frac{1}{16}b_4^3=\frac{5}{16}A^3-\frac{3}{4}Aa_2-\frac{1}{2}h,$$

$$S_8=\frac{1}{2}b_9-\frac{1}{4}b_4b_7=-\frac{3}{4}Aa_3=\frac{1}{8}A\lambda, \quad S_9=\frac{1}{2}b_{10}-\frac{1}{4}b_4b_8-\frac{1}{8}b_6^2+\frac{3}{16}b_4^2b_6-\frac{5}{128}b_4^4,$$

$$=\frac{35}{128}A^4+\frac{3}{32}\lambda x_1A^2-\frac{3}{4}Ah+\frac{1}{480}\lambda^2x_1^2. \tag{15}$$

In this manner we have defined in (14) a new transcendental function which, when $\lambda=0$, is identical with $\mathrm{sn}(Cv)$. Since λ appears only to positive powers, we can write $S(z)$, $z=Cv$, as the sum of $\mathrm{sn}(z)$ and a function in z and λ, as shown in (5). It is this second term, which is the new transcendent of Painlevé.

9. Determination of the Parameter h

An examination of the coefficients S_i, defined explicitly in (15) of Section 8, shows that the parameter h has not yet been evaluated. Since C and x_1 have been assumed to be the arbitrary constants of the solution, h cannot be independently chosen also.

For the purpose of determining h, which appears explicitly in both S_7 and S_9, we shall let $\lambda=0$ and then equate S_7 to the explicit coefficient of v^7 in the expansion of $\mathrm{sn}(Cv)$. It will be recalled (Section 13, Chapter 6) that $\mathrm{sn}(z,k)$ has the following development:

$$\mathrm{sn}(z,k)=z-(1+k^2)\frac{z^3}{3!}+(1+14k^2+k^4)\frac{z^5}{5!}-(1+135k^2+135k^4+k^6)\frac{z^7}{7!}$$

$$+(1+1{,}228k^2+5{,}478k^4+1{,}228k^6+k^8)\frac{z^9}{9!}-(1+11{,}069k^2$$

$$+165{,}826k^4+165{,}826k^6+11{,}069k^8+k^{10})\frac{z^{11}}{11!}+\dots. \tag{1}$$

Referring now to the values given in (15) of Section 8, we set $\lambda=0$, and, recalling that $B=C^2$ [see (7), Section 6, Chapter 7], we obtain the equation:

$$\sqrt{B}S_7=C\left[\frac{5}{16}A^3-\frac{1}{2}h\right]=-\frac{C^7}{7!}(1+135k^2+135k^4+k^6). \tag{2}$$

Since $k^6=-1$ and $1-k^2+k^4=0$ (see Section 6, Chapter 7), the right-hand member of (2) reduces to $-3C^7(2k^2-1)/(2^4\cdot7)$. Hence, solving for h, we obtain:

$$h=\frac{3C^6}{56}(2k^2-1)+\frac{5}{8}A^3. \tag{3}$$

But from (7) of Section 6, Chapter 7, we have $A=-C^2(1+k^2)/3$, whence $A^3=-C^6(2k^2-1)/9$. Introducing this value into (3), we thus obtain

$$h=-\frac{27}{56}A^3+\frac{5}{8}A^3=\frac{1}{7}A^3. \tag{4}$$

We observe that h also appears explicitly in the coefficient of S_9. It is instructive to verify (4) by comparing S_9 with the coefficient of z^9 in (1). Since we have, when $\lambda=0$,

$$S_9=\left(\frac{35}{128}-\frac{3}{28}\right)A^4=\frac{149}{7\cdot9\cdot128}C^8(k^2-1),$$

the identification of S_9/C^8 with the coefficient of z^9 leads to establishing the identity:

$$5\cdot9\cdot149(k^2-1)=1+1{,}228k^2+5{,}478k^4+1{,}228k^6+k^8,$$

where $k^6=-1$.

Replacing k^8 by $-k^2$, k^6 by -1 and k^4 by k^2-1, we see that the right-hand member is reduced to $6{,}705=5\cdot9\cdot149$ multiplied by k^2-1.

10. Generalization of $S(v)$

From the results of the preceding sections, it is now possible to define a function, which we shall denote by $S(v,k;\lambda)$, which reduces to $\operatorname{sn}(v,k)$ when $\lambda=0$, and to $S(v)$ when $k^6=-1$.

In order to define this function, let us write $\operatorname{sn}(v)$ as follows:

$$\operatorname{sn}(v)=v+A_3v^3+A_5v^5+A_7v^7+A_9v^9+\ \ldots, \tag{1}$$

where the A_i are the functions of k^2 given explicitly in (1) of Section 9.

Let us now, in (14) of Section 8, set $C=1$, which we can do since C is an arbitrary constant, from which it follows that B is also equal to 1. The coefficients S_i are now replaced by their explicit values as given by (15) of Section 8 and we thus obtain the following expansion:

$$S(v)=v+\frac{1}{2}Av^3+\frac{3}{8}A^2v^5+\frac{27}{112}A^3v^7+\frac{149}{896}A^4v^9+\ \ldots$$

$$+\lambda\left[\frac{1}{20}x_1v^5+\frac{1}{12}v^6+\frac{3}{40}x_1Av^7+\frac{1}{8}Av^8+\left(\frac{3}{32}x_1A^2+\frac{1}{480}\lambda x_1^2\right)v^9+\ \ldots\right]. \tag{2}$$

Let us now observe that $A_3 \doteq -(1+k^2)/3!$ reduces to $A/2$, when $k^6 = -1$. Thus, if a new function were formed in which A in $S(v)$ is replaced by $2A_3$, we would have a function which, with respect to the coefficients in which A occurs, would reduce to $S(v)$ when $k^6 = -1$.

Similarly, A_5 reduces to $3A^2/8$ and hence A^2 can be replaced by $8A_5/3$. We obtain in this way the following table of equivalents:

$$A_3 \text{ reduces to } \frac{1}{2} A, \qquad\qquad A \text{ is replaced by } 2A_3;$$

$$A_5 \text{ reduces to } \frac{3}{8} A^2, \qquad\qquad A^2 \text{ is replaced by } \frac{8}{3} A_5;$$

$$A_7 \text{ reduces to } \frac{27}{112} A^3, \qquad A^3 \text{ is replaced by } \frac{112}{27} A_7;$$

$$A_9 \text{ reduces to } \frac{149}{896} A^4, \qquad A^4 \text{ is replaced by } \frac{896}{149} A_9;$$

$$A_{11} \text{ reduces to } \frac{201}{1792} A^5, \qquad A^5 \text{ is replaced by } \frac{1792}{201} A_{11}. \qquad (3)$$

When these substitutions are made we obtain the following function:

$$S(v,k;\lambda) = \mathrm{sn}(v,k) + \lambda\phi(v,k;\lambda), \qquad (4)$$

where $\phi(v,k;\lambda)$ has the explicit expansion:

$$\phi(v,k;\lambda) = \frac{1}{20} x_1 v^5 + \frac{1}{12} v^6 + \frac{3}{40} x_1 v^7 + \frac{1}{4} A_3 v^8 + \left(\frac{1}{4} A_5 x_1 + \frac{1}{480} \lambda x_1^2\right) v^9 + \ldots$$

$$(5)$$

If we set $x_1 = 0$, the expansion of this function reduces to the following:

$$\phi(v,k;\lambda) = \frac{1}{12} v^6 \left[1 + 3A_3 v^2 + 5A_5 v^4 + \frac{1}{6} \lambda v^5 + \frac{1121}{135} A_9 v^6 + \cdots \right]. \qquad (6)$$

We thus see that *the difference between* $S(v,k;\lambda)$ *and* $\mathrm{sn}(v,k)$ *is of the order of* v^6. The function defined by (6) we shall call the *first transcendent of Painlevé.*

Although the function $S(v,k;\lambda)$ defined in (4) is real for real values of k, it is actually complex when it is introduced into the solution of the Painlevé equation, since $A^2 = k^2/3$, where $k^2 = 1 + \omega$ and ω is a complex cube root of 1.

Some interest may attach to the expansion of $S(v)$ in this form. The problem reduces to that of the evaluation of powers of A, the first six of which are found to be the following:

$$A = -(2+\omega)/3, \quad A^2 = (1+\omega)/3, \quad A^3 = -(1+2\omega)/9, \quad A^4 = \omega/9,$$

$$A^5 = -(1+\omega)/27, \quad A^6 = -1/27.$$

It is readily shown that A^n has the following value:

$$A^n = \frac{(-1)^n}{3^n} \left[\{2^n - {}_nC_2 \; 2^{n-2} + {}_nC_3 \; 2^{n-3} - {}_nC_5 \; 2^{n-5} + {}_nC_6 \; 2^{n-6} + \ldots\} \right.$$
$$\left. + \omega\{{}_nC_1 \; 2^{n-1} - {}_nC_2 \; 2^{n-2} + {}_nC_4 \; 2^{n-4} - {}_nC_5 \; 2^{n-5} + \ldots\} \right]. \quad (7)$$

When these values are substituted in equation (2) and x_1 set equal to zero, the following expansion results:

$$S(v) = \left[v - \frac{1}{3} v^3 + \frac{1}{8} v^5 - \frac{3}{112} v^7 + \frac{67}{16128} v^{11} + \ldots \right.$$
$$+ \frac{\lambda}{12} \left(v^6 - v^8 + \frac{5}{8} v^{10} + \frac{9\lambda}{88} v^{11} - \frac{1121}{5040} v^{12} + \ldots \right) \Big]$$
$$+ \omega \left[\left(-\frac{1}{6} v^3 + \frac{1}{8} v^5 - \frac{3}{56} v^7 + \frac{149}{8064} v^9 - \frac{67}{16128} v^{11} + \ldots \right) \right.$$
$$\left. - \frac{\lambda}{24} \left(v^8 + \frac{5}{4} v^{10} - \frac{2242}{2745} v^{12} + \ldots \right) \right]. \quad (8)$$

11. The Second Painlevé Transcendent

The equation which defines the second Painlevé transcendent can be conveniently written in the form

$$\frac{d^2y}{dx^2} = 3y^3 + xy + \mu, \quad (1)$$

where μ is an arbitrary parameter. As in the case of the first Painlevé equation, two expansions will be of interest to us, one valid in the neighborhood of the pole: $x = x_1$, and the second in the neighborhood of a regular point: $x = x_0$ at which values of y_0 and y_0' are specified. The first of these expansions can be written in the form:

$$y = \frac{a_{-1}}{v} + a_0 + a_1 v + a_2 v^2 + a_3 v^3 + a_4 v^4 + \ldots, \quad v = x - x_1. \quad (2)$$

When y is substituted in (1) and coefficients of like powers are equated, values of a_n are obtained of which the first seven are the following:

$$a_{-1} = 1, \quad a_0 = 0, \quad a_1 = -x_1/6, \quad a_2 = -\frac{1}{4}(1+\mu), \quad a_3 = h,$$

$$a_4 = \frac{1}{72} x_1(1+3\mu), \quad a_5 = \frac{1}{3024}(27 + 108\mu - 216hx_1 + 81\mu^2 - 2x_1^3), \quad (3)$$

where both x_1 and h are arbitrary constants.

The explicit values of a_n through $n=15$ are given in *Appendix* 4. Others can be computed from the following recursion formula:

$$a_{n+2} = \frac{1}{(n+1)(n+2)} \left[2 \sum_{i=-1}^{n+2} a_i \sum a_j a_{n-(i+j)} + a_{n-1} + a_n x_1 \right], \quad n > -1, \quad (4)$$

where $a_m = 0$, when $m < -1$.

For example, when $n=2$ in (4), we get

$$a_4 = \frac{1}{12} (6a_{-1}^2 a_4 + 12a_{-1} a_0 a_3 + 12a_{-1} a_1 a_2 + 6a_0^2 a_2 + 6a_0 a_1^2 + a_1 + a_2 x_1).$$

When the values for a_i, $i = -1, 0, 1\,2, 3$, as given in (3), are substituted and a_4 solved for, the coefficient given in (3) is obtained.

The second expansion of the solution of (1) is the following Taylor's series about the point: $x = x_0$:

$$y = y_0 + y_0'(x - x_0) + \frac{y_0''}{2!} (x - x_0)^2 + \frac{y_0^{(3)}}{3!} (x - x_0)^3 + \ldots, \quad (5)$$

where y_0 and y_0' are specified arbitrarily, y_0'' is obtained from the differential equation, and $y_0^{(3)}$ and higher derivatives are evaluated from the successive derivatives of (1).

The first few of these derivatives are the following:

$$y^{(3)} = 6y^2 y' + xy' + y,$$

$$y^{(4)} = 6y^2 y'' + 12yy'^2 + xy'' + 2y' = 12y^5 + 8xy^3$$
$$+ 6\mu y^2 + x^2 y + \mu x + 12yy'^2 + 2y',$$

$$y^{(5)} = 12y'^3 + 36yy'y'' + 6y^2 y^{(3)} + xy^{(3)} + 3y''. \quad (6)$$

The values of $y^{(n)}$ through $n=10$ are given in *Appendix* 4.

As in the case of the first equation of Painlevé, the relationship between expansions (2) and (5) is readily seen if x in (2) is replaced by x_0 and y by y_0. Equation (2) is differentiated and x and y' assume respectively the values x_0 and y_0'. Since $v_0 = x_0 - x_1$, we now replace x_1 by $x_0 - v_0$, and thus obtain the following equations:

$$-y_0 = \frac{A_{-1}}{v_0} + A_1 v_0 + A_2 v_0^2 + A_3 v_0^3 + A_4 v_0^4 + A_5 v_0^5 + A_6 v_0^6 + \ldots,$$

$$-y_0' = -\frac{A_{-1}}{v_0^2} + A_1 + 2A_2 v_0 + 3A_3 v_0^2 + 4A_4 v_0^3 + 5A_5 v_0^4 + 6A_6 v_0^5 + \ldots, \quad (7)$$

where the coefficients have the following values:

$$A_{-1}=1, \quad A_1=-\frac{1}{6}x_0, \quad A_2=-\frac{1}{12}(1+3\mu), \quad A_3=h, \quad A_4=\frac{1}{72}(x_0+3\mu),$$

$$A_5=\frac{1}{3024}(-15-18\mu+216\mu x_0+81\mu^2-2x_0^3),$$

$$A_6=-\frac{1}{6024}(2x_0^2+21\mu x_0^2+756\mu h+72h). \tag{8}$$

Negative values of y and y' are used in (7) on the assumption that $x_0 < x_1$ and hence v_0 is negative. This will be the case if the initial point in (5) lies between the origin and the first pole of the solution.

In applying equations (7) to the numerical evaluation of x_1 and h, values of y and y' are determined for some value of x sufficiently near to x_1 so that $v=x-x_1$ is less than one in absolute value. To accomplish this we begin with an initial set of values of y and y' at some specified value of x, let us say, $x=0$. Since series (5) converges very slowly when x is in the neighborhood of x_1, its usefulness is limited and we have recourse to the method of continuous analytic continuation, which is described in Chapter 9, or to some other method of approximation. Having finally obtained the desired values, we denote them by x_0, y_0, y_0' and substitute them in equations (7).

Since h appears first in A_3, terms beyond this value in (7) are neglected and the resulting equations are then used to approximate h and x_1. Since h appears linearly, it can be eliminated. The resulting equation for the approximation of v_0 is the following cubic:

$$(1+3\mu)v_0^3+4(x_0+3y_0')v_0^2-36y_0v_0-48=0. \tag{9}$$

From the value of v_0 thus determined we compute the pole, x_1. In order to obtain h we substitute v_0 in equations (7) and solve either of them for h.

By means of the method of analytic continuation, tables have been computed for y and y' corresponding to $\mu=0, 1, 2, 3, 4, 5$ and the boundary conditions: $x_0=0$, $y_0=1$, $y_0'=0$. These values have been recorded in Table II in the *Appendix*. The graphical representations of y and y' for $\mu=0, 1, 3,$ and 5 are found in Figures 3 and 4.

It will be evident from the graphs that the polar values of the solutions tend to move toward the origin with increasing μ. Making use of the values of y and y' in the neighborhood of $x=+1$, we obtain estimates of x_1 for several values of μ by solving equation (9). These estimates are as follows:

For $\mu=0$, $x_1=1.26$; for $\mu=1$, $x_1=1.16$; for $\mu=5$, $x_1=0.95$.

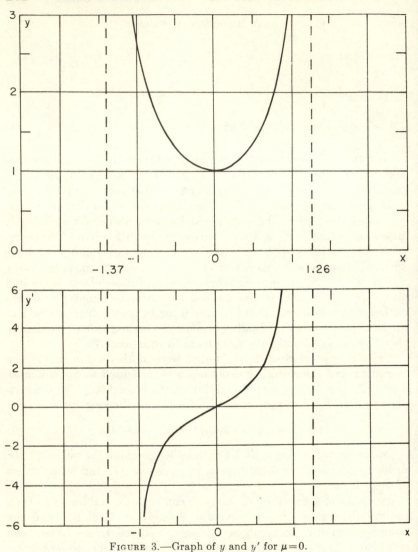

FIGURE 3.—Graph of y and y' for $\mu=0$.

12. The Boutroux Transformation of the Second Painlevé Equation

A transformation similar to that which we have described in Section 7 for the first Painlevé equation was also given by Boutroux for the equation:

$$\frac{d^2y}{dx^2}=2y^3+xy+\mu.$$ (1)

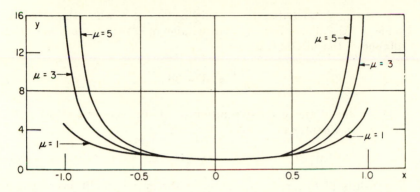

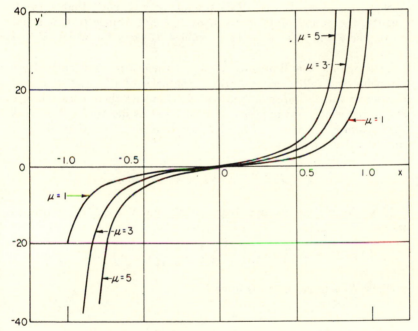

FIGURE 4.—Graph of y and y' for $\mu=1$, 3, and 5.

Applying to (1) the transformation:

$$y=x^{1/2}w, \quad t=\frac{2}{3}x^{3/2}, \tag{2}$$

we obtain the following equation:

$$\frac{d^2w}{dt^2}=2w^3-2w-\frac{1}{t}\frac{dw}{dt}+\frac{1}{9}\frac{w}{t^2}+\frac{2}{3}\frac{\mu}{t}. \tag{3}$$

Boutroux showed that for certain regions of t, when the absolute value of t is sufficiently large, the solution of (3) is asymptotic to the solution of the elliptic equation:

$$\frac{d^2W}{dt^2} = 2W^3 - 2W. \tag{4}$$

One form of the general solution of (4) can be written [see (7), Section 10, Chapter 7]:

$$W = C/\mathrm{sn}(Cu,k), \quad k^2 = 2/C^2 - 1, \quad u = t - t_0. \tag{5}$$

The values of C and t_0 depend upon the initial conditions imposed upon W.

Since, for every k, there exists a quantity Ω, such that $\mathrm{sn}(z+\Omega) = \mathrm{sn}(z)$, the period of W is given by $\Omega_1 = \Omega/C$. Since $t = t_0$ is a pole of W, there will exist a series of values: $t_0 + m\Omega_1$ for which $W(t)$ is infinite.

Thus within the Boutroux regions, when x is sufficiently large, the poles of y will be asymptotic to the poles of W. Let T_1 and T_2 be successive poles of w on the real axis, and let them correspond to real initial conditions imposed at $x = 0$. Let us denote by X_1 and X_2 the asymptotic images of T_1 and T_2 respectively. It is then observed from (2) that we can write

$$T_2 - T_1 = \frac{2}{3}(X_2^{3/2} - X_1^{3/2}) \sim \Omega_1. \tag{6}$$

If we denote the distance between X_1 and X_2 by δ, then from (6) we get

$$(X_1 + \delta)^{3/2} - X_1^{3/2} = X_1^{3/2}\left(1 + \frac{3}{2}\frac{\delta}{X_1} + \frac{3}{8}\frac{\delta^2}{X_1^2} + \cdots\right) - X_1^{3/2} \sim \frac{3}{2}\Omega_1.$$

Since X_1 is large, we thus have

$$\delta \sim \Omega_1 X_1^{-1/2}. \tag{7}$$

The conclusion is thus reached that the distance between the real poles of the solutions of (1) approaches zero asymptotically as the square-root of the distance of the poles from the origin.

One difference between the asymptotic behavior of δ for the cases of the first and second Painlevé transcendents is to be observed. In the first case Ω_1 depends upon the parameter λ of the equation, but in the second case Ω_1 is independent of the parameter μ.

13. Methods of Analytic Continuation

The problem of extending the solutions of the Painlevé transcendents beyond their first singular points on the axis of reals is one of some interest. Two methods are available. The first of these, which we shall call the method of "pole-vaulting," is made practical by the existence in the neighborhood of the pole of both a Taylor's series and a Laurent series with a single infinite term. The second method makes use of an analytic continuation around the singular point.

We shall illustrate these methods by making an extension into the segment beyond its first pole of the first Painlevé transcendent when $\lambda=1$. The solution is initially defined at $x_0=0$ by the values: $y_0=1$, $y_0'=0$.

In applying the method of "pole-vaulting" we first compute the value of x_1 (the first pole) by means of equation (11), Section 6. For the initial values we take from Table I in the *Appendix* the following: $x_0=0.95$, $y_0=15.16$, $y_0'=118$. Substituting these in equation (11), we find without difficulty the following solution of the quintic: $v_0=0.2568$, and hence compute:

$$x_1=x_0+v_0=1.2068.$$

This enables us to obtain new initial values y_0 and y_0' at the arbitrarily selected point: $x_0=1.5$ by means of equation (10), Section 6. We thus find: $y_0=11.62$, $y_0'=-79.46$. With these as initial conditions, the solution is now extended by continuous analytic continuation (see Chapter 9) to $x=2.80$, where we obtain the values: $y=11.23$, $y'=75.47$. Entering these values in equation (11), and solving the quintic as before, we are able to determine the second pole. This is found to be: $x_2=3.0982$. It is obvious that this method can be indefinitely continued.

The technique of the second method is described in Section 11 of Chapter 9. By means of it, and employing a rectangular path, the solution of the differential equation is analytically extended into the complex plane and back again to the axis of reals at $x=1.5$. The values thus obtained are $y=11.78$ and $y'=-79.56$, which, when one considers the complex nature of the computation, agree very well with those obtained by the first method. The graphical representation of the curve thus found is shown in Figure 5.

It is interesting to compare the extended function with a series of functions computed respectively at the points $x=10$, $x=20$, and $x=30$, but each satisfying at its respective point the same initial conditions, namely, $y=1$, $y'=0$. These graphs are shown in Figure 6, which may be compared with those in Figure 5. The former show the contraction between successive poles as x increases.

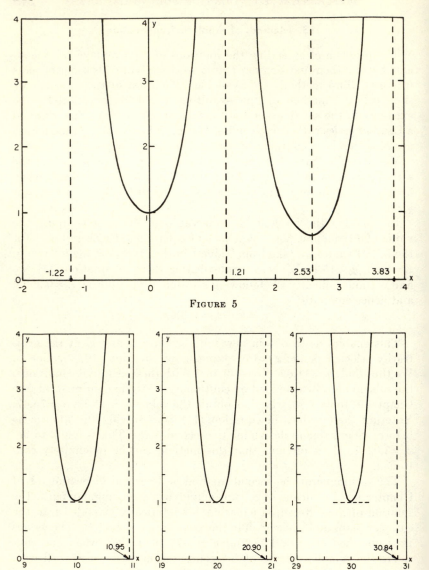

FIGURE 5

FIGURE 6

Chapter 9

Continuous Analytic Continuation

1. Introduction

ONE OF THE ADVANTAGES enjoyed by linear equations over nonlinear ones resides in the relative simplicity of the algorithms available for the computation of their solutions. For the time ultimately arrives in most investigations when it becomes necessary to exhibit in numerical or graphical form the function which solves an equation.

In preceding chapters we have given three types of existence theorems, which, in a sense, provide algorithms for the construction of the integral of an equation. But a survey of these methods will make it clear that there is a great deal of difference between proving the existence of a solution and the actual attaining of it in a graphical or numerical form. In general, complexities increase with each step.

Let us consider, for example, Cauchy's method of limits, which, in its essential feature, is merely the construction of an appropriate Taylor's series which satisfies formally the differential equation and converges, together with its derivatives, to preassigned values at a given point $x=x_0$. But the computation of the coefficients of the series is usually a monumental task, since the evaluation of higher derivatives from the original equation often increases exponentially in difficulty. The convergence of the series is limited by the radius $r=|x_0-x_1|$, where x_1 is the nearest singular point. And even when r is very large, the error in the approximation soon increases beyond practical limits with an increasing value of x_0-x.

Therefore, the need exists for a method of approximation which has the following features:

(a) The method should be what we shall call *linearly iterative*. That is to say, each successive step should be connected with its immediate predecessor by an algorithm which involves associations that do not increase in complexity at each step.

(b) The method should be applicable to approximations in the complex domain. It should not be limited to approximations on the real axis.

(c) It should be capable of extension to the neighborhood of any point in the complex plane, which is not excluded by a natural bound.

(d) The error in the approximation should increase linearly at most. That is to say, the error reached in the nth step should not be greater than nK, where K is a preassigned constant.

A method which has these properties, and which we are about to describe, will be called the method of *continuous analytic continuation*. It can be used for the approximation of solutions of equations of any order, although in our discussion we shall consider its application to differential equations of second order.

As an introduction to the method, it will be instructive to describe another one related to it, which we shall call the *method of curvature*. This approximation has already been discussed in Section 2 of Chapter 2. The method, however, is much more limited in its scope than that of continuous analytic continuation, since its applicability is to real values of the variables, and it is adapted only to differential equations of first and second orders. It does, however, have the advantage that it can be applied graphically to the construction of an integral curve.

2. The Method of Curvature

Let us assume that the equation to be solved can be written in the form

$$y'' = f(x,y,y'), \tag{1}$$

and that the desired integral passes through the point $P_0 = (x_0, y_0)$ with slope y_0'.

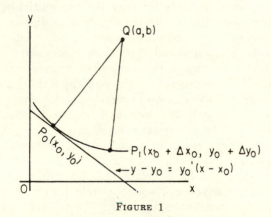

FIGURE 1

By means of equation (1) we now determine y_0'' and thus can compute the value of the radius of curvature, R_0, from the formula:

$$R^2 = \frac{(1+y'^2)^3}{y''^2}. \tag{2}$$

Referring to Figure 1, let us denote by $Q=(a,b)$ the center of curvature. Explicitly the coordinates a and b are computed from the formulas:

$$a=x_0-\frac{y_0'R_0}{\sqrt{1+y_0'^2}}, \quad b=y_0+\frac{R_0}{\sqrt{1+y_0'^2}}. \tag{3}$$

The line through P_0, which is tangent both to the integral curve and to the circle of curvature, is given by

$$y-y_0=y_0'(x-x_0). \tag{4}$$

Let us now consider a second point $P_1=(x_1,y_1)$ on the circle of curvature, which is derived from P_0 by adding an increment Δx to x_0; that is,

$$P_1=(x_0+\Delta x, \, y_0+\Delta y).$$

To a second approximation we shall have

$$y_1=y_0+\Delta y=y_0+y_0'\Delta x+\frac{1}{2}y_0''(\Delta x)^2. \tag{5}$$

Since P_1 lies on the circle of curvature, we can now compute the derivative y' at P_1 by means of formulas (3) as follows:

$$\left(\frac{dy}{dx}\right)_{P_1}=-\left(\frac{x_1-a}{y_1-b}\right)=-\left(\frac{x_0+\Delta x-a}{y_0+\Delta y-b}\right),$$

$$=-\left\{\frac{y_0''\Delta x+y_0'(1+y_0'^2)}{y_0''\left[y_0'\Delta x+\frac{1}{2}y_0''(\Delta x)^2\right]-(1+y_0'^2)}\right\}. \tag{6}$$

Denoting this value by y_1', we now substitute x_1,y_1,y_1' in (1) to find y_1''. The computation is then repeated to obtain $P_2=(x_2,y_2)$ and y_2'. Other points are similarly determined and an approximation to the integral curve is thus attained.

An estimate of the error involved in this method can be obtained by considering the difference between y_1' as given by (6) and y_1' computed from the derivative of (5). Denoting this difference by D, we thus find its value to be

$$D=\frac{1}{2}(y_0'')^2(\Delta x)^2\left[\frac{3y_0'+y_0''\Delta x}{1+\frac{3}{2}y_0'^2-\frac{1}{2}(y_0'+y_0'\Delta x)^2}\right]. \tag{7}$$

Neglecting differentials of third and higher powers, we have the following approximation for D:

$$D\sim\frac{3}{2}\frac{y_0'(y_0'')^2}{1+y_0'^2}(\Delta x)^2. \tag{8}$$

3. Analytic Continuation

We shall assume that $f(z)$ is a function of the complex variable z and that it is analytic throughout a simply connected region A except at a finite number of points in which it has singularities. Let us now connect two points a and b of the region A by a simple continuous curve, L, which neither intersects one of the singular points nor encloses one of them. The situation is shown in Figure 2, where the P_1 are the singular points of $f(z)$.

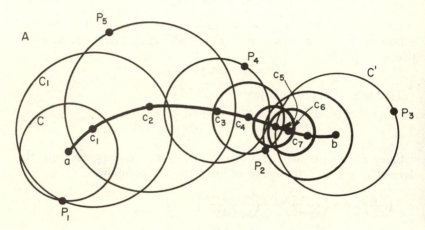

FIGURE 2

Since $f(z)$ is analytic in A except at the points of singularity, it is analytic at $z=a$ where it has derivatives of all orders. Therefore, $f(z)$ can be expanded in the neighborhood of a in the form of a Taylor's series,

$$f(z)=f(a)+(x-a)f'(a)+\frac{(x-a)^2}{2!}f''(a)+\ldots+\frac{(x-a)^n}{n!}f^{(n)}(a)+\ldots. \quad (1)$$

This series converges within a circle of radius R about $z=a$, where R is the distance from a to the nearest singular point. If the situation is that described in Figure 2, then $R=|a-P_1|$, and series (1) converges within the circle C.

The problem proposed here is that of finding the value of $f(x)$ at the point $z=b$, which lies outside of the circle C. This is the problem which is solved by *analytic continuation*.

To accomplish this we select some point c_1 which lies on L and is also within the circle C. Since $f(z)$ is analytic at c_1, both the value of $f(c_1)$ and the values of all of its derivatives can be computed at c_1 by means of the convergent series (1). From these values a new Taylor's series is now constructed which this time converges within the circle C_1, the radius of which is equal to the distance between c_1 and P_1. Part of this circle lies outside of C so we have now enlarged the domain within which we can define $f(z)$ by means of a Taylor's series; or, in other words, we have *analytically extended* $f(z)$ beyond the region enclosed by C and into the region interior to C_1.

Since the singular points are finite in number and isolated, and since L does not include or enclose any of them, it is clear that by a sufficient number of repetitions of the process just described we shall eventually reach the point $z=b$ with a Taylor's expansion that converges within the circle C'.

It should be observed that the circles of convergence vary in size, since their radii depend upon the position of the singular points with respect to the line L. If one is actually computing the values of $f(z)$ at points along L, this matter of the size of the circles of convergence is an important one, since, in general, the rate of convergence of the Taylor's series depends upon the size of the radius of the circle of convergence. The smaller the radius the larger the number of terms that will be required to attain values of $f(z)$ within the limits of a prescribed error.

It has been assumed that the path L does not enclose any of the singular points. The importance of this assumption is readily seen, for if one of the points, let us say P_1, is a branch point of $f(z)$, then any path which encloses P_1 will carry the function to another sheet of the Riemann surface and we should not be able to reach the final point $z=b$.

4. The Method of Continuous Analytic Continuation

The method of analytic continuation which we have described in the preceding section is especially well adapted to the computation of a function described by a differential equation. Let us assume that the function is a solution of the differential equation

$$y'' = f(x,y,y'),\tag{1}$$

and that at the point x_0 it has the value y_0 and its first derivative has the value y_0'.

Let us assume that we are interested in obtaining the value of y along a path L between the initial point $a = (x_0, y_0)$ and a second point b. We shall assume that the path is so chosen that the function $y(x)$ is analytic throughout its length. It should be observed, however, that the choice of such a path is not an obvious consequence of the form of the differential equation.

Take, for example, the simple equation: $y'' = 6y^2$, which we have studied in Chapter 7. Cauchy's existence theorem assures us that a solution with the prescribed boundary conditions exists in the neighborhood of every point in the finite plane. But there is no obvious reason to infer from the form of the equation that if L is the real axis and the boundary conditions are real, the path L will contain an infinite number of poles of the solution, as it actually does.

Under the conditions just assumed there will exist a solution of equation (1) in the neighborhood of the point a, which can be written in the following form:

$$y(x)=y_0+y_0'\Delta x+\frac{y_0''}{2!}(\Delta x)^2+\frac{y_0^{(3)}}{3!}(\Delta x)^3+\ldots+\frac{y_0^{(n)}}{n!}(\Delta x)^n+R_n, \quad (2)$$

where $\Delta x=x-x_0$, and

$$R_n=y_p^{(n+1)}(\Delta x)^{n+1}/(n+1)!, \quad x_p=x_0+\theta(\Delta x), \quad 0\leqq\theta\leqq 1.$$

Similarly, for the computation of $y'(x)$, we have

$$y'(x)=y_0'+y_0''\Delta x+\frac{y_0^{(3)}}{2!}(\Delta x)^2+\ldots+\frac{y_0^{(n)}}{(n-1)!}(\Delta x)^{n-1}+R_n', \quad (3)$$

where we write

$$R'=y^{(n+1)}(\Delta x)^n/n!, \quad x_p=x_0+\theta(\Delta x), \quad 0\leqq\theta\leqq 1. \quad (4)$$

The values of $y_0^{(m)}$ for $m=2, 3, \cdots$, m are computed from the differential equation and its successive derivatives. Thus, for example, we have

$$y^{(3)}=f_x+f_yy'+f_{y'}y''=f_x+f_yy'+f f_{y'}, \quad (5)$$

where the subscripts indicate partial derivatives. Let us take note of the fact that $y^{(3)}$ has been expressed in terms of x, y, and y' only.

We now observe that each one of the coefficients of $(\Delta x)^m$ in (2) and (3) can be expressed in terms of the initial values: $P_0=(x_0,y_0,y_0')$. Let us write these coefficients as follows:

$$f(x_0,y_0,y_0')=f_2(P_0),$$

$$f_x(P_0)+y_0 f_y(P_0)+f(P_0) f_{y'}(P_0)=f_3(P_0), \quad (6)$$

and so on to higher derivatives.

Equations (2) and (3) then assume the form:

$$y_1=y_0+y_0'\Delta x+\frac{1}{2!}f_2(P_0)(\Delta x)^2+\frac{1}{3!}f_3(P_0)(\Delta x)^3+\ldots+\frac{1}{n!}f_n(P_0)(\Delta x)^n+R_n,$$

$$y_1'=y_0'+f_2(P_0)\Delta x+\frac{1}{2!}f_3(P_0)(\Delta x)^2+\ldots+\frac{1}{(n-1)!}f_{n-1}(P_0)(\Delta x)^{n-1}+R_n',$$

$$(7)$$

where y_1 and y_1' are the values respectively of $y(x)$ and $y'(x)$ at the point $x_1 = x_0 + \Delta x$.

In this manner, beginning with an initial set of values, it is now possible to compute successive values of y and y', the errors in each approximation being largely controlled by the size of Δx. Thus, denoting the successive values by y_i and y_i', we obtain as approximating equations the following:

$$y_{i+1} = \dot{y}_i + y_i' \Delta x + \frac{1}{2!} f_2(P_i)(\Delta x)^2 + \ldots + \frac{1}{n!} f_n(P_i)(\Delta x)^n,$$

$$y_{i+1}' = y_i' + f_2(P_i)\Delta x + \frac{1}{2!} f_3(P_i)(\Delta x)^2 + \ldots + \frac{1}{(n-1)!} f_{n-1}(P_i)(\Delta x)^{n-1}.$$

$$(8)$$

The errors, which are obviously cumulative, will be of the order of mR and mR', where m is the number of iterations and R and R' are the absolute values of the respective maxima of R_i and R_i' along the path L.

The nature of the continuation is readily understood from Figure 3, which shows the successive circles that carry the computation from the point a to the point b along the path L. The center of each circle lies on L and each radius is equal to Δx. The error in each approximation will vary from circle to circle, this variation depending upon the proximity of the center of the circle to the nearest singular point of the solution which is being computed. Thus the maximum values of R_i and R_i' will, in general, be those corresponding to the circle which lies closest to a singular point. It is to be observed, however, that Δx is not necessarily the same value throughout the length of L, but can be varied according to the proximity of the singular points.

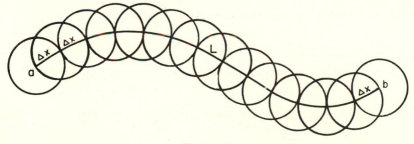

FIGURE 3

The great power of continuous analytic continuation as a computing device resides in two factors. In the first place the P_i, and consequently the coefficients $f_n(P_i)$, are functions only of x_i, y_i, and y_i'. No higher derivatives enter explicitly into the computation. In the second place, even for complicated equations, the iterations defined by (8) are relatively simple since one rarely needs to go beyond the fourth derivative, that is to say, beyond the term containing $f_4(P_i)$. The approximation, like a turtle, carries its house with it.

5. An Elementary Example Illustrating the Method of Continuous Analytic Continuation

As a simple, and somewhat amusing, illustration of the method described in the preceding section, let us consider the solution by this means of the following linear equation:

$$y''+y=0. \tag{1}$$

Since the derivatives form the following sequences:

$$y''=-y,\ y^{(4)}=y,\ y^{(6)}=-y,\ .\ .\ .,\ y^{(2n)}=(-1)^n y,$$

$$y^{(3)}=-y',y^{(5)}=y',y^{(7)}=-y',\ .\ .\ .,y^{(2n+1)}=(-1)^n y', \tag{2}$$

equations (8) of Section 4 reduce to the following:

$$y_{i+1}=y_i+y_i'\Delta x+\frac{1}{2!}(-y_i)(\Delta x)^2+\frac{1}{3!}(-y_i')(\Delta x)^3+\frac{1}{4!}y_i(\Delta x)^4+\ .\ .\ .,$$

$$=y_i\cos\Delta x+y_i'\sin\Delta x; \tag{3}$$

$$y_{i+1}'=y_i'-y_i\Delta x+\frac{1}{2!}(-y_i')(\Delta x)^2+\frac{1}{3!}y_i(\Delta x)^3+\frac{1}{4!}y_i'(\Delta x)^4-\ .\ .\ .,$$

$$=-y_i\sin\Delta x+y_i'\cos\Delta x. \tag{4}$$

It is thus observed that we have the relationship,

$$y_{i+1}^2+y_{i+1}'^2=y_i^2+y_i'^2. \tag{5}$$

Let us now solve equations (3) and (4) for sin Δx and cos Δx. We thus obtain:

$$\sin\Delta x=\frac{y_i'y_{i+1}-y_iy_{i+1}'}{y_i^2+y_i'^2},\quad \cos\Delta x=\frac{y_iy_{i+1}+y_i'y_{i+1}'}{y_i^2+y_i'^2}. \tag{6}$$

Substituting these values into the equations,

$$y_{i+2}=y_{i+1}\cos\Delta x+y'_{i+1}\sin\Delta x,$$

$$y'_{i+2}=-y_{i+1}\sin\Delta x+y'_{i+1}\cos\Delta x, \tag{7}$$

we thus get the following iteration free from the increment Δx:

$$y_{i+2}=\frac{y_i(y_{i+1}^2-y_{i+1}'^2)+2y_{i+1}y_i'y_{i+1}'}{y_i^2+y_i'^2},$$

$$y'_{i+2}=\frac{y_i'(y_{i+1}'^2-y_{i+1}^2)+2y_iy_{i+1}y_{i+1}'}{y_i^2+y_i'^2}. \tag{8}$$

It should be observed, however, that an initial choice must be made of four values, let us say, y_0, y_0' and y_1, y_1', before the sequence, y_2, y_2'; y_3, y_3'; etc., can be computed. The first pair of values are those which determine the initial conditions of the original differential equation (1). The second pair, arbitrarily chosen except for the one restriction given by (5), determine the increment Δx of the iteration.

It is interesting to observe that if the four initial values are chosen to be rational fractions, then the iterated values will also be rational fractions. Each pair of fractions will form a set of Pythagorean fractions, that is to say, rational pairs which satisfy the equation:

$$x^2+y^2=y_0^2+y_0'^2.$$

The number of such sets is enumerably infinite. The iteration (8), however, picks from the totality of Pythagorean fractions a set which forms separately the ordinates at fixed intervals of two harmonic curves.

For example, if $y_0=1$, $y_0'=0$; $y_1=3/5$, $y_1'=-4/5$, then we compute by means of (8) the following subsequent sets:

$$y_2=-7/25,\ y_2'=-24/25;\ y_3=-117/125,\ y_3'=-44/125;$$

$$y_4=-527/625,\ y_4'=336/625;\ y_5=-237/3125,\ y_5'=3116/3125;$$

$$y_6=11753/15625,\ y_6'=10296/15625.$$

Substituting the first two pairs of values in equations (6), we get $\sin\Delta x=4/5$, $\cos\Delta x=3/5$, and thus determine: $\Delta x=0.927295$ radians.

The solution of the original equation (1) corresponding to the initial conditions: $y_0=1$, $y_0'=0$ at $x=0$, is $y=\cos x$ and its derivative is $y'=-\sin x$. The graphs of these two functions showing the location on them of the rational values just computed are given in Figure 4.

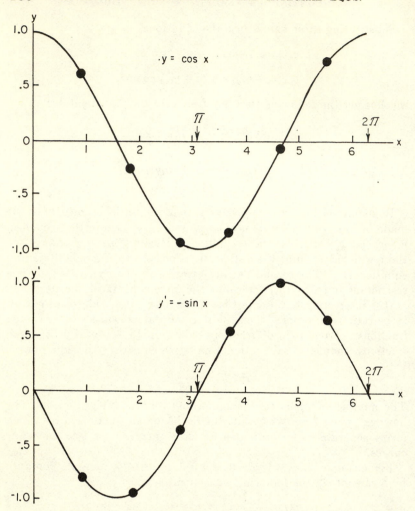

FIGURE 4.—Graphs of $y = \cos x$ and $y' = -\sin x$, showing positions of rational points on the curves.

6. The Solution of $y'' = 6y^2$ by the Method of Continuous Analytic Continuation

As an illustration of the usefulness of the method of continuous analytic continuation when approximating into the neighborhood of a singular point, we shall apply the method to the equation:

$$y'' = 6y^2, \tag{1}$$

prescribing the following initial values: $y(0) = 1$, $y'(0) = 0$.

From the extensive discussion of this equation given in Chapter 7, we know that its solution, corresponding to the initial conditions, is a doubly periodic function with a real period equal to 2Ω, where $\Omega = 1.2143254$. At the point $x = \Omega$ the solution has a simple polar singularity. Both a table of values between $x = 0$ and $x = 1$ and the graphical representation of the solution are given in Section 9 of Chapter 7.

Recognizing the fact that our application of the method of analytic continuation will carry us speedily into the neighborhood of the pole, where the error increases rapidly, we shall illustrate the essential power of the method by applying it to this unfavorable case.

We shall assume that $\Delta x = 0.01$, but in the approximations [(7), Section 4] of y and y' we shall let $n = 4$. That is to say, our approximations of the two functions are made respectively with quartic and cubic polynomials. To obtain the explicit expansions we first compute

$$y^{(3)} = f_3(P) = 12yy', \quad y^{(4)} = f_4(P) = 12(6y^3 + y'^2). \tag{2}$$

From these values we then obtain the following approximating equations:

$$y_{i+1} = y_i + y_i' 10^{-2} + (3y_i^2)10^{-4} + (2y_i y_i')10^{-6} + \left(3y_i^3 + \frac{1}{2}y_i'^2\right)10^{-8}, \tag{3}$$

$$y_{i+1}' = y_i' + (6y_i^2)10^{-2} + (6y_i y_i')10^{-4} + (12y_i^3 + 2y_i'^2)10^{-6}. \tag{4}$$

The first iteration yields the values: $y_1 = 1.00030003$ and $y_1' = 0.060012$, both of which are correct to the places indicated. The values obtained through 100 iterations are given in the table in Section 9 of Chapter 7.

Since the computation into the neighborhood of a polar singularity places a great strain upon the method, it is instructive to ascertain with some precision just what accuracy has been attained. For $x = 1$ it is possible to obtain an accurate value of y from formula (4) of Section 7, Chapter 7, since, for the boundary values assumed above, we know that $h = 1/7$ and $x_1 = 1.2143254$, correct to the last figure. We thus find that y has the value 21.7711 and y' the value 203.181 at $x = 1$, these approximations being correct to the last figure. We now compare these values with those obtained by the method just used after 100 iterations, that is, $y = 21.7444$ and $y' = 202.8241$. The errors are thus seen to be respectively 2.7×10^{-2} and 3.7×10^{-1}.

It is not usually possible to obtain a precise value for the error of the approximations of $y(x)$ and $y'(x)$ after m iterations, since there is always a small error in the initial conditions for each iteration. But if E_m and E_m' are respectively the absolute values of the errors in y

and y' after m iterations, then these values will usually satisfy the following inequalities:

$$mR \leqq E_m \leqq mS,$$
$$mR' \leqq E'_m \leqq mS', \qquad (5)$$

where R and R' are the quantities defined in Section 4, corresponding to derivatives of order $n+1$ and S and S' are similar values corresponding to derivatives of order n. The assumption is made that $S>R$ and $S'>R'$.

Thus, in the problem of this section, $n=4$, from which we have

$$y^{(4)}/4! = \left(3y^3 + \frac{1}{2}y'^2\right), \quad y^{(5)}/5! = 3y^2 y'.$$

Substituting the maximum observed values, namely, $y=21.7444$ and $y'=202.8241$, in these formulas, we obtain

$$y^{(4)}/4! = 92543.32, \quad y^{(5)}/5! = 287697.20.$$

Computing R and S for $m=100$, $\Delta x=0.01$, we obtain the following inequality:

$$2.88 \times 10^{-3} < E < 9.25 \times 10^{-2},$$

which is quite satisfactory.

The estimate for the error of the derivative is not as good, but is clearly conservative, since we find

$$1.44 < E' < 37.0.$$

7. The Numerical Evaluation of the First Painlevé Transcendent

The first Painlevé equation, which has been described at some length in Chapter 8, is a natural generalization of the equation discussed in Section 6. From its form,

$$\frac{d^2 y}{dx^2} = 6y^2 + \lambda x, \qquad (1)$$

we see that the elliptic equation is the special case $\lambda=0$.

The numerical solution of this equation is readily attained by means of the method of continuous analytic continuation. From the table of derivatives given in *Appendix* 3, we obtain the following Taylor's series to five terms:

$$y = y_0 + y'_0 \Delta x + \left(3y_0^2 + \frac{1}{2}\lambda x_0\right)(\Delta x)^2 + \left(2y_0 y'_0 + \frac{1}{6}\lambda\right)(\Delta x)^3$$
$$+ \left(3y_0^3 + \frac{1}{2}\lambda x_0 y_0 + \frac{1}{2}y_0'^2\right)(\Delta x)^4 + R, \quad \Delta x = x - x_0, \quad (2)$$

where we write

$$R = \frac{1}{10} (30y_a^2 y_a' + \lambda y_a + 3\lambda x_a y_a')(\Delta x)^5, \quad x_a = x_0 + \theta \Delta x, \ 0 < \theta < 1.$$

Similarly, for the computation of y', we have

$$y' = y_0' + (6y_0^2 + \lambda x_0)\Delta x + \left(6y_0 y_0' + \frac{1}{2}\lambda\right)(\Delta x)^2$$
$$+ (12y_0^3 + 2\lambda x_0 y_0 + 2y_0'^2)(\Delta x)^3 + R', \quad (3)$$

where $R' = 5R/\Delta x$.

In the actual computation Δx was set equal to 0.01. Thus, denoting the successive values of y and y' respectively by y_i and y_i', we obtain as approximating equations the following:

$$y_{i+1} = y_i + y_i' 10^{-2} + \left(3y_i^2 + \frac{1}{2}\lambda x_i\right) 10^{-4} + \left(2y_i y_i' + \frac{1}{6}\lambda\right) 10^{-6}$$
$$+ \left(3y_i^3 + \frac{1}{2}\lambda x_i y_i + \frac{1}{2}y_i'^2\right) 10^{-8}, \quad (4)$$

$$y_{i+1}' = y_i' + (6y_i^2 + \lambda x_i)10^{-2} + \left(6y_i y_i' + \frac{1}{2}\lambda\right) 10^{-4}$$
$$+ (12y_i^3 + 2\lambda x_i y_i + 2y_i'^2)10^{-6}. \quad (5)$$

Beginning with the initial values: $y_0 = 1$, $y_0' = 0$, $x_0 = 0$, values of y_i and y_i' have been computed for $\lambda = 1$, 2, 3, 4, 5 over the range between -1.00 and $+1.00$, except in the immediate vicinity of $+1.00$, where the growth of y and y' is very rapid. These values have been recorded in Table 1 in the *Appendix*. The graphical representation of the solution has been given in Chapter 8.

The errors in the approximation have been estimated at $x = 0.50$ and found to be less than 4×10^{-7} and 2×10^{-4} respectively for y and y'. Five-figure values are given for y and y' to $x = \pm 0.50$, but from this point to ± 0.80 four-figure values are given for y'. Thereafter the approximation for y is reduced to four figures and for y' to three figures.

A direct check on the approximations is difficult, since there is no alternative way to compute the values of the function and its derivative in the upper part of the range. But for $x = 0.1$, $\lambda = 5$, the Taylor's series converges sufficiently fast to permit a 9-decimal approximation. Thus we find the value: $y = 1.03114\ 1446$, correct to the last place. By the method of analytic continuation through ten iterations we obtain: $y = 1.03114\ 1419$. The error is thus 3.7×10^{-8}.

By means of the method described in the preceding section we find that the error, E, has the following bounds:

$$2.44\times10^{-9}<E<3.86\times10^{-7},$$

which is quite satisfactory.

8. The Numerical Evaluation of the Second Painlevé Transcendent

The equation defining the second Painlevé transcendent, referring to Chapter 8, is the following:

$$\frac{d^2y}{dx^2}=2y^3+xy+\mu. \tag{1}$$

From the table of derivatives given in *Appendix* 4 we obtain the following Taylor's series to five terms:

$$y=y_0+y_0'\Delta x+\left(y_0^3+\frac{1}{2}x_0y_0+\frac{1}{2}\mu\right)(\Delta x)^2+\left(y_0^2y_0'+\frac{1}{6}x_0y_0'+\frac{1}{6}y_0\right)(\Delta x)^3$$

$$+\frac{1}{24}[(2y_0^3+x_0y_0+\mu)(6y_0^2+x_0)+12y_0y_0'^2+2y_0'](\Delta x)^4+R,\ \Delta x=x-x_0, \tag{2}$$

where we write

$$R=\frac{1}{120}[12y_a'^3+(2y_a^3+x_ay_a+\mu)(36y_ay_a'+3)$$

$$+(6y_a^2y_a'+x_ay_a'+y_a)(6y_a^2+x_a)](\Delta x)^5, \tag{3}$$

in which $x_a=x_0+\theta\Delta x$, $0<\theta<1$.

Similarly, for the computation of y', we have

$$y'=y_0'+(2y_0^3+x_0y_0+\mu)\Delta x+\left(3y_0^2y_0'+\frac{1}{2}x_0y_0'+\frac{1}{2}y_0\right)(\Delta x)^2$$

$$+\frac{1}{6}[(2y_0^3+x_0y_0+\mu)(6y_0^2+x_0)+12y_0y_0'^2+2y_0'](\Delta x)^3+R', \tag{4}$$

where $R'=5R/\Delta x$.

Beginning with an initial set of values, x_0, y_0, y_0', we now compute successive values of y and y', where in the actual computation Δx was set equal to 0.01. Thus, denoting the successive values of y and y' respectively by y_i and y', we obtain as approximating equations the following:

$$y_{i+1}=y_i+y_i'(10)^{-2}+\left(y_i^3+\frac{1}{2}x_iy_i+\frac{1}{2}\mu\right)(10)^{-4}+\left(y_i^2y_i'+\frac{1}{6}x_iy_i'+\frac{1}{6}y_i\right)(10)^{-6}$$

$$+\frac{1}{24}[(2y_i^3+x_iy_i+\mu)(6y_i^2+x_i)+12y_iy_i'^2+2y_i'](10)^{-8}, \tag{5}$$

$$y'_{i+1}=y'_i+(2y_i^3+x_iy_i+\mu)(10)^{-2}+\left(3y_i^2y'_i+\frac{1}{2}x_iy'_i+\frac{1}{2}y_i\right)(10)^{-4}$$

$$+\frac{1}{6}(2y_i^3+x_iy_i+\mu)(6y_i^2+x_i)+12y_iy'^2_i+2y'_i](10)^{-6}. \quad (6)$$

Beginning with the initial values: $x_0=0$, $y_0=1$, $y'_0=0$, computations were made of y and y' over the range between $x=-1.00$ and $x=+1.00$ at intervals of 0.01 corresponding to $\mu=0, 1, 2, 3, 4, 5$. As in the case of the first Painlevé functions, it was found necessary to limit the range for a few values near both ends of the interval, where y and y' become unusually large. Graphs are given in Chapter 8.

A direct estimate of the errors in the approximation was made by means of a Taylor's expansion about the origin. For $x=0.20$, after 20 iterations, the value obtained by analytical continuation corresponding to $\mu=5$ was 1.14439 6449 and by the Taylor's series was 1.14439 65698. Hence, the error was 1.2×10^{-7}.

If we compute the bounding errors as described in Section 6, we obtain for the limits of the theoretical error E the following inequality:

$$9.4\times10^{-9}<E<8.9\times10^{-7},$$

which is in agreement with what has just been observed.

The value of y for $\mu=5$ was computed at $x=0.40$ by using terms to x^{10} in the Taylor's series. The resulting estimate, correct only to five decimal places, was found to be in agreement to this order of approximation with the value obtained by continuous analytic continuation.

In Table 2 in the Appendix the values of y and y' are given to four decimal places between $x=0.00$ and $x=0.50$. Thereafter, to $x=0.80$, the values of y' are reduced to four significant figures. Beyond $x=0.80$ the values of y are reduced to four significant figures and those of y' to three significant figures.

9. The Analytic Continuation of the Van der Pol Equation

It will be useful to express the Van der Pol equation (see Section 2, Chapter 7), that is,

$$\frac{d^2y}{dx^2}-\epsilon(1-y^2)\frac{dy}{dx}+ay=0, \quad (1)$$

in the form of a continuous analytic continuation.

The third and fourth derivatives of y, expressed in terms of y and y', are found to be the following:

$$y^{(3)} = a\epsilon(-y+y^3) + (\epsilon^2 - a - 2\epsilon^2 y^2 + \epsilon^2 y^4)y' - 2\epsilon yy'^2, \qquad (2)$$

$$y^{(4)} = a(a-\epsilon^2)y + 2a\epsilon^2 y^3 - a\epsilon^2 y^5 + \epsilon[\epsilon^2 - 2a + (8a - 3\epsilon^2)y^2$$
$$+ 3\epsilon^2 y^4 - \epsilon^2 y^6]y' + 8\epsilon^2(-y+y^3)y'^2 - 2\epsilon y'^3. \qquad (3)$$

Denoting by y_i and y_i' the successive values of y and y' corresponding to an increment Δx, we obtain the following iterative equations for the analytic continuation of equation (1):

$$y_{i+1} = y_i + y_i' \Delta x + \frac{1}{2}\left[-ay_i + \epsilon(1-y_i^2)y_i'\right](\Delta x)^2$$
$$+ \frac{1}{6}\left\{a\epsilon(-y_i + y_i^3) - [a - \epsilon^2(1-y_i^2)^2]y_i' - 2\epsilon y_i y_i'^2\right\}(\Delta x)^3$$
$$+ \frac{1}{24}\left\{\bar{a}(a-\epsilon^2)y_i + 2a\epsilon^2 y_i^3 - a\epsilon^2 y_i^5 + [2a\epsilon(-1+4y_i^2)\right.$$
$$\left. + \epsilon^3(1-y_i^2)^3]y_i' + 8\epsilon^2(-y_i + y_i^3)y_i'^2 - 2\epsilon y_i'^3\right\}(\Delta x)^4 + R; \qquad (4)$$

$$y_{i+1}' = y_i' + \left[-ay_i + \epsilon(1-y_i^2)y_i'\right]\Delta x + \frac{1}{2}\left\{a\epsilon(-y_i + y_i^3)\right.$$
$$\left. - [a - \epsilon^2(1-y_i^2)^2]y_i' - 2\epsilon y_i y_i'^2\right\}(\Delta x)^2 + \frac{1}{6}\left\{a(a-\epsilon^2)y_i\right.$$
$$+ 2a\epsilon^2 y_i^3 - a\epsilon^2 y_i^5 + [2a\epsilon(-1+4y_i^2) + \epsilon^3(1-y_i^2)^3]y_i'$$
$$\left. + 8\epsilon^2(-y_i + y_i^3)y_i'^2 - 2\epsilon y_i'^3\right\}(\Delta x)^3 + R'. \qquad (5)$$

10. The Analytic Continuation of Volterra's Equation

In Section 4 of Chapter 5 Volterra's equation has been given in the following form:

$$y\frac{d^2y}{dx^2} = \left(\frac{dy}{dx}\right)^2 + c(-y+y^2)\frac{dy}{dx} + ac(y^2 - y^3). \qquad (1)$$

In order to obtain the analytic continuation for this equation, we first compute the third and fourth derivatives of y as follows:

$$y^{(3)} = -ac^2 y(1-y)^2 + [(3ac+c^2) - (4ac+2c^2)y + c^2 y^2]y'$$
$$+ (-3c+4cy)y'^2/y + y'^3/y^2, \qquad (2)$$

$$y^{(4)} = a^2 c^2 y(1-y)(3-4y) + ac^3 y(1-y)^3 + [-5ac^2(1-y)(2-3y)$$
$$- c^3(1-y)^3]y' + [(6ac+7c^2) - (11ac+18c^2)y + 11c^2 y^2]y'^2/y$$
$$+ (11cy - 6c)y'^3/y^2 + y'^4/y^3. \qquad (3)$$

Denoting by y_i and y_i' the successive values of y and y' corresponding to an increment Δx, we obtain the following iterative equations for the analytical continuation of (1):

$$y_{i+1} = y_i + y_i'\Delta x + \frac{1}{2}[acy_i(1-y_i) - c(1-y_i)y_i' + y_i'^2/y_i](\Delta x)^2$$

$$+ \frac{1}{6}\{-ac^2y_i(1-y)^2 + [(3ac+c^2) - (4ac+2c^2)y_i + c^2y_i^2]y_i'$$

$$+ c(-3+4y_i)y_i'^2/y_i + y_i'^3/y_i^2\}(\Delta x)^3 + \frac{1}{24}\{a^2c^2y_i(1-y_i)(3-4y_i)$$

$$+ ac^3y_i(1-y_i)^3 + [-5ac^2(1-y_i)(2-3y_i) - c^3(1-y_i)^3]y_i'$$

$$+ [(6ac+7c^2) - (11ac+18c^2)y_i + 11c^2y_i^2]y_i'^2/y_i + (11cy_i-6c)y_i'^3/y_i^2$$

$$+ y_i'^4/y_i^3\}(\Delta x)^4 + R; \qquad (4)$$

$$y_{i+1}' = y_i' + [acy_i(1-y_i) - c(1-y_i)y_i' + y_i'^2/y_i]\Delta x + \frac{1}{2}\{-ac^2y_i(1-y_i)^2$$

$$+ [(3ac+c^2) - (4ac+2c^2)y_i + c^2y_i^2]y_i' + c(-3+4y_i)y_i'^2/y_i$$

$$+ y_i'^3/y_i^2\}(\Delta x)^2 + \frac{1}{6}\{a^2c^2y_i(1-y_i)(3-4y_i) + ac^3y_i(1-y_i)^3$$

$$+ [-5ac^2(1-y_i)(2-3y_i) - c^3(1-y_i)^3]y_i' + [(6ac+7c^2)$$

$$- 11)ac+18c^2)y_i+11c^2y_i^2]y_i'^2/y_i + (11cy_i-6c)y_i'^3/y_i^2 + y_i'^4/y_i^3\}(\Delta x)^3 + R'.$$

$$(5)$$

Values of y and y' have been computed by means of these formulas for $a=2$, $c=1$, with boundary conditions $y(0)=1$, $y'(0)=-4$, from $x=0.00$ to $x=6.00$. These are recorded in Table IV of the *Appendix*.

11. The Technique of Continuous Analytic Continuation Around a Singular Point

Although the usual application of continuous analytic continuation will be to integrations along the real azis, the need occasionally arises to extend the method to a path in the complex plane. The general considerations involved in this have been given in Section 4; but one may be puzzled to know more specifically how such a continuation is to be accomplished. In order to explain the technique we shall confine ourselves to an example, which has already been described in Section 13 of Chapter 8.

The Painlevé transcendent defined by the equation

$$y''=6y^2+\lambda y, \quad y(0)=1, \quad y'(0)=0, \tag{1}$$

has a polar singularity (when $\lambda=1$), at $x_1=1.2068$. The problem is to continue analytically the solution from the origin to some point x_0 lying between 0 and x_1, then into the plane of the complex variable, and finally back again to a second point X on the real axis beyond x_1.

While almost any continuous curve can be used as the path of integration, it has been found most convenient to employ a path constructed from linear segments parallel respectively to the imaginary axis or the axis of reals.   One then begins with the following iterative equations:

$$y_{i+1}=y_i+y_i'\Delta x+\left(3y_i^2+\frac{1}{2}\lambda x_i\right)(\Delta x)^2+\left(2y_iy_i'+\frac{1}{6}\lambda\right)(\Delta x)^3$$

$$+\left(3y_i^3+\frac{1}{2}\lambda x_iy_i+\frac{1}{2}y_i'^2\right)(\Delta x)^4,$$

$$y_{i+1}'=y_i'+(6y_i^2+\lambda x_i)\Delta x+\left(6y_iy_i'+\frac{1}{2}\lambda\right)(\Delta x)^2$$

$$+(12y_i^3+2\lambda x_iy_i+2y_i'^2)(\Delta x)^3. \tag{2}$$

We now introduce the following complex values:

$$y=u+jv,\ y'=u'+jv',\ x=w+jz,\ \Delta x=\Delta w+j\Delta z,\ j=\sqrt{-1}. \tag{3}$$

These quantities are now substituted into system (2) and the real and imaginary expressions thus obtained form separate iterations. Employing for simplicity the abbreviations: $\Delta w=\Delta$, $\Delta z=\delta$, we obtain, after a tedious calculation, the following iterative system:

$$u_{i+1}=u_i+(u_i'\Delta-v_i'\delta)+\left(3u_i^2-3v_i^2+\frac{1}{2}\lambda w_i\right)(\Delta^2-\delta^2)-(12u_iv_i+\lambda z_i)\Delta\delta$$

$$+\left(2u_iu_i'-2v_iv_i'+\frac{1}{6}\lambda\right)(\Delta^3-3\Delta\delta^2)-(u_iv_i'+u_i'v_i)(3\Delta^2\delta-\delta^3)$$

$$+\left[3u_i^3-9u_iv_i^2+\frac{1}{2}\lambda(u_iw_i-z_iv_i)+\frac{1}{2}u_i'^2-\frac{1}{2}v_i'^2\right](\Delta^4-6\Delta^2\delta^2+\delta^4)$$

$$-[36u_i^2v_i-12v_i^3+2\lambda(u_iz_i+w_iv_i)+4u_i'v_i'](\Delta^3\delta-\Delta\delta^3);$$

$$v_{i+1}=v_i+(v_i'\Delta+u_i'\delta)+\left(6u_iv_i+\frac{1}{2}\lambda z_i\right)(\Delta^2-\delta^2)+(6u_i^2-6v_i^2+\lambda w_i)\Delta\delta$$

$$+\left(2u_iu_i'-2v_iv_i'+\frac{1}{6}\lambda\right)(3\Delta^2\delta-\delta^3)+(u_iv_i'+u_i'v_i)(\Delta^3-3\Delta\delta^2)$$

$$+[12u_i^3-36u_iv_i^2+2\lambda(u_iw_i-z_iv_i)+2u_i'^2-2v_i'^2](\Delta^3\delta-\Delta\delta^3)$$

$$+\left(9u_i^2v_i-3v_i^3+\frac{1}{2}\lambda(u_iz_i+w_iv_i)+u'v'\right)(\Delta^4-6\Delta^2\delta^2+\delta^4);$$

$$u'_{i+1}=u'_i+(6u_i^2-6v_i^2+\lambda w_i)\Delta-(12u_iv_i+\lambda z_i)\delta$$

$$+\left(6u_iu'_i-6v_iv'_i+\frac{1}{2}\lambda\right)(\Delta^2-\delta^2)-(6u_iv'_i+6u'_iv_i)\Delta\delta$$

$$+[12u_i^3-36u_iv_i^2+2\lambda(u_iw_i-z_iv_i)+2u_i'^2-2v_i'^2](\Delta^3-3\Delta\delta^2)$$

$$-[36u_i^2v_i-12v_i^3+2\lambda(u_iz_i+w_iv_i)+4u'_iv'_i](3\Delta^2\delta-\delta^3);$$

$$v'_{i+1}=v'_i+(6u_i^2-6v_i^2+\lambda w_i)\delta+(12u_iv_i+\lambda z_i)\Delta$$

$$+(12u_iu'_i-12v_iv'_i-\lambda)\Delta\delta+(3u_iv'_i+3u'_iv_i)(\Delta^2-\delta^2)$$

$$+[12u_i^3-36u_iv_i^2+2\lambda(u_iw_i-z_iv_i)+2u_i'^2-2v^12v_i'^2](3\Delta^2\delta-\delta^3)$$

$$+[36u_i^2v_i-12v_i^3+2\lambda(u_iz_i+w_iv_i)+4u'_iv'_i](\Delta^3-3\Delta\delta^2).$$

We now apply this iterative scheme to the path illustrated in Figure 5. Along the segment AB, with increments equal to 0.01, we have $z_i=0$, $w_i=i\times10^{-2}$, $\Delta=10^{-2}$, $\delta=0$, $v_i=v'_i=0$. The initial values for u are $u_0=1$, $u'_0=0$. This computation has already been described in Section 7 and the values are recorded in Table 1 in the *Appendix*. From this table we find for the terminal values of u at B the following: $u=2.0226$, $u'=5.4606$.

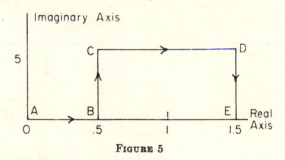

FIGURE 5

We now integrate along the path (BC) where we have the values: $z_i=i\times10^{-2}$, $w_i=0.5$, $\Delta=0$, $\delta=10^{-2}$, $v_0=v'_0=0$. The variation in this path is shown graphically in Figure 6. At the point C we have the following terminal values:

$$u=0.3289,\ u'=-0.7011,\ v=1.2119,\ v'=3.4670.$$

The next integration is along the segment (CD) where we have the values: $z_i=0.5$, $w_i=0.5+i\times10^{-2}$, $\Delta=10^{-2}$, $\delta=0$. The terminal values for the path (BC) now become the initial values for this integration. The variations in u and v are shown in Figure 6. At the point D we obtain the following terminal values:

$$u=-1.4147,\ u'=9.9821,\ v=-2.7136,\ v'=0.0155.$$

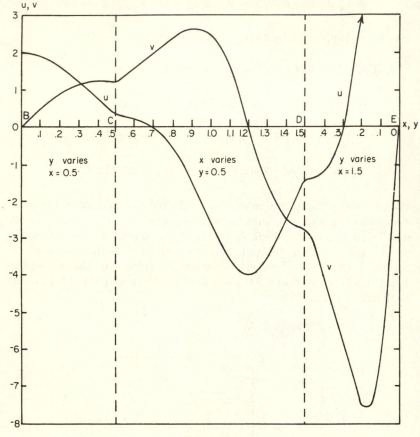

FIGURE 6

The final integration is along the path (DE) where we have the values: $z_i=-i\times10^{-2}$, $w_i=1.5$, $\Delta=0$, $\dot{\delta}=-10^{-2}$. The terminal values for the segment (CD) are the initial values for this computation. The variations of u and v are shown in Figure 6. At the point E we have the final terminal values:

$$u=11.7796,\ u'=-79.5588,\ v=0.1152,\ v'=0.5880.$$

The magnitude of the error in this computation is measured by the final values of v and v', both of which should be zero. But when the values of u and u' are compared with those previously obtained by the method of "pole-vaulting" (Section 13, Chapter 8), respectively 11.62 and -79.46, the agreement is seen to be quite satisfactory.

Chapter 10

The Phase Plane and Its Phenomena

1. Introduction

PHYSICAL PHENOMENA to the interpretation of which the theory of differential equations has been so abundantly applied are often of an oscillatory character. In many instances the observed oscillation is nearly periodic, and hence its mathematical description may differ little from an appropriate sum of sine and cosine terms. The astronomers were pioneers in the investigation of such oscillatory phenomena and it is not an exaggeration to say that much of our current knowledge of nonlinear equations had its origin in the difficulties of celestial mechanics.

Fortunately for astronomy, and also for those who dwell upon the surface of the earth, the solar system is dominated by a single massive object which contains 99.87 percent of the entire known matter within the effective reach of its gravitational influence. This lucky circumstance provided nearly elliptical orbits and almost periodic motions for the observation and study of the early astronomers. But it is also true that the existence of a single massive planet, Jupiter, with more than 70 percent of the total planetary mass of the system, has been the source of trouble with its strong perturbing effect upon the other planets. In order to take account of the influence of Jupiter, as well as that of the other members of the system, the astronomers were forced to make long and difficult computations in order to obtain satisfactory approximations of the solutions of the essentially nonlinear system of equations involved. But since the paths were nearly elliptical and the motions almost periodic the difficulties were not insuperable.

Closely associated with the problem of oscillation is that of stability. Poincaré's classical studies in the field of nonlinear equations originated in his investigation of stability conditions associated with the trajectories defined by the equations of celestial mechanics. The question of whether or not the solar system itself is a stable configuration resolves itself into the question of whether or not the nearly periodic motions of the planets can be described by a convergent series of periodic functions. Since the equations of the problem of three bodies cannot be integrated in terms of the elementary functions, the problem of the stability of the planetary system cannot be solved by an examination of an explicit solution. Poincaré and others attempted to find methods by means of which the problems of stability could be answered by an implicit study of the defining differential equations.

2. The Phase Plane and Limit Cycles

Applied problems not infrequently appear in the following form:

$$\frac{dy}{dt}=P(x,y), \quad \frac{dx}{dt}=Q(x,y), \tag{1}$$

where $P(x,y)$ and $Q(x,y)$ are functions which share some common domain of the x,y-plane.

If the first equation is divided by the second, we obtain the following differential equation of first order:

$$\frac{dy}{dx}=\frac{P(x,y)}{Q(x,y)}, \tag{2}$$

which we shall assume has a solution conveniently written as follows:

$$f(x,y)=0. \tag{3}$$

But it is also possible to attain this solution in another way. Differentiating the first equation in (1), we have three equations in the variables x and dx/dt, from which, in theory at least, we can obtain an equation of second order in y which is independent of x and dx/dt. The solution of this equation we shall denote by: $y=y(t)$. Proceeding similarly with the second equation in (1), we ultimately obtain another function: $x=x(t)$. The pair of equations:

$$x=x(t), \quad y=y(t) \tag{4}$$

are the parametric equivalent of the solution given by (3).

It is customary to speak of the function $f(x,y)=0$ as a function in the *phase plane*, that is to say, the plane of the variables x, y. Since the function contains an arbitrary constant, its graphical representation will be a series of curves in the phase plane. These are called *phase trajectories*.

It is possible to extend this idea in various ways. Let us, for example, consider the following differential equation of second order:

$$\frac{d^2x}{dt^2}=F(t,x,x'), \tag{5}$$

which we can write as the following system:

$$\frac{dx}{dt}=y, \quad \frac{dy}{dt}=F(t,x,y).\tag{6}$$

The solution of this system can then be written as the pair of equations given by (4), which are now the parametric representation of a function $f(x,y)=0$. In this simple manner we have associated with (5) a phase plane and a set of phase trajectories.

It is also possible to generalize to more variables. To do this we replace (1) by the system:

$$\frac{dx}{dt}=P(x,y,z), \quad \frac{dy}{dt}=Q(x,y,z), \quad \frac{dz}{dt}=R(x,y,z).\tag{7}$$

The parametric solution:

$$x=x(t), \quad y=y(t), \quad z=z(t),\tag{8}$$

now represents a two-parameter set of curves in a *phase space* of three dimensions.

Returning now to equations (4), let us assume that the functions are defined as t varies from $-\infty$ to $+\infty$. A curve is thus traced which may be a complex configuration, such as those shown in Figure 1. The arrows show the direction in which the point $P=(x,y)$ describes the trajectory.

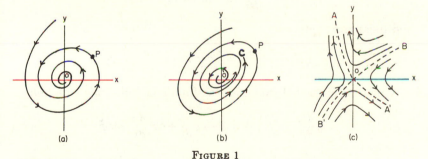

(a) (b) (c)

FIGURE 1

In the first curve (a) the point P approaches the origin as a limit, as t varies continuously from $-\infty$ to $+\infty$. This limit is called a *focal point* and the motion is characterized as *stable*.

In the second figure (b) we observe a different phenomenon. The trajectories are observed to approach asymptotically a fixed curve C. This curve is called a *limit cycle*. Trajectories of this type were observed by Poincaré and others and have been the object of much study. As one may intuitively infer, the functions defined by (4) will approach limiting forms which are periodic. Since the direction of the arrows shows that the point P is approaching the limit cycle, the motion is *stable*. If, however, the arrows pointed outward, then the motion would be *unstable* outside of the limit cycle and would approach the origin as a contracting spiral inside of the cycle.

The third figure (c) represents an *unstable* motion about the origin, since the trajectories are all hyperbolic curves which approach the origin, but then depart from it. The origin is called a *saddle point*. The significance of this term will be explained later. The most interesting feature of the figure is found in the curves AA' and BB', toward which the trajectories move asymptotically. Since they separate the plane into regions which contain different motions, each curve is called a *separatrix*.

Since our interest in this chapter will be largely that of the phenomena associated with oscillatory solutions of differential equations, our attention is focused upon the problem of the existence of limit cycles in the phase plane. When such a cycle exists, then the solutions of the differential equation, although they may not be periodic, will approach periodicity. Hence the subject of limit cycles and that of periodic functions are closely related.

Earlier in the book we examined certain differential equations which had periodic solutions. One of these was the equation of the simple pendulum,

$$\frac{d^2y}{dt^2} + n^2 \sin y = 0, \tag{9}$$

and its approximation,

$$\frac{d^2y}{dt^2} + n^2y - \frac{n^2}{6}y^3 = 0. \tag{10}$$

A more complicated example was found in Volterra's equation:

$$y\frac{d^2y}{dt^2} = \left(\frac{dy}{dt}\right)^2 + acy^2 - cy\frac{dy}{dt} + cy^2\frac{dy}{dt} - acy^3. \tag{11}$$

Another equation which will interest us, both for its comparative simplicity, the extensive attention which has been given to it, and the limit cycle to which its solutions are asymptotic, is that of Van der Pol,

$$\frac{d^2y}{dt^2} - \epsilon(1-y^2)\frac{dy}{dt} + ay = 0. \tag{12}$$

The connection between this equation and the earlier one of Lord Rayleigh has already been discussed in Chapter 7. In a later chapter it will be shown that its solutions are oscillatory and that they approach periodicity as t increases.

It is interesting to observe that of the four equations just given only the first has a harmonic term in y. In the coefficients of none of them does the independent variable appear explicitly, which means that their solutions can be written in the form: $y=f(t+p)$, where p is an arbitrary constant.

It is instructive also to consider the physical systems which led to the equations and the reasons for the existence of periodic, or quasi-periodic, solutions. In the problem of the pendulum, from which we derived equations (9) and (10), the assumption was made that there was no frictional coefficient present, which, of course, is a situation never met in the physical world. Every clock runs at the expense of an energy input. The solutions of (9) and (10) are periodic, but this is by virtue of a false assumption. Realism can be restored, however, by the addition of a resistance factor to the left-hand member and a force function to the right-hand member. Thus, equation (10) could be written:

$$\frac{d^2y}{dt^2}+k\frac{dy}{dt}+n^2y-\frac{n^2}{6}y^3=f\cos\omega t. \tag{13}$$

This is Duffing's equation, to which reference has been made in Chapter 7. Its solution presents many difficulties, which are complicated by the fact that an explicit function of t appears in the equation. Its solution will be discussed at some length in a later chapter.

Equation (11) has already been examined extensively in Chapter 4. It describes the variations in the numbers of a population which is in conflict with the members of a second population. As we have seen, there exists a periodic solution of the equation. The population grows and decreases in a regular manner, which approximates a sinusoidal motion. It resembles the oscillations of a pendulum, but with this interesting difference, that the motion does not need to be sustained by any force function extrinsic to the conflicting populations. Equation (11) thus defines a function, which, like that describing the motion of the frictionless pendulum, is not subjected to a damping factor and thus maintains its sinusoidal character.

The fourth equation defines the current in an electrical circuit which contains a triode oscillator. The circuit, which we are about to describe, is sometimes referred to as a feed-back circuit and under certain conditions is capable of maintaining an oscillating current even though the driving potential is not itself an alternating electromotive force (e.m.f.).

In Figure 2, T is a triode vacuum tube, which contains a plate P (the anode), a filament F (the cathode), and a grid G. If P is positively charged the resulting electric field in the tube causes a current to flow from the filament and the current is controlled by the grid

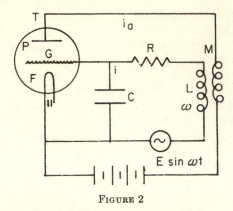

FIGURE 2

potential. It can be shown that the differential equation which describes the current i is the following:

$$L\frac{di}{dt}+Ri+\frac{1}{C}\int^{t} idt-M\frac{di_a}{dt}=E\sin\omega t, \tag{14}$$

where M is the mutual inductance and L, R, and C respectively the inductance, resistance, and capacity of the circuit.

It is a matter of experimental observation that the grid current i_a can be represented by a cubic polynomial of the form,

$$i_a=Au-Bu^3, \tag{15}$$

where we have

$$u=k\int_0^t idt.$$

If i_a, as given by (15), is now substituted in (14), we obtain the following equation:

$$L\frac{d^2u}{dt^2}-[(kMA-R)-3MBku^2]\frac{du}{dt}+\frac{1}{C}u=kE\sin\omega t. \tag{16}$$

This equation can be simplified in appearance if we write,

$$\epsilon=(kMA-R)/L,\quad \eta=3MBk/L, \tag{17}$$

from which we have

$$\frac{d^2u}{dt^2}-\epsilon\left(1-\frac{\eta}{\epsilon}u^2\right)\frac{du}{dt}+\frac{1}{LC}u=\frac{kE}{L}\sin\omega t. \tag{18}$$

Finally, if we write: $y=(\eta/\epsilon)^{1/2}u$, equation (18) assumes the desired form:

$$\frac{d^2y}{dt^2}-\epsilon(1-y^2)\frac{dy}{dt}+ay=E_0\sin\omega t, \tag{19}$$

where $a=1/LC$ and $E_0=kE(\eta/\epsilon)^{1/2}/L$.

In the study of the self-excited oscillations due to the feedback characteristics of the circuit, the alternating e.m.f. is set equal to zero and we have the Van der Pol equation (12). For this phenomenon it is necessary that ϵ be positive, which leads to the condition from (8) that $kMA>R$. The reason why the solution is oscillatory under this condition is readily seen from the equation. If y is initially zero, then the multiplier of dy/dt is negative and thus acts like a negative resistance. Hence y increases; but when y exceeds 1, the resistance becomes positive and the function diminishes. The oscillation is the result of these two opposing actions. In recent literature it has been customary to refer to such motions as *relaxation oscillations*, a term introduced by Von der Pol in 1926.

3. Phase Curves and Forcing Functions

Some of the complexities which we find in the phase trajectories associated with nonlinear differential equations can be illustrated by examples taken from the domain of linear differential equations, when these equations contain forcing functions. That such should be the case is readily seen from such a simple example as the following:

$$\frac{dx}{dt}+x=v, \tag{1}$$

where v, the forcing function, is a solution of the equation

$$\frac{dv}{dt}=v^2. \tag{2}$$

If equation (1) is differentiated and v eliminated between the three equations, the following nonlinear equation of second order is obtained:

$$x''+x'-x'^2-2xx'-x^2=0. \tag{3}$$

Hence the complexities of the simple linear equation are actually those of the nonlinear equation (3).

Because of the relative simplicity of linear equations, it will be convenient to examine some of the phenomena of phase trajectories by the device which we have just illustrated.

Let us first consider the following equation:

$$A \frac{d^2x}{dt^2} + B \frac{dx}{dt} + Cx = E \cos qt, \tag{4}$$

which is familiar to everyone from its connection with various physical problems. In one of its most obvious applications it describes the charge (x) in a simple electric circuit containing an impressed alternating e.m.f. represented by the term: $E \cos qt$.

The solution is readily found to be

$$x(t) = Ke^{-at} \sin (\omega t + p) + L \cos (qt - \theta), \ \omega \neq q, \tag{5}$$

where K and p are arbitrary constants, and the other parameters are defined as follows:

$$a = \frac{B}{2A}, \ \omega = \frac{1}{2A} \sqrt{4AC - B^2}, \ \tan \theta = \frac{Bq}{C - Aq^2}, \ L = \frac{E}{\sqrt{(C - Aq^2)^2 + B^2q^2}}. \tag{6}$$

The derivative of $x(t)$, which we shall denote by $y(t)$, is seen to be

$$x'(t) = y(t) = Ke^{-at}[-a \sin (\omega t + p) + \omega \cos (\omega t + p)] - Lq \sin (qt - \theta). \tag{7}$$

We shall now discuss the phase trajectories defined parametrically by the equations:

$$x = x(t), \quad y = y(t). \tag{8}$$

Several special cases are to be recognized as follows:

I. When a is positive and $L = 0$, the curve defined by (8) is a spiral, which approaches zero as a limit point as $t \to \infty$. From the criteria established in Section 2, the motion would be described as *stable*.

II. When a is positive and $L = 0$, the curve is a spiral, which approaches the ellipse

$$q^2x^2 + y^2 = q^2L^2, \tag{9}$$

as its limit cycle. The motion is thus *stable* and ultimately becomes periodic.

III. When $a = 0$, $L = 0$, the curve is the ellipse

$$\omega^2x^2 + y^2 = K^2\omega^2, \tag{10}$$

The motion is thus *stable* and *periodic*.

IV. When $a = 0$, $L \neq 0$, the curve is given parametrically by the equations

$$x = K \sin (\omega t + p) + L \cos (qt - \theta),$$

$$y = K \omega \cos (\omega t + p) - Lq \sin (qt - \theta), \tag{11}$$

provided $\omega \neq q$. The curve is in general a complex configuration which lies within the rectangle

$$-M \leqq x \leqq +M, \quad -N \leqq y \leqq +N, \qquad (12)$$

where we write: $M = |K| + |L|$, $N = |K\omega| + |Lq|$.

If $\omega = kq$, where k is a rational fraction, the trajectory will be closed and the motion is periodic. The case where k is irrational will be discussed in the next section.

V. If $a = 0$, $E = 0$, and $\omega = q$, the solution given by (5) is replaced by the following:

$$x(t) = K \sin (qt + p) + Qt \sin qt, \quad Q = \frac{1}{2}\sqrt{E^2/AC}; \qquad (13)$$

from which we get for $y(t)$:

$$y(t) = Kq \cos (qt + p) + Q(\sin qt + qt \cos qt). \qquad (14)$$

The solution of the original equation is now seen to be *unstable*, since $x(t)$ increases with t. The phase trajectories for sufficiently large values of t are asymptotic to the curve defined parametrically by the equations:

$$x = Qt \sin qt, \quad y = qQt \cos qt. \qquad (15)$$

Since, for any fixed time t_1, x and y, given by (15), lie on the ellipse

$$q^2 x^2 + y^2 = q^2 Q^2 t_1^2, \qquad (16)$$

the phase trajectories are elliptical spirals, the axes of which increase linearly with t. In this case, dynamical and electrical systems are said to exhibit *resonance*.*

A few examples will serve to illustrate some of the cases described above:

Example 1. Given the differential equation:

$$\frac{d^2x}{dt^2} + 2\frac{dx}{dt} + (4\pi^2 + 1)x = E \, F(t), \qquad (17)$$

where $F(t) = (3\pi^2 + 1) \cos \pi t - 2\pi \sin \pi t$, we shall discuss the motion which it defines.

Solution: If we set $E = 0$, we have Case I above. The particular solution and its derivative, which correspond to the initial conditions: $x = 0$, $y = x' = 2\pi$ at $t = 0$, are found to be the following:

$$x = e^{-t} \sin 2\pi t, \quad y = e^{-t}(-\sin 2\pi t + 2\pi \cos 2\pi t). \qquad (18)$$

---

*By *resonance* we mean the phenomenon in which large vibrations are caused by small forces. Thus a large ship will sometimes roll heavily in a light sea, when the period of the waves is equal to the natural period of the ship. Similarly, a bridge may be badly damaged by a column of marching men. For a more extended account of this phenomenon see A. G. Webster: *The Dynamics of Particles.* 3d ed., 1949, pp. 152–155, 175.

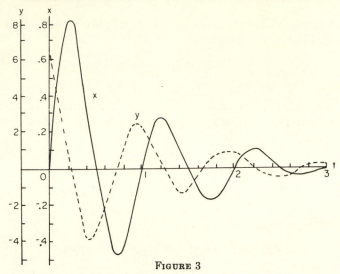

FIGURE 3

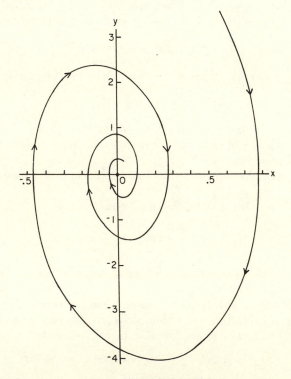

FIGURE 4

These are graphically represented in Figure 3. As $t \to +\infty$, both functions tend to zero. The phase curve, $x=x(t)$, $y=y(t)$ is seen from Figure 4 to be a spiral which continuously approaches the focal point $(0,0)$. The motion is thus to be characterized as stable.

If we now assume that $E=1$, which brings us to Case II, we find as a particular solution and its derivative the following functions:

$$x=e^{-t} \sin 2\pi t + \cos \pi t, \quad y=e^{-t}(-\sin 2\pi t + 2\pi \cos 2\pi t) - \pi \sin \pi t. \quad (19)$$

In this case, as $t \to +\infty$, $x \to \cos \pi t$ and $y \to -\pi \sin \pi t$, which means that the phase curve is asymptotic to the ellipse:

$$\pi^2 x^2 + y^2 = \pi^2. \quad (20)$$

The graphs of $x=x(t)$ and $y=y(t)$ are shown in Figure 5 and the phase curve in Figure 6. It is interesting to observe that the approach of the phase curve to its limiting ellipse is not uniformly an exterior one, since at one point it actually traverses the ellipse. But this is not actually an exceptional phenomenon, since we have already observed an example in the problem of pursuit when the path of the pursued was a circle. (See Figure 12, Chapter 5.)

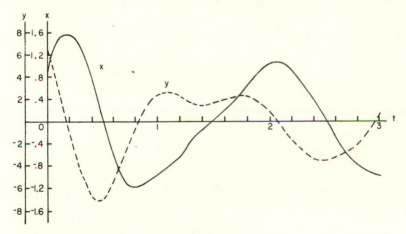

FIGURE 5

In the example just given the boundary conditions for $t=0$ are $x=1$, $y=2\pi$. If these conditions are varied, more complicated phase curves may result. Thus, if we let $x=0$, $y=\frac{1}{2}(3\pi^2+1)$, when $t=0$, the graphs of $x(t)$ and $y(t)$ shown in Figure 7 and that of the phase trajectory shown in Figure 8 were obtained. The graph of the forcing function, $F(t)$, is also shown in Figure 7. These curves were made by an analogue computer. In constructing these graphs by this method

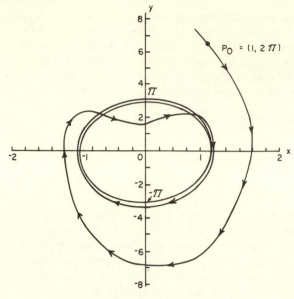

FIGURE 6

it was found convenient to replace equation (17) by the following equivalent system:

$$\frac{d^2x}{dt^2}+2\frac{dx}{dt}+(4\pi^2+1)x=z, \quad \frac{d^2z}{dt^2}+\pi^2z=0, \tag{21}$$

where $z(0)=(3\pi^2+1)$, $z'(0)=-2\pi^2$.

If z is eliminated between the two equations, we obtain the following linear differential equation of fourth order satisfied by x:

$$\frac{d^4x}{dt^4}+2\frac{d^3x}{dt^3}+(5\pi^2+1)\frac{d^2x}{dt^2}+2\pi^2\frac{dx}{dt}+\pi^2(4\pi^2+1)x=0. \tag{22}$$

Example 2. We shall now discuss the motion defined by the following equation:

$$\frac{d^2x}{dt^2}+4\pi^2x=3\pi^2\cos\pi t. \tag{23}$$

Solution: Corresponding to the initial conditions: $x=1$, $y=x'=2\pi$, for $t=0$, the solution of this equation and its derivative are readily found to be:

$$x=\sin 2\pi t+\cos\pi t, \quad y=2\pi\cos 2\pi t-\pi\sin\pi t. \tag{24}$$

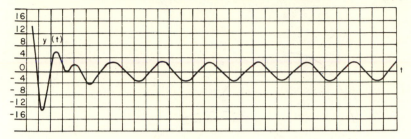

Time Scale: 1cm = 1 sec.

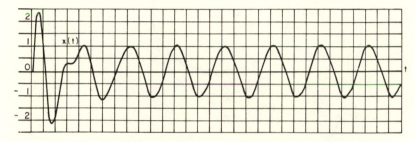

F(t)

FIGURE 7

The graphs of $x=x(t)$ and $y=y(t)$ are shown in Figure 9. The phase trajectory is shown in Figure 10. It will be observed that both x and y are periodic functions of t and thus the phase curve is closed. The motion may thus be described as a *vortex motion*. Although the origin of coordinates is not a limit point, the *limit in the mean** (l.i.m.) of both x and y is zero. It is natural, therefore, to say that the vortex is about this point.

This example illustrates Case IV for the special condition that the periods of the forcing function and of the solution of the homogeneous equation are commensurable. When this condition is not satisfied the situation is very different. This case will be considered in the next section.

---

*The limit in the mean of a function $f(t)$ is defined as follows: $\text{l.i.m.} f(t) = \lim\limits_{\lambda \to \infty} \dfrac{1}{2\lambda} \int_{-\lambda}^{\lambda} f(t)\,dt$.

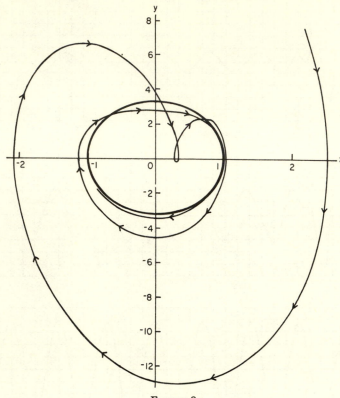

FIGURE 8

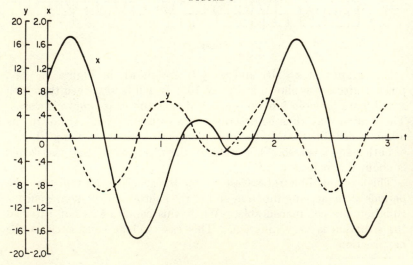

FIGURE 9

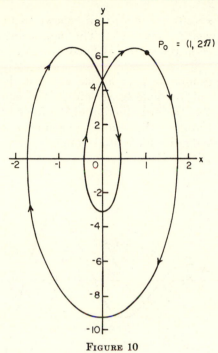

$P_0 = (1, 2\pi)$

FIGURE 10

If the right-hand member of equation (23) is set equal to zero, we have an example of Case III. The general solution is

$$x = A \sin 2\pi(t+a), \tag{25}$$

and the phase curve is the ellipse:

$$4\pi^2 x^2 + y^2 = A^2. \tag{26}$$

The solution is thus a periodic function of t and the motion in the phase-plane is a vortex motion.

Example 3. Given the differential equation

$$\frac{d^2x}{dt^2} + 4x = 4 \cos 2t, \tag{27}$$

we shall discuss the motion which it defines.

Solution: We are now in the circumstances of Case V above, since $w = q = 2$. The solution and its derivative, which both vanish at $t = 0$, are seen to be the following:

$$x = t \sin 2t, \quad y = x' = 2t \cos 2t + \sin 2t. \tag{28}$$

The graphs of $x=x(t)$ and $y=y(t)$ are shown in Figure 11 and the phase curve is given in Figure 12. The motion, which exhibits the effects of resonance in the system defined by the equation, is clearly an unstable one.

The surprising number of phenomena, which we have been able to exhibit by means of a simple linear differential equation of second order, is the result of the introduction of the force function $E \cos qt$. It will be found, however, that similar phenomena are discovered in the solution of nonlinear differential equations where such a force function is not explicitly observed.

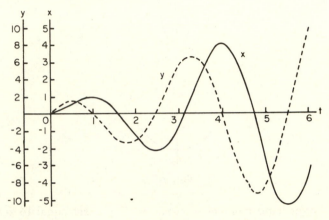

FIGURE 11

As a result of the special analysis which we have given above, let us classify the solutions which we have examined. In cases I and V the motion is nonperiodic. But in the former it is stable, since the point $x=0$, $y=0$ is approached as $t \to \infty$, while it is unstable in the latter, since the phase trajectory expands indefinitely as t increases.

In Case II the motion is nonperiodic, but it approaches a periodic solution asymptotically as t increases. In this case the solution does not·approach a limit point, but the motion has a limit cycle in the phase-space.

The only completely periodic motion was found in Cases III and IV, but in the latter only under certain special circumstances, when the natural period of the solution of the homogeneous equation is commensurate with the period of the forcing function.

It is thus apparent that *periodic motion must be regarded generally as a very special kind of motion.* Periodicity cannot be regarded *a priori* as an inherent property of the motion of a dynamical system, even when its phase curve is limited to a closed region of space. Thus, for example, the motion of the moon about the earth cannot be assumed to be periodic in spite of the fact that its path intertwines with a mean orbit which is stable. One satisfactory proof of periodicity in such a dynamical system is to exhibit the motion in a convergent Fourier series and to show that such a series accounts for all the observed variations.

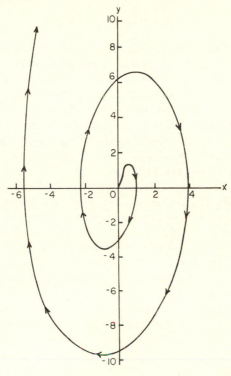

FIGURE 12

In a classical paper of great elegance G. W. Hill (1838–1914) reduced the determination of the motion of the lunar perigee to a differential equation of the form:

$$\frac{d^2y}{dt^2}+A(t)y=0, \tag{29}$$

where $A(t)$ is a periodic function. Since this equation was found to have a periodic solution, the problem of the periodicity of the motion associated with the perigee was thus established in a very satisfactory manner. But in most cases the proof of periodicity is elusive.

In the next section we shall consider an example from Case IV, which shows how deep-seated the problem of periodicity may become and how difficult it is in the case of special equations to determine whether or not the solutions possess this important property.

4. Nonperiodic Solutions in a Closed Area

A good many years ago C. Jordan (1838–1922), while reflecting on the properties of a curve defined by the parametric equations:

$$x=x(t),\ y=y(t),\quad a\leqq t\leqq b,\tag{1}$$

asked the question: Can a continuous curve be defined which will fill a space? That is to say, can the functions x and y of (1), assumed to be continuous in t, be found such that there would be a one-to-one correspondence between the points of the continuum: $|x|\leqq A$, $|y|\leqq B$, and the points on the segment (a,b)? The answer to this question was given in the affirmative in 1890 by G. Peano (1858–1932), who exhibited such a continuous "space-filling curve" now known to mathematicians as *Peano's curve*. Other examples were forthcoming from E. H. Moore and David Hilbert.

Now the construction of such a curve might well have been considered something of a pathological example, illustrating the ingenuity of mathematicians in probing the depths inherent in a definition of continuity, were it not for the fact that the physicists and astronomers had reached something resembling this same problem. Investigating the question of the equipartition of energy in the kinetic theory of gases, J. Clerk Maxwell, L. Boltzmann, Lord Rayleigh and others had been led to a consideration of the possible configurations of a swarm of particles and the distribution of their velocities within a bounded space. In other words, they were interested in the distribution of energy in a phase-space of $2n$ dimensions given by the coordinates:

Position coordinates: $q_i=q_i(t),\quad i=1, 2,\ .\ .\ .\ , n,$

Velocity coordinates: $\dot{q}_i=\dot{q}_i(t),\quad i=1, 2,\ .\ .\ .\ , n.\tag{2}$

We shall not consider here the details of the attempts made to partition the total energy of the system among the individual constituent energies which comprised it. But one of the proposed hypotheses was very intriguing. This assumed that at some time, any specified configuration would be attained to as close a degree of approximation as one desired.

During this same period of time the astronomers found that their problems led to a similar question. Given a dynamical system with a fixed constant energy, such, for example, as that of the planetary system, does there exist an infinite number of periodic motions in the neighborhood of a given periodic motion of general stable type? This question in turn led to another: Does the dynamical system approach within a specified error any preassigned configuration? This problem was one of the principal concerns of H. Poincaré.

whose researches led him to concepts of stability and limit cycles in the solution of differential equations. The name *ergodic* (*ergon*= energy, *hodos*=path) was given to the problem and led ultimately through the researches of G. D. Birkhoff to a statement of the "ergodic theorem" in a form acceptable to modern mathematics.

As a simple illustration of the general problem, consider the motion of a billiard ball moving on a frictionless table with perfectly elastic cushions. If the direction of the ball is such that the point of contact with the cushion divides the length of the cushion into incommensurable parts, then it is reasonable to believe that the ball in the course of time will pass through any preassigned area of the table. The proof of this proposition is readily attained by an application of the ergodic theorem.*

The existence of paths which, like that traced by the billiard ball, pass through any preassigned small part of a given closed area does not pose an entirely philosophical problem. Consider, for example, the path of the moon about the earth. Perturbed by both the earth and the sun, and to a lesser degree by the planets, the motion of the moon has presented a problem of great complexity. To note this one needs merely to examine the heroic computations of E. W. Brown, who devoted a lifetime to the problem, or to the theory of Charles Delaunay, which contains one equation 170 pages in length, or to the classical papers of G. W. Hill, who reduced the problem of the lunar perigee to the evaluation of a determinant of infinite order. The mean path of the moon referred to the earth is a circle of radius 239,000 miles, but the actual path lies within an annulus with radii equal respectively to 222,000 and 253,000 miles. It is probably true, however difficult it would be to prove the proposition, that, given sufficient time, the moon would traverse any given small area within this annulus.

No name appears to have been given to this space-filling path, but one might possibly refer to the area traversed as an *ergchorad*, or energy-area (*ergon*=energy, *chora*=area). Such energy spaces are not uncommon in the theory of nonlinear problems, nor, for that matter, as will soon be demonstrated, in linear equations as well. In fact, pure harmonic motion is a much rarer phenomenon. As any operator of an analogue computer knows, the tracing needle of the machine appears much more ready to describe an ergchorad in phase-space, than to follow the highly specialized path of a curve defined by the sum of harmonic terms.

The significance of the time factor in the ergodic problem may be comprehended from the following philosophical considerations. Let us assume that a very small ball lies upon the floor of an otherwise empty room of dimensions *a*, *b*, *c*. Let us assume further that the ball is

*See, for example, G. D. Birkhoff: "What is the Ergodic Theorem?" *American Math. Monthly*, Vol. 49, 1942, pp. 222–226. *Collected Works*, 1950, pp. 713–717.

given a linear velocity equal to v, which, of course, is to endow it with a total energy of $\frac{1}{2} mv^2$, where m is its mass. If the ball is perfectly elastic and if the walls of the room are sufficiently rough so that reflection from any point of impingement is random, the energy of the ball will be distributed uniformly throughout the room and the ball itself in the course of time will make contact with any prescribed area of the wall to any specified order of approximation.

Since the square of the greatest distance between any two points in the room is $a^2+b^2+c^2$, and since the area of the walls is $A=2(ab+bc+ac)$, the number of contacts made by the ball per unit of surface in unit time will be greater than n defined as follows:

$$n=\frac{v}{A\sqrt{a^2+b^2+c^2}}. \tag{3}$$

If anyone attempted to enter the room by way of a door of dimensions: p,q, he would encounter a shower of particles, which in each unit of time would exert a pressure in excess of $kpqn$, where k is the pressure exerted at each contact of the ball. If v is very large, this total pressure would be very great and one would assert that the empty room was, in fact, a solid.

Returning from these philosophical matters, let us examine an actual curve which passes as near as one wishes to every point in an elliptically shaped space within which its values are defined. For this purpose, let us consider the linear differential equation:

$$\frac{d^2x}{dt^2}+2x=-2\cos 2t. \tag{4}$$

The solution which corresponds to the boundary conditions: $x(0)=1$, $x'(0)=\sqrt{2}$ is the following function:

$$x(t)=\sin \sqrt{2}\, t+\cos 2t, \tag{5}$$

and its derivative is

$$x'(t)=y(t)=\sqrt{2}\,\cos \sqrt{2}\, t-2\sin 2t. \tag{6}$$

We now consider the phase trajectory defined by the equations:

$$x=x(t), \quad y=y(t). \tag{7}$$

Since the periods of the constituent harmonic terms of (5) are incommensurable, the phase curve generated when t varies between $-\infty$ and $+\infty$ will not be periodic, but will wander in a series of loops throughout the interior of a nearly elliptical region with semi-axes equal to the respective periods. The structure of this curve is seen from Figure 13, which shows that portion generated when t varies from -11.6 to $+11.6$.

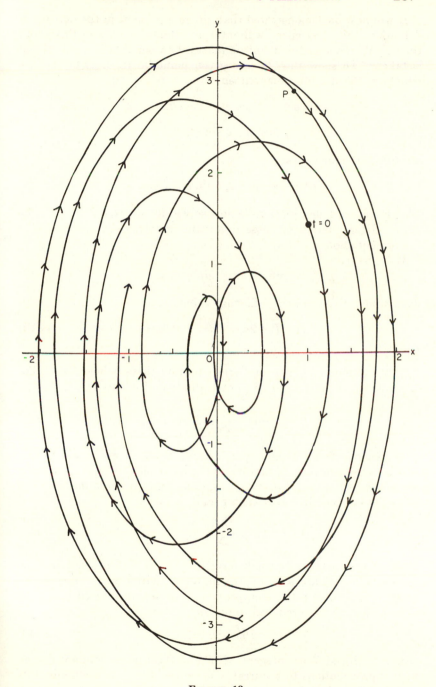

FIGURE 13

It will now be demonstrated that, given any point in the domain of definition, the phase curve will enter an infinite number of times into the neighborhood of this point no matter how small the neighborhood is taken. To show this, let the given point be $P=(x_0, y_0)$. To the equations (5) and (6), now written,

$$x_0 = \sin pt + \cos qt, \quad p = \sqrt{2}, \quad q = 2,$$

$$y_0 = p \cos pt - q \sin qt,$$

we adjoin the identities:

$$\sin^2 pt + \cos^2 pt = 1, \quad \sin^2 qt + \cos^2 qt = 1. \tag{8}$$

This system of four equations is now solved for $\sin pt$ and $\sin qt$. The actual determination of these quantities involves the solution of a quartic equation.

We thus have:

$$\sin pt = s_0, \quad \sin qt = s_1, \tag{9}$$

from which we obtain the following values:

$$pt = T_0 + 2n\pi, \quad qt = T_1 + 2m\pi, \tag{10}$$

where m and n are integers.

Solving equations (10) for t and equating these values, we obtain the following equations for the determination of m and n:

$$pm - qn = \frac{1}{2\pi}(qT_0 - pT_1). \tag{11}$$

In general there will not exist integers m and n which will satisfy this equation, since p, q, and the right-hand member will usually be irrational numbers. However, given any irrational number θ, there will exist integers M and N such that

$$\left| \theta - \frac{M}{N} \right| < \epsilon, \tag{12}$$

where ϵ is an arbitrarily small positive number. If all the irrational numbers in (11) are thus approximated to within some preassigned error, and if these rational approximations are introduced into (11), this equation will be reduced to one of the form

$$Pm - Qn = R, \tag{13}$$

where P, Q, and R are integers. If P and Q contain a common factor, new approximations to the irrational numbers in (11) are made until a P and Q are found which are relatively prime.

Equation (12) is a Diophantine equation, which has the following solution:

$$m = Rm_0 - Qs, \quad n = Rn_0 - Ps, \tag{14}$$

where s is any integer and m_0 and n_0 satisfy the equation:

$$Pm_0 - Qn_0 = 1. \tag{15}$$

Since P and Q are relatively prime, values of m_0 and n_0 exist and are readily found by means of continued fractions as follows:

Let P/Q be expanded into the following continued fraction:

$$P/Q = a_0 + \frac{1}{a_1 +} \frac{1}{a_2 +} \frac{1}{a_3 +} \cdots + \frac{1}{a_h}, \tag{16}$$

which will terminate for some finite value of h, since P and Q are integers.

The last term in (16) is now removed and the resulting fraction evaluated. The numerator and denominator of this fraction, with proper signs, will be respectively the desired values n_0 and m_0.

Hence, any of the infinitely many values of m and n defined by (14), when substituted in equations (10), will yield an infinite set of values of t for which the phase curve passes through the neighborhood of the given point.

As an illustration, let us consider the following example:

Example: Find the value of t for which the phase trajectory defined by equations (5) and (6) passes through the neighborhood of the rational point: $x = 0.8212$, $y = 2.8978$, where the neighborhood is defined to be a circle of radius equal to 0.0001. The point is denoted by P on the graph shown in Figure 13.

Solution: To the indicated degree of accuracy, we find the following values of $\sin pt$ and $\sin qt$ corresponding to x and y:

$$\sin pt = 0.1609, \quad \sin qt = -0.7510, \quad p = \sqrt{2}, \quad q = 2. \tag{17}$$

From these we get, $T_0 = 0.1616$, $T_1 = -0.8496$. Using the approximations: $\pi = 3.1416$, $p = 1.4142$, we obtain the following equation for the evaluation of m and n:

$$7071m - 10000n = 1213. \tag{18}$$

The expansion of $7071/10000$ as a continued fraction is found to be

$$7071/10000 = 0 + \frac{1}{1+} \frac{1}{2+} \frac{1}{2+} \frac{1}{2+} \frac{1}{2+} \frac{1}{2+} \frac{1}{3+} \frac{1}{14+} \frac{1}{2}. \tag{19}$$

Deleting the last term and evaluating the resulting fraction, we obtain 3416/4831, from which we find: $m_0=4831$, $n_0=3146$. Substituting these values in (14) we get

$$m=586003-10000s, \quad n=4143608-7071s, \tag{20}$$

from which we have, when $s=586$, the values $m=3$, $n=2$. These values, when substituted in (10), give 9.0 as one value of t. An infinite set of other values is provided by the one-parameter set of values of m and n obtained from (20). Finally, to verify the correctness of the analysis, we substitute $t=9$·in (5) and (6) and thus obtain the values: $x=0.821167$, $y=2.897774$, which are observed to be within the neighborhood of the prescribed point.

The functions $x(t)$ and $y(t)$ which we have just examined belong to what is called the class of *almost periodic functions*, the theory of which was initiated by Harold Bohr in 1924.*

By an almost periodic function is meant a function which satisfies the equation

$$f(x+\tau)=f(x), \tag{21}$$

within an error that can be made arbitrarily small, where τ denotes any number of an infinite set of values "spread over the whole range from $-\infty$ to $+\infty$ in such a way as not to leave empty intervals of arbitrarily great length." These values of τ are called *translation numbers*.

For example, if $f(x)=\sin px+\cos qx$, where p and q are incommensurable numbers, then τ will be a translation number provided integers m and n can be found such that

$$|p\tau-2n\pi|<\epsilon_1, \text{ and } |q\tau-2m\pi|<\epsilon_2, \tag{22}$$

where ϵ_1 and ϵ_2 are arbitrarily small positive numbers. If $\epsilon<\epsilon_1$ and ϵ_2, we can write

$$p\tau=2n\pi+\epsilon, \quad q\tau=2m\pi+\epsilon, \tag{23}$$

from which we get

$$\tau=2n\pi/p+\epsilon/p=2m\pi/q+\epsilon/q. \tag{24}$$

If π/p, π/q, ϵ/p, and ϵ/q are now represented by rational fractions to any desired degree of approximation, then, as we have already seen above, an infinite number of integral values of m and n can be found which satisfy equation (24).

---

*H. Bohr: "Zur Theorie der fastperiodischen Funktionen," *Acta Mathematica*, Vol. 45, 1924, pp. 29–127, Vol. 46, 1925, pp. 101–214, Vol. 47, 1926, pp. 237–281.

A comprehensive survey of the subject will be found in A. S. Besicovitch: *Almost Periodic Functions*. Cambridge, 1932, xiii+180 pp.

There will thus exist an infinite sequence of translation numbers for which we have

$$f(x+\tau)=\sin\ p(x+\tau)+\cos\ q(x+\tau)=\sin\ (px+\epsilon)+\cos\ (qx+\epsilon),$$
$$=\sin\ px+\cos\ qx+2\theta\epsilon,\ \text{where}\ |\theta|<1.$$

More generally, an almost periodic function can be expanded into a series of the following form:

$$f(x)=\sum_{n=0}^{\infty}A_{n}e^{i\lambda_{n}x},\qquad (25)$$

where $\sum|A_n|^2$ converges.   This is a special case of a Dirichlet series, which in its turn includes Fourier series as a special case.

Conversely, it can be shown that the series (25) is the representation of an almost periodic function, provided $\sum|A_n|^2$ converges.

PROBLEMS

1. Find the first time after $t=0$ when the phase curve defined by equations (5) and (6) passes through the neighborhood of the rational point: $x=1.7565$, $y=1.1551$, where the neighborhood is the interior of the circle of radius 0.0001 about the point. Verify that $\sin\ pt=0.6651$ and $\sin\ qt=-0.0495$. *Answer:* $t=9.4$.

2. Verify that $t=25$ is the first time after $t=0$ when the phase curve defined by equations (5) and (6) passes through the neighborhood of the rational point: $x=0.2492$, $y=-0.4627$. Assume that the neighborhood is the interior of the circle of radius 0.0001 about the point. Verify that $\sin\ pt=-0.7158$, $\sin\ qt=-0.2624$.

5. The Pendulum Problem as a Fourier Series

As an example of a nonlinear equation which has a periodic solution, let us return to the equation of the simple pendulum, which has already been discussed at some length in Section 10 of Chapter 7.

It will be convenient to write the equation in the form

$$\frac{d^2y}{dt^2}=Ay+By^3,\qquad (1)$$

and to write the solution in the form

$$y=C\ \text{sn}(\lambda x,k),\quad x=t+p,\qquad (2)$$

where λ and p are arbitrary constants and k and C are defined as follows:

$$k^2=-\frac{(\lambda^2+A)}{\lambda^2},\quad C^2=-\frac{2(\lambda^2+A)}{B}.\qquad (3)$$

If we assume that C is known, as well as A and B, then both λ and k can be computed from (3). Let us denote by K and K' the complete and complementary elliptic integrals corresponding to k. If in (2) we now set $p=0$, and let $\lambda t = u$, then the solution of (1) assumes the form

$$y = C \operatorname{sn}(u, k). \tag{4}$$

But from Section 19 of Chapter 6, we know that this function can be written as the following Fourier series:

$$y = C \frac{\pi}{Kk} \sum_{n=0}^{\infty} A_n \sin (2n+1)z, \tag{5}$$

where we abbreviate:

$$A_n = 1/\sinh \left[\left(n+\frac{1}{2} \right) \pi K'/K \right], \quad z = \frac{1}{2} (\pi u/K) = \frac{1}{2} (\lambda \pi t/K). \tag{6}$$

Although we thus seem to have obtained the solution of the nonlinear equation (1) in linear form as the sum of simple harmonic terms, the apparent linearization is soon seen to be an illusion. For both the individual amplitudes, A_n, and the multiplier of t are functions of k, and thus also functions of the parameter C.

It will now be of interest to write equation (5) in a second form. To achieve this we introduce the symbol: $q = \exp (-\pi K'/K)$, and observe from (14), Section 18, Chapter 6, the expansion:

$$\frac{2\pi}{Kk} = q^{-1/2} (1 + q^2 + q^6 + q^{12} + q^{20} + \ldots)^{-2}, \tag{7}$$

It now follows from equation (7) of Section 19, Chapter 6 that (5) above can be written as follows:

$$y = C \sum_{n=0}^{\infty} a_n \sin (2n+1)z, \tag{8}$$

where we abbreviate:

$$a_n \doteq \frac{q^n}{1 - q^{2n+1}} (1 + q^2 + q^6 + q^{12} + \ldots)^{-2}, \tag{9}$$

or, in expanded form:

$$a_n = q^n (1 - 2q^2 + 3q^4 - 6q^6 + 11q^8 - 18q^{10} + 28q^{12} + \ldots)$$
$$\times (1 + q^{2n+1} + q^{4n+2} + q^{6n+3} + \ldots). \tag{10}$$

Explicitly we have for the first three coefficients the following expansions:

$$a_0 = 1 + q - q^2 - q^3 + 2q^4 + 2q^5 - 4q^6 - 4q^7 + 7q^8 + 7q^9 - 11q^{10} - 11q^{11}$$
$$+ 17q^{12} + \ldots;$$
$$a_1 = q(1 - 2q^2 + q^3 + 3q^4 - 2q^5 - 5q^6 + 3q^7 + 9q^8 - 5q^9 - 15q^{10} + 9q^{11}$$
$$+ 23q^{12} + \ldots);$$
$$a_2 = q^2(1 - 2q^2 + 3q^4 + q^5 - 6q^6 - 2q^7 + 11q^8 + 3q^9 - 17q^{10} - 6q^{11}$$
$$+ 26q^{12} + \ldots). \tag{11}$$

It will now be convenient to write equation (1) in the following form:

$$\frac{d^2y}{dt^2} + y + ry^3 = 0, \tag{12}$$

and to express the solution (8) in terms of a and r, where

$$a = Ca_0. \tag{13}$$

Since, in equation (3), we have $A = -1$, $B = -r$, if we eliminate λ, we obtain

$$k^2 = \frac{-x}{2+x}, \quad k'^2 = \frac{2+2x}{2+x}, \tag{14}$$

where $x = rC^2$ and $k^2 + k'^2 = 1$.

Observing that a_0 in (13) is a function of q, we now seek a relationship between q and x, which will enable us to express a as a series in x. This objective will be attained by way of the function

$$\epsilon = \frac{1}{2}\left(\frac{1-s}{1+s}\right), \quad s^2 = k'. \tag{15}$$

This function, by (14), Section 20, Chapter 6, is connected with q through the following equation:

$$q = \epsilon + 2\epsilon^5 + 15\epsilon^9 + 150\epsilon^{13} + 1707\epsilon^{17} + \ldots. \tag{16}$$

By (14) we have

$$s = \left(1 + \frac{1}{2}v\right)^{\frac{1}{4}}, \qquad v = \frac{x}{1 + \frac{1}{2}x}, \tag{17}$$

and from (15),

$$\epsilon = \frac{1}{2}\left(\frac{1-s}{1+s}\right) = \frac{1}{2}\frac{(1-s)^2}{1-s^2} = \frac{1}{2}\frac{(1-s)^2(1+s^2)}{1-s^4}, \tag{18}$$

that is,

$$2(1-s^4)\epsilon = 1 - 2s + 2s^2 - 2s^3 + s^4. \tag{19}$$

We now observe the following expansions:

$$s=\left(1+\frac{1}{2}v\right)^{\frac{1}{4}}=1+\frac{1}{8}v-\frac{3}{128}v^2+\frac{7}{1024}v^3-\frac{77}{32768}v^4+\cdots,$$

$$s^2=\left(1+\frac{1}{2}v\right)^{\frac{1}{2}}=1+\frac{1}{4}v-\frac{1}{32}v^2+\frac{1}{128}v^3-\frac{5}{2048}v^4+\cdots, \tag{20}$$

$$s^3=\left(1+\frac{1}{2}v\right)^{\frac{3}{4}}=1+\frac{3}{8}v-\frac{3}{128}v^2+\frac{5}{1024}v^3-\frac{45}{32768}v^4+\cdots,$$

$$s^4=1+\frac{1}{2}v.$$

When these values are substituted in (19) there results

$$\epsilon=-\frac{1}{32}v+\frac{1}{128}v^2-\frac{21}{8192}v^3+\cdots. \tag{21}$$

From (17) we now obtain the following expansions:

$$v=x\left(1-\frac{1}{2}x+\frac{1}{4}x^2-\frac{1}{8}x^3+\cdots\right),$$

$$v^2=x^2\left(1-x+\frac{3}{4}x^2-\frac{1}{2}x^3+\cdots\right), \tag{22}$$

$$v^3=x^3\left(1-\frac{3}{2}x+\frac{3}{2}x^2-\frac{5}{4}x^3+\cdots\right),$$

and when these are substituted in (21) there is finally obtained:

$$\epsilon=-\frac{1}{32}x+\frac{3}{128}x^2-\frac{149}{8192}x^3+\cdots. \tag{23}$$

Observing from (16) that, to the degree of approximation given by (23), we have

$$q=\epsilon, \tag{24}$$

it is now possible to express Ca_i in terms of x. We thus obtain

$$Ca_0=C\left(1-\frac{1}{32}x+\frac{23}{1024}x^2-\frac{547}{32768}x^3+\cdots\right),$$

$$=C\left(1-\frac{1}{32}rC^2+\frac{23}{1024}r^2C^4-\frac{547}{32768}r^3C^6+\cdots\right),$$

$$Ca_1=C\left(-\frac{1}{32}rC^2+\frac{3}{128}r^2C^4-\frac{297}{16384}r^3C^6+\cdots\right), \tag{25}$$

$$Ca_2=C\left(\frac{1}{1024}r^2C^4-\frac{3}{2048}r^3C^6+\cdots\right).$$

These equations show explicitly the manner in which the arbitrary constant C and the parameter r enter into the coefficients of the sine terms in the expansion (8).

We return now to the original problem, that of expressing the coefficients of (8) in terms of a as defined by (13). To achieve this we first square (13), multiply by r, and use the abbreviation: $\mu = ra^2$. We thus have $\mu = x\, a_0^2$, whence

$$x = \mu/a_0^2 = \mu(1 - 2q + 5q^2 - 8q^3 + \ldots),$$

$$= \mu \left(1 + \frac{1}{16}\, x - \frac{43}{1024}\, x^2 + \frac{15}{512}\, x^3 - \ldots \right). \tag{26}$$

This equation is now inverted by the method of Lagrange* and we obtain the following expansion of x in terms of μ:

$$x = \mu + \frac{1}{16}\, \mu^2 - \frac{39}{1024}\, \mu^3 + \ldots. \tag{27}$$

This is now substituted in (23) and note taken of (24), from which we have:

$$\epsilon = q = -\frac{1}{32}\, \mu + \frac{11}{512}\, \mu^2 - \frac{461}{32768}\, \mu^3 + \ldots. \tag{28}$$

This value of q is now substituted in the expansions:

$$Ca_1 = a(a_1/a_0) = a(q - q^2 + q^4 + \ldots),$$
$$Ca_2 = a(a_2/a_0) = a(q^2 - q^3 + \ldots), \tag{29}$$

and we thus obtain the desired coefficients:

$$Ca_0 = a, \quad Ca_1 = a \left(-\frac{1}{32}\, \mu + \frac{21}{1024}\, \mu^2 - \frac{461}{32768}\, \mu^3 + \ldots \right),$$

$$= -\frac{1}{32}\, ra^3 \left(1 - \frac{21}{32}\, ra^2 + \frac{461}{1023}\, r^2 a^4 + \ldots \right),$$

$$Ca_2 = a \left(\frac{1}{1024}\, \mu^2 - \frac{43}{32768}\, \mu^3 + \ldots \right),$$

$$= \frac{1}{1024}\, r^2 a^5 \left(1 - \frac{43}{32}\, ra^2 + \ldots \right). \tag{30}$$

It will be of interest also to express the multiplier of t in (6) in terms of both x and μ. We adopt the notation: $\omega = \frac{1}{2}\, (\lambda\pi/K)$.

*See H. T. Davis: *Tables of the Higher Mathematical Functions*, Vol. 1, p. 79.

We first observe from (3) that

$$\lambda^2 = 1 + \frac{1}{2} rC^2 = 1 + \frac{1}{2} x, \tag{31}$$

and from equation (14), Section 18, Chapter 6, that

$$\frac{2K}{\pi} = (1 + 2q + 2q^4 + 2q^9 + \ldots)^2. \tag{32}$$

We thus have

$$\omega^2 = \lambda^2 (1 + 2q + 2q^4 + 2q^9 + \ldots)^{-4},$$

$$= \left(1 + \frac{1}{2} x\right)(1 - 8q + 40q^2 - 160q^3 + \ldots). \tag{33}$$

Observing from (24) that to the desired approximation $q = \epsilon$, we replace q in this expression by (23) from which we obtain the following:

$$\omega^2 = 1 + \frac{3}{4} x - \frac{3}{128} x^2 + \frac{9}{512} x^3 + \ldots,$$

$$= 1 + \frac{3}{4} rC^2 - \frac{3}{128} r^2 C^4 + \frac{9}{512} r^3 C^6 + \ldots. \tag{34}$$

The equivalent expansion in terms of μ is obtained by replacing x in this expression with its expansion in μ as given by (27). We thus obtain

$$\omega^2 = 1 + \frac{3}{4} \mu + \frac{3}{128} \mu^2 - \frac{57}{4096} \mu^3 + \ldots,$$

$$= 1 + \frac{3}{4} ra^2 + \frac{3}{128} r^2 a^4 - \frac{57}{4096} r^3 a^6 + \ldots. \tag{35}$$

As a numerical example, let us consider the case of the pendulum described in Section 10 of Chapter 7. The equation in the notation of this section is

$$\frac{d^2y}{dt^2} + y - \frac{1}{6} y^3 = 0, \tag{36}$$

and the value of C is $\pi/3$.

Noting that $r = -1/6$, we introduce C and r into formulas (25) and thus obtain:

$$Ca_0 = 1.05407, \quad Ca_1 = 0.00692, \quad Ca_2 = 0.000044. \tag{37}$$

Substituting C and r into formula (34), we find

$$\omega^2 = 0.862022, \tag{38}$$

from which we get: $\omega = 0.928451$.

The desired Fourier series is thus explicitly the following:

$$y = 1.05407 \sin \omega t + 0.00692 \sin 3\omega t + 0.000044 \sin 5\omega t + \ldots \quad (39)$$

The convergence of the series which we have given above is not great as is shown by a more exact determination of the constants in (39). The coefficients of the series will be found to be too small by two units in the last place and the value of ω too large by four units in the last place.

6. Periodic Solutions

Before proceeding to other particular equations it is necessary to discuss the important question: How does one determine when a given differential equation has a periodic solution? This is obviously a difficult question to answer and the best efforts of great analysts have been devoted to the problems derived from it. For example, it is not easy to see, a priori, why the two equations

$$y'' + y - y^3/6 = 0, \qquad y'' + \sin y = 0, \qquad (1)$$

should have real solutions with real periods, nor why the equations:

$$y'' + (y^2 - 1)y' + y = 0, \qquad y'' + x \sinh (y/x) = 0, \qquad (2)$$

should have real solutions which are not periodic, but which approach periodicity as x increases along the positive real axis.

That periodicity is a very special property is seen from the fact that if $y(x)$ is a periodic function, it must satisfy the following equation:

$$y(x + a) = y(x), \qquad (3)$$

where a is a constant. Moreover, all the derivatives of $y(x)$, if they exist, are also periodic, as we see by taking the nth derivative of (3).

Since equation (3) can be written as the following linear differential equation of infinite order:

$$y' + \frac{a}{2!} y'' + \frac{a^2}{3!} y^{(3)} + \frac{a^3}{4!} y^{(4)} + \ldots = 0, \qquad (4)$$

we see that every periodic analytic function must be a solution of this equation. We thus reach the curious conclusion that every solution of a nonlinear differential equation, which is both analytic and periodic, is also the solution of the linear differential equation (4).

Equation (4) has the following infinite set of solutions:

$$y=e^{pix}, \qquad p=2\pi n/a, \qquad n=0, \pm 1, \pm 2, \ldots, \qquad (5)$$

and thus any finite sum of the following form:

$$y(x)=\frac{1}{2} A_0 + \sum_{m=1}^{n} A_m \cos \frac{m\pi x}{a} + \sum_{m=1}^{n} B_m \sin \frac{m\pi x}{a}, \qquad (6)$$

where A_m and B_m are arbitrary, is also a solution.

We can now establish certain general criteria, which assure the existence of a solution of the following differential equation of second order:

$$\frac{d^2y}{dx^2}=f(x,y,y'). \qquad (7)$$

If a set of values A_m and B_m exists for all integral values of m, so that series (6) formally satisfies (7) as $n \to \infty$, and if the first two derivatives of (6) exist for values of $|x| \leqq a$, then (7) has a periodic solution of period a.

From the fact that if $y(x)$ is a periodic solution of (7), and if its deriviative exists, then $y'(x)$ must also be periodic, we see that the corresponding phase-trajectory is closed. Conversely, *if any phase-trajectory of (7) is closed, then the equation has a periodic solution.*

If equation (7) has a periodic solution within a domain R, and if the function $f(x,y,y')$ is analytic for all values of x,y,y' in R, then every derivative of y will exist and be a periodic function. This follows from the fact that every derivative of (7) can be expressed in terms of y and y' and the argument of the calculus of limits (Section 2, Chapter 4) insures the existence of the derivatives within R.

If one is willing to relax the requirements of analyticity and substitute the limiting processes of summability, the domain of periodic solutions of differential equations can be considerably enlarged. An instructive example is supplied by the following equation.

$$\frac{d^2y}{dx^2}=0, \qquad (8)$$

which has as a solution the continuous periodic function shown in Figure 14(A). Within the interval; $-a \leqq x \leqq a$, y is given by the formula:

$$y=\begin{cases} 1+x/a, & -a \leqq x \leqq 0, \\ 1-x/a, & 0 \leqq x \leqq a. \end{cases} \qquad (9)$$

This function meets the first criterion given above, since it is represented by the following Fourier series:

$$y(x) = \frac{1}{2} + \frac{4}{\pi^2}\left[\cos\frac{\pi x}{a} + \frac{1}{9}\cos\frac{3\pi x}{a} + \frac{1}{25}\cos\frac{5\pi x}{a} + \ldots\right], \tag{10}$$

which converges for all real values of x.

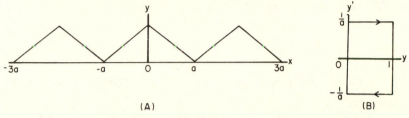

(A)   (B)

FIGURE 14

It is clear that the function defined by (9) is a solution of the differential equation, but this is not true in a strictly analytical sense for (10), since we have $y'' = -(4/a^2)\ C(\theta)$, $\theta = \pi x/a$, where we abbreviate:

$$C(\theta) = \cos\theta + \cos 3\theta + \cos 5\theta + \ldots, \tag{11}$$

and this series is clearly divergent for almost all values of θ.

However, except at the points $x = ma$, that is for $\theta = m\pi$, where m is any positive or negative integer, the series is summable to zero in the Fejér-Cesàro sense,[*] and thus may be said to satisfy the differential equation except at these critical points.

To see this we form the Cesàro sums:

$$S_1 = \cos\theta = \frac{\sin 2\theta}{2\sin\theta}, \quad S_2 = \cos\theta + \cos 3\theta = \frac{\sin 4\theta}{2\sin\theta}$$

$$S_3 = \cos\theta + \cos 3\theta + \cos 5\theta = \frac{\sin 6\theta}{2\sin\theta}, \ldots, \quad S_n(\theta) = \frac{\sin 2n\theta}{2\sin\theta},$$

from which we get

$$S = S_1 + S_2 + \ldots + S_n = \frac{\sin(n+1)\theta\sin n\theta}{2\sin^2\theta}. \tag{12}$$

It thus follows that we have

$$\lim_{n\to\infty}\frac{S}{n} = 0,$$

except when $\theta = m\pi$.

[*]L. Fejér: "Untersuchungen über Fourische Reihen," *Mathematische Annalen*, Vol. 58, 1904, pp. 51-69.

The phase-trajectory is shown in Figure 14 (B). The perpendicular lines of the diagram through $y=0$ and $y=1$ may be regarded as limiting forms of the transition curves of the nth segments of the Fourier series for y'. Strictly speaking y' has only the value 0 when $y=0$ and $y=1$.

It is instructive to compare the first equation of (1) with the example just given. Referring to Section 5, we see that its solution can be written as the Fourier series

$$y=\sum_{n=0}^{\infty} B_n \sin (2n+1)z, \qquad (13)$$

where $z=ax$, and $B_n=P/\sinh \left[b\left(n+\frac{1}{2}\right)\right]$, in which a, P, and b are constants.

It is obvious that the series which defines $y^{(m)}$ converges uniformly for all values of m and for any point on the real axis of z.

7. Additional Aspects of Periodicity—Floquet's Theory

Deeper problems connected with periodicity than those which we have described were encountered by the astronomers. They were disturbed, for example, by the intrusion of *secular terms* in the description of phenomena, which were fundamentally periodic. That is to say, the solution of their equations led to expansions of the following form:

$$y=\sum(a_n+b_nt+c_nt^2+ \ldots) \cos (nt+\gamma_n), \qquad (1)$$

where t appears explicitly in the multiplier of the harmonic terms.

Such terms are called secular, since they represented variations which progress in one direction for long periods of time, even though ultimately they may prove to be periodic. Such, for example, are movements in the line of nodes, the line of apsides, the inclination and eccentricity of the planets. But these variations in astronomy are relatively small and cause small disturbance to the harmonic terms which they multiply. But in an electrical system, where many vibrations occur in a short space of time, the appearance of a factor of the form $t \cos nt$ would immediately introduce the phenomenon of resonance.

On the other hand, the coefficient of the cosine term in (1) may actually be a periodic function and thus, even for rapid changes in t, would produce no resonance effect. An elementary example is furnished by the following function:

$$y=\cos pt \cos qt, \qquad (2)$$

where p is assumed to be very much smaller than q, as is the case in the secular variations of astronomy. If the values of y in which we are interested are limited to a fixed interval of t, let us say, $0 \le t \le T$, then (2) might well be replaced by the function:

$$y = \left(1 - \frac{p^2 t^2}{2!} + \frac{p^4 t^4}{4!} \right) \cos qt, \tag{3}$$

which, while not actually a periodic function, would vary little from one within the assumed interval of t.

How to handle these secular terms became a problem of great concern to the astronomers and a number of ingenious methods were developed. We shall describe one of these in Chapter 12.

A second problem which interested the astronomers was that of determining conditions under which a differential equation with periodic coefficients would have periodic solutions.

Let us first observe that the general solution of a differential equation may not be a periodic function, but that the equation may nevertheless have particular solutions which are. Thus the linear equation

$$A \frac{d^2 y}{dt^2} + B \frac{dy}{dt} + Cy = 0, \tag{4}$$

where the coefficients are constants, has as a general solution the function:

$$y(t) = \alpha e^{r_1 t} + \beta e^{r_2 t}, \tag{5}$$

where r_1 and r_2 are roots of the equation: $Ar^2 + Br + C = 0$.

If $y(t)$ is periodic, then there must exist a constant ω such that

$$y(t+\omega) = y(t), \tag{6}$$

which reduces to the following condition:

$$e^{r_1 \omega} [\alpha \, e^{r_1 t} + \beta \, e^{r_2 t} \, e^{(r_2 - r_1) \omega}] = \alpha \, e^{r_1 t} + \beta \, e^{r_2 t}. \tag{7}$$

If $r_1/r_2 = p/q$, where p and q are integers, then $\omega = 2q\pi i/r_2$ is seen to be a period. But if this condition is not satisfied, that is, if r_1/r_2 is not rational, the particular solutions $\alpha e^{r_1 t}$ and $\beta e^{r_2 t}$ are separately periodic with periods equal respectively to $2\pi i/r_1$ amd $2\pi i/r_2$.

This elementary analysis was extended by G. Floquet[*] to a linear equation in which the coefficients are periodic functions of the inde-

[*]*Annales de l'École Normale Supérieure*, Sup. 2, Vol. 12, 1883, p. 47. Accounts of the Floquet theory will also be found in E. T. Whittaker and G. N. Watson: *Modern Analysis*, 4th ed. Cambridge, 1927, pp. 412–413; E. L. Ince: *Differential Equations*, London, 1927, pp. 381–384; N. Minorsky, *Nonlinear Mechanics*, Ann Arbor, 1947, pp. 357–360.

pendent variable. With sufficient generality this equation can be written in the following form:

$$\frac{d^2y}{dt^2}+A(t)\ y=0,\tag{8}$$

where $A(t)$ denotes the following series:

$$A(t)=a_0+\sum_{n=1}^{\infty}2a_n\ \cos 2nt.\tag{9}$$

This equation is called Hill's equation after G. W. Hill (1838–1914), who introduced it in his study of the motion of the lunar perigee.[*] A celebrated special case is the differential equation of Mathieu, namely,

$$\frac{d^2y}{dt^2}+(a-2b\ \cos 2t)\ y=0,\tag{10}$$

which was introduced by E. L. Mathieu (1835–90) in a discussion of the vibrations of an elliptic membrane.[†]

Floquet's theory, with which we shall be mainly concerned, is designed to establish the existence of a periodic solution of a linear differential equation of any order with coefficients which are all periodic functions of a fixed period ω. The theory is sufficiently explained if we limit its application to a differential equation of second order, such, for example, as equation (8) above.

Let $u_1(t)$ and $u_2(t)$ be any linearly independent solutions of the equation, from which it follows that

$$U(t)=Au_1(t)+Bu_2(t),\tag{11}$$

where A and B are arbitrary constants, is the general solution.

Since the coefficients of the equation are periodic functions of period ω, it is clear that both $u_1(t+\omega)$ and $u_2(t+\omega)$ are also solutions of the equation. Hence these functions can be expressed linearly in terms of the fundamental set and we have

$$u_1(t+\omega)=a_1u_1(t)+a_2u_2(t),\quad u_2(t+\omega)=b_1u_1(t)+b_2u_2(t).\tag{12}$$

The general solution can then be written:

$$U(t+\omega)=(Aa_1+Bb_1)\ u_1(t)+(Aa_2+Bb_2)\ u_2(t).\tag{13}$$

[*] "On the Part of the Motion of the Lunar Perigee, which is a function of the Mean Motion of the Sun and Moon," Cambridge, Mass., 1877.   Reprinted in *Acta Mathematica*, Vol. 8, 1886, pp. 1-36.   Also Hill's *Collected Works*, Vol. 1, 1905, pp. 243-270.   A bibliography and an account of the solution of this equation is given in H. T. Davis: *Theory of Linear Operators*, Bloomington, Ind., 1936, pp. 436-440.

[†] E. L. Mathieu: *Journal de Mathematiques*, Vol. 13 (2), 1868, p. 137.

If we now write

$$U(t+\omega)=k\,U(t), \tag{14}$$

then A and B must satisfy the following set of equations:

$$Ak=Aa_1+Bb_1, \quad Bk=Aa_2+Bb_2. \tag{15}$$

Since these are homogeneous equations in A and B, the necessary and sufficient condition for the existence of values other than zero assumes the following form:

$$\begin{vmatrix} a_1-k & b_1 \\ a_2 & b_2-k \end{vmatrix} =0. \tag{16}$$

If k is one of the roots of this equation, then the general solution of the differential equation will satisfy (14). Let us now write $k=e^{\lambda\omega}$ and define the function

$$W(t)=e^{-\lambda t}U(t). \tag{17}$$

We then have from (14)

$$W(t+\omega)=e^{-\lambda(t+\omega)}U(t+\omega)=e^{-\lambda t}e^{-\lambda\omega}U(t+\omega)=e^{-\lambda t}U(t)=W(t). \tag{18}$$

The differential equation thus has a solution of the form

$$U(t)=e^{\lambda t}W(t), \tag{19}$$

where $W(t)$ is a periodic function.

The principal difficulty in the actual solution of the original equation is found in the problem of determining λ. If this constant is zero, then the solution is periodic, but otherwise it is either stable, if $\lambda<0$, or unstable if $\lambda>0$. The equation of Mathieu has been extensively investigated from this point of view and a series of periodic functions have been determined for it, which are called *Mathieu functions*. The elegant researches of Hill also make use of this theory and lead to the evaluation of a determinant of infinite order. Since both of these problems have been extensively treated in the references cited above, further discussion will be omitted.

8. Periodicity as a Phenomenon of the Phase-Plane

Since the theory of Floquet, described in Section 7, makes fundamental use of the property of linearity, it cannot be applied to the periodicity problem of nonlinear equations. One must either explore the possibility of finding a convergent Fourier expansion which satisfies the differential equation, or investigate the behavior of solutions in phase-space. If a closed trajectory is found, then by the criterion of Section 6, the existence of a periodic solution is automatically proved.

If a differential equation

$$\frac{d^2y}{dt^2}=f(y,y'),\tag{1}$$

is derived by the elimination of x from a system of equations:

$$\frac{dx}{dt}=P(x,y),\qquad \frac{dy}{dt}=Q(x,y),\tag{2}$$

then the existence of a closed cycle in the phase-plane (x,y) carries with it the periodicity of y as a function of t. It was in this way that the existence of periodic solutions of Volterra's equation

$$yy''=(y')^2+ay(1-y)y'+acy^2(1-y),\tag{3}$$

was established in Chapter 5. The length of the period was determined by the interval: $T=t_2-t_1$, where t_1 is the time corresponding to a point: $P_1=(x_1,y_1)$ on the phase-trajectory and t_2 the next value of t when the curve reenters the point P_1.

It was an adaptation of this method that A. Liénard used in establishing criteria for the existence of a periodic solution for the following equation:

$$\frac{d^2y}{dt^2}+f(y)\frac{dy}{dt}+g(y)=0,\tag{4}$$

and that N. Levinson and O. K. Smith applied similarly to the more general equation:*

$$\frac{d^2y}{dt^2}+f(y,v)\frac{dy}{dt}+g(y)=0,\qquad v=\frac{dy}{dt}.\tag{5}$$

The theorem of Liénard is as follows:

THEOREM A. *The functions* f(y) *and* g(y) *in equation* (4) *are continuous and integrable and satisfy the following additional conditions:*

(a) f(y) *is an even function, and* g(y) *is an odd function such that* y g(y) >0;

(b) *The functions*

$$F(y)=\int_0^y f(y)\,dy,\quad G(y)=\int_0^y g(y)\,dy,$$

tend toward ∞ *as* y→∞ .

---

*See *Bibliography* for references.

(c) $F(y)$ *has a single positive zero*, $y = y_0$. *In the interval* $(0, y_0)$, $F(y)$ *is negative, but for* $y > y_0$ $F(y)$ *is positive and increases monotonically.*

Under these conditions there exists a periodic solution of (4), *which is unique to within a simple translation of the variable* t.

To prove this theorem we first introduce the Liénard variable

$$z = y' + F(y), \tag{6}$$

from which we have, by reference to (4),

$$\frac{dz}{dt} = y'' + f(y)y' = -g(y). \tag{7}$$

Moreover, since $y' = dy/dt = z - F(y)$, we obtain the derivative

$$\frac{dz}{dy} = -\frac{g(y)}{z - F(y)}. \tag{8}$$

We now consider the trajectories in the Liénard-plane (y, z), which are geometrically simpler than those in the phase-plane (y, y'), since they are symmetrical with respect to the origin. Thus, if y is replaced by $-y$ and z by $-z$ in (8), the equation is unchanged, since both $g(y)$ and $F(y)$ are odd functions of y.

This is a very useful property, since it means that if a closed trajectory passes through the point $(0, z_0)$ it will also pass through the point $(0, -z_0)$; and conversely, if a trajectory passes through these two points, it must necessarily be closed. We shall use this fact to show that there must exist one and only one closed trajectory for equation (4) in the phase-plane and thus the equation has a unique periodic solution. This follows immediately from the observation that when $y = 0$, we have $z_0 = y'_0$, and hence a closed trajectory in one plane implies a closed trajectory in the other.

For this purpose we introduce the function

$$\lambda(y, z) = \frac{1}{2} z^2 + G(y), \tag{9}$$

which reduces to $z^2/2$ when $y = 0$. Thus, in order to establish the existence of a periodic solution of equation (4), it is merely necessary to exhibit a trajectory for which $\lambda(0, z_0) = \lambda_A$ equals $\lambda(0, -z_0) = \lambda_B$.

To achieve this, we write

$$d\lambda = z\, dz + g(y)\, dy = -\frac{g \cdot F}{z - F(y)}\, dy, \tag{10}$$

from which it follows that

$$\int_A^B d\lambda = \lambda_B - \lambda_A = -\int_A^B \frac{g \cdot F}{z - F}\, dy = \frac{1}{2}(z_1^2 - z_0^2) = \frac{1}{2}(y_1'^2 - y_0'^2). \tag{11}$$

Our problem thus reduces to showing that there exists a unique trajectory for which we have

$$I = -\int_A^B \frac{g \cdot F}{z-F} dy = 0. \tag{12}$$

The argument now proceeds geometrically and to understand it we refer to Figure 15, which shows a phase-trajectory: $y'=y'(y)$, a Liénard-trajectory, $z=z(y)$, and the function $F(y)$. Although these graphs have actually been constructed from the Van der Pol equation [equation (12), Section 2] for the case where $\epsilon=1$, $a=1$, that is, for $f(y)=y^2-1$, $g(y)=y$, the argument based upon them is perfectly general.

From the figure we see that I can be expressed as the sum of three integrals, the first (I_1), extending over the interval from A to A', the second (I_2), from A' to A'', and the third (I_3) from A'' to B. Since $F(y)$ is negative from $y=0$ to $y=y_0$, and thereafter increases monotonically, I_1 and I_3 are positive, but I_2 is negative.

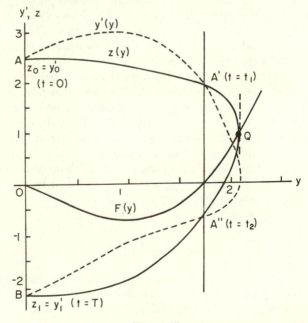

Figure 15

We now observe that if A is moved upward along the z-axis, the values of I_1 and I_3 will not increase. On the other hand, the absolute value of I_2 will increase, since the point Q will move upward along the curve of $F(y)$ and hence the integrand of (12) becomes greater since $F(y)$ is monotonically increasing.

It is thus evident from this geometrical argument that by adjusting the point A along the z-axis, a position will be reached where I is zero. Moreover this position is unique and thus there will exist one and only one periodic solution of the equation.

The situation thus described is readily seen if the variable of integration in (12) is changed to t. Since the origin of t can be arbitrarily chosen, we let $t=0$ at A. If the values of t at A', A'', and B are denoted respectively by t_1, t_2, and T, then the integral I can be written:

$$I=-\int_0^T gF\,dt=I_1+I_2+I_3, \tag{13}$$

where the limits of I_1, I_2, I_3 are respectively $(0, t_1)$, $(t_1,t_2,)$, (t_2,T).

Returning to the Van der Pol equation, we compute values of the integrand as a function of T, corresponding to an initial value of $z_0=2.5$. These are shown graphically in Figure 16, from which one can readily ascertain that the value of $-I_2$ exceeds the sum of I_1 and I_3, and thus the value of I is negative. From this we conclude that the initial choice of z_0 was too great, and when it is reduced to 2.18, we find that $I=0$.

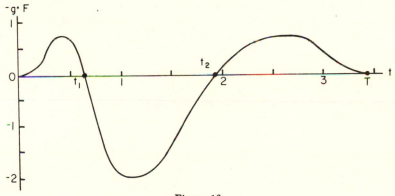

Figure 16

As we have already said above, the more general equation (5) was the object of study by Levinson and Smith. We shall state their principal theorems, but shall not give proofs of them. The first theorem, which abandons the condition of uniqueness, is as follows:

THEOREM B. *The functions* f(y,v) *and* g(y) *are continuous and integrable and satisfy the following conditions:*

(a) y g(y)$>$0 *for* |y|$>$0. *Moreover* G(y) *tends toward infinity as* y$\rightarrow\pm\infty$.

(b) $f(0,0)$ *is negative. There exists a positive value of* y, *namely,* y = y_0, *such that* $f(y,v) \geqq 0$ *for* $|y| \geqq y_0$, *and a positive value* M, *such that* $f(y,v) \geqq -M$ *for* $|y| \leqq y_0$.

(c) *There exists a positive value* $y_1 y_0 >$, *such that*

$$\int_{v_0}^{v_1} f(y,v)\, dy \;\geqq\; 10 M y_0 \tag{14}$$

where v>0 *is an arbitrarily decreasing function of* y.

Under these conditions there exists at least one periodic solution of equation (5).

A theorem somewhat more general than that of Liénard, but less general than Theorem B, restores the uniqueness of the period solution. This is achieved by means of the function $\lambda(y,v)$, defined by (9) above, in which z has been replaced by v. This theorem is as follows:

THEOREM C. *Let* R_1 *denote the region in the* $(y,v) = plane$ *where* $f(y,v)$ *is negative and* R_2 *the region where* $f(y,v)$ *is positive. Furthermore, let* $R_1(c)$ *denote that part of the curve* $\lambda(y,v) = c$ *which lies in* R_1 *and let* $R_2(c)$ *denote that part which lies in* R_2.

To the requirements of Theorem B, the following condition is added:
For every value of c *the minimum value of*

$$F(y,v) = \frac{1}{v^2} + \frac{1}{v f(y,v)} \frac{\partial f(y,v)}{\partial v} \tag{15}$$

on $R_2(c)$ *is positive and exceeds the maximum of* $F(y,v)$ *on* $R_1(c)$.

Under these conditions equation (5) *has a periodic solution, which is unique to within a simple translation of the variable* t.

Chapter 11

Nonlinear Mechanics

1. Introduction

THE TERM *nonlinear mechanics* has been applied in recent years to a series of investigations in the field of nonlinear differential equations, which have had their origin for the most part in applications to physical phenomena. The independent variable is time. The systems considered are reducible in general to ordinary differential equations of second order. Investigations subsumed under the generic title of nonlinear mechanics are concerned principally with phase spaces, with expressions which represent energy terms, with phenomena included under the subject of relaxation oscillations, stability and instability points, and with certain loci called limit cycles.

The literature of the subject is now very extensive. Originating in the classical researches of Poincaré, Bendixson, Liapounoff, and others around the beginning of the 20th century, the ideas remained for a time relatively fallow. But the pressure of problems arising in various technical processes finally turned the attention of scientists to the subject and to a resurvey of the methods contained in the original memoirs. We have already given some of the historical details earlier in the book.

In the chapter we shall be concerned principally with a differential system of the following form:

$$\frac{dx}{dt} = P(x,y), \quad \frac{dy}{dt} = Q(x,y). \tag{1}$$

Unless otherwise specified, $P(x,y)$ and $Q(x,y)$ shall be functions analytic within a domain D of the x,y-plane. Thus, by the theory of limits, there will exist functions:

$$x = x(t), \quad y = y(t), \tag{2}$$

for a range T of the variable t, which satisfy equations (1).

Equations (2) define parametrically a curve in the x,y-plane. Since t appears only implicitly in the equations of the original system, the differential equation of this curve is obtained by dividing one equation in (1) by the other. We thus get the following equation of first order·

$$\frac{dy}{dx}=\frac{P(x,y)}{Q(x,y)}. \tag{3}$$

Assuming the existence of a solution of this equation throughout some domain D, let us denote it by the following function:

$$f(x,y)=0. \tag{4}$$

It is customary in nonlinear mechanics to refer to the x,y-plane as the *phase plane*, and any graphical representation of equation (4) as a *phase trajectory*.

It is readily proved that both $x(t)$ and $y(t)$ are the solutions of differential equations of second order. To see this, let us assume that the first equation in (1) has been solved explicitly for x in terms of y and $\dot{y}$, and the second equation has been solved for y in terms of x and $\dot{x}$, that is

$$x=\psi(y,\dot{y}), \quad y=\phi(x,\dot{x}), \tag{5}$$

where we now use the customary notation of mechanics: $dx/dt=\dot{x}$, $dy/dt=\dot{y}$.

Let us now differentiate the first equation in (1). We thus get

$$\ddot{y}=P_x(x,y)\dot{x}+P_y(x,y)\dot{y}, \tag{6}$$

where $P_x(x,y)$ and $P_y(x,y)$ denote the partial derivatives of $P(x,y)$. If we now replace $\dot{x}$ by $Q(x,y)$ from (1) and replace x by $\psi(y,\dot{y})$ from (5), we obtain the following differential equation of second order of which $y=y(t)$ is a solution:

$$\ddot{y}=P_y(\psi,y)\dot{y}+P_x(\psi,y)Q(\psi,y). \tag{7}$$

A similar procedure for the second equation in (1) leads to the following differential equation satisfied by $x=x(t)$:

$$\ddot{x}=Q_x(x,\phi)\dot{x}+Q_y(x,\phi)P(x,\phi). \tag{8}$$

As an example we shall derive the equation in y from the following system:

$$\dot{y}=-y+xy=P(x,y), \quad \dot{x}=x-xy=Q(x,y).$$

We thus compute:

$$x=1+\dot{y}/y=\psi(y.\dot{y}), P_x=y, \quad P_y=1-x=-\dot{y}/y, \quad Q(\psi,y)=(y+\dot{y})(1-y)/y.$$

When these values are substituted in (7), we obtain the following equation:

$$y\ddot{y}=\dot{y}^2+(y-y^2)\dot{y}+y^2-y^3. \tag{9}$$

2. A Preliminary Example

It will be instructive to investigate in some detail the following set of equations, which presents many of the problems encountered in more general systems:

$$\frac{dx}{dt}=Cx+Dy, \quad \frac{dy}{dt}=Ax+By. \tag{1}$$

The multipliers of x and y are constants.

If $dy/dt=\dot{y}$ is divided by $dx/dt=\dot{x}$, we obtain the differential equation of the phase trajectories as follows:

$$\frac{dy}{dx}=\frac{Ax+By}{Cx+Dy}, \quad BC-AD\neq 0. \tag{2}$$

This equation has been extensively discussed in Chapter 2, but it will now be important to examine the relationship of its solution to that of system (1).

For this purpose we now obtain the differential equations satisfied separately by $x(t)$ and $y(t)$, namely, those given by (7) and (8) of Section 1. For system (1) these equations are found to be identical. Thus we find that $x(t)$ is a solution of the linear equation:

$$\ddot{x}-(B+C)\dot{x}+(BC-AD)x=0. \tag{3}$$

Since y, as well as x, is also a solution of (3), we thus obtain the following explicit expressions for the two functions:

$$x=ae^{\lambda_1 t}+be^{\lambda_2 t}, \quad y=ce^{\lambda_1 t}+de^{\lambda_2 t}, \tag{4}$$

where the multipliers are arbitrary constants and λ_1 and λ_2 are roots of the quadratic equation:

$$\begin{vmatrix} C-\lambda & A \\ D & B-\lambda \end{vmatrix} \equiv \lambda^2-(B+C)\lambda+BC-AD=0. \tag{5}$$

Equation (5), called the *characteristic equation* of system (1), has already been encountered in Chapter 2. It now assumes a major position in the definition of the categories into which the solution of equation (2) can be placed.

If equations (4) are now solved for $\exp(\lambda_1 t)$ and $\exp(\lambda_2 t)$ in terms of x and y and logarithms taken of both sides of the solutions, we get

$$\lambda_1 t=\log k(dx-by), \quad \lambda_2 t=\log k(ay-cx), \quad k^{-1}=ad-bc=\Delta. \tag{6}$$

These equations define the phase trajectories in terms of the parameter t. If t is eliminated, then we obtain the equation of the trajectories in the following form:

$$(dx - by) = K(ay - cx)^{\lambda_1/\lambda_2}, \quad K = \Delta^{1-\lambda_1/\lambda_2}. \tag{7}$$

This equation is identical with the one obtained in Chapter 2 by the direct solution of equation (2). The apparent existence of four arbitrary constants is an illusion, since the general solution of (2) introduces only one arbitrary parameter. The constant K is a function of k and hence of the four parameters: a, b, c, and d, but it will not be important in our discussion at this point to indicate the explicit relationships between the parameters.

Let us now write equation (7) in the following form:

$$w = Kz^{\lambda_1/\lambda_2}, \tag{8}$$

where we abbreviate:

$$w = dx - by, \quad z = ay - cx. \tag{9}$$

If we assume that $ad - bc = 1$, then equations (9) define a rotation of axes. The w,z-axes will be orthogonal to the x,y-axes if we impose the further restrictions that $ab + cd = 0$, but this we shall not assume. It will be observed that both axes pass through the origin, which is a singular point of equation (2).

We shall now consider the character of the curve defined by (8) as it is related to the roots of the characteristic equation. We shall consider several special cases as follows:

CASE I. *Both roots real, unequal, and negative.* Referring to equation (4) we see that as t moves from $-\infty$ to $+\infty$, x and y decrease and approach zero as a limit. The motion thus described we shall call *stable*. Equation (8) represents a family of curves with a common nodal point at the origin. At this point the curves are tangent to the z-axis, which means that the corresponding curves in the x,y-plane are tangent to the line: $ay = cx$.

An example, illustrating this case, is shown in Figure 1. The function $w = Kz^{4/3}$ is represented for various values of K. Since the characteristic equation is assumed to have the roots: $\lambda_1 = -4$ and $\lambda_2 = -3$, the motion is stable and the arrows indicate that the point (x,y) moves toward the origin as $t \to \infty$.

CASE II. *Both roots real, unequal, and positive.* In this case we see from equation (4) that x and y are zero when $t = -\infty$ and increase without limit as $t \to +\infty$. The motion is thus characterized as *unstable*. But as in Case I equation (8) represents a family of curves with a common *nodal point* at the origin.

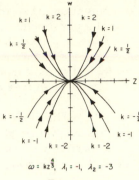

$$\omega = kz^{\frac{4}{3}}, \ \lambda_1 = -1, \ \lambda_2 = -3$$

FIGURE 1

Figure 1, which shows curves under Case I, can equally well be used to illustrate Case II. If we assume that $\lambda_1 = +4$ and $\lambda_2 = +3$, the same curves are obtained. But in this case the arrows will be reversed and the point (x,y) moves away from the origin as $t \to \infty$.

Another example is supplied by Figure 2, which shows two trajectories originating at the nodal point 0. Their equations are given by

$$y - 2x = K(y - x)^{4/3}, \tag{10}$$

where $K=1$, and -1 respectively. The differential equation is

$$\frac{dy}{dx} = \frac{2x + 2y}{5x - y}, \tag{11}$$

and the roots of the characteristic equation are 3 and 4. The motion is thus unstable, as is indicated by the direction of the arrows.

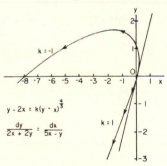

FIGURE 2

The equations which define the motion in the two cases are as follows:

$$K=1: \quad \begin{aligned} x &= -(e^{4t} + e^{3t}), \\ y &= -(e^{4t} + 2e^{3t}); \end{aligned} \qquad K=-1: \quad \begin{aligned} x &= e^{3t} - e^{4t}, \\ y &= 2e^{3t} - e^{4t} \end{aligned} \tag{12}$$

It will be seen from these equations that the two trajectories are tangent to the line $y=2x$ at the nodal point.

CASE III. *Both roots real, but differing in sign.* It is clear from (4) that in general both x and y will increase toward infinity as t approaches either $+\infty$ or $-\infty$. Hence the motion is *unstable*. Equation (8) now represents a family of curves of hyperbolic type for which w is infinite when $z=0$.

Figure 3 illustrates Case III. The curves are representations of the function $w=Kz^{-4/3}$ for various values of K. The roots of the characteristic equation are respectively -4 and 3 and the motion is unstable. The point 0 is called a *saddle point*.

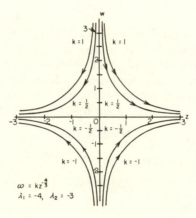

FIGURE 3

CASE IV. *The roots are conjugate complex numbers, the real part of which is positive, that is to say,* $\lambda_1=\lambda+\mu i$, $\lambda_2=\lambda-\mu i$, $\lambda>0$. Referring now to Section 5, Chapter 2, we see that equation (8) is replaced by the following:

$$\mu \log r=\lambda \arctan\frac{v}{u}+k, \qquad (13)$$

where we write: $u=Cx+Dy-\lambda x$, $v=\mu x$, $r^2=u^2+v^2$. Equation (13) thus represents a family of spirals. Equations (4) are now replaced by the following:

$$x=ae^{\lambda t}\cos(\mu t+b), \quad y=ce^{\lambda t}\cos(\mu t+d). \qquad (14)$$

The motion is observed to be *unstable* since the amplitudes increase toward $+\infty$ as $t\to+\infty$. In this case and the next the singular point is often called a *focus*.

Figure 4 illustrates this case. The curves shown in the figure for two values of K have the equation:

$$r = Ke^{p\theta}, \quad p = 1/2\pi, \tag{15}$$

where $\lambda_1 = p+i$, $\lambda_2 = p-i$. Since p is a positive number, the motion is unstable, and the spirals unwind from the origin.

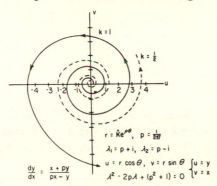

FIGURE 4

CASE V. *The roots are conjugate complex numbers the real part of which is negative.* Referring to Case IV, we see that the phase trajectories are spirals, but in this case the motion is *stable*, since $\lambda < 0$, and the amplitudes of the motion described by equations (14) diminish toward zero as $t \to +\infty$.

This case is also illustrated by Figure 4 and the curves defined by equation (15). If p is now a negative number, let us say, $p = -1/2\pi$, then the spirals wind in toward the origin as $t \to \infty$, and the motion is stable.

CASE VI. *The roots are pure imaginaries, that is to say,* $\lambda_1 = \mu i$, $\lambda_2 = -\mu i$. Referring to Section 5, Chapter 2, we see that the phase trajectories are a family of ellipses given by the equation:

$$Ax^2 + 2Bxy - Dy^2 = K, \quad -AD > B^2. \tag{16}$$

The motion is thus a stable one about the origin, which is now called a *vortex point*.

This case is illustrated by Figure 5, which represents graphically the ellipses:

$$x^2 - 2xy + 3y^2 = K, \tag{17}$$

derived as solutions of the equation:

$$\frac{dy}{dx} = \frac{x-y}{x-3y}.$$

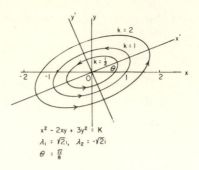

$x^2 - 2xy + 3y^2 = K$
$\lambda_1 = \sqrt{2}i, \quad \lambda_2 = -\sqrt{2}i$
$\theta = \frac{\pi}{8}$

FIGURE 5

The characteristic roots are $\pm\sqrt{2}\,i$ and the major axis makes an angle: $\theta=\pi/8$ with the x-axis. The origin is a vortex point.

CASE VII. *The roots are equal.* In this case equations (4) are replaced by the following:

$$x=e^{\lambda t}(a+bt), \quad y=e^{\lambda t}(c+dt). \tag{18}$$

The motion is thus seen to be *stable* if $\lambda=\frac{1}{2}(B+C)<0$ and *unstable* if $\lambda>0$. The phase trajectories may be straight lines, parabolas, or somewhat complicated logarithmic curves. These various cases are described in Section 5 of Chapter 2. Since the roots are equal, the constants in equation (2) must satisfy the condition:

$$(B-C)^2+4AD=0. \tag{19}$$

Figure 6 illustrates this case. The equation of the trajectories is the following:

$$y=x(K+\log x), \tag{20}$$

which is derived from equation (2) by assuming: $A=B$, $B=C$, $D=0$.

The two roots are equal to B. The equations defining the motion are respectively:

$$\frac{dx}{dt}=Ax, \quad \frac{d^2y}{dt^2}-2B\frac{dy}{dt}+B^2y=0. \tag{21}$$

If B is assumed to be negative, the motion is stable as shown in the figure. But if B is positive, the motion is unstable.

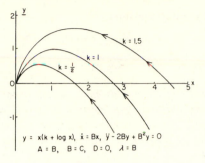

$$y = x(k + \log x), \quad \dot{x} = Bx, \quad \ddot{y} - 2By + B^2 y = 0$$
$$A = B, \quad B = C, \quad D = 0, \quad \lambda = B$$

FIGURE 6

3. The Stability Theorem

The significance of the preliminary example given in the preceding section becomes apparent when we consider what is called *the stability theorem* of nonlinear mechanics.

We shall consider the following system:

$$\frac{dy}{dt} = P(x,y), \quad \frac{dx}{dt} = Q(x,y), \tag{1}$$

where $P(x,y)$ and $Q(x,y)$ are functions analytic in a domain D of the x,y-plane.

We now make the transformation: $x = \xi + p$, $y = \eta + q$, and thus obtain

$$\frac{d\eta}{dt} = P(\xi + p, \ \eta + q) = P(p,q) + \xi P_x(p,q) + \eta P_y(p,q)$$

$$+ \frac{1}{2!} [\xi^2 P_{xx}(p,q) + 2\xi\eta P_{xy}(p,q) + \eta^2 P_{yy}(p,q)] + \ldots,$$

$$\frac{d\xi}{dt} = Q(\xi + p, \ \eta + q) = Q(p,q) + \xi Q_x(p,q) + \eta Q_y(p,q)$$

$$+ \frac{1}{2!} [\xi^2 Q_{xx}(p,q) + 2\xi\eta Q_{xy}(p,q) + \eta^2 Q_{yy}(p,q)] + \ldots. \tag{2}$$

The *singular points* of (1) are the intersections of the curves

$$P(p,q) = 0, \quad Q(p,q) = 0. \tag{3}$$

If one of these singular points is the point $P_0 = (p_0, q_0)$, then the *characteristic equation* of the system with respect to this point is the following:

$$\Delta(\lambda) \equiv \begin{vmatrix} C - \lambda & A \\ D & B - \lambda \end{vmatrix} = \lambda^2 - (B + C)\lambda + BC - AD = 0, \tag{4}$$

where we adopt the abbreviations:

$$A = P_x(p_0, q_0), \quad B = P_y(p_0, q_0), \quad C = Q_x(p_0, q_0), \quad D = Q_y(p_0, q_0).$$

The *stability theorem** then asserts:

In the neighborhood of a singular point P_0 of system (1), the stability characteristics of the solution are determined by the characteristic roots of equation (4) in the same sense as they characterize the solutions of the linear system as set forth in the various cases described in Section 2.

To prove this let us write equation (2) as follows:

$$\frac{d\eta}{dt} = A\xi + B\eta + \phi(\xi, \eta),$$

$$\frac{d\xi}{dt} = C\xi + D\eta + \psi(\xi, \eta). \tag{5}$$

We now introduce the transformation:

$$\eta = -Du - (C - \lambda)v,$$

$$\xi = -(B - \mu)u - Av, \tag{6}$$

where λ and μ are assumed to be distinct roots of $\Delta(\lambda) = 0$, defined by (4).

When this transformation is made on (5), we obtain:

$$D\dot{u} + (C - \lambda)\dot{v} = [CD + D(B - \mu)]u + [C(C - \lambda) + AD]v - \phi^*(u, v),$$

$$(B - \mu)\dot{u} + A\dot{v} = [AD + B(B - \mu)]u + [A(C - \lambda) + AB]v - \psi^*(u, v), \tag{7}$$

where $\phi^*(u, v)$ and $\psi^*(u, v)$ are respectively the functions $\phi(\xi, \eta)$ and $\psi(\xi, \eta)$ in the new variables and where $\dot{u}$ and $\dot{v}$ are derivatives with respect to t.

Eliminating $\dot{v}$, and making use of the relationships: $\lambda + \mu = B + C$, $\lambda\mu = BC - AD$, we obtain:

$$\Delta(\lambda, \mu)\dot{u} = \lambda\Delta(\lambda, \mu)u - A\Delta(\lambda)v - A\phi^* + (C - \lambda)\psi^*, \tag{8}$$

where we have

$$\Delta(\lambda, \mu) = AD - (C - \lambda)(B - \mu) = B\lambda + C\mu - 2\lambda\mu,$$

$$= \mu^2 + B(\lambda - \mu) - \lambda\mu = (\lambda - \mu)(B - \mu) = (\lambda - \mu)(\lambda - C). \tag{9}$$

***

*This is also known as the theorem of Liopounoff.

Since $\Delta(\lambda)=0$, we get from (8) and (9):

$$\dot{u}=\lambda u+\frac{1}{\lambda-\mu}\left[-\frac{A}{\lambda-C}\phi^*-\psi^*\right]. \tag{10}$$

By a similar analysis, in which we eliminate $\dot{u}$ from (7) and reduce the resulting equation, we obtain the following expression for $\dot{v}$:

$$\dot{v}=\mu v+\frac{1}{\lambda-\mu}\left[\phi^*-\frac{D}{\lambda-C}\psi^*\right]. \tag{11}$$

If we now multiply (10) by u and (11) by v and add the two equations, we obtain an equation which can be written conveniently as follows:

$$\frac{d\rho^2}{dt}=2(\lambda u^2+\mu v^2)+f(u,v)=F(u,v), \tag{12}$$

where $\rho^2=u^2+v^2$ and $f(u,v)$ is a function which vanishes together with its derivatives of first and second orders at the point $P_0=(0,0)$.

Let us now consider the surface,

$$z=F(u,v).$$

Computing the first and second derivatives of z at P_0, we have

$$z_u=z_v=0,\ z_{uu}=2\lambda,\ z_{vv}=2\mu,\ z_{uv}=0,$$

whence it follows that

$$\Delta=z_{uu}z_{vv}-(z_{uv})^2=4\lambda\mu. \tag{13}$$

We shall now consider five cases: I. *Both roots real and positive.* II. *Both roots real and negative.* III. *Both roots real, but differing in sign.* IV. *Both roots conjugate complex numbers.* V. *Both roots pure imaginaries.*

I. When both roots are real and positive we see from (13) that z has a minimum value at P_0. Let us now represent the point (u,v) in polar coordinates:

$$u=\rho\cos\theta,\ v=\rho\sin\theta, \tag{14}$$

where both ρ and θ are functions of t, and let us consider the motion thus defined in the neighborhood of the origin. It will be convenient to enclose this neighborhood within a circle C of radius ξ_0. The trajectory is schematically shown in Figure 7.

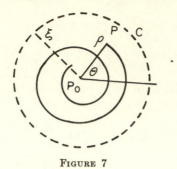

FIGURE 7

Since our interest is primarily in the stability of the motion, we shall consider only the radial velocity: $v_\rho = d\rho/dt$. When the transformation defined by (14) is made in (12), the following equation is obtained:

$$\frac{d\rho}{dt} = \lambda\rho(1 - k\sin^2\theta) + \rho g(\rho,\theta), \qquad (15)$$

where $k = 1 - \mu/\lambda$ and $g(\rho,\theta)$ vanishes at least to the first degree as $\rho \to 0$.

Integrating (15) from some initial point C, we now have

$$\int_{\rho_0}^{\rho}\frac{d\rho}{\rho} = \log\rho - \log\rho_0 = \int_{t_0}^{t}[\lambda(1 - k\sin^2\theta) + g(\rho,\theta)]\,dt. \qquad (16)$$

Since λ and μ have been assumed to be distinct roots, one is greater than the other. If we assume that λ is the larger value, and if both have the same sign, then $k < 1$. Since $g(\rho,\theta)$ vanishes at least to the first degree in ρ, the radius ξ of C can be so chosen that $g(\rho,\theta)$ remains arbitrarily small within the circle: From the theorem of mean value for integrals we then have

$$\log\rho - \log\rho_0 = [\lambda(1 - k\sin^2\theta_1) + \epsilon](t - t_0), \qquad (17)$$

where θ_1 is some value between 0 and 2π, and ϵ is arbitrarily small.

From this it is clear that when $\lambda > 0$, as was assumed, ρ increases as t increases and will ultimately reach the boundary of C. The motion is thus to be characterized as an unstable spiral about the singular point P_0.

II. When both roots are real and negative, then from (13) we see that z has a minimum value at P_0. Without changing the argument given in Case I, we arrive at equation (17). Now, however, we assume that $\lambda < 0$. As t increases, the right-hand member of (17) approaches $-\infty$, from which we conclude that $\rho \to 0$. Thus the motion is stable and P_0 is a stable nodal point.

III. When both roots are real, but differ in sign from one another, then at the point P_0 the surface $z = F(u,v)$ has a *saddle point*.*

In order to investigate the trajectory within the circle C of Figure 7, we now take the derivative of (12) from which we have

$$\frac{d^2\rho^2}{dt^2} = 4\left(\lambda u \frac{du}{dt} + \mu v \frac{dv}{dt}\right) + f_u(u,v)\frac{du}{dt} + f_v(u,v)\frac{dv}{dt}. \tag{18}$$

Replacing du/dt and dv/dt by their values from (10) and (11) respectively, we then get

$$\frac{d^2\rho^2}{dt^2} = 4(\lambda^2 u^2 + \mu^2 v^2) + G(u,v), \tag{19}$$

where $G(u,v)$ is a function whose first and second derivatives vanish at P_0.

Observing that

$$\frac{d^2\rho^2}{dt^2} = 2\rho\frac{d^2\rho}{dt^2} + 2\left(\frac{d\rho}{dt}\right)^2,$$

we can compute $d^2\rho/dt^2$ from (19) and (15). A simple calculation yields the following:

$$\frac{d^2\rho}{dt^2} = \rho\lambda^2[(1-k\sin^2\theta)^2 + 2k^2\sin^2\theta\cos^2\theta] + \rho\, h(\rho,\theta), \tag{20}$$

where $h(\rho,\theta)$ vanishes at least to the first degree as $\rho \to 0$.

If we now consider a point within the circle C of Figure 7 and observe that $h(\rho,\theta)$ can be made arbitrarily small by an appropriate choice of ξ, then $d^2\rho/dt^2$ is positive except at P_0 where it vanishes. By an argument similar to that already given in Case I, we can now show that $d\rho/dt$ is also positive. Hence the radial distance of P from P_0 will increase as t increases. Moreover, the rate of this increase will be positive. We thus see that the motion is unstable. In this case, P_0 is called a saddle point.

IV. If μ and λ are conjugate complex numbers, let us say $a+bi$ and $a-bi$ respectively, we transform equations (10) and (11) by writing

$$u = U+iV, \quad v = U-iV. \tag{21}$$

---

*If P_0 is a minimum point on the surface $z=F(u,v)$, the surface will lie above its tangent plane in the neighborhood of the point, that is to say, the surface is concave up. But if P_0 is a maximum point, then the surface will lie below its tangent plane and is concave down. In both cases, at the point, $z_u = z_v = 0$, and Δ, given by (13), is positive. But if $z_u = z_v = 0$, and Δ is negative, then the surface will lie partly above and partly below its tangent plane at P_0. It is thus neither concave up nor concave down, but *saddle shaped*. The point is thus referred to as a *saddle point*.

We thus get
$$\dot{U}+\dot{V}i=aU-bV+i(aV+bU)+\ \ldots,$$

$$\dot{U}-\dot{V}i=aU-bV-i(aV+bU)+\ \ldots. \tag{22}$$

When these equations are first added and then subtracted there results

$$\dot{U}=aU-bV+F_1(U,V), \quad \dot{V}=aV+bV+F_2(U,V), \tag{23}$$

where F_1 and F_2 are functions which vanish, together with their first derivatives, at P_0.

Forming the sum: $U\dot{U}+V\dot{V}$, and using the abbreviation: $R^2 = U^2+V^2$, we obtain the equation:

$$\frac{dR^2}{dt}=2aR^2+UF_1(U,V)+VF_2(U,V). \tag{24}$$

If we now make the transformation: $U=R$ cos ϕ, $V=R$ sin ϕ, equation (24) reduces to

$$\frac{dR}{dt}=aR+RH(R,\phi), \tag{25}$$

where $H(R,\phi)$ vanishes at least to the first degree as $R\to 0$.

The argument of Case I is now applicable and the stability of the solution is seen to depend upon the sign of a. When a is negative, the solution is stable and when a is positive, it is unstable. The singular point is called a *focus*.

V. If both λ and μ are pure imaginaries, then a in equation (25) is zero and we have

$$\frac{dR}{dt}=RH(R,\phi). \tag{26}$$

Since the right-hand member can be made arbitrarily small within the circle C of Figure 7 by proper choice of ξ, R will be nearly constant. Thus as t increases the trajectory remains within C and a vortex motion is defined. This, however, is very far from assuming that the solution will be periodic, as we shall show later. When the trajectories are closed the vortex point is often called a *center*.

4. An Application of the Stability Theorem

We shall now consider the system (1) of Section 3 in which $P(x,y)$ and $Q(x,y)$ are general quadratic functions. We thus write

$$\frac{dy}{dt}=P(x,y), \qquad \frac{dx}{dt}=Q(x,y), \tag{1}$$

where we have explicitly

$$P(x,y) = E + Ax + By + Lx^2 + Mxy + Ny^2,$$
$$Q(x,y) = F + Cx + Dy + Gx^2 + Hxy + Ky^2. \tag{2}$$

In these functions E, F, and the coefficients of the variable terms are assumed to be constants.

Although we have already introduced this system in Chapter 5, we shall examine it more critically here. By means of the transformation:

$$x = \xi + p, \quad y = \eta + q, \tag{3}$$

we obtain the following system:

$$\frac{d\eta}{dt} = E + Ap + Bq + Lp^2 + Mpq + Nq^2 + (A + 2Lp + Mq)\xi$$
$$+ (B + 2Np + Mq)\eta + L\xi^2 + M\xi\eta + N\eta^2,$$
$$\frac{d\xi}{dt} = F + Cp + Dq + Gp^2 + Hpq + Kq^2 + (C + 2Gp + Hq)\xi$$
$$+ (D + Hp + 2Kq)\eta + G\xi^2 + H\xi\eta + K\eta^2. \tag{4}$$

The singular points are given by the intersections of the conics

$$Lp^2 + Mpq + Nq^2 + Ap + Bq + E = 0,$$
$$Gp^2 + Hpq + Kq^2 + Cp + Dq + F = 0. \tag{5}$$

There are thus, in general, four singular points, but not all of these may be real. In certain degenerate cases, the number of points may be less than four. Let us denote the singular points by $P_i = (p_i, q_i)$, $i = 1, 2, 3, 4$.

In the discussion which follows we shall limit ourselves to the case where the singular points are real. From the stability theorem of the preceding section, the nature of the solution in the neighborhood of the singular points is determined by the roots of the equation:

$$\begin{vmatrix} C' - \lambda & A' \\ D' & B' - \lambda \end{vmatrix} = \lambda^2 - (B' + C')\lambda + B'C' - A'D' = 0, \tag{6}$$

where we write:

$$A' = A + 2Lp_i + Mq_i, \qquad B' = B + Mp_i + 2Nq_i, \tag{7}$$
$$C' = C + 2Gp_i + Hq_i, \qquad D' = D + Hp_i + 2Kq_i.$$

We shall now examine several special cases from the point of view of the stability theorem.

Example 1. We shall now consider Volterra's equation, the stable, periodic solution of which was given in Chapter 5. It will be sufficient for our purpose to consider the following special case:

$$\frac{dx}{dt}=2x-2xy, \quad \frac{dy}{dt}=-y+xy. \tag{8}$$

Equations (5) reduce to the degenerate conics,

$$2p-2pq=0, \quad -q+pq=0, \tag{9}$$

which, as shown in Figure 8, intersect in the two points: $P_1=(0,0)$ and $P_2=(1,1)$.

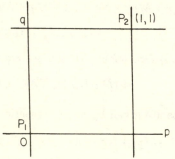

Figure 8

At P_1 equation (6) reduces to $(\lambda+1)(\lambda-2)$. Since both roots are real, but differ in sign, this singular point is an unstable saddle point.

At P_2 we have $A'=1$, $B'=C'=0$, $D'=-2$, and equation (6) becomes: $\lambda^2+2=0$. Since both roots are imaginary, this singular point is a vortex point.

The solution in the neighborhood of P_2 has been adequately discussed in Chapter 5. It is graphically represented in the first quadrant of Figure 9. We turn, therefore, to an examination of the solution about P_1.

It will be recalled from the previous treatment that the equation of the phase trajectories was found explicitly to be

$$\eta=C\xi, \tag{10}$$

where we wrote: $\eta=e^x/x$, $\xi=(ye^{-2})^2$, and C is an arbitrary constant. The function $\eta(x)$ has a minimum value of e at $x=1$ and $\xi(y)$ has a maximum value of e^{-2} at $y=1$. Let us denote these respectively by η_0 and ξ_0.

The functions $\eta(x)$ and $\xi(y)$ are represented graphically in Figure 10. From (a) and (b) of that figure we see that real positive values of x and y, that is to say, points in the first quadrant of Figure 9, will satisfy equation (10) if and only if $C \geqq \eta_0/\xi_0 = e^3 = 20.09$.

If, however, the point (x,y) is in the second quadrant of Figure 9 $(x<0, y>0)$, it is evident from (b) and (c) of Figure 10, that equation (10) will have no real solution unless $C<0$. It is also clear that in this case there will be two values of y corresponding to each value of x on each phase trajectory.

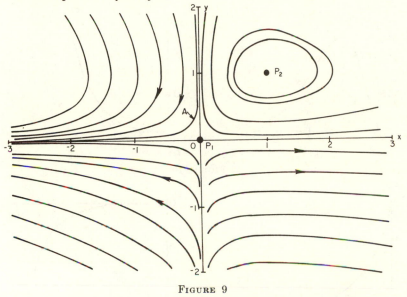

FIGURE 9

When the point (x,y) is in the third quadrant $(x<0, y<0)$, it is observed from (c) and (d) of Figure 10 that (10) will have no real solutions unless $C<0$. In this case there will correspond only one value of y to each value of x.

Finally, when (x,y) is in the fourth quadrant $(x>0, y<0)$, it is clear from (a) and (d) of Figure 10 that real solutions of (10) will exist only if $C>0$. It is also seen that for every value of x there will correspond a unique value of y.

The phase trajectories in each of the four quadrants are shown in Figure 9. The integral curves: $y=y(t)$, $x=x(t)$, and their first derivatives, which correspond to the phase trajectory marked A in Figure 9, are shown in Figure 11. The numerical computation of these curves is tedius, but can be made by the methods described earlier in Chapter 5. It has been found simpler to obtain the trajectories given in Figure 9 and the integral curves in Figure 11 by use of an analogue computer.

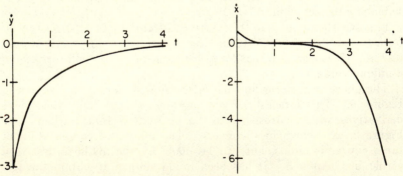

FIGURE 10

FIGURE 11

If we now examine Figure 9 as a whole, we observe that the phase curves are separated into four distinct families by the x- and y-axes. As we have said earlier in Chapter 10, such a separation is commonly observed in many nonlinear problems. It is convenient to refer to such a curve (or curves), which thus separate the families of solutions, as a *separatrix*. The equations of the separatrix are usually very difficult to determine, since they depend upon critical values of the arbitrary constant. In the present instance, the equations $x=0$ and $y=0$, which separate the families of solutions, correspond respectively to $C=\infty$ and $C=0$.

Example 2. We shall consider next the following system:

$$\frac{dx}{dt}=x+y+x^2+y^2, \quad \frac{dy}{dt}=x-y-x^2+y^2. \tag{11}$$

Equations (5) reduce to a circle and two intersecting straight lines, namely:

$$\left(p+\frac{1}{2}\right)^2+\left(q+\frac{1}{2}\right)^2=\frac{1}{2}, \quad (p-q)(p+q-1)=0. \tag{12}$$

Their intersections give two real points: $P_1=(0,0)$, $P_2=(-1,-1)$ and two complex points: $P_3=(-\omega,-\omega^2)$, $P_4=(-\omega^2,-\omega)$, where ω is a complex cube root of 1. Since our interest is in real trajectories, we shall not be concerned with P_3 and P_4.

At P_1 equation (6) reduces to $\lambda^2-2=0$, and since the roots are real, but differ in sign, the motion in the neighborhood of P_1 is unstable and of hyperbolic type.

At P_2 we have $A'=3$, $B'=-3$, $C'=-1$, $D'=-1$, from which we derive equation (6): $\lambda^2+4\lambda+6=0$. Since the roots, $\lambda=-2\pm\sqrt{2}i$, are conjugate imaginaries with real part negative, the motion in the neighborhood of P_2 is a stable spiral.

These conclusions are seen to be verified in Figure 12, which shows the phase trajectories in the neighborhood of the two singular points. These trajectories were obtained by the use of an analogue computer.

The *separatrix* in this case is of considerable interest. This curve appears to consist of three branches presumably emerging from the neighborhood of P_1, two of which separate the spirals from the hyperbolic curves, and one of which separates the unstable trajectories which emerge respectively from the second and third quadrants.

The equations of motion, that is, $x=x(t)$, $y=y(t)$, together with their first derivatives, are graphically represented in Figure 13. The curves denoted by (a) correspond to the phase trajectory marked (1) in Figure 12 and those denoted by (b) correspond to the phase trajectory marked (2). The origins of time in both cases are at points outside of the area shown in Figure 12.

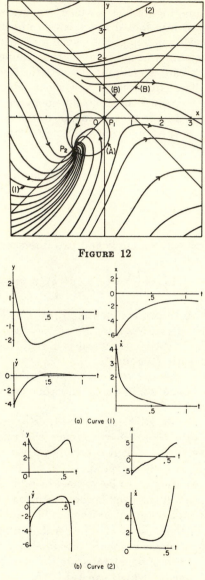

FIGURE 12

(a) Curve (1)

(b) Curve (2)

FIGURE 13

The graphical representations of the equations given in (12) are shown in Figure 12, where they are denoted respectively by (A) and (B). The circle (A) defines a locus at every point of which the slopes of the phase curves are infinite. Similarly, the two lines (B) define loci at each point of which the slopes of the phase curves are zero.

Exceptions, of course, are the singular points, P_1 and P_2, which lie on each locus.

Example 3. The following system

$$\frac{dx}{dt} = -5x + 6y + x^2 - 3xy + 2y^2, \quad \frac{dy}{dt} = -7x - 14y + 2x^2 - 5xy + 4y^2, \quad (13)$$

has the following four real singular points:

$$P_1 = (0,0),$$
$$P_2 = (-4,-6),$$
$$P_3 = (3\sqrt{2}-2, \sqrt{2}-3),$$
$$= (2.2426, -1.5858),$$
$$P_4 = (-3\sqrt{2}-2, -\sqrt{2}-3),$$
$$= (-6.2426, -4.4142).$$

These points are shown graphically in Figure 14, where they appear as the intersections of the ellipse (A), defined by the equation:

$$2x^2 - 5xy + 4y^2 - 7x + x14y = 0, \quad (14)$$

with the hyperbola (B), defined by

$$x^2 - 3xy + 2y^2 - 5x + 6y = 0. \quad (15)$$

The characteristic equations, computed from (6) and (7), which correspond to the four points, are the following:

At P_1: $\lambda^2 - 9\lambda - 28 = 0$; $\lambda = 4.5 \pm \frac{1}{2}\sqrt{193}$;

At P_2: $\lambda^2 + 9\lambda - 28 = 0$; $\lambda = 4.5 \pm \frac{1}{2}\sqrt{193}$;

At P_3: $\lambda^2 + 4\sqrt{2}\lambda + 28 = 0$; $\lambda = -2\sqrt{2} \pm 2\sqrt{21}i$;

At P_4: $\lambda^2 - 4\sqrt{2}\lambda + 28 = 0$; $\lambda = 2\sqrt{2} \pm 2\sqrt{21}i$. $\quad (16)$

An inspection of the cases enumerated in Section 2 shows us that the four singular points have the following characteristics, which determine the form of the phase trajectories in their neighborhoods:

P_1: An unstable saddle point, since the characteristic roots are real, but differ in sign. (Case III.)

P_2: An unstable saddle point, since the roots are real, but differ in sign. (Case III.)

P_3: A stable spiral point, since the roots are complex numbers the real part of which is negative.   (Case V.)

P_4: An unstable spiral point, since the roots are complex numbers the real part of which is positive.   (Case IV.)

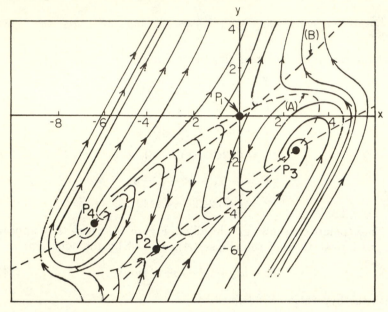

FIGURE 14

It will be seen from an inspection of Figure 14 that these characteristics are confirmed by the trajectories.   One also observes that the slopes of the trajectories are zero where they intersect the ellipse (A) and infinite where they intersect the hyperbola (B).

Example 4.   The following system:

$$\frac{dx}{dt}=-2x+3y+8x^2-5xy-4y^2, \quad \frac{dy}{dt}=-2x+2y+5x^2+xy-6y^2, \quad (17)$$

has two complex singular points and two real singular points: $P_1=(0,0)$ and $P_2=(1,1)$, which are the intersections of the hyperbola defined by the equation:

$$8x^2-5xy-4y^2-2x+3y=0, \qquad (18)$$

and the intersecting lines given by

$$5x^2+xy-6y^2-2x+2y=(5x+6y-2)(x-y)=0. \qquad (19)$$

Their graphs, shown in Figure 15, are denoted respectively by (A) and (B).

The characteristic equations, computed by (6) and (7), are found to be the following: At P_1: $\lambda^2+2=0$; at P_2: $\lambda^2+9=0$.

Since the characteristic roots are pure imaginaries in both cases, we have Case V of Section 3 and thus the points are vortex points. This conclusion is justified by the phase trajectories shown in Figure 15.

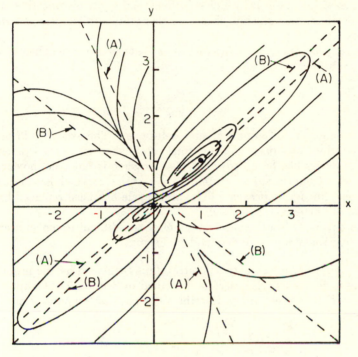

But we observe an interesting phenomenon. The vortices are not closed as they were in Example 1 and the motion thus described is not periodic. It is thus evident that other conditions must be imposed upon the coefficients of $P(x,y)$ and $Q(x,y)$ if periodicity is to be a property of the motion defined by equations (1).

5. Limit Cycles

Although the theory which we have described above is sufficient to determine the singular points of the system

$$\frac{dy}{dt}=P(x,y), \quad \frac{dx}{dt}=Q(x,y), \tag{1}$$

and to characterize the stability or instability of the solution in the neighborhood of these points, our information about the solution is still very far from complete.

A conspicuous property of nonlinear equations is the existence in the phase plane in certain cases of *limit cycles*. These important configurations we have already introduced, although somewhat superficially, in Chapter 12. They are stable closed curves, independent of initial conditions, toward which solutions tend in an asymptotic sense, or from which they unwind, as it were, as t tends toward plus or minus infinity.

Limit cycles are of much interest, because they are either themselves closed solutions of the equation

$$\frac{dy}{dx} = \frac{P(x,y)}{Q(x,y)}, \tag{2}$$

or are the asymptotic limits of such solutions. The existence of periodic solutions of equations (1) is thus established when the existence of limit cycles has been proved. The converse is not true, however, as one sees from the simple system: $x'=y$, $y'=-x$, which has periodic solutions, but no limit cycle. The limit cycle has been replaced by a system of vortex cycles instead.

The nature of limit cycles and some of the difficulties encountered in finding them can be understood by examples.

Example 1. An instructive illustration is furnished by the problem of pursuit in the circular case, as described in Section 9 of Chapter 5.

The differential equations describing the path of the pursuer are the following:

$$\frac{d\phi}{d\theta} = \frac{a}{\rho} \cos \phi - 1, \quad \frac{d\rho}{d\theta} = a \sin \phi - ka, \tag{3}$$

where the variables are those shown in Figure 16 and k is the ratio of the velocity of the pursuer to that of the pursued. If this ratio is less than 1, then the path of the pursuer is asymptotic to a circle of radius ka about the origin, independent of the origin of pursuit. This is illustrated by the two cases shown in Figure 16, where (a) the pursuer starts within the circle of the pursued at 0, and where (b) he starts outside the circle at 0'. The asymptotic limit of the two paths is a limit cycle.

That the limit cycle is actually itself a solution of the system can be shown readily. The differential equation satisfied by ρ [see (14), Section 9, Chapter 5] is the following:

$$\rho \frac{d^2\rho}{d\theta^2} + \rho\sqrt{\Delta} - \Delta = 0, \text{ where } \Delta = a^2 - \left[\left(\frac{d\rho}{d\theta}\right)^2 - ka\right]^2. \tag{4}$$

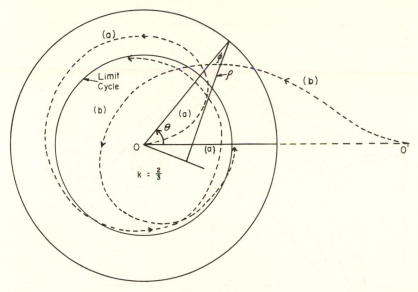

FIGURE 16

But if the circle of radius ka is the path of the pursuer, then ρ is a constant and equal to half of the length of the chord tangent to the circle, that is,

$$\rho = a\sqrt{1-k^2}.$$

This value of ρ is observed to be a solution of (4).

It is instructive to discuss the problem of pursuit from the point of view of this chapter. From equations (3), in which we set $a=1$, $k=2/3$, we obtain the singular points from the intersections of the curves:

$$\rho(\sin \phi - 2/3) = 0, \tag{5}$$

$$\rho - \cos \phi = 0. \tag{6}$$

The first curve (5) consists of the ϕ-axis and the lines,

$$\phi = \arcsin 2/3 = 0.7297 = \phi_0, \ \pi - \phi_0, -\pi - \phi_0, \text{ etc.}$$

The second curve is the graph of the cosine, which intersects (5) at the points P_0, P_1, P_2, and P_3 shown in Figure 17.

For the point $P_0 = (\pi/2, 0)$ the characteristic equation reduces to

$$(\lambda + 1)(\lambda - 1/3) = 0.$$

Hence this point is unstable and in its neighborhood the phase curves will be of hyperbolic type. This is also the case for the point P_3.

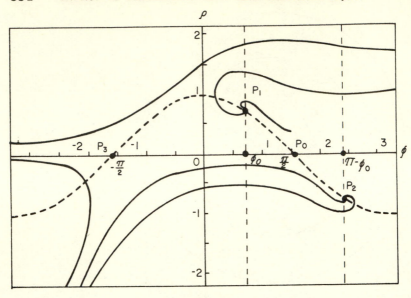

FIGURE 17

For the point $P_1 = (0.7297, 0.7454)$ the characteristic equation is

$$9\lambda^2 - 6\lambda + 5 = 0,$$

with the roots: $\lambda = -1/3 \pm (2/3)i$. Hence P_1 is a stable point and in its neighborhood the phase trajectories will be stable spirals. The point P_2 shares these characteristics with P_1.

A confirmation of this analysis is offered by the curves shown in the phase diagram (Figure 17).

Example 2. In his extensive treatise on the subject of nonlinear equations H. Poincaré gave a number of examples of limit cycles.* The following, somewhat modified with respect to notation, is typical of his analysis:

$$\frac{dy}{dt} = x - y + x(x^2 + y^2) + y(x^2 + y^2), \quad \frac{dx}{dt} = -x - y + x(x^2 + y^2) - y(x^2 + y^2).$$

$$(7)$$

It is clear that the origin is a singular point and that the solution is stable in its neighborhood, since the characteristic equation reduces to: $(1 + \lambda)^2 + 1 = 0$.

Introducing polar coordinates: $r^2 = x^2 + y^2$, $\theta = \arctan(y/x)$, we have upon differentiating:

$$r\,r' = x\,x' + y\,y', \quad r^2\theta' = x\,y' - y\,x'. \tag{8}$$

---

*See *Bibliography*, Poincaré (1). In particular, Chapter 7.

If the values of x' and y' as given by (7) are now substituted in (8), the system is seen to reduce to the following:

$$r'=r(r^2-1), \quad \theta'=r^2+1. \tag{9}$$

The solutions of these equations are readily found to be

$$r=(1+Ke^{2t})^{-1/2}, \quad \theta=t+t_0+\frac{1}{2}\log(1-r^2), \tag{10}$$

where K and t_0 are arbitrary constants. We shall assume that K is positive.

As $t\to\infty$, $r\to0$, thus confirming the stability of the singular point; and as $t\to-\infty$, $r\to1$, which shows that the unit circle is a limit cycle. We also observe that the limit cycle is a solution of the system in phase-space.

An examination of the second equation of (9) shows that as $t\to+\infty$, $\theta\to+\infty$, and as $t\to-\infty$, $\theta\to-\infty$. The phase trajectories are thus spirals.

A similar analysis, assuming that K is negative, shows that the trajectories are spirals exterior to the unit circle, which approach the circle as a limit cycle as $t\to-\infty$.

Example 3. As a final example, showing that a system may have more than one limit cycle, let us consider the following equations:

$$\frac{dy}{dt}=x+y\left[ac\sqrt{x^2+y^2}+(bc+ad)+\frac{bd}{\sqrt{x^2+y^2}}\right],$$

$$\frac{dx}{dt}=-y+x\left[ac\sqrt{x^2+y^2}+(bc+ad)+\frac{bd}{\sqrt{x^2+y^2}}\right], \tag{11}$$

where a, b, c, d are constants such that $D=ad-bc\neq0$.

Converting to polar coordinates and substituting in equations (8), we reduce (11) to the following system:

$$\frac{dr}{dt}=(ar+b)(cr+d), \quad \frac{d\theta}{dt}=1. \tag{12}$$

The solution of (12) is found to be

$$r=-\frac{(adKe^{Dt}-bc)}{ac(Ke^{Dt}-1)}, \quad \theta=t+t_0, \tag{13}$$

where K and t_0 are arbitrary constants.

For the special case: $a=b=-c=d=1$, $D=2$, we have

$$r=\frac{Ke^{2t}+1}{Ke^{2t}-1},\tag{14}$$

which represents a series of spiral curves that approach the unit circle both from outside and from within the circle. There thus exists a single limit cycle, namely, $r=1$.

If, however, we assume the values: $a=1$, $b=2$, $c=-1$, $d=1$, $D=3$, then equation (13) becomes

$$r=\frac{Ke^{3t}+2}{Ke^{3t}-1},\tag{15}$$

and we have a curve which approaches asymptotically both the circle $r=1$ and the circle $r=2$. There thus exist two limit cycles.

The graphs of equation (15) for the cases where K is respectively positive and negative have been represented schematically in Figure 18. It is thus seen that the limit cycle: $r=1$ is stable, but that the limit cycle: $r=2$ is unstable.

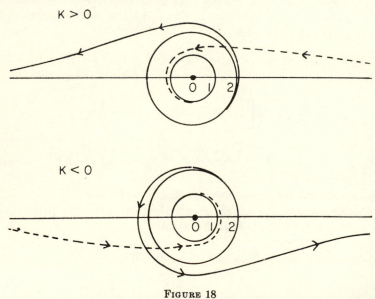

FIGURE 18

6. Some Further Comments About Limit Cycles

The existence of a limit cycle carries with it the existence of a set of solutions of the differential system, which, while not actually periodic themselves, are asymptotic to a periodic function. If the limit cycle is itself a solution of the system in phase-space, then there exists a periodic solution of the differential equations. The importance of establishing the existence of limit cycles is evident from these facts.

The manner in which one can make use of a limit cycle is readily illustrated by Example 3 of the preceding section. One solution of the original system of equations was obtained in the following form:

$$r(t) = \frac{Ke^{2t}+1}{Ke^{2t}-1}, \quad \theta = t + t_0, \tag{1}$$

where, for simplicity, we shall assume that K is positive.

The existence of the limit cycle, $r=1$, proves the existence of a set of almost periodic functions, which we shall assume can be written as follows:

$$x(t) = A(t) \cos (t+t_0), \quad y(t) = A(t) \sin (t+t_0), \tag{2}$$

where the amplitude, $A(t)$, is to be determined. We know, however, that its limiting value is unity.

If the functions $x(t)$ and $y(t)$ are substituted in the original equations, then $A(t)$ is found to satisfy the following equation:

$$\frac{dA}{dt} = 1 - A^2, \tag{3}$$

or, as we could have found more readily from (1),

$$A(t) = r(t). \tag{4}$$

We thus see that the solution of the original system consists of functions x and y, which are harmonic except for the damping factor $r(t)$. This factor approaches 1 as its limiting value.

This example, simple as it is, provides a guide to more general problems. It suggests that when the existence of a limit cycle has been established, the solution in the neighborhood of the limit cycle can be represented by functions of the form:

$$x = A(t)S(t), \quad y = A(t)C(t), \tag{5}$$

where $S(t)$ and $C(t)$ are harmonic functions with a common period Ω.

As an illustration of a somewhat more general situation than that of the example just given, let us consider the system of Example 2 of Section 5 for which the following solution was obtained:

$$r(t) = \frac{1}{\sqrt{1+Ke^{2t}}}, \quad \theta = t + t_0 + \frac{1}{2} \log (1-r^2). \tag{6}$$

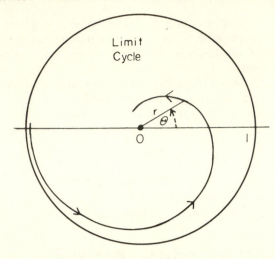

FIGURE 19

Let us now assume a solution of the form:

$$x=A(t) \cos (t+\phi), \quad y=A(t) \sin (t+\phi), \tag{7}$$

where the amplitude $A(t)$ and the phase $\phi=\phi(t)$ are functions of t to be determined.

When x and y are substituted in the original equations and proper simplifications made, A and ϕ will be found to be solutions respectively of the following equations:

$$\frac{dA}{dt}=-A+A^3, \quad \frac{d\phi}{dt}=A^2, \tag{8}$$

from which we get

$$A(t)=r(t), \quad \phi(t)=t_0+\frac{1}{2} \log (1-r^2). \tag{9}$$

If K is assumed to be positive, we see that $r(t)$ varies from 1 to 0 as t varies from $-\infty$ to $+\infty$. The motion is thus a damped harmonic motion with a variable phase. It is unstable with respect to the limit cycle, $r=1$, but is stable with respect to the singular point: $x=0$, $y=0$.

The phase trajectory is shown in Figure 19, where it has been assumed that $K=1$, $t_0=0$. The graphs of $x=x(t)$ and $y=y(t)$ are given in Figure 20, where t varies between $-\pi/2$ and $+\pi/2$.

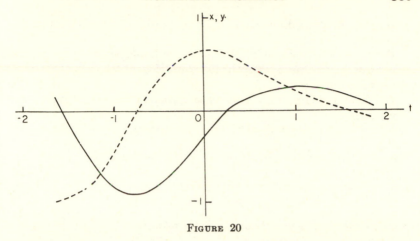

FIGURE 20

Although the examples which we have just given are elementary, they illustrate, nevertheless, one mode of approach to more complicated problems. When a limit cycle exists, the solutions in its neighborhood can often be approximated by the determination of asymptotic forms for the functions $A(t)$, $S(t)$, and $C(t)$ given in (5). S. Poisson was one of the pioneers in using such approximations and his methods have been generalized and extended by others, conspicuous among whom was H. Poincaré. Although an extensive account of this problem is beyond the scope of this book, we shall return to it in a more general example in Section 3 of Chapter 12.

7. Periodic Solutions—The Homogeneous Polynomial Case

In this and the next section we shall consider the question of the existence of periodic solutions for the following system:

$$\frac{dy}{dt}=P(x,y), \quad \frac{dx}{dt}=Q(x,y), \tag{1}$$

where $P(x,y)$ and $Q(x,y)$ are polynomials in x and y.

The system where P and Q are linear functions has been discussed in Section 2 and from the number of special cases exhibited there one can readily infer the complexities which appear when the degrees of the polynomials exceed 1. Although the stability theorem applies to all polynomial systems, it gives limited information with respect to the existence of periodic solutions in the neighborhood of the singular points. As we have seen, it is necessary that the singularity should be a vortex point, but this is not a sufficient condition except in the linear case. In fact, as we shall see in Section 8, the problem of finding sufficient conditions when the polynomials are of degree two is one of great difficulty and almost nothing is known for higher exponents.

But when the polynomials P and Q are homogeneous functions of x and y of degree $2n-1$, then the following elegant theorem (*Frommer's theorem*) can be proved:

If P(x,y) *and* Q(x,y) *are homogeneous polynomials of odd degree, and if we write*

$$p(u)=P(1,u), \quad q(u)=Q(1,u), \tag{2}$$

then a necessary and sufficient condition that the equation

$$\frac{dy}{dx}=\frac{P(x,y)}{Q(x,y)}, \tag{3}$$

have a closed cycle about the origin is the following:

$$I=\int_{-\infty}^{\infty}\frac{q}{p-uq}\,du=0. \tag{4}$$

The proof follows from the fact that the solution of (3), as we have seen in Section 4 of Chapter 2, can be written in the form

$$\log kx=\int\frac{q}{p-uq}\,du. \tag{5}$$

If x_1 and x_2 are successive intersections of an integral curve with the line $y=u_1x$, then one has

$$\log kx_2-\log kx_1=\int_{u_1}^{\infty}\frac{q}{p-uq}\,du+\int_{-\infty}^{u_1}\frac{q}{p-uq}\,du+\int_{-\infty}^{\infty}\frac{q}{p-uq}\,du,$$

that is,

$$\log\frac{x_2}{x_1}=2\int_{-\infty}^{\infty}\frac{q}{p-uq}\,du.$$

If the integral curve is to form a closed trajectory, then for every u^1 we must have $x_1=x_2$. It is clear that the necessary and sufficient condition for this is the vanishing of the integral.

Example 1. Let us first apply this theorem to the linear case where we have

$$P(x,y)=Ax+By, \quad Q(x,y)=Cx+Dy.$$

Since $p=A+Bu$, $q=C+Du$, we have

$$I=\int_{-\infty}^{\infty}\frac{C+Du}{X(u)}\,du,$$

where we abbreviate: $X(u)=A+(B-C)u-Du^2$.

We thus obtain

$$I = \frac{1}{2}(B+C) \int_{-\infty}^{\infty} \frac{du}{X(u)} - \frac{1}{2} \log X(u) \Big|_{-\infty}^{\infty}.$$

It is clear that the first term is zero if $B+C=0$, and that the second term will vanish if $X(u)$ has no zero in the finite interval. The condition for this is that $BC-AD=-B^2-AD>0$. Referring to Case VI of Section 2, we see that these are the conditions for the existence of an elliptical trajectory about the origin.

Example 2. We now consider the case of *Frommer's curve*, where $P(x,y)$ and $Q(x,y)$ are homogeneous cubical polynomials, that is,

$$P(x,y)=x^3+ax^2y+bxy^2+cy^3, \quad Q(x,y)=ax^3+bx^2y+cxy^2-y^3.$$

We now have the following integral:

$$I=\int_{-\infty}^{\infty} \frac{a+bu+cu^2-u^3}{1+u^4} du,$$

$$=(a-c)\frac{\sqrt{2}}{2}\log\left(\frac{u^2+\sqrt{2}u+1}{u^2-\sqrt{2}u+1}\right)-(a+c)\frac{\sqrt{2}}{4} \text{ arc tan } \frac{\sqrt{2}u}{u^2-1}$$

$$+\frac{b}{2} \text{ arc tan } u^2+\frac{1}{4}\log(u^4+1) \Big|_{-\infty}^{\infty}.$$

Since all terms except the second are zero, it is clear that the necessary and sufficient condition for a closed cycle is merely that

$$a+c=0. \tag{6}$$

The "shoemaker's last" shown in Figure 21 is the special case where $a=b=-c=1$, and the initial condition is $x_0=1/3$, $y_0=0$.

Since Frommer's theorem refers only to equations of odd degree, it is interesting to examine the case where $P(x,y)$ and $Q(x,y)$ are homogeneous polynomials of second degree and to enquire whether equation (3) can have closed trajectories. That this is, indeed, the case is seen from the fact that the conic:

$$ax^2+2bxy+cy^2-K(ax+by)=0, \tag{7}$$

where K is arbitrary, is a solution of the equation:

$$\frac{dy}{dx}=-\frac{a^2x^2+2ab xy+(2b^2-ac)y^2}{abx^2+(2ac-b^2)xy}. \tag{8}$$

When $ac-b^2>0$, the conic is an ellipse.

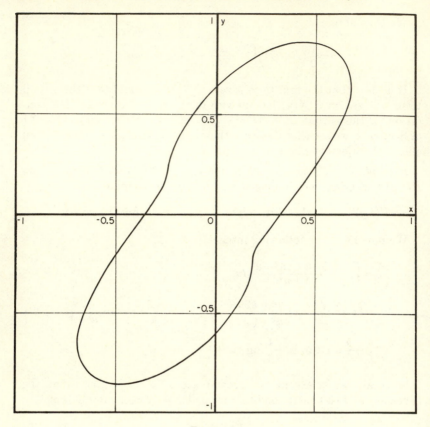

FIGURE 21

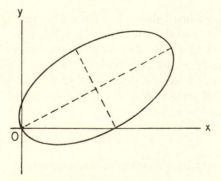

FIGURE 22

But comparing Figure 22 with Figure 21, we see that there is a big difference between the two cases. The trajectory obtained from the cubic is a true vortex cycle, since it includes the singular point in its interior; but in the quadratic case, the trajectory passes through the singular point.

The significance of this is seen if we consider equations (1), which in the first case will have periodic solutions, but in the second case will not, since an infinite time will be required for a point to trace a complete cycle.

This is illustrated by the following simple example:

$$\frac{dy}{dt}=x^2-y^2, \quad \frac{dx}{dt}=-2xy. \tag{9}$$

The phase trajectory is the circle: $x^2+y^2-Kx=0$. If one now eliminates x from these equations, he obtains the following equation in y:

$$y''+6yy'+4y^3=0. \tag{10}$$

This is recognized as a generalized Riccati equation, which, by the method described in Section 10, Chapter 3, is found to have the solution:

$$y=\frac{1}{2}\frac{w'}{w},$$

where $w=a+bt+ct^2$. Since $x'/x=-2y$, we also find that $x=1/w$.

It is thus evident that the origin is entered by x and y only as $t\to\infty$. Hence, the circle is not a closed trajectory in t-space.

8. Periodic Solutions—The General Quadratic Equation

We shall now consider the problem of establishing sufficient conditions for the existence of periodic solutions of the system:

$$\frac{dy}{dt}=P(x,y), \quad \frac{dx}{dt}=Q(x,y), \tag{1}$$

where $P(x,y)$ and $Q(x,y)$ are the following quadratics:

$$P(x,y)=Ax+By+Lx^2+Mxy+Ny^2,$$

$$Q(x,y)=Cx+Dy+Gx^2+Hxy+Ky^2. \tag{2}$$

This problem is one of considerable complexity and cannot be said to have reached final form at the present time. The first investigations appear to have been made by H. Dulac in 1908 and 4 years later by W. Kapteyn, but the problem languished and was not revived until 1934 when M. Frommer, generalizing the methods of Dulac, produced an extensive memoir which contained numerous examples of periodic motion for system (1) with specification of

sufficiency criteria. This system of equations was critically examined
in 1952 by N. N. Bautin and again in 1955 by I. G. Petrovskii and
E. M. Landis. A reexamination of the problem was recently undertaken by J. E. Faulkner and the author.

Some of the difficulties are readily understood from examples.
In the linear case, that is to say, when the quadratic terms are deleted
from (2), a closed cycle is obtained in the phase plane if and only
if the origin is a vortex point. But one readily sees that this criterion
is not sufficient for the general quadratic system by examining Example 4 of Section 4. Both of the real singular points are vortex
points, but about neither of them is there a closed trajectory in
the phase plane.

The following example, due to Faulkner, also shows another
aspect of the situation:

$$\frac{dy}{dt}=-10x-y+4x^2+2xy+4y^2, \quad \frac{dx}{dt}=6y-2y^2. \tag{3}$$

This system has two real singular points: $P_1=(0,0)$ and $P_2=$
$(2.5,0)$. About the first the phase trajectories are stable spirals,
but about the second they are hyperbolic and unstable as shown in
Figure 23. But the significant aspect of the example is found in the
fact that the separatrix (S) is the circle:

$$(x-\tfrac{1}{2})^2+y^2=1, \tag{4}$$

which satisfies the phase equation:

$$\frac{dy}{dx}=\frac{P(x,y)}{Q(x,y)}. \tag{5}$$

One observes that it is the limit cycle for both the interior and the
exterior spirals as $x \rightarrow -\infty$.

Another example, taken from Frommer, shows that it is possible
to have closed trajectories about two vortex points for the same
equation. The following system

$$\frac{dy}{dt}=x-2xy, \quad \frac{dx}{dt}=-y+x^2+y^2, \tag{6}$$

has four real singular points, namely,

$$P_1=(0,0), \quad P_2=(0,1), \quad P_3=\left(\frac{1}{2},\frac{1}{2}\right), \quad P_4=\left(-\frac{1}{2},\frac{1}{2}\right). \tag{7}$$

The first two points are readily shown to be vortex points and the
last two unstable saddle points. From the graphical representation
of x,y in the phase plane, as exhibited in Figure 24, it is clear that
the cycles are closed about the points P_1 and P_2.

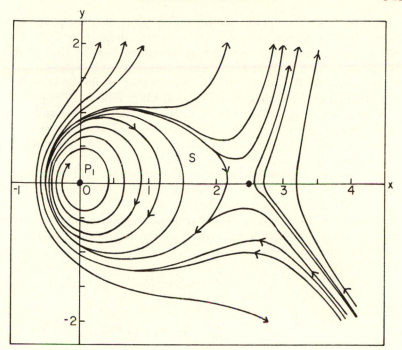

FIGURE 23

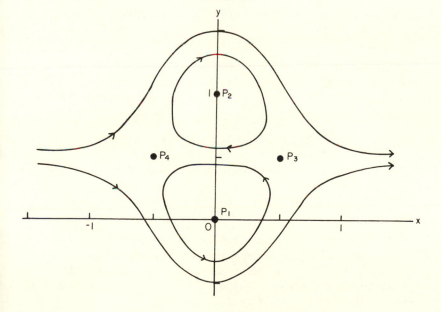

FIGURE 24

Several important questions are raised by these examples. How many limit cycles are possible for the system defined by (1)? How many vortex points can one set of equations have? Under what conditions will the trajectories about a vortex point be closed?

According to Bautin the maximum number of limit cycles is 3, but his proof of this statement is very difficult to follow. The problem was reexamined in 1955 by I. G. Petrovskii and E. M. Landis (see *Bibliography*), who reaffirmed the correctness of Bautin's theorum, but their analysis is also long and intricate. The first example of a system having one limit cycle was given by Frommer.

The second question is answered by the theorem that *at most only two of the singular points can be vortex points*. The following proof is due to Faulkner:

Referring to Section 4, we see that if the singular point $P=(p,q)$ is to be a vortex point, then the roots of the following characteristic equation must be pure imaginaries:

$$\begin{vmatrix} C'-\lambda & A' \\ D' & B'-\lambda \end{vmatrix} \equiv \lambda^2 - (B'+C')\lambda + B'C' - A'D' = 0, \tag{8}$$

where we write:

$$A' = A + 2Lp + Mq, \quad B' = B + Mp + 2Nq,$$
$$C' = C + 2Gp + Hq, \quad D' = D + Hp + 2Kq. \tag{9}$$

For the roots of equation (8) to be pure imaginaries we must have

$$B' + C' = 0, \quad B'C' - A'D' > 0. \tag{10}$$

We shall examine first the case where three of the critical points are collinear. Then $P(x,y)$ and $Q(x,y)$ are degenerate conics and will have a common linear factor so that we can write them as follows:

$$P(x,y) = (ax+by+c)(mx+ey+f),$$
$$Q(x,y) = (ax+by+c)(gx+hy+k). \tag{11}$$

We thus see that the equation in the phase plane reduces to the following linear case:

$$\frac{dy}{dx} = \frac{(gx+hy+k)}{(mx+ey+f)}, \tag{12}$$

which we know from Section 2 can have only one vortex point.

We now examine the nondegenerate case to see whether it can have more than the two vortex points which we have already exhibited in the example given above. The quantities p and q must satisfy the first equation in (10), which we shall now write

$$B + C + (M+2G)p + (2N+H)q = 0, \tag{13}$$

and also the definitive equations:

$$P(p,q)=0, \quad Q(p,q)=0. \tag{14}$$

The only way in which (13) and (14) can be satisfied by three sets of values of p and q in the nondegenerate case is for (13) to be identically zero, from which we have,

$$B+C=0, \quad M+2G=0, \ 2N+H=0. \tag{15}$$

Since any three noncollinear points can be transformed into any other three noncollinear points by a nonsingular linear transformation, we can take $(0,0)$, $(1,0)$, and $(0,1)$ as the three singular points. Substituting these in (14), we have

$$E=F=0, \ C+G=0, \ A+L=0, \ D+K=0, \ B+N=0. \tag{16}$$

It is now possible to write all the coefficients A through N in terms of three coefficients, namely, A, B, and D. These relations are

$$A=A, \ B=B, \ C=-B, \ D=D, \ E=0, \ F=0, \ G=B,$$
$$H=2B, \ K=-D, \ L=-A, \ M=-2B, \ N=-B. \tag{17}$$

When these are substituted in (9), there results

$$A'=A-2Ap-2Bq, \qquad B'=B-2Bp-2Bq,$$
$$C'=-B+2Bp+2Bq, \qquad D'=D+2Bp-2Dq. \tag{18}$$

The inequality in (10) now yields three inequalities corresponding to the three singular points as follows:

$$-AD-B^2>0, \tag{19}$$
$$AD+2AB-B^2>0, \tag{20}$$
$$AD-2BD-B^2>0. \tag{21}$$

Assuming that $B=0$, we get from (19) that $-AD>0$, and from (20) or (21), that $AD>0$. Thus B cannot be zero and we can write:

$$A=\alpha B, \ D=-\beta B. \tag{22}$$

Adding (19) and (20), we get

$$2(AB-B^2)=2B^2(\alpha-1)>0, \tag{23}$$

and consequently $\alpha>0$.

Adding (19) and (21), we have

$$2(-BD-B^2)=2B^2(\beta-1)>0, \tag{24}$$

and consequently $\beta>0$.

Substituting (22) in (19), (20), and (21), we obtain the following inequalities:

$$\alpha\beta-1>0, \quad -\alpha\beta+2\alpha-1>0, \quad -\alpha\beta+2\beta-1>0. \tag{25}$$

It has already been shown that for all three of these inequalities to be satisfied it is necessary for α and β to be positive. In this case the first inequality in (25) is satisfied only to the right of the curve

$$\alpha\beta-1=0. \tag{26}$$

The second inequality is satisfied only to the right of the curve

$$-\alpha\beta+2\alpha-1=0, \tag{27}$$

and the third inequality to the left of the curve

$$-\alpha\beta+2\beta-1=0. \tag{28}$$

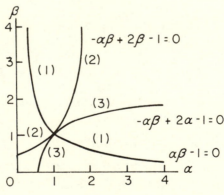

FIGURE 25

These curves are shown in Figure 25. Since the curves (28) and (29) are tangent at the point $\alpha=\beta=1$, it is not possible to satisfy the last two inequalities of (25) for positive values of α and β. But this is necessary if the first inequality of (25) is also to be satisfied. From this we reach the conclusion that there can be at most two vortex points.

We come finally to the third question asked above. Under what conditions will the trajectories about a vortex point be closed? In this case the vortex point is called a *center*. For simplicity we shall assume that the origin is such a point, an assumption that imposes no restriction, since any such point can be transferred to the origin by a linear transformation. Referring to equations (1) and (2), we see that the coefficients of the linear terms must satisfy the conditions:

$$B+C=0, \quad BC-AD>0, \quad \text{or} \quad -AD>C^2. \tag{29}$$

It is convenient first to express system (1) in canonical form. To achieve this we begin with the following transformation:

$$x=-\frac{1}{A}X-\frac{C}{kA}Y, \quad y=-\frac{1}{k}Y, \tag{30}$$

where $k^2=-AD-C^2$.

System (1) then assumes the following form:

$$\frac{dY}{dt}=kX+kp(X,Y), \quad p(X,Y)=L'X^2+H'XY+N'Y^2,$$

$$\frac{dX}{dt}=-kY+kq(X,Y), \quad q(X,Y)=G'X^2+M'XY+K'Y^2. \tag{31}$$

When we divide by k and write $\tau=kt$, we obtain the system

$$\frac{dY}{d\tau}=X+p(X,Y), \quad \frac{dX}{d\tau}=-Y+q(X,Y). \tag{32}$$

Finally we introduce the rotation

$$X=x'\cos\theta-y'\sin\theta, \quad Y=x'\sin\theta+y'\cos\theta, \tag{33}$$

which leaves the form of (32) unchanged, but provides a parameter θ. This is now determined so that

$$L'+N'=0. \tag{34}$$

Dropping the primes on x' and y', we now have the original system in the desired canonical form as follows:

$$\frac{dy}{d\tau}=x+p(x,y), \quad \frac{dx}{d\tau}=-y+q(x,y), \tag{35}$$

where we write

$$p(x,y)=ax^2+(2b+\alpha)xy-ay^2,$$

$$q(x,y)=-bx^2+(2a-\beta)xy-dy^2. \tag{36}$$

The phase equation thus becomes

$$\frac{dy}{dx}=-\frac{x+ax^2+(2b+\alpha)xy-ay^2}{y+bx^2-(2a-\beta)xy+dy^2}. \tag{37}$$

By arguments of considerable complexity it can be shown that *periodic solutions exist for system (35) and closed cycles in phase space for (37) provided the coefficients satisfy any one of the following four conditions:*

I. $b+d=0$.

II. $a=\beta=0$.

III. $\alpha=\beta=0$.

IV. $\beta=\alpha+5(b+d)=0, \quad a^2+bd+2d^2=0$.   (38)

Let us examine several special cases as follows:

Example 1. Volterra's system:

$$\frac{dy}{dt}=-cy+cxy, \quad \frac{dx}{dt}=ax-axy.$$ (39)

By the linear transformation: $x=x'+1, y=y'+1$, the vortex point $(1,1)$ is transformed to the origin. Dropping primes, we can write (39) as follows:

$$\frac{dy}{dt}=c(x+xy), \quad \frac{dx}{dt}=-a(y+xy).$$

Since $A=c, B=C=0, D=-a$, we have $k^2=ac$ and transformation (30) becomes:

$$x=-\frac{1}{c}X, \quad y=-\frac{1}{k}Y,$$

from which we obtain

$$\frac{dY}{d\tau}=X-\frac{1}{k}XY, \quad \frac{dX}{d\tau}=-Y+\frac{1}{c}XY, \quad \tau=kt.$$

It is clear that this system is included under Case I where $b+d=0$.

Example 2. Frommer's system:

$$\frac{dy}{dt}=x-2xy, \quad \frac{dx}{dt}=-y+x^2+y^2.$$ (40)

Since this is already in standard form we compare the right-hand members of the equations with (36) and thus obtain: $a=\alpha=\beta=0$, $b=d=-1$, from which we see that we have either Case II or Case III.

For the second vortex point, we make the successive transformations:

$$x=x', \quad y=y'+1; \quad x'=X, \quad y'=-Y,$$

which reduces the system to the same form as (40).

Example 3. As a final example we shall consider the system given in Example 4, Section 5, namely,

$$\frac{dy}{dt}=-2x+2y+5x^2+xy-6y^2, \quad \frac{dx}{dt}=-2x+3y+8x^2-5xy-4y^2. \quad (41)$$

It will be recalled that this system has two vortex points: $P_1=(0,0)$ and $P_2=(1,1)$, but that the phase trajectories about neither of them are closed. We shall now examine the point at the origin to see why this is so.

For this purpose we first make the transformation (30), which in this case is explicitly the following:

$$x=\frac{1}{2}X-\frac{1}{\sqrt{2}}Y, \quad y=-\frac{1}{\sqrt{2}}Y. \quad (42)$$

System (41) then assumes the following form:

$$\frac{dY}{d\tau}=X-\frac{5}{4}X^2+\frac{11}{4}\sqrt{2}XY,$$

$$\frac{dX}{d\tau}=-Y+\frac{3}{4}\sqrt{2}\ X^2-\frac{1}{2}\sqrt{2}\ Y^2, \tau=\sqrt{2}t. \quad (43)$$

We now apply the rotation (33) to this system and in order to satisfy condition (34), we find the following values for sin θ and cos θ:

$$\sin\theta=-\frac{5}{27}\sqrt{27}, \cos\theta=\frac{1}{27}\sqrt{54}.$$

In terms of these values the parameters of the canonical polynomials (36) are found to be:

$$a=-680\gamma,\ b=512\gamma,\ d=868\gamma,\ \alpha=-1130\gamma,\ \beta=-(1615\sqrt{2}+1360)\gamma,$$

where we use the abbreviation: $\gamma=\sqrt{27}/2916$.

Since none of the criteria given in (38) is satisfied by these quantities, it is now clear why the trajectories in the neighborhood of the origin were not closed curves.

9. Topological Considerations—Poincaré's Index—Bendixson's Theorem

In the earlier history of nonlinear mechanics, before the advent of the great computing devices of the present time, the study of nonlinear systems was more frequently advanced by topological arguments than by the quantitative methods of analysis and the use of numerical integration. But these theorems still retain their usefulness in many cases and often serve as guides to more exact determinations of the motions defined by the systems of differential equations. We shall now give a brief introduction to certain useful ideas advanced by Poincaré and to a related theorem of Bendixson.*

In order to describe the nature of a singular point Poincaré introduced what he called an *index*. In order to understand this term, let us consider a set of trajectories defined by the following system:

$$\frac{dy}{dt}=P(x,y), \quad \frac{dx}{dt}=Q(x,y). \tag{1}$$

As we have seen from the examples given earlier in this chapter, these trajectories may form closed cycles, they may unwind from a singular point and ultimately approach infinity in the form of an unstable spiral, they may approach a limit cycle, or behave in any of the other ways which we have previously described. It has been found useful in qualitative descriptions of these trajectories to divide them into semitrajectories by designating an arbitrary point P_0 on each of them. As t advances from $-\infty$, a variable point on one of them will ultimately reach P_0, and as t continues to $+\infty$ the second half of the trajectory is then described. The nature of the motion, whether it is periodic, stable, or unstable, is then determined by the behavior of the moving point in each half of the trajectory.

This general picture can now be made more precise by attaching to each point a vector which shows both the direction of the motion and its magnitude. The topological arguments of Poincaré apply to the direction of the vector, rather than to the magnitude, since the latter is determined by quantitative measurement. But the direction of the field yields qualitative information, which depends upon the functions $P(x,y)$ and $Q(x,y)$ in (1), as we shall soon see, and thus does not require a knowledge of the integrals of the system.

We shall find it necessary to distinguish between the *index of a closed curve* and the *index of a point*, although both of these are related. Let us first consider a vector field, which contains within its boundary a single singular point S and a simple closed curve C that does not pass through S. At each intersection of C with a trajectory (it may actually itself be a trajectory), there will be a direction angle of the vector field, which we shall denote by ϕ. If a point now moves in a counterclockwise direction along C, the angle ϕ will vary and after the completion of a circuit, ϕ will have the value $2\pi I$, where I is an integer, since the direction angle of the field has returned to its initial value. The quantity I is the index of the curve C.

---

*For these and other topological arguments the reader is referred to the extensive work of S. Lefschetz (*see Bibliography*), in particular *Differential Equations—Geometric Theory*, New York, 1957.

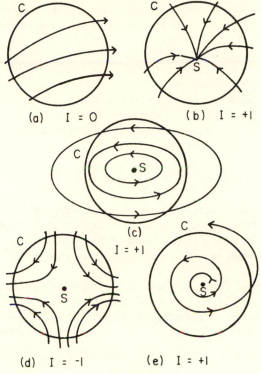

(a)   I = 0

(b)   I = +I

(c)   I = +I

(d)   I = -I

(e)   I = +I

FIGURE 26

The situation is shown graphically in Figure 26 for these singular points, which we have discussed earlier in this chapter. If there is no singular point in the interior of C, as in (a) of the figure, then the net variation in ϕ is zero and we have $I=0$. But if S is a nodal point, as in (b), then ϕ changes by 2π and the index equals $+1$. This is also true for the vortex center (c) and for the spiral point (e). But in the case of the saddle point, shown in (d), ϕ changes by -2π and we have $I=-1$.

The following properties of indexes can be proved without great difficulty, although they are not obvious:

1. The index of a closed curve which contains several singularities (of the nodal, focal, or saddle point type) is the algebraic sum of their indexes.

A simple example is furnished by the phase diagram of Volterra's problem shown in Figure 9. One can readily show that if a closed curve encloses the points P_1 and P_2, then its index is 0, since P_1 is a saddle point and P_2 a center.

2. The index of a closed trajectory is $+ 1$; and, conversely, if C is a closed trajectory, then it must contain at least one singular point the index of which is $+1$.

3. A closed trajectory must always contain $2n+1$ singular points, and the number of singular points of index 1 must exceed by one the number of saddle points.

Since the index is thus seen to be determined without knowing the solution of system (1), it should be possible to find it analytically from the functions $P(x,y)$ and $Q(x,y)$. That this is, indeed, the case is found in the following formula:*

$$I=\frac{1}{2\pi}\int_C d\arctan\frac{P}{Q}=\frac{1}{2\pi}\int_C\frac{P\,dQ-Q\,dP}{P^2+Q^2}, \tag{2}$$

where the path of integration is the closed curve C given above.

If C is chosen sufficiently small so that only the singular point S is enclosed by it, then I is the *index of the singular point*. But if C encloses more than one such point, then we call I the *index of the curve*.

The theorem of Bendixson provides a criterion by means of which, in certain regions, one can show that no closed trajectory exists and that system (1) has no periodic solution.

This criterion depends upon the following identity of Green:

$$\int_C Q\,dy-P\,dx=\int\int_A\left(\frac{\partial Q}{\partial x}+\frac{\partial P}{\partial y}\right)dx\,dy,$$

where C is a simply connected curve in the x,y-plane, which encloses the area A.

If C is a closed trajectory, then the line integral is zero since we have

$$Q\,dy-P\,dx=x'\,dy-y'\,dx=(x'y'-y'x')dt=0.$$

Hence the area integral also vanishes and we thus derive the theorem *that the function*

$$J=\frac{\partial Q}{\partial x}+\frac{\partial P}{\partial y}$$

must either be zero or change sign in A if C is a closed trajectory.

---

*For a systematic discussion of this integral the reader is referred to E. Picard: *Traité d'Analyse*, Paris, 1891, Vol. 1, Chap. 3, Sec. 4, where the theorem is proved that *the integral I, taken along C in the positive sense, is equal to the excess of the number of roots of the system*

$$P(x,y)=0, \quad Q(x,y)=0,$$

for which the functional determinant $(P_xQ_y-P_yQ_x)$ *is positive, over the number of roots for which this determinant is negative.*

If one now applies this theorem to the system: $P=Ax+By$, $Q=Cx+Dy$, he will find that $I=+1$, when $\Delta=BC-AD>0$ and $I=-1$, when $\Delta<0$. Referring to the characteristic equation (5), Section 2, we see that the first condition gives us either a nodal point or a focal point, and the second condition gives us a saddle point.

As an example, consider the system

$$y' = -y + xy, \ x' = x - xy,$$

from which we compute: $J = x - y$.

The, line, $x = y$, passes through the two singular points: (0,0) and (1,1) and divides the plane into two parts in neither of which it is possible to have a closed trajectory. Referring to Figure 9, we see that this conclusion is confirmed. However, this example also reveals the weakness of the theorem, since a closed trajectory does exist in the first quadrant, a fact not disclosed by the analysis just given. For this reason Bendixson's theorem is frequently referred to as a negative criterion.

Chapter 12

Some Particular Equations

1. Introduction

In this chapter we shall consider the solution of several particular equations, which have arisen naturally in the study of physical phenomena. Some of these have been mentioned in Chapter 1. As was the case in the solution of other equations which we have described in earlier pages, each requires special treatment.

By this time the reader has doubtless observed that the term "solved," when applied to any equation, has a varying degree of uncertainty about it. In the words of Poincaré, most problems are never actually solved, but only "more or less solved." In the case of differential equations the reduction of the solution to a function contained in the classical corpus of functions is usually considered a highly satisfactory achievement. In other cases the expression of a solution in the form of an infinite series, or in terms of some other convergent algorithm, will suffice. In practical applications, however, those who seek the solution of an equation will not be satisfied until the function has been reduced to tabular form. In this evaluation a specified order of approximation must be attained throughout a prescribed range of satisfactory size. Others who are interested primarily in the theoretical aspects of a solution will be content with an enumeration of the critical properties of the solving function, such, for example, as its zeros, its maxima and minima, its infinities, and other types of singularities. In other cases, where these desirable attainments cannot be achieved, recourse may be had to various approximations and to graphical methods. Several useful schemes have been devised to achieve this, the use of isoclines, for example, and the application of analogue computers where these machines are available.

In order to illustrate the various techniques by means of which equations may be "more or less solved," we shall examine several classical equations. The first of these is the equation of Van der Pol, to which reference has already been made in earlier pages. This equation was originally solved by the method of isoclines.

2. The Equation of Van der Pol

The equation

$$\frac{d^2y}{dt^2}-\epsilon(1-y^2)\frac{dy}{dt}+ay=0,\ \epsilon,\ a>0,\tag{1}$$

when it was first discussed by B. Van der Pol in 1926, probably attracted more attention because of the curious nature of its phase diagram than because of its explanation of the behavior of the triode oscillator. The equation in its phase plane provides an excellent example of a limit cycle, which is approached both from within and without by the phase trajectories.

Let us first write (1) in the form

$$\frac{dy}{dt}=x,\ \frac{dx}{dt}=\epsilon(1-y^2)x-ay=\epsilon x-ay-\epsilon xy^2.\tag{2}$$

If we make the transformation:

$$x=\xi+p,\ y=\eta+q,$$

then system (2) becomes

$$\dot{\eta}=p+\xi,$$

$$\dot{\xi}=\epsilon p-aq-\epsilon pq^2+\epsilon(1-q^2)\xi-(a+2\epsilon pq)\eta-\epsilon(\xi\eta^2+2q\xi\eta+p\eta^2).\tag{3}$$

Setting the constant terms equal to zero, that is, $p=0$, $\epsilon p-aq-\epsilon pq^2=0$, we see that system (2) has only one singular point in the finite plane, namely, the origin.

From the characteristic equation

$$\begin{vmatrix}\epsilon-\lambda & 1\\ -a & -\lambda\end{vmatrix}\equiv\lambda^2-\epsilon\lambda+a=0,\tag{4}$$

we have: $\lambda=\mu\pm\sqrt{\mu^2-a}$, where $\mu=\frac{1}{2}\epsilon$.

If $\mu^2<a$, then λ_1, λ_2 are conjugate complex numbers with positive real parts. Hence the origin is an *unstable focal point*. If $\mu^2>a$, the roots are unequal positive numbers and the origin is thus an *unstable nodal point*. This is also the case if $\mu^2=a$.

By the theorem of Liénard (Chapter 10, Section 8), equation (1) has a unique periodic solution, which has its image in the plase plane as a single closed configuration. This means that the phase trajectories emerging from the neighborhood of the origin must approach this curve as a limit cycle and that the solutions of (1) associated with them tend asymptotically to periodic functions.

If we examine the region exterior to the limit cycle, we see that the point at infinity is also a singular point of the equation. If we bring this point into the interior of the limit cycle by means of the transformation: $t=1/z$, equation (1) assumes the form:

$$z^4\frac{d^2y}{dz^2}+2z^3\frac{dy}{dz}+\epsilon(1-y^2)z^2\frac{dy}{dz}+ay=0. \tag{5}$$

If $\epsilon=0$ the solutions of (5) are $\exp(\pm i\sqrt{a}/z)$, which have essential singularities at the origin. This is a property shared by the solutions of the nonlinear equation when $\epsilon\neq0$. It is easy to show that the solution of (5) is unstable in the neighborhood of $z=0$ by the simple expedient of testing the solution of equation (1) for points exterior to the limit cycle. All phase trajectories will be found to approach the limit cycle.

Various methods have been used to solve equation (1), the first being that of isoclines which Van der Pol applied in his initial integration both to obtain the phase trajectories and the graphical representation of the solutions. Since from (2) we have

$$\frac{dx}{dy}=\frac{\epsilon(1-y^2)x-ay}{x}, \tag{6}$$

the isoclines are given by the following one-parameter cubic:

$$(m-\epsilon)x+ay+\epsilon xy^2=0. \tag{7}$$

Since equation (1) is readily integrated by an analogue computer, this method, when available, is superior to the more tedious application of the method of isoclines. Both methods are graphical, however, and the accuracy attained by them is thus necessarily limited. The construction of the integral of (1) from the phase trajectory for any specified set of initial conditions follows the method already illustrated in the integration of Volterra's equation given earlier in the book. (Section 4, Chapter 5.)

If one desires greater numerical accuracy, the method of continuous analytic continuation is available to him. The pertinent formulas for the Van der Pol equation have already been given in Section 9 of Chapter 9.

Various solutions of equation (1) are illustrated in the several figures of this section. Figure 1 shows the limit cycle for the case where $\epsilon=0.1$, together with phase trajectories, one originating inside the limit cycle and one originating outside of the cycle. In deriving these curves the constant a in (1) has been set equal to unity.

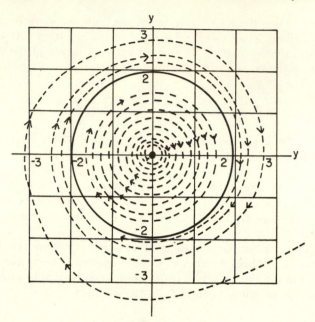

FIGURE 1

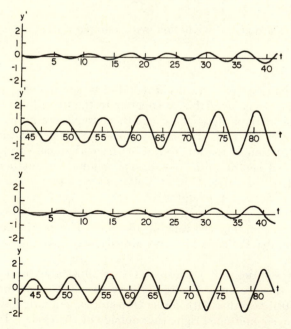

FIGURE 2

In Figure 2 the graphs of $x=y'(t)$ and $y=y(t)$ are shown, where the approach to the limit cycle is from an interior point. The slow growth of both of these functions is to be observed from the graphs. This, of course, is a consequence of the many loops of the spiral within the limit cycle.

In Figure 3, also for the case where $\epsilon=0.1$, the graphs of $x=y'(t)$ and $y=y(t)$ are shown, but now the origin is a point exterior to the limit cycle. The approach to the harmonic solution is observed to be much more rapid in this case than in the preceding one.

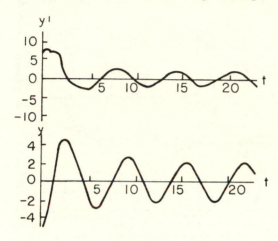

FIGURE 3

We observe that for a small value of ϵ the departure of the solution from the linear case, namely, where $\epsilon=0$, is not great. This is particularly to be noticed for the limit cycle, which is nearly circular. In the linear case, which was discussed at some length in Section 3 of Chapter 5, the limit cycle is replaced by a vortex cycle. But the vortex cycle cannot be derived simply from the limit cycle by merely letting $\epsilon\to0$, for, as will become clear in the next section, the breadth of the limit cycle is always equal to 4 whatever the value of ϵ. But the vortex cycle, corresponding to the linear equation derived from (1) by setting $\epsilon=0$, can have a diameter of any length.

As we increase ϵ, the shapes of both the limit cycle and the phase curves undergo considerable alteration. This is illustrated in Figure 4, which shows phase trajectories, both interior and exterior, together with the limit cycle, for the case where $\epsilon=1$. We note that the limit cycle has increased in length, but not in breadth. We also observe the approach of the phase curves to the lines A and A', which form separatrices for the phase curves.

The shapes of the derivative curve, $x = y'(t)$, and of the integral curve, $y = y(t)$ depart considerably from normal harmonic motion. This departure is graphically shown in Figure 5, where the approach to the limit cycle is from an exterior point.

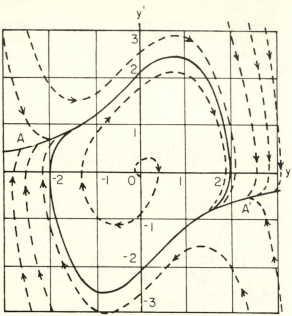

FIGURE 4

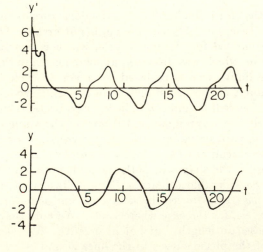

FIGURE 5

As ϵ is further increased the shape of the limit cycle changes rapidly. Its breadth remains equal to 4, but it gains in length and develops sharp corners at $y=2$ and $y=-2$. This is shown in Figure 6 ($\epsilon=5$), in which is exhibited one interior and one exterior phase trajectory.

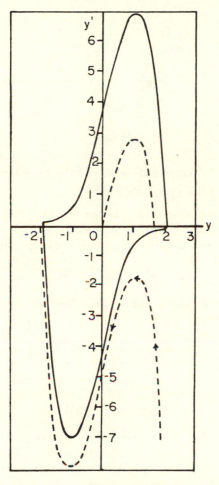

FIGURE 6

The integral tends to a more rectangular shape, while the derivative curve flattens out except for a series of long spikes near the corners of the graph of $y(t)$. These peculiarities are shown in Figure 7, where the initial point was taken interior to the limit cycle.

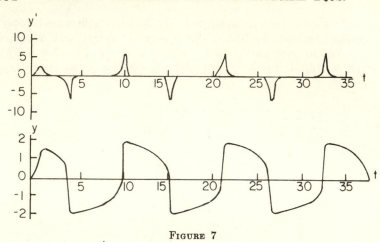

3. An Analytical Approximation to the Solution of the Van der Pol Equation

It will be observed that if in equation (1) of Section 2 we replace t by ωt, then determine ω from the equation $a\omega^2=1$, and finally replace $\omega\epsilon$ by ϵ, we can reduce the equation to the following simpler form:

$$y''-\epsilon(1-y^2)y'+y=0. \tag{1}$$

From the phase-diagrams which we have given in Section 2, we observe that if the initial values of the point (y,y') lie within the limit cycle, then the amplitude of y slowly increases to a limiting value of 2, which is independent of the parameter ϵ. Moreover, the period remains almost constant and equal to 2π. This suggests that the solution of (1) might be approximated by a function of the following form:

$$y=A(t)\sin[t+\phi(t)], \tag{2}$$

where $A(t)$ approaches 2 as t increases and $\phi(t)$ is a slowly varying function of t. This variation in both $A(t)$ and $\phi(t)$ is also a function of the parameter ϵ, since both reduce to constants as $\epsilon\to0$.

It will be convenient to consider a somewhat more general equation written as follows:

$$y''+\lambda^2y+\epsilon F(y,y')=0, \tag{3}$$

the solution of which will be assumed to be:

$$y=A\sin(\lambda t+\phi), \tag{4}$$

where both A and ϕ are functions of t as in (2).

It will be further assumed that, as a first approximation, the derivative of y can be written:

$$y' = A\lambda \cos (\lambda t + \phi). \tag{5}$$

The following analysis follows that given by N. Kryloff and N. Bogoliuboff in 1937 in their treatise on *Nonlinear Mechanics* (see *Bibliography*).

Let us first write for simplicity: $\theta = \lambda t + \phi$. We now differentiate (4) and thus obtain:

$$y' = A' \sin \theta + A\lambda \cos \theta + A\phi' \cos \theta, \tag{6}$$

which, by means of (5), reduces to the following:

$$A' \sin \theta + A\phi' \cos \theta = 0. \tag{7}$$

Differentiating (5), we get

$$y'' = A'\lambda \cos \theta - A\lambda^2 \sin \theta - A\lambda\phi' \sin \theta. \tag{8}$$

When this value of y'' is substituted in the original equation (3) and proper simplifications made, one obtains

$$A'\lambda \cos \theta - A\lambda\phi' \sin \theta = - \epsilon F(A \sin \theta, a\lambda \cos \theta). \tag{9}$$

Equations (7) and (9), being linear functions of A' and ϕ', can now be solved readily for these quantities. We thus get

$$A' = -\frac{\epsilon}{\lambda} F(A \sin \theta, A\lambda \cos \theta) \cos \theta,$$

$$\phi' = \frac{\epsilon}{A\lambda} F(A \sin \theta, A\lambda \cos \theta) \sin \theta. \tag{10}$$

In order to obtain a first approximation to A and ϕ, we now expand the functions in the right-hand members of (10) as Fourier series, that is, we write

$$F(A \sin \theta, A\lambda \cos \theta) \cos \theta = \frac{1}{2} A_0 + \sum_{n=1}^{\infty} [A_n \cos n\theta + B_n \sin n\theta],$$

$$F(A \sin \theta, A\lambda \cos \theta) \sin \theta = \frac{1}{2} A_0' + \sum_{n=1}^{\infty} [A_n' \cos n\theta + B' \sin n\theta], \tag{11}$$

where we have the customary integrals:

$$A_n = \frac{1}{\pi} \int_0^{2\pi} F(A \sin \theta, A\lambda \cos \theta) \cos \theta \cos n\theta \, d\theta,$$

$$B_n = \frac{1}{\pi} \int_0^{2\pi} F(A \sin \theta, A\lambda \cos \theta) \sin \theta \sin n\theta \, d\theta, \tag{12}$$

with similar integrals for A_n' and B_n'.

Let us now integrate A' between t and $t+T$, where T is a period of $\sin \theta$ and $\cos \theta$. We thus have

$$\int_t^{t+T} \frac{dA}{dt}\,dt = A(t+T) - A(t) = TA'(t+pT),\ \ 0 < p < 1. \tag{13}$$

Since we have assumed that $A(t)$ varies little over a cycle, the last term in (13) can be replaced approximately by $TA'(t)$. Moreover, since the integration of (11) between t and $t+T$ reduces to $\frac{1}{2}A_0 T$, we obtain the following approximate equation:

$$\frac{dA}{dt} = -\frac{\epsilon}{2\lambda} A_0. \tag{14}$$

By a similar analysis we also have for the approximation of ϕ:

$$\frac{d\phi}{dt} = \frac{\epsilon}{2A\lambda} A_0'. \tag{15}$$

If we now make application to the equation of Van der Pol, we write

$$F(y,y') = -(1-y^2)y' = y'y^2 - y',$$

from which we get

$$A_0 = \frac{1}{\pi} \int_0^{2\pi} (A^3 \sin^2 \theta \cos^2 \theta - A \cos^2 \theta)d\theta = \frac{1}{4} A^3 - A,$$

$$A_0' = \frac{1}{\pi} \int_0^{2\pi} (A^3 \sin^3 \theta - A \sin \theta \cos \theta)d\theta = 0. \tag{16}$$

Substituting the first of these values in (14), we obtain the equation

$$\frac{dA}{dt} = \frac{1}{2} \epsilon A \left(1 - \frac{1}{4} A^2\right). \tag{17}$$

This is an elementary form of Abel's equation [see (5), Section 9, Chapter 3], which has the solution:

$$A(t) = \frac{2ke^{\epsilon t/2}}{\sqrt{1+k^2 e^{\epsilon t}}}, \tag{18}$$

in which $k^2 = a_0^2/(4-a_0^2)$, where a_0 is the value of $A(t)$ when $t=0$.

From (15) and (16) we see that $\phi = \phi_0$, a constant. Hence the approximate solution of equation (1) has the following form:

$$y = A(t) \sin (t+\phi_0). \tag{19}$$

For small values of ϵ this interesting result preserves some of the observed features of the solution of equation (1). Thus $A(t) \to 2$ as $t \to \infty$, and the period is approximately equal to 2π. But as ϵ increases the departure is apparent in the derivative of (19):

$$y' = A'(t) \sin (t+\phi_0) + A(t) \cos (t+\phi_0). \tag{20}$$

As $t \to \infty$, $A(t) \to 2$ and consequently $A'(t) \to 0$, as we see from (17). Thus the amplitude of y' approaches 2, which is obviously far from the case as one sees from an inspection of the phase trajectories given above.

The process which we have just described is sometimes called the *equivalent linearization* of a nonlinear equation, since equation (3) has in effect been replaced by a linear equation. Thus, in equation (8), if we replace $\cos \theta$ by y'/A and $A \sin \theta$ by y, we obtain the following linear equation of second order:

$$y'' + \lambda \left(\frac{A'}{A}\right) y' + (\lambda^2 + \lambda\phi')y = 0, \tag{21}$$

where A' and ϕ' are given respectively by (14) and (15).

PROBLEMS

1. Apply the method just described to the equation:

$$\frac{d^2y}{dt^2} + y + ry^3 = 0, \tag{22}$$

and show that in this manner one obtains the first two terms in the expansion of ω^2 given by formula (35), Section 5, Chapter 10.*

2. By the method given in (D), Section 3, Chapter 7, establish the equivalence between the Van der Pol equation and the following Rayleigh equation:

$$\frac{d^2z}{dt^2} - \epsilon \left[1 - \frac{1}{3}\left(\frac{dz}{dt}\right)^2\right]\frac{dz}{dt} + z = 0. \tag{23}$$

Show that the periods of $y(t)$ and $z(t)$ are equal.

3. By a very elaborate argument it can be shown that the period, T, of the periodic solutions of (1) and (23) has the following asymptotic form:

$$T = A\epsilon + 2B\epsilon^{-1/3} - \frac{22}{9}\frac{\log \epsilon}{\epsilon} + C\epsilon^{-1} + O(\epsilon^{-4/3}), \tag{24}$$

where $A = 3 - \log 4 = 1.6138$, $2B = 7.0143$, $C = 0.0087$. From the values given in Table III of the *Appendix*, estimate the value of T for $\epsilon = 5$ and compare this value with the one obtained from (24).

---

*Kryloff and Bogoliuboff extended their method to higher approximations, although technical difficulties increased rapidly. When the method was applied to equation (22), they succeeded in obtaining approximations for the first three coefficients of the Fourier expansion of the solution. These coefficients were equivalent to Ca_0, Ca_1 to two terms, and Ca_3 to one term as given by formula (30), Section 5, Chapter 10. Their value of ω^2 was equivalent to three terms in expansion (35) of that section. With respect to this extension, they state: "Generally speaking, the higher approximations provide quantitative rather than new qualitative information. In view of this and of the difficulty of computing the higher approximation, it is usually quite sufficient to obtain the first approximation."

For a discussion of this problem consult the following (see *Bibliography*): Dorodnicyn, Haag (1) and (2), Cartwright (3), and Stoker, p. 141.

4. By an ingenious argument Cartwright has shown that the constant B in Problem 3 has the following value:

$$B = u(0) + \int_0^\infty \frac{dx}{u(x)}, \qquad (25)$$

where $u(x)$ is that solution of the equation

$$u \frac{du}{dx} = 2xu + 1 \qquad (26)$$

for which $u(x) \to 0$ as $x \to -\infty$. Show that if $u = x^2 + v$, then equation (26) is replaced by the Riccati: $dx/dv = x^2 + v$. Show also that if $x = -\phi'/\phi$, where $\phi' = d\phi/dv$, then ϕ is a solution of the equation:

$$\frac{d^2\phi}{dv^2} + v\phi = 0. \qquad (27)$$

5. Referring to Problem 4, we know that one solution of (27) is given by Airy's integral, that is, $\phi(v) = \mathrm{Ai}(-v)$, where

$$\mathrm{Ai}(x) = \frac{1}{\pi} \int_0^\infty \cos\left(\frac{1}{3}t^3 + xt\right) dt.$$

Given the following facts: (a) that

$$\lim_{x \to \infty} \mathrm{Ai}(x) = 0;$$

(b) that $d\,\mathrm{Ai}(x)/dx$ has a maximum value at $x = -1.01879$; and (c) that $\mathrm{Ai}(x)$ has a zero value at $x = -2.33811$, evaluate the constant B defined by (25).

4. Stellar Pulsation as a Limit-Cycle Phenomenon

In a paper bearing the title of this section W. S. Krogdahl, employing a generalization of the Van der Pol equation, has had considerable success in explaining the shape of velocity curves observed in the pulsation of variable stars of the Cepheid type.* Such a velocity curve is shown in Figure 8 for the Cepheid RT Aurigae as observed by J. C. Duncan.

We let $a = r_0$ be the mean or equilibrium radius of the star for some layer and $r = r(t)$ be the variable radius. For convenience the time t is replaced by a new variable τ, connected with t by the equation:

$$\tau = \left(\frac{4}{3} \pi G \rho_0\right)^{1/2} t, \qquad (1)$$

where G is the gravitational constant and ρ_0 is the mean density of the star in its equilibrium position.

Astrophysical Journal, Vol. 122, 1955, pp. 43–51.

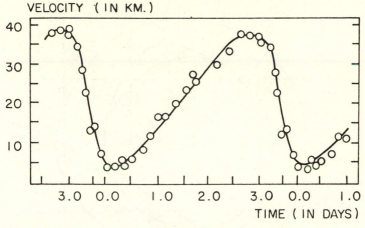

VELOCITY (IN KM.)

TIME (IN DAYS)

FIGURE 8

In normal gas spheres the pressure p is usually assumed to conform to the adiabatic relationship:

$$p = p_0 (\rho/\rho_0)^\gamma, \qquad (2)$$

where γ is an empirical constant, chosen in this case to have the value 5/3. But in the case of the Cepheids it is assumed that equation (2) should be augmented by another term, which is a function of both r and r', where the indicated derivative in r' is taken with respect to τ.

The variable radius is then written in the form

$$r(\tau) = a[1 + q(\tau)]^{1/3}, \qquad (3)$$

from which we have the variable velocity:

$$v(\tau) = r'(\tau) = \frac{1}{3} a q' / [1 + q(\tau)]^{2/3} = \frac{a^3}{3r^2} q'. \qquad (4)$$

By an argument which we shall not give here, but which involves assumptions concerning the term that is added to (2), and the replacement of $q(\tau)$ by $\lambda Q(\tau)$, where λ is an empirical constant, the author obtains the following equation for the determination of Q:

$$Q'' = -Q + \frac{2}{3} \lambda Q^2 - \frac{14}{27} \lambda^2 Q^3 + \mu(1 - Q^2) Q' + \frac{2}{3} \lambda(1 - \lambda Q) Q'^2, \qquad (5)$$

in which μ is a second empirical constant.

If $\lambda = 0$, this equation is seen to reduce to that of Van der Pol. One may assume, therefore, that if λ is not too large the equation will

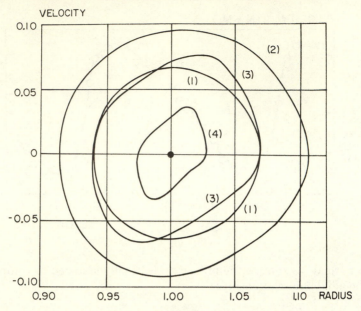

FIGURE 9.—Phase-plane limit cycle solutions of equation (5) in terms of $r(\tau)$ and $dr/d\tau$.

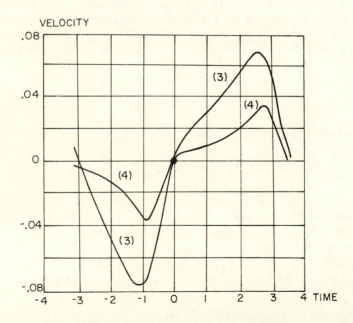

FIGURE 10.—Velocity curves from the solution of equation (5). The time is in units of τ and the velocity in units of $-dr/d\tau$.

have limit-cycle solutions, which are characterized by the parameters λ and μ. This is, indeed, the case, and the author has exhibited several of these cycles in the velocity-radius plane for the following values of the parameters:

(1) $\mu=0.01$, $\lambda=0.1$; (2) $\mu=0.1$, $\lambda=0.15$; (3) $\mu=0.5$, $\lambda=0.1$;

(4) $\mu=1.0$, $\lambda=0.04$; (5) $\mu=0.1$, $\lambda=0.1$.

The limit cycles in the first four cases are shown in Figure 9. From these diagrams the velocity curves have been constructed, those for (3) and (4) being shown in Figure 10. Comparing the general form of these curves with that of the Cepheid velocity curve shown in Figure 8, the general argument of the author seems to be confirmed.

5. Emden's Equation

Another classical nonlinear equation, which has been the object of much study and which will interest us here, is *Emden's equation*. This equation has the following form:

$$\frac{d^2y}{dx^2}+\frac{2}{x}\frac{dy}{dx}+y^n=0, \tag{1}$$

where n is a constant parameter. The boundary conditions, which are of most interest, are the following:

$$y=1, \; y'=0, \text{ when } x=0. \tag{2}$$

This equation was first studied by the German astrophysicist, Robert Emden, in his work on *Gas Kugeln* (1907), which considered the thermal behavior of a spherical cloud of gas acting under the mutual attraction of its molecules and subject to the classical laws of thermodynamics. Since the application of the equation to this problem throws light upon its solution, we shall review the theory from which it was derived.

Let us consider a spherical cloud of gas (Figure 11) and denote its hydrostatic pressure at a distance r from the center by P. Let $M(r)$ be the mass of the sphere of radius r, ϕ the gravitational potential of the gas, and g the acceleration of gravity.

From these we have the usual equations:

$$g=\frac{GM(r)}{r^2}=-\frac{d\phi}{dr}, \tag{3}$$

where G (the gravitational constant) $= 6.668 \times 10^{-8}$ cgs units.

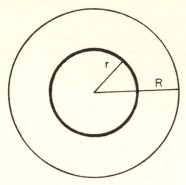

FIGURE 11

Three conditions are assumed for the determination of ϕ and P, namely,

(a) $dP = -g\rho dr = \rho d\phi$, where ρ is the density of the gas;   (4)

(b) $\nabla^2\phi = \dfrac{d^2\phi}{dr^2} + \dfrac{2}{r}\dfrac{d\phi}{dr} = -4\pi G\rho;$ (5)

(c) $P = K\rho^\gamma$, where γ and K are empirical constants.   (6)

From (4) and (6), together with the assumption that $\phi = 0$, when $\rho = 0$, that is, at the surface of the sphere, where $r = R$ as shown in Figure 11, we readily find that

$$\rho = L\phi^n, \text{ where } n = 1/(\gamma-1),\ L = \{(n+1)K\}^{-n}.\tag{7}$$

When this value of ρ is introduced into equation (5), we obtain

$$\nabla^2\phi = -a^2\phi^n,\tag{8}$$

where $a^2 = 4\ LG$.

If we now let $\phi = \phi_0 y$, where ϕ_0 is the value of ϕ at the center of the sphere, and let

$$r = x/[a\phi_0^{\frac{1}{2}(n-1)}],$$

then (8) reduces to Emden's equation as given in (1). The boundary conditions specified in (2) merely state that, at the center of the sphere, ϕ reduces to ϕ_0 and that $g = -d\phi/dr$ is zero there.

Since we have assumed that $\phi = 0$ at the surface of the sphere, when $y = 0$, and $\phi = \phi_0$ at the center of the sphere, when $y = 1$, it is clear that we are interested in those values of the solution of equation (1) between 0 and 1. We also observe that the radius R of the gas sphere and its total mass, $M = M(R)$, are given by the following equations:

$$R = (r)_{y=0},\quad GM = \left(-r^2\frac{d\phi}{dr}\right)_{y=0}\tag{9}$$

Let us now consider the solution of equation (1). General solutions corresponding to $n=0$ and $n=1$ are readily obtained and are found to be the following:

$$n=0: \quad y=a+\frac{b}{x}-\frac{x^2}{6}; \tag{10}$$

$$n=1: \quad y=a\,\frac{\sin x}{x}+b\,\frac{\cos x}{x}. \tag{11}$$

For $n=5$ a particular solution, involving one arbitrary constant, is found to be the following:

$$n=5: \quad y=[3a/(x^2+3a^2)]^{\frac{1}{2}}. \tag{12}$$

For the boundary conditions prescribed in (2) these three cases reduce respectively to the following functions:

$$n=0: \quad y=1-\frac{x^2}{6}; \quad n=1: \quad y=\frac{\sin x}{x}; \quad n=5: \quad y=\left(1+\frac{x^2}{3}\right)^{-\frac{1}{2}}. \tag{13}$$

It is clear from (7) that the first solution must be discarded since it corresponds to an infinite value of γ. Similarly, the third solution is of doubtful usefulness, since by (9) it corresponds to a sphere of infinite radius. Acceptable solutions, therefore, are those for which n lies between 0 and 5.

In the general case the solution of (1) contains two arbitrary constants, but one of these is what is called a "constant of homology." By this we mean that if $y=y(x)$ is a particular solution, then

$$y=A^\beta y(Ax), \quad \beta=2/(n-1), \quad n\neq 1, \tag{14}$$

is also a solution, where A is an arbitrary constant. But this fact does not help in finding the general solution, since (14) still contains only a single arbitrary constant.

In order to obtain a solution of equation (1) for values of n between 0 and 5 and over an adequate range of the variable x, recourse was had to a Taylor's expansion about $x=0$ and an analytic continuation of the series. The following solution, satisfying conditions (2), was obtained by J. R. Airey:[*]

$$y=1-\frac{x^2}{3!}+n\frac{x^4}{5!}+(5n-8n^2)\frac{x^6}{3\cdot 7!}+(70n-183n^2+122n^3)\frac{x^8}{9\cdot 9!}$$

$$+(3150n-1080n^2+12642n^3-5032n^4)\frac{x^{10}}{45\cdot 11!}+\cdots \tag{15}$$

---

[*]For this and other details, including tables, see *Mathematical Tables*, Vol. 2, British Association for the Advancement of Science, 1932.

A number of tables have been computed corresponding to values of n from 0 to 6, but the most important and accurate are those published by the British Association for the Advancement of Science in 1932. The graphical representation of five of these functions is shown in Figure 12.

Table I gives the values of x for which $y(x)=0$ and the corresponding values of $-y'(x)$ and $-x^2y'(x)$.

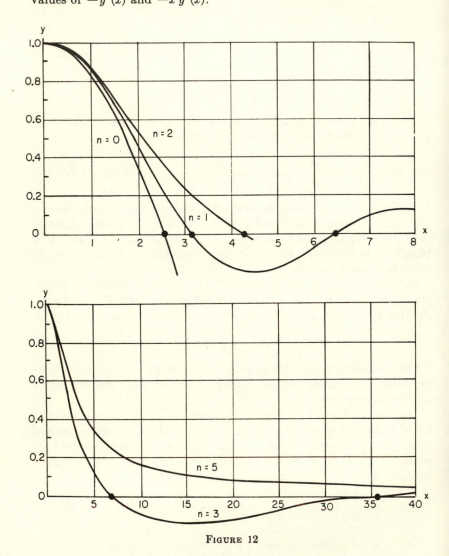

FIGURE 12

Astrophysicists use the table of Emden's functions to estimate the density and the internal temperature of stars. For this purpose one introduces the following auxiliary quantities:

$$R^*=x_0, \quad M^*=(-x^2y')_{x=x_0}, \quad D(x)=-x/(3y'), \quad D^*=D(x_0), \quad (16)$$

where x_0 is the first zero of $y(x)$.

Let R be the observed radius of the star and M its mass. Then by (9)

$$R=kR^*, \quad GM=kM^*\phi_0, \quad (17)$$

where $k=[a\phi_0^{(n-1)/2}]^{-1}$.

We then have

$$\frac{GMR^*}{M^*R}=\frac{kM^*\phi_0 R^*}{M^*kR^*}=\phi_0. \quad (18)$$

Let ρ_m be the mean density of the star, that is, $\rho_m=M/\left(\frac{4}{3}\pi R^3\right)$, and let ρ_r be the density of an interior sphere of radius r. We then have

$$\rho_r=\frac{M_r}{\frac{4}{3}\pi r^3}=\frac{3}{4\pi G r^3}\left(-r^2\frac{d\phi}{dr}\right)=\frac{\phi_0}{4\pi k^2 G}\left(-\frac{3y'}{x}\right)=\frac{\phi_0}{4\pi k^2 G}\frac{1}{D(x)}.$$

Since $D(x)=1$ when $x=0$, and since $\rho_m=\rho_R$, we get as the ratio between the mean density and the density at the center of the star the following:

$$\rho_0/\rho_m=D(x_0). \quad (19)$$

TABLE I

$n=0$	$n=1$	$n=1.5$	$n=2$
$x_0=2.44949$	$x_0=3.14159$	$x_0=3.65375$	$x_0=4.35287$
$-y_0'=0.81650$	$-y_0'=0.31831$	$-y_0'=0.20330$	$-y_0'=0.12725$
$-x_0^2y_0'=4.89898$	$-x_0^2y_0'=3.14159$	$-x_0^2y_0'=2.71406$	$-x_0^2y_0'=2.41105$
$n=2.5$	$n=3$		
$x_0=5.35528$	$x_0=6.89685$	$x_0=35.96194$	$x_0=102.60285$
$-y_0'=0.07626$	$-y_0'=0.04243$	$-y_0'=0.00480$	$-y_0'=0.00119$
$-x_0^2y_0'=2.18720$	$-x_0^2y_0'=2.01824$	$-x_0^2y_0'=6.2040$	$-x_0^2y_0'=12.551$
$n=3.5$	$n=4$	$n=4.5$	$n=5$
$x_0=9.53581$	$x_0=14.97155$	$x_0=31.83646$	$x_0=\infty$
$-y_0'=0.02079$	$-y_0'=0.00802$	$-y_0'=0.00171$	$-y_0'=0$
$-x_0^2y_0'=1.89056$	$-x_0^2y_0'=1.79723$	$-x_0^2y_0'=1.73780$	$-x_0^2y_0'=1.73205$

Formulas for central pressure (P_0) and central temperature (T_0) are similarly obtained. Thus, integrating (4), and noting that ϕ is zero at the surface of the star, we have

$$P_0 = \int_0^{\phi_0} \rho\, d\phi = \int_0^{\phi_0} L\phi^n d\phi = \frac{L\phi_0^{n+1}}{n+1} = \frac{\rho_0\phi_0}{n+1}. \tag{20}$$

We now introduce the formula: $P = C\rho T$, where C is a technical constant equal to $\kappa/(H\mu\beta)$ in which κ is the Boltzmann constant $= 1.380 \times 10^{-16}$ erg. deg.$^{-1}$, H is the mass of the hydrogen molecule $= 1.673 \times 10^{-24}$ gm., μ is the molecular weight of the gas in terms of hydrogen and β the ratio of radiation pressure to total pressure. Since, in stars of an order of size between 10^{33} and 10^{34} gm., $\mu\beta$ is assumed to have a value of 2, we find that $C = 4.124 \times 10^7$. Hence, from (20), we get

$$T_0 = \frac{P_0}{C\rho_0} = 2.425 \times 10^{-8} \frac{\phi_0}{n+1}. \tag{21}$$

The following application (from A. S. Eddington*) illustrates the use of these formulas and the significance of the Emden equation in the study of stellar phenomena.

For the bright component of Capella we have: $M = 8.30 \times 10^{33}$ gm. and $R = 9.55 \times 10^{11}$ cm., from which we compute: $\rho_m = 0.00227$. Assuming that $\gamma = 4/3$ in (6), we find from (19) that

$$\rho_0 = D(x_0)\rho_m = 54.18\rho_m = 0.1230 \text{ gm. per cc.}$$

Since, for $n = 3$, $R^* = 6.90$ and $M^* = 2.018$, we compute from (18)

$$\phi_0 = \frac{6.668 \times 10^{-8} \times 8.30 \times 10^{33} \times 6.90}{2.018 \times 9.55 \times 10^{11}} = 1.981 \times 10^{15}.$$

Hence, using (20), we get

$$P_0 = \frac{1}{4}(0.1234 \times 1.981 \times 10^{15}) = 6.11 \times 10^{13} \text{ dynes per cm.}^2,$$

and from (21),

$$T_0 = \frac{1}{4}(2.425 \times 10^{-8} \times 1.981 \times 10^{15}) = 1.20 \times 10^7 \text{ degrees.}$$

In more recent years the theory of Emden has been modified to take account of new knowledge with respect to radiation pressure and the so-called "guillotine factor", τ, which is related to density through the equation: $\tau = \tau_0\rho^\alpha$. This factor depends upon the ionizing

---

The Internal Constitution of the Stars. Cambridge, 1926, viii + 407 pp. See, in particular, Chap. 4.

potentials of the constituent elements of the star and enters into the coefficient of energy absorption. This theory tends to increase the value of the ratio ρ_0/ρ_m, which, for the sun, under the Emden theory is 23.41 and under the new theory is 79.1. For an extensive description of this theory and the differential system, which replaces the Emden equations, the reader is referred to an article by S. Chandrasekhar: "The Structure, the Composition, and the Source of Energy of Stars" in a *Symposium on Astrophysics* (1951), edited by J. A. Hynek.

PROBLEM

Given the following values for the sun: $M = 1.991 \times 10^{33}$ gm. and $R = 6.960 \times 10^{10}$ cm., show that $\rho_m = 1.410$ gm. per cm.[3]. Assuming that $\gamma = 7/5$, show that $\rho_0 = 23.41 \ \rho_m$. Also compute ϕ_0, P_0, and T_0.

6. The Differential Equation of Isothermal Gas Spheres

In the problem considered in the previous section, the solution of the Emden equation for $n = 5$ had no zero in the finite plane. This implied an infinite distribution of the gas sphere and directed attention to problems associated with such distributions, which are perhaps approximated in the case of the giant stars. In the isothermal case, where the temperature, T, remains constant, n is infinite. This fact makes it necessary to modify the arguments given in the preceding section.

Referring to Section 5, we see that when $n = \infty$, $\gamma = 1$ and $P = K\rho$. Hence, since $dP = \rho d\phi$, we get by integration: $\phi = K \log (\rho/\rho_0)$, that is, $\rho = \rho_0 \exp (\phi/K)$. If ρ_0 is the central density, then ϕ_0 must be zero, a change from the condition in the previous case where ϕ was zero only at the boundary of the sphere. The constant temperature is connected with K by the formula: $T = K\mu\beta/\kappa$, where the constants have the same significance as in Section 5.

Poisson's equation is now replaced by

$$\nabla^2\phi = -a^2 e^{\phi/K}, \quad a^2 = 4\pi\rho_0 G, \tag{1}$$

which is generally known as *Liouville's equation*. It has already been described in Section 7 of Chapter 1.

Assuming spherical symmetry as before, equation (1) in polar coordinates reduces to the following:

$$\frac{d^2\phi}{dr^2} + \frac{2}{r}\frac{d\phi}{dr} + a^2 e^{\phi/K} = 0, \tag{2}$$

which replaces equation (8) of Section 5.

If we let $\phi = Ky$, $r = (\sqrt{K}/a)x$, then (2) becomes

$$\frac{d^2y}{dx^2} + \frac{2}{x}\frac{dy}{dx} + e^y = 0, \tag{3}$$

which is to be solved subject to the boundary conditions: $y(0) = y'(0) = 0$.

It will be found by tedious computation that the solution about the origin can be expressed by the following series:

$$y = -\frac{1}{6}x^2 + \frac{1}{5 \cdot 4!}x^4 - \frac{8}{21 \cdot 6!}x^6 + \frac{122}{81 \cdot 8!}x^8 - \frac{61 \cdot 67}{495 \cdot 10!}x^{10} + \cdots,$$

$$= -0.16666\ 66667\ x^2 + 0.00833\ 3333\ x^4 - 0.00052\ 91006\ x^6$$

$$+ 0.00003\ 73555\ x^8 - 0.00000\ 22753\ x^{10}. \tag{4}$$

Using approximation methods, Emden computed a table of values of y from $x = 0$ to $x = 2{,}000$, which are graphically represented in Figure 13.

Only one particular solution of equation (3) is known, namely, the function $y = \log 2 - 2 \log x$.

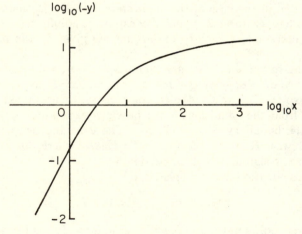

FIGURE 13

The counterpart of equation (3) in which e^y is replaced by e^{-y} appears in O. W. Richardson's theory of thermionic currents when one seeks to determine the density and electric force of an electron gas in the neighborhood of a hot body in thermal equilibrium.*

---

*See Richardson: *The Emission of Electricity from Hot Bodies.* 2d ed. London, 1921, 320 pp. In particular, pp. 47–55.

Richardson states that "the general condition for this equilibrium at constant temperature is that the force on the electron in any element of volume arising from the electric field should balance the force on the same element of volume arising from the pressure gradient."

This is equivalent to having the electric intensity E satisfy the following nonlinear partial differential equation:

$$E \text{ div } E + \frac{KT}{e_0} \nabla^2 E = 0, \tag{5}$$

with the additional condition:

$$\text{grad log } n = \frac{e_0}{\kappa T} E, \tag{6}$$

where κ is Boltzmann's constant, T the constant temperature, e_0 the charge on the electron, and n the number of electrons per unit volume.

To obtain the differential equation satisfied by v, the volume of unit mass, one first computes the work done when unit mass is moved from a point A in the field to a second point B against a pressure p. Denoting this work by w, we get

$$w = \int_A^B p \, dv. \tag{7}$$

But since the gas is in equilibrium the work done on the electron by the electric force is equal to w. Denoting this work by w', we have

$$w' = -\int_A^B N_0 e_0 \frac{dV}{ds} \, ds, \tag{8}$$

where V is the potential, N_0 the number of electrons per unit mass, and ds is the element of the path from A to B.

Equating (7) and (8), we have

$$\int_A^B p \, dv + \int_A^B N_0 e_0 \frac{dV}{ds} \, ds = 0. \tag{9}$$

Since $p = RT/v$, we can write (9) as follows:

$$\int_A^B \frac{RT}{v} \frac{dV}{ds} \, ds + \int_A^B N_0 e_0 \frac{dV}{ds} \, ds = 0,$$

from which we have the equation:

$$\frac{RT}{v} \frac{dV}{ds} + N_0 e_0 \frac{dV}{ds} = 0. \tag{10}$$

The potential V satisfies Poisson's equation:

$$\nabla^2 V = 4\pi\rho = 4\pi N_0 e_0/v. \tag{11}$$

If the thermionic emission is from a flat plate of infinite extent, then $ds = dx$ and equation (11) reduces to

$$\frac{d^2V}{dx^2} = 4\pi\rho. \tag{12}$$

Making use of this equation, we eliminate V from (10) and thus obtain the following equation which Richardson used in his study of thermionic distribution in the neighborhood of flat surfaces:

$$\frac{d^2v}{dx^2} - \frac{1}{v}\left(\frac{dv}{dx}\right)^2 + C = 0, \quad C = 4\pi N_0^2 e_0^2/RT. \tag{13}$$

If the emission is from a spherical surface with radial symmetry, then equation (11) must be expressed in polar coordinates and we have

$$\frac{1}{r^2}\frac{d}{dr}\left(r^2\frac{dV}{dr}\right) = 4\pi\rho. \tag{14}$$

Using this to eliminate V from (10) we obtain the following nonlinear equation:

$$\frac{d^2v}{dr^2} - \frac{1}{v}\left(\frac{dv}{dr}\right)^2 + \frac{2}{r}\frac{dv}{dr} + C = 0. \tag{15}$$

By means of the transformation:

$$v = e^y, \quad r = \alpha x, \quad \alpha^2 C = 1,$$

equation (15) assumes the following form:

$$\frac{d^2y}{dx^2} + \frac{2}{x}\frac{dy}{dx} + e^{-y} = 0. \tag{16}$$

PROBLEMS

1. Show that equation (13) has a particular solution of the form:

$$v = (\alpha x + \beta)^2.$$

Make the transformation $\log v = y$ and obtain the general solution of (13).

2. Obtain a series solution of (16) which satisfies the conditions:

$$y(0) = y'(0) = 0.$$

7. Equations of Emden Type

By an equation of Emden type, we shall mean any equation of the following form:

$$\frac{d^2y}{dx^2}+\frac{2}{x}\frac{dy}{dx}+f(y)=0, \tag{1}$$

where $f(y)$ is some given function of y. In the preceding two sections $f(y)=y^n$ and e^y.

We shall now consider eight additional cases by specializing $f(y)$ as follows:

$$f(y) = \pm\sin y,\ \pm\cos y,\ \pm\sinh y,\ \pm\cosh y. \tag{2}$$

But if x and y are variables in the complex plane, then an Emden equation involving any one of these eight functions can be reduced to an Emden equation involving any other. To see this, let us first observe that under the linear transformation

$$y=u+p, \tag{3}$$

where p is a constant, equation (1) assumes the form:

$$\frac{d^2u}{dx^2}+\frac{2}{x}\frac{du}{dx}+f(u+p)=0. \tag{4}$$

If $f(y)$ is a periodic function, and if p is any period, then equation (1) is unchanged by the transformation. This means that *any solution of (1) which is increased by a period of* f(y) *is also a solution.* Since all the functions in (2) are periodic, this property is possessed by any equation of which they are a member.

Let us now denote any member of set (2) by $f_m(y)$ and any other member by $f_n(y)$. We then observe that real or imaginary values of p and q exist, such that

$$f_m(qu+p)=kf_n(u), \tag{5}$$

where k is a constant. For example, $\cosh(u+\frac{1}{2}\pi i)=i\sinh u$ and $\cosh(iu+\frac{1}{2}\pi i)=\sin u$.

If we now subject the equation

$$\frac{d^2y}{dx^2}+\frac{2}{x}\frac{dy}{dx}+f_m(y)=0, \tag{6}$$

to the transformation:

$$y=qu+p,\quad x=ct, \tag{7}$$

where $c^2 = kq$, then (6) transforms into the following equation:

$$\frac{d^2u}{dt^2} + \frac{2}{t}\frac{du}{dt} + f_n(u) = 0. \tag{8}$$

Thus the proposition is established.

As in the cases previously considered, an analytic solution of (1) is always possible in the neighborhood of $x = 0$ for the boundary conditions:

$$y(0) = y_0, \quad y'(0) = 0. \tag{9}$$

If we denote by f_0, f_0', f_0'', etc., the values of $f(y)$ and its derivatives when $y = y_0$, then the expansion of the solution to four terms in the neighborhood of $x = 0$ is given by the following series:

$$y = y_0 - f_0\frac{1}{3!}x^2 + f_0 f_0'\frac{1}{5!}x^4 - (5f_0^2 f_0'' + 3f_0 f_0'^2)\frac{1}{3 \cdot 7!}x^6 + \ldots \tag{10}$$

The term in y' in equation (1) can be removed by means of the transformation:

$$y = \frac{1}{x}z(x). \tag{11}$$

We thus obtain

$$\frac{d^2z}{dx^2} + xf(z/x) = 0. \tag{12}$$

We shall now consider separately the eight cases defined by (2), where we shall assume the boundary conditions: $y(0) = 1$, $y'(0) = 0$.

(a) $f(y) = \sin y$.

The phase trajectory, that is, $y' = F(y)$, is shown in Figure 14 and the values of y and y' as functions of x are graphed in Figure 15. The motion is seen to be a damped oscillation.

(b) $f(y) = -\sin y$.

This case can be referred to (a) by setting $p = -\pi$ in equation (3). The focal point, which was zero in the first equation, is now transferred to π, as one observes from Figure 16. But the boundary conditions are now different, for they correspond to $y(0) = 1 - \pi$, $y'(0) = 0$, in the first case.

(c) $f(y) = \cos y$.

This problem can be referred to (a) by setting $p = \pi/2$ in equation (3) and choosing the boundary conditions: $y(0) = 1 + \pi/2$. In the phase-trajectory, Figure 18, a focal point appears at $y = -\pi/2$, $y' = 0$. The motion is again a damped oscillation.

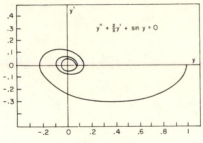

FIGURE 14

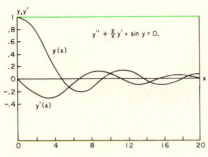

FIGURE 15

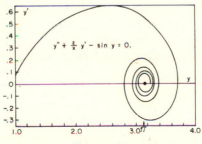

FIGURE 16

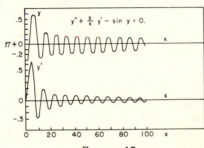

FIGURE 17

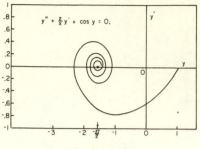

FIGURE 18

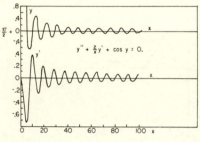

FIGURE 19

(d) $f(y) = -\cos y$.

This problem is referred back to case (a) by setting $p = -\pi/2$ and using the boundary conditions: $y(0) = 1 - \pi/2$, $y' = 0$. The phase diagram and the curves $y = y(x)$, $y' = y'(x)$ are shown in Figure 20.

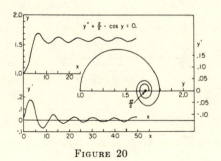

FIGURE 20

(e) $f(y) = \sinh y$.

The phase trajectory for this case is shown in Figure 22 and the graphs of $y = y(x)$ and $y' = y'(x)$ in Figure 23. We see what appears at first sight to be a surprising fact, namely, that this case differs very little from (a) where $f(y) = \sin y$.

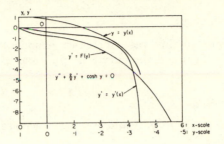

FIGURE 21

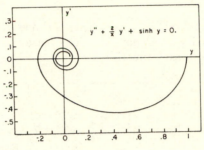

FIGURE 22

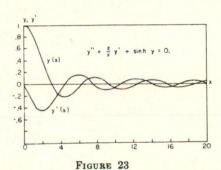

FIGURE 23

An explanation is readily obtained from equation (12), however, for if we expand $f(z/x)$ as a power series in z/x for the two cases, we obtain the following two equations:

$$\text{Case (a):} \quad z'' + z - \frac{1}{3!}\frac{z^3}{x^2} + \frac{1}{5!}\frac{z^5}{x^4} - \ldots = 0,$$

$$\text{Case (e):} \quad z'' + z + \frac{1}{3!}\frac{z^3}{x^2} + \frac{1}{5!}\frac{z^5}{x^4} + \ldots = 0. \tag{13}$$

If $|z|$ remains less than or equal to 1, then, as x increases, both equations are asymptotic to $z'' + z = 0$.

(f) $f(y) = -\sinh y$.

The solution in this case increases exponentially. The phase trajectory, $y' = F(y)$ and the curves $y = y(x)$ and $y' = y'(x)$ are shown in Figure 24.

(g) $f(y) = \cosh y$.

In this case both $y = y(x)$ and $y' = y(x)$ decrease exponentially, as shown in Figure 21.

(h) $f(y) = -\cosh y$.

This case resembles closely case (f) as is shown by Figure 25, which may be compared with Figure 24. The reason is obvious since the dominating term in each function is $-e^y$.

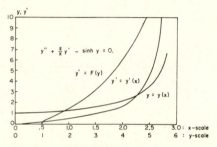

FIGURE 24

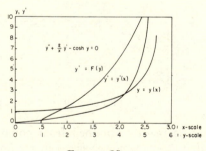

FIGURE 25

8. The Duffing Problem

The Duffing problem concerns the integration of the following equation:

$$\frac{d^2y}{dt^2} + ay + by^3 = K \sin \Omega t, \quad a > 0, \tag{1}$$

which is that of a simple pendulum moved by a driving force $K \sin \Omega t$. This equation was the subject of a small book by G. Duffing in 1918 and has been investigated by many others since that time.

Without essential loss of generality it can be written in the simpler form

$$y'' + y + ry^3 = K \sin \Omega t, \tag{2}$$

since we can make the transformation: $t = ct'$ and choose c so that $a = 1$. It should also be observed that the driving force could equally well be $K \cos \Omega t$, since the equation is unchanged by the linear translation: $t = t' + p$.

Although equation (1) looks comparatively simple, this simplicity is entirely specious. Its solution is a matter of great difficulty and only limited progress has been made in the mathematical understanding of it. The reason for this is readily understood, for, in the comparatively simple linear case, where $r = 0$, we have already shown in Chapter 10 the numerous complexities introduced by simple variations of the forcing function. These complexities are multiplied severalfold in (2), where, for example, the period of the solution of the null equation

$$y'' + y + ry^3 = 0, \tag{3}$$

does not depend alone upon r, but also upon the amplitude of the motion. In the linear case we have the phenomenon of resonance when the period of the forcing function is equal to the period of the solution of the null (homogeneous) equation. It is independent of K. But this is not the case for (2), where resonance is a function of r, Ω, and K.

The analytical difficulties of the problem are perhaps most readily apprehended if we write equation (2) as the following system:

$$y'' + y + ry^3 = z, \quad z'' + \Omega^2 z = 0, \tag{4}$$

which we can do, since the forcing function is a solution of the second equation.

If the first equation is differentiated twice and z eliminated, we obtain the following nonlinear equation of fourth order:

$$y^{(4)} + (1 + \Omega^2 + 3ry^2)y'' + 6ry(y')^2 + \Omega^2 y + r\Omega^2 y^3 = 0, \tag{5}$$

which is the null-equivalent of (1). If one refers to Example 1, Section 3, Chapter 10, he will be instructed as to the complexities inherent in the solution of (5) by observing those introduced in the same manner in the much simpler case of a linear system.

The method which we shall use here is to express the solution of (1) as the following Fourier series:

$$y = a_1 \sin \omega t + a_3 \sin 3\omega t + a_5 \sin 5\omega t + \ldots, \tag{6}$$

where ω is a period of the null equation (3).

This, of course, gives us considerable latitude in the choice of ω, since, as we have shown in Section 5, Chapter 10, this quantity depends upon an arbitrary constant and can be written in the following form:

$$\omega^2 = 1 + \frac{3}{4}rC^2 - \frac{3}{128}r^2C^4 + \frac{9}{512}r^3C^6 + \cdots, \tag{7}$$

where C is arbitrarily given. When $r = -1/6$ (the case of the pendulum), C is the value of the maximum excursion of the bob.

In general, the period of the driving force will be different from ω. To surmount this difficulty we shall express $K \sin \Omega t$ as a Fourier series in ω, that is

$$K \sin \Omega t = K_1 \sin \omega t + K_3 \sin 3\omega t + K_5 \sin 5\omega t + \cdots, \tag{8}$$

where the coefficients are derived from the formula:

$$K_{2n+1} = \frac{2}{\pi} \frac{K \sin \mu}{\mu^2 - (2n+1)^2}, \quad \mu = \Omega/\omega. \tag{9}$$

We have already shown in Section 5, Chapter 10, that the solution of (3) can be written as the following Fourier series:

$$u = u_1 \sin \omega t + u_3 \sin 3\omega t + u_5 \sin 5\omega t + \cdots, \tag{10}$$

and explicit values have been derived for the parameters.

We shall now assume that the solution of (2) can be written in the form

$$y = u + v, \tag{11}$$

and shall seek to show that v can be expanded in a Fourier series similar to (10), namely,

$$v = v_1 \sin \omega t + v_3 \sin 3\omega t + v_5 \sin 5\omega t + \cdots. \tag{12}$$

If y as given by (11) is now forced into equation (2), the following equation is obtained for the determination of v:

$$v'' + v + 3r \, u^2 v + 3r \, uv^2 + r \, v^3 = K \sin \Omega t. \tag{13}$$

The question now, which is far from obvious, is to determine whether or not v can have the form specified in (12). If (12) is now substituted in (13) and the multiplications $u^2 v$, uv^2, and v^3 performed, it will be found that the resulting series will contain terms of the form: S_3^3, $S_1 S_5^2$, $S_3 S_5 S_7$, etc., where we use the abbreviations:

$$S_1 = \sin \omega t, \quad S_3 = \sin 3\omega t, \quad S_5 = \sin 5\omega t, \text{ etc.} \tag{14}$$

But an easy calculation shows that

$$S_3^3 = \frac{3}{4} S_3 - \frac{1}{4} S_9, \quad S_1 S_5^2 = \frac{1}{2} S_1 + \frac{1}{4} S_9 - \frac{1}{4} S_{11},$$

$$S_3 \, S_5 \, S_7 = \frac{1}{4} \, (S_1 + S_5 + S_9 - S_{15}), \text{ etc.} \quad (15)$$

It thus follows that the left-hand member of (13) reduces to a linear combination of S_1, S_3, S_5, S_7, etc., the coefficients of which will be defined by a system of algebraic equations obtained by setting the coefficients of S_1, S_3, S_5, S_7, etc., equal respectively to K_1, K_3, K_5, K_7, etc., obtained from equation (8).

In this manner we have shown that the solution of the original equation can be expressed formally as a Fouriers series of the specified type defined by (6). We now substitute (6) in equation (2). After a somewhat tedious calculation it will be found that the first four parameters in (6) must satisfy the following system of equations:

$$(1-\omega^2)a_1 + \frac{3}{4} \, r(a_1^3 + 2a_1a_3^2 + 2a_1a_5^2 + 2a_1a_7^2 - a_1^2a_3 - 2a_1a_3a_5 - 2a_1a_5a_7$$

$$+ 2a_3a_5a_7 + a_3^2a_5 - a_3^2a_7) = K_1,$$

$$(1-9\omega^2)a_3 + \frac{1}{4} \, r(-a_1 + 6a_1^2a_3 - 3a_1^2a_5 + 6a_1a_3a_5 - 6a_1a_3a_7 + 6a_1a_5a_7 + 3a_3^3$$

$$+ 6a_3a_5^2 + 6a_3a_7^2 + 3a_5^2a_7) = K_3, \quad (16)$$

$$(1-25\omega^2)a_5 + \frac{3}{4} \, r(a_1a_3^2 - a_1^2a_3 + 2a_1^2a_5 - a_1^2a_7 + 2a_1a_3a_7 + 2a_3a_5a_7$$

$$+ 2a_3^2a_5 + a_5^3 + 2a_5a_7^2) = K_5,$$

$$(1-49\omega^2)a_7 + \frac{3}{4} \, r(-a_1a_3^2 - a_1^2a_5 + 2a_1^2a_7 + 2a_1a_3a_5 + 2a_3a_5^2$$

$$+ 2a_3^2a_7 + 2a_5^2a_7 + 2a_7^3) = K_7.$$

From this obviously complicated set of equations we can obtain at least one simple result. Thus, if $r=0$, we have

$$a_{2n+1} = \frac{K_{2n+1}}{[1-(2n+1)^2\omega^2]}, \quad n=0, 1, 2, \text{ etc.} \quad (17)$$

provided $\omega \neq 1/(2n+1)$. Otherwise we shall have resonance in the linear system such as that exhibited in Example 3, Section 3, Chapter 10. Since ω, as defined by (12), equals 1 when $r=0$, it is clear that we may expect difficulties with the solution if ω has a value close to unity.

If we assume that the solution of (1) is given by series (6), it is clear that at $t=0$ we have $y_0=0$. Moreover, for any arbitrarily chosen value of y', let us say y_0', we obtain the following equation:

$$y_0' = \omega(a_1 + 3a_3 + 5a_5 + 7a_7 + \ . \ . \ .). \tag{18}$$

This is a linear relationship between the parameters, which must be added to the nonlinear set (16) above.

If $r \neq 0$, and if we neglect all harmonic terms beyond the third, then system (16) reduces to the following:

$$4(1-\omega^2)a_1 + 3r \ a_1^3 + 6r \ a_1a_3^2 + 6r \ a_1a_5^2 - 3r \ a_1^2a_3 - 6r \ a_1a_3a_5 + 3r \ a_3^2a_5 = 4K_1$$

$$4(1-9\omega^2)a_3 - ra_1^3 + 6r \ a_1^2a_3 - 3r \ a_1^2a_5 + 6r \ a_1a_3a_5 + 3r \ a_3^3 + 6r \ a_3a_5^2 = 4K_3. \tag{19}$$

Adjoining to this system the linear equation,

$$y_0' = \omega(a_1 + 3a_3 + 5a_5), \tag{20}$$

by means of which one of the three parameters can be removed, let us say a_5, we obtain a set of two cubic equations in two variables which will, in general, lead to nine solutions.

There is no simple way in which this system can be solved for a_1 and a_3, although the eliminant for either variable can be formed as a determinant of sixth order by *Sylvester's dialytic method* (See Section 6, Chapter 13). Since the resulting equation, in the general case, is one of ninth degree there will be at least one real solution. This can usually be located graphically without great difficulty and computed to any desired degree of accuracy.

As an example we shall consider the equation:

$$y'' + y - \frac{1}{6} y^3 = 2 \sin 3\omega t. \tag{21}$$

The null equation:

$$u'' + u - \frac{1}{6} u^3 = 0, \tag{22}$$

we have already solved in Section 5, Chapter 10. Corresponding to an initial displacement of $\pi/3$, we found the values: $\omega^2=0.86202$, $\omega=0.92845$, and this we shall assume defines the value of ω in (21). Since 3ω is considerably larger than the critical value of unity, we anticipate no difficulty with the convergence of series (6). Further reasons for this will be given in the next section. We shall assume as initial values: $y_0=y_0'=0$.

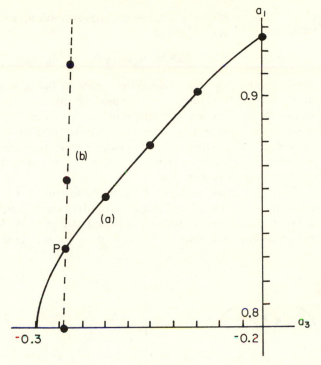

FIGURE 26

Since the constants in (19) are the following: $\omega^2=0.86202$, $r=-1/6$, $K_1=0$, $K_3=2$, we have, after the elimination of a_5, the following set of equations:

$$1.10384a_1-1.08a_1^3-3.72a_1a_3^2+0.12a_1^2a_3+0.6a_3^3=0, \qquad (23\text{-a})$$

$$-162.1963a_3+0.4a_1^3-6.84a_1^2a_3-2.16a_1a_3^2+5.16a_3^3=48 \qquad (23\text{-b})$$

Proceeding graphically as shown in Figure 26, we determine approximately the real point (P) of the intersection of the curves represented by equations (23). These are denoted respectively by (a) and (b) in the figure. The coordinates of this point are then determined more exactly by interpolation and are found to be:

$$a_1=0.8333, \quad a_3=-0.2878. \qquad (24)$$

The value of a_5 is then computed from (20) in which $y_0'=0$, from which we have: $a_5=0.0060$.

Therefore, to a satisfactory degree of approximation, we obtain the solution of (21) in the following form:

$$y = 0.8333 \sin \omega t - 0.2878 \sin 3\omega t + 0.0060 \sin 5\omega t. \qquad (25)$$

The phase diagram for this equation is shown in (a) of Figure 27, which shows a simple configuration consisting of two symmetric ovals. This we have called the diagram of the *double egg*. The graphical representations of y and its first two derivatives are given in (b), (c), and (d) of the figure. These same curves through several cycles were obtained by the analogue computer in order to confirm the analysis and are shown in Figure 28.

Intrigued by the great variety of figures obtained in the phase-plane for various choices of K and Ω, the computers made a number of these. A rather characteristic pattern is shown in Figure 29, together with the component functions y, y', and y''. In this case $K=2$, $\Omega = 5\omega$.

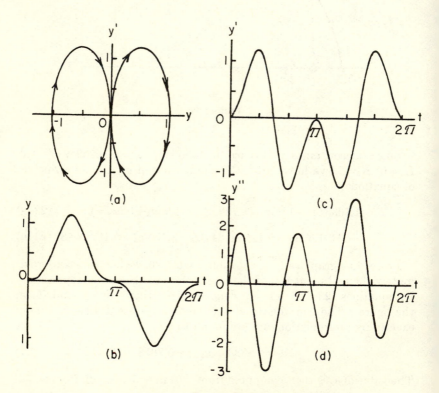

FIGURE 27

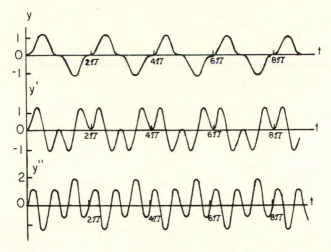

FIGURE 28

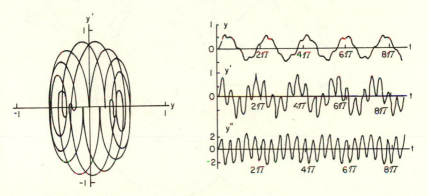

FIGURE 29

The beautiful configuration shown in Figure 30 was discovered by Norriss Hetherington and has been called the *Norriss Heart*. It is given by the values: $K=1$, $\Omega=2\omega$. The pattern shown in Figure 31, corresponding to $K=3$, $\Omega=3\omega$, was named *Murphy's Eyeballs*, in recognition of the titular diety who presides over error in computing laboratories.*

---

*Murphy is credited with the discovery of three propositions: (1) If anything can go wrong, it will; (2) Things when left alone can only go from bad to worse; (3) Nature sides with the hidden flaw. During this investigation Murphy's eyes were constantly upon these computations.

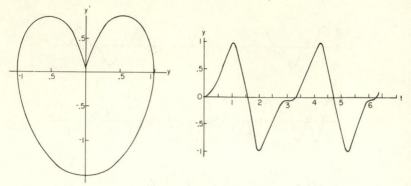

FIGURE 30

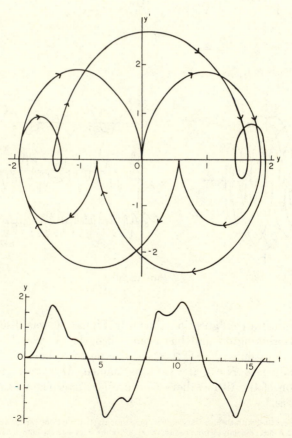

FIGURE 31

9. Nonlinear Resonance—The Jump Phenomenon

We turn next to a consideration of the more difficult case of the Duffing equation, that is,

$$y'' + y + ry^3 = K \sin \Omega t, \tag{1}$$

where Ω is a value in the neighborhood of ω, a period of the null equation.

We have shown in an empirical way that the system of equations given by (16) in Section 8 can have stable solutions for appropriately chosen values of Ω and K_1. But an existence theorem assuring stable solutions would be very difficult to give because of the obvious analytical difficulties presented by the nonlinear character of the equations. In spite of this, however, it is clear from Cauchy's existence theorem that a local solution of (1) must exist in the neighborhood of $t=0$ for every value of Ω and K. But it is a different matter to determine whether or not this solution can be expressed in a convergent series of the form assumed in the preceding section.

If we examine the case where $r=0$ and $\Omega=1$, we find that the solution of (1), for which $y_0 = y_0' = 0$, has the form

$$y = \frac{1}{2} K \sin t - \frac{1}{2} Kt \cos t. \tag{2}$$

While the solution is thus oscillatory it is not periodic. On the contrary, it is unstable and approaches infinity with an amplitude which increases as $Kt/2$. This solution, of course, is well known, and when it appears, the phenomenon which it describes is said to exhibit *resonance*.

That resonance should also be observed in solutions of equation (1) when $r \neq 0$ is to be expected, but the conditions under which it appears are by no means easy to state. One very curious aspect of such unstable solutions is what has been called the "jump" phenomenon. A solution $y=y(t)$ will appear, through a considerable interval of time, to be a stable oscillation. Then, without obvious reason, it will become unstable in such a sudden and abrupt manner that the motion which it describes appears very much like a jump.

The phenomenon of resonance as exhibited by equation (1) differs from that in the linear case through a functional relationship between K and Ω which divides the region of stability from that of instability. It is our purpose to discover something about this function.

For this purpose we shall make use of the method of continuous analytic continuation described in Chapter 9 as an empirical probe and shall seek by means of it to determine the boundary between the regions of resonance and nonresonance. The results thus obtained are then checked by applying the differential analyzer to the same problem. It will be convenient in this investigation to formulate the results in terms of a known period of the null equation and for this purpose we shall assume the values $r=-1/6$, $\omega=0.92845$, $\omega^2=0.86202$. This choice of ω, it will be recalled from Section 5, Chapter 10, corresponds to the case of a pendulum with an initial displacement of $\pi/3$. The boundary conditions will be $y_0=y_0'=0$ for $t=0$.

In order to illustrate the method we shall apply it first to the case where $K=2$, $\Omega=\omega$. Since a local solution exists in the neighborhood of the origin, we find its expansion to $(\Delta t)^7$ by evaluating successive derivatives of $y(t)$. The following series is thus obtained:

$$y=\frac{2}{3!}\,\Omega\,(\Delta t)^3 -\frac{2}{5!}\,(\Omega+\Omega^3)\,(\Delta t)^5 +\frac{2}{7!}\,(\Omega+\Omega^3+\Omega^5)\,(\Delta t)^7+ \ldots. \tag{3}$$

The radius of convergence, of course, is unknown. That the series is entire is very doubtful, since the elliptic function which solves the null equation has polar singularities in the finite plane.

Since equation (3) is inadequate for the evaluation of y except in the immediate vicinity of the origin, we now introduce the equations by means of which the solution can be analytically continued. In spite of the complex configurations to which they are to be applied, sufficient accuracy will be attained by using derivatives only to fourth order. We thus obtain the following:

$$y_{i+1}=y_i+y_i'\Delta t+\frac{K}{2!}\,\sin\,\Omega t_i(\Delta t)^2+\frac{K}{3!}\,\cos\,\Omega t_i\,(\Delta t)^3$$

$$-\frac{K\Omega^2}{4!}\,\sin\,\Omega t_i\,(\Delta t)^4+\frac{1}{2}\,(-y_i-ry_i^3)\,(\Delta t)^2+\frac{y_i'}{3!}\,(-1-3ry_i^2)\,(\Delta t)^3$$

$$+[-6ry_iy_i'^2+(1+3ry_i^2)\,(y_i+ry_i^3-K\,\sin\,\Omega t_i]\,\frac{(\Delta t)^4}{4!}, \tag{4}$$

$$y'_{i+1}=y_i'+K\,\sin\,\Omega t_i\Delta t+\frac{K\Omega}{2!}\,\cos\,\Omega t_i\,(\Delta t)^2-\frac{K\Omega^2}{3!}\,\sin\,\Omega t_i\,(\Delta t)^3$$

$$+(-y_i-ry_i^3)\Delta t+\frac{y_i'}{2!}\,(-1-3ry_i^2)\,(\Delta t)^2+[-6ry_iy_i'^2$$

$$+(1+3ry_i^2)\,(y_i+ry_i^3-K\,\sin\,\Omega t_i]\,\frac{(\Delta t)^3}{3!}.$$

With increments of $\Delta t=0.1$, the solution was extended analytically to $t=\pi$ and was found to increase monotonically. The results of this computation are shown in Figure 32. The instability of the solution is readily seen from the phase diagram.

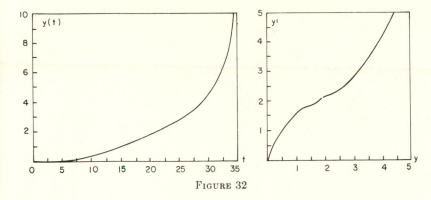

FIGURE 32

A systematic study was now initiated to determine the approximate line,

$$F(K,\Omega)=0,$$

which defines the boundary between the stable and the unstable solutions. Curves were first constructed in the phase-plane (y',y) for various values of K and $\Omega=n\omega$ until two regions were roughly outlined, one in which there was resonance and the other in which there was no resonance. The boundary between the two regions was then more exactly obtained by computing y and y' by means of formulas (4) for a systematic set of values of K and $n\omega$, where K and n varied by small increments.

The results of this investigation are shown in Figure 33, where values of K and Ω in the shaded area were found to produce unstable solutions and those in the unshaded area stable solutions. The cross in the shaded area corresponds to the resonance pattern shown in Figure 32 and the dots in the unshaded region correspond to the stable solutions given in Figures 27 to 31. In order to define the boundary line AB a least-squares polynomial was fitted to the midpoints between neighboring positions of stability and instability. The equation of the polynomial thus obtained is the following:

$$k=3.3864(n-1)-0.8811(n-1)^2-0.0052(n-1)^3+0.0701(n-1)^4$$

$$-0.0109(n-1)^5, \tag{5}$$

where n varies between 1 and 5.

A special investigation was made of the small region of stability between 0 and ω, since it was here that one encountered the curious jump phenomenon which has already been described above. A

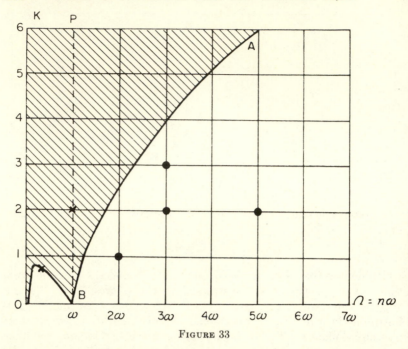

FIGURE 33

characteristic jump is shown in Figure 34, which gives the graphs of $y(t)$ and $y'(t)$ corresponding to $K=0.8$, $\Omega=0.27\omega$. The phase-diagram appears in (a) of Figure 35.

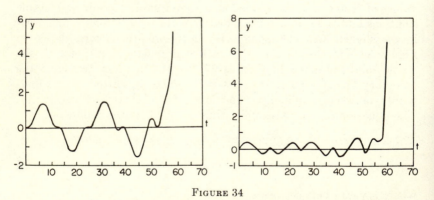

FIGURE 34

The problem of defining the boundary between the stable and un-stable regions was much more difficult than in the previous case, since very complex patterns were encountered and for very small values of Ω long runs on both the digital and analogue computers

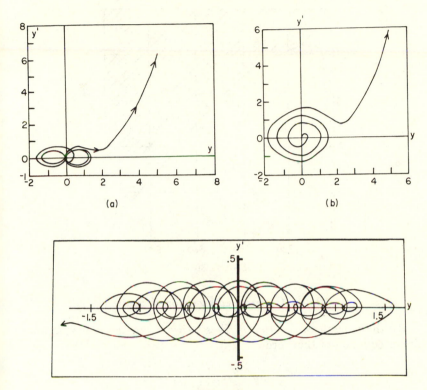

were necessary before resonance appeared. This is shown by the phase graphs in (b) and (c) of Figure 35, which correspond respectively to the points $K=0.25$, $\Omega=0.81\omega$ and $K=0.9$, $\Omega=0.05\omega$. The results of this investigation are shown in Figure 36, where the boundary CD is well defined, but the boundary OC is to be regarded as somewhat conjectural. As in the previous case, a least-square polynomial was fitted to the midpoints of contiguous positions of stability and instability. The equation of the polynomial thus obtained is given as follows:

$$\frac{K}{10}=0.22642(1-n)-1.0570(1-n)^2+3.5602(1-n)^3$$

$$-4.8358(1-n)^4+2.2435(1-n)^5, \quad (6)$$

where n varies between 0.2 and 1.

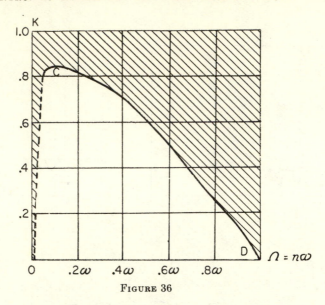

FIGURE 36

10. The Generalized Equation of Blasius

By the *generalized equation of Blasius* we mean the following non-linear equation of third order:

$$\frac{d^3y}{dx^3} + ay\,\frac{d^2y}{dx^2} = \beta\left[\left(\frac{dy}{dx}\right)^2 - 1\right] \tag{1}$$

where a and β are arbitrary constants. But a can be set equal to 1 without loss of generality, as one sees from the transformation: $y = pz$, $x = pt$, where $p^2 a = 1$.

The boundary conditions of greatest interest (for $a = 1$) are the following:

$$y(0) = y'(0) = 0, \quad y'(x) \rightarrow k, \quad \text{as } x \rightarrow \infty, \tag{2}$$

where k is a constant. When $\beta \geqq 0$, these conditions are sufficient to insure a unique solution of the equation, but this uniqueness fails when $\beta < 0$.

The equation for the case where $\beta = 0$ was originally solved by H. Blasius, who introduced it in studying the laminar flow of a fluid. The more general problem, where $\beta \neq 0$. has been the object of study by Goldstein, Howarth, Falkner and Skan, Hartree, and others. (See *Bibliography*).

Since the equation occupies an important place in the boundary layer problem of hydrodynamics, a short account of its origin may prove instructive.* One considers the flow of a fluid which streams past a plate placed edgeways in it. We shall denote the velocity of the fluid by U, which is assumed to be constant except for the disturbance of the plate. The coefficient of viscosity and the density are respectively ν and ρ. The Reynolds number is assumed to be sufficiently small so that the motion is without turbulence.

The equations of laminar flow with a pressure gradient are the following:

$$u\frac{\partial u}{\partial x}+ v\frac{\partial u}{\partial y}=\nu\frac{\partial^2 u}{\partial y^2}-\frac{1}{\rho}\frac{\partial p}{\partial x},$$

$$\frac{\partial u}{\partial x}+\frac{\partial v}{\partial y}=0, \qquad (3)$$

where x is the direction of flow, y the direction of the normal to the plate, u and v the components of the velocity in the directions x and y respectively, and p the static pressure in the boundary layer. The situation is shown graphically in Figure 37.

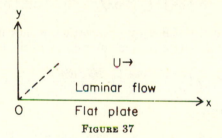

FIGURE 37

The physical assumptions, that there is no slip at the boundary and that the boundary is a stream-line, are expressed by the conditions that $u=v=0$, when $y=0$. A third condition requires that $u/U\to1$ as $y\to\infty$.

If we introduce a stream function ψ, where

$$u=-\frac{\partial\psi}{\partial y}, \quad v=\frac{\partial\psi}{\partial x}, \qquad (4)$$

then equations (3) reduce to the following single equation:

$$\frac{\partial\psi}{\partial y}\frac{\partial^2\psi}{\partial x\partial y}-\frac{\partial\psi}{\partial x}\frac{\partial^2\psi}{\partial y^2}=-\nu\frac{\partial^3\psi}{\partial y^3}-\frac{1}{\rho}\frac{\partial p}{\partial x}. \qquad (5)$$

*See, for example, Sir Horace Lamb: *Hydrodynamics.* Cambridge, p. 684 *et seq.* An extensive and admirable account of the entire problem is to be found in Shih-I Pai: *Viscous Flow Theory: I-Laminar Flow.* Princeton, 1956, xiii + 384 pp. In particular, Chaps. 8 and 9.

Introducing the assumption that

$$p + \frac{1}{2}\rho v^2 = \text{a constant, and } v = kx^m,$$

we have

$$\frac{\partial p}{\partial x} = \rho v \frac{\partial v}{\partial x} = \rho\, m\, v^2/x,$$

and equation (5) becomes

$$\frac{\partial \psi}{\partial y}\frac{\partial^2 \psi}{\partial x \partial y} - \frac{\partial \psi}{\partial x}\frac{\partial^2 \psi}{\partial y^2} + \nu \frac{\partial^3 \psi}{\partial y^3} - mv^2/x = 0. \tag{6}$$

If we now transform to the new variables

$$\phi = \psi/\sqrt{xv\nu}, \quad z = yv/\sqrt{xv\nu}, \tag{7}$$

equation (6) assumes the following form:

$$x\left[\frac{\partial \phi}{\partial z}\frac{\partial^2 \phi}{\partial x \partial z} - \frac{\partial^2 \phi}{\partial z^2}\frac{\partial \phi}{\partial x}\right] + \left[m\left(\frac{\partial \phi}{\partial z}\right)^2 - \frac{m+1}{2}\phi\frac{\partial^2 \phi}{\partial z^2} + \frac{\partial^3 \phi}{\partial z^3} - m\right] = 0. \tag{8}$$

The assumption, justified by observation, is now made that ϕ is a function of z alone, and equation (8) reduces to the following:

$$m\left(\frac{d\phi}{dz}\right)^2 - \frac{m+1}{2}\phi\frac{d^2\phi}{dz^2} + \frac{d^3\phi}{dz^3} - m = 0. \tag{9}$$

If we now write

$$\eta = Az, \quad f = -A\phi, \quad \text{where } A^2 = (m+1)/2,$$

equation (9) reduces to

$$f^{(3)} + ff'' = \beta(f'^2 - 1), \quad \beta = 2m/(m+1),$$

that is to say, with proper change of variables, to (1).

We shall now consider the solution of the original equation of Blasius, that is to say, where $\beta = 0$, subject to the boundary conditions (2). But we can, without loss of generality, set $a = 1$ and $k = 1$. To see this we subject equation (1), $\beta = 0$, to the transformation:

$$y = pz, \quad x = qt, \tag{10}$$

and thus obtain:

$$\frac{d^3 z}{dt^3} + apq\, z\frac{d^2 z}{dt^2} = 0, \quad \frac{dy}{dx} = \frac{p}{q}\frac{dz}{dt}. \tag{11}$$

The conditions are fulfilled if we set $apq=1$, $p/q=k$, that is, when

$$p=\sqrt{k/a}, \quad q=\sqrt{1/(ka)}. \tag{12}$$

The problem is thus reduced to solving the equation:

$$y^{(3)}+y\,y^{(2)}=0, \tag{13}$$

subject to the boundary conditions:

$$y(0)=y'(0)=0, \quad y'\to1, \; x\to\infty. \tag{14}$$

In order to achieve this we now observe that y can be written as follows:

$$y=\alpha^{1/3}\phi(\alpha^{1/3}x), \tag{15}$$

where α is an arbitrary parameter.

When ϕ is expanded as a power series and note taken of the first two conditions in (14), the following solution is achieved:

$$y=\frac{1}{2!}\,a_0\alpha x^2-\frac{1}{5!}\,a_1\alpha^2x^5+\frac{1}{8!}\,a_2\alpha^3x^8-\ldots+(-1)^n\,\frac{1}{(3n+2)!}\,a_n\alpha^{n+1}x^{3n+2}+\ldots, \tag{16}$$

where the a_n are determined from the following formula

$$a_n=\sum_{m=0}^{n-1}{}_{3n-1}C_{3m}a_ma_{n-m-1}, \tag{17}$$

in which ${}_qC_s$ is the binomial coefficient.

From this formula we obtain the following values:

$$a_0=a_1=1, \; a_2=11, \; a_3=375, \; a_4=27{,}897, \; a_5=3{,}817{,}137,$$
$$a_6=865{,}874{,}115, \; a_7=298{,}013{,}289{,}795.$$

In order to solve the problem one now selects arbitrarily some value of α and by a process of continuation evaluates the limiting value of y' obtained from the derivative of (16). If this limit is k, then the value of α which belongs to the conditions (14), let us say α_1, is given by the formula:

$$\alpha_1=k^{-3/2}\alpha. \tag{18}$$

In the numerical solution of this problem Howarth computed a five-decimal table of y, y', and y'' with $k=2$ and Hartree a four-decimal table of y' for $k=1$. The values of α found respectively by the two computers were 1.32824 and 0.4696, but it is seen from (18) that the ratio of the two numbers is $2\sqrt{2}=2.828427$. One also sees that the values in the two tables are related through the following transformation, where y is Howarth's variable and z that of Hartree:

$$y = \sqrt{2}z, \quad \frac{dy}{dx} = 2\frac{dz}{dt}, \quad \frac{d^2y}{dx^2} = 2\sqrt{2}\frac{d^2z}{dt^2}, \quad x = \tfrac{1}{2}\sqrt{2}t.$$

The following values are those computed by Howarth:

x	y	y'	y''	x	y	y'	y''
0.0	0.00000	0.00000	1.32824	2.3	2.88826	1.96537	0.11793
0.1	0.00664	0.13282	1.32795	2.4	3.08534	1.97558	0.08748
0.2	0.02656	0.26553	1.32589	2.5	3.28329	1.98309	0.06363
0.3	0.05974	0.39788	1.32033	2.6	3.48189	1.98849	0.04537
0.4	0.10611	0.52942	1.30957	2.7	3.68094	1.99231	0.03171
0.5	0.16557	0.65957	1.29204	2.8	3.88031	1.99496	0.02173
0.6	0.23795	0.78756	1.26637	2.9	4.07990	1.99675	0.01459
0.7	0.32298	0.91253	1.23147	3.0	4.27964	1.99795	0.00961
0.8	0.42032	1.03352	1.18666	3.1	4.47948	1.99873	0.00620
0.9	0.52952	1.14953	1.13173	3.2	4.67938	1.99922	0.00392
1.0	0.65003	1.25954	1.06701	3.3	4.87931	1.99954	0.00243
1.1	0.78120	1.36263	0.99345	3.4	5.07928	1.99973	0.00148
1.2	0.92230	1.45798	0.91237	3.5	5.27926	1.99984	0.00088
1.3	1.07252	1.54492	0.82582	3.6	5.47925	1.99991	0.00051
1.4	1.23099	1.62303	0.73603	3.7	5.67924	1.99995	0.00029
1.5	1.39682	1.69210	0.64544	3.8	5.87924	1.99997	0.00017
1.6	1.56911	1.75218	0.55651	3.9	6.07923	1.99999	0.00009
1.7	1.74696	1.80354	0.47151	4.0	6.27923	1.99999	0.00005
1.8	1.92954	1.84666	0.39234	4.1	6.47923	2.00000	0.00003
1.9	2.11605	1.88224	0.32050	4.2	6.67923	2.00000	0.00001
2.0	2.30576	1.91104	0.25694	4.3	6.87923	2.00000	0.00001
2.1	2.49806	1.93392	0.20208	4.4	7.07923	2.00000	0.00000
2.2	2.69238	1.95174	0.15589				

The generalized equation of Blasius was investigated by Hartree, who discovered that when β is positive or zero, then conditions (2) are sufficient to define a unique solution; but when β is negative, the property of uniqueness disappears. It is thus necessary to replace the third condition in (2). This Hartree did by assuming that y' approaches 1 from below as rapidly as possible, that is to say, by making $y''(0)$ as large as possible, subject to the condition $y' \leqq 1$. If y' approaches 1 from above, this would mean that $y(x)$ has a maximum value, a situation which would occur only if there was a reversal of the normal gradiant of the tangential velocity in the boundary layer. This does not appear to be a probable physical condition.

It was found that the new boundary condition imposed limits upon β and that it could not be fulfilled if β was less than -0.199. The solutions attained for $\beta = -0.198$ and $\beta = 1$ are compared with the Blasius solution ($\beta = 0$) in Figure 38, based upon the Hartree tables.

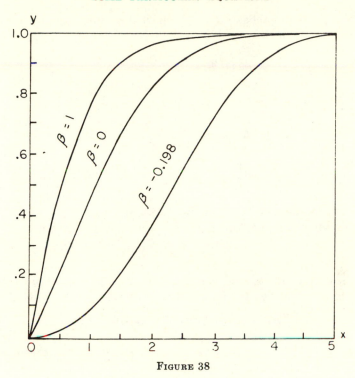

y

FIGURE 38

11. Miscellaneous Examples

In this section we shall consider briefly a few examples of nonlinear equations, which have appeared in various physical problems.

(a) The first of these is the *Thomas-Fermi equation:*

$$\frac{d^2y}{dx^2} = \frac{1}{\sqrt{x}} y^{3/2} \tag{1}$$

which appears in the problem of determining the effective nuclear charge in heavy atoms.* The solution is defined for the boundary values:

$$y(0) = 1, \quad y(x) \to 0, \quad \text{as } x \to \infty. \tag{2}$$

The differential equation itself belongs to equations of Emden type, since the transformation $y = xz$ leads to the following:

$$z'' + \frac{2}{x} z' - z^{3/2} = 0, \tag{3}$$

but the boundary conditions are no longer simple ones.

* For this reference see the *Bibliography.*

By graphical methods Fermi obtained the following approximation for values of x in the neighborhood of the origin:

$$y = 1 - Bx + \frac{4}{3} x^{3/2}, \quad B \sim 1.58, \tag{4}$$

a result which E. D. Baker extended in the following series:[*]

$$y = 1 - Bx + \frac{1}{3} x^3 - \frac{2}{15} B x^4 + \cdots$$
$$+ x^{3/2} \left[\frac{4}{3} - \frac{2}{5} Bx + \frac{3}{70} B^2 x^2 + \frac{4}{63} \left(\frac{2}{3} + \frac{B^3}{16} \right) x^3 + \cdots \right], \tag{5}$$

with the more accurate value: $B = 1.588588$.

Observing that equation (1) has the particular solution:

$$y_1(x) = 144/x^3, \tag{6}$$

which satisfies the second condition in (2) but not the first, S. A. Sommerfeld achieved the following interesting approximation:[*]

$$y = y_1(x) \{ 1 + [y_1(x)]^{\lambda_1/3} \}^{\lambda_2/2}, \tag{7}$$

where $\lambda_1 = 0.772$ is the positive root and $\lambda_2 = -7.772$ is the negative root of the equation:

$$\lambda^2 + 7\lambda - 6 = 0. \tag{8}$$

To obtain this the following transformation is first made:

$$x = 1/t, \quad y = w/t, \tag{9}$$

which reduces equation (1) to the form:

$$t^4 \frac{d^2 w}{dt^2} = w^{3/2}. \tag{10}$$

The particular solution (6) now becomes: $w_1 == 144\ t^4$ and the boundary conditions (2) are the following:

$$w \sim t \text{ as } t \to \infty, \quad w(0) = 0. \tag{11}$$

The solution of (10) is now assumed to be of the form

$$w = w_1 (1 + \alpha t^\lambda), \tag{12}$$

---

[*]For these references see the *Bibliography*.

and when this is substituted in (10), we obtain the following expansion:

$$12^3 t^6 \left[1 + \frac{\alpha}{12}(4+\lambda)(3+\lambda)t^\lambda + \ldots \right] = 12^3 t^6 \left(1 + \frac{3}{2}\alpha t^\lambda + \ldots \right).$$

When the coefficients of α are equated, it is found that λ must satisfy equation (8).

Since the second boundary condition in (11) is satisfied, equation (12) must now be adapted so that the first boundary condition is also satisfied. For this purpose, one now writes

$$W = w_1(1+\beta t^\lambda)^n = [144t^3(1+\beta t^\lambda)^n]t. \tag{13}$$

It is thus clear that the first boundary condition will be satisfied if β and n are chosen so that

$$144t^3(1+\beta t^\lambda)^n = 144t^3\beta^n t^{\lambda n}(1+\beta^{-1}t^{-\lambda})^n \sim 1 \tag{14}$$

as $t \to \infty$.

Since the condition is satisfied provided $\lambda n + 3 = 0$, $144\beta^n = 1$, and since $\lambda_1\lambda_2 = -6$, the desired approximation is attained if $\lambda = \lambda_1$, $n = -3/\lambda_1 = \lambda_2/2$. Returning to the original variables x and y, we obtain equation (7).

The numerical integration of equation (1) has been made by Fermi and by V. Bush and S. H. Caldwell, the latter using a differential analyzer for that purpose. For small values of x the Baker series is adequate, but not for values much beyond $x=1$. The Sommerfeld approximation gives surprisingly good estimates when x is large, but underestimates y near the origin, as one sees from the following table:

x	y(Fermi)	y(Bush-Caldwell)	y(Sommerfeld)
0. 0	1. 000	1. 000	1. 000
0. 5	0. 607	0. 607	0. 556
1. 0	0. 425	0. 425	0. 385
2. 0	0. 244	0. 247*	0. 221
3. 0	0. 157	0. 156*	0. 143
4. 0	0. 108	0. 106*	0. 099
5. 0	0. 079	0. 0788	0. 072
10	0. 024	0. 0244	0. 0228
20	0. 0056	0. 0058	0. 0056
30	0. 0022	0. 0022	0. 0022
100	0. 00010		0. 00010

*Interpolated values.

(b) The second equation of interest to us in this section is what is called the "White-dwarf" equation, namely,

$$x\frac{d^2y}{dx^2}+2\frac{dy}{dx}+x(y^2-C)^{3/2}=0, \tag{15}$$

which S. Chandrasekhar introduced in his study of the gravitational potential of these degenerate (white-dwarf) stars.* The boundary conditions are the following:

$$y(0)=1, \quad y'(0)=0.$$

This equation is one of Emden type (See Section 7), where $f(y)=(y^2-C)^{3/2}$. In fact, it reduces to Emden's equation with index $n=3$ when $C=0$.

An expansion exists in the neighborhood of the origin, which Chandrasekhar gives as follows:

$$y=1-\frac{q^3}{6}x^2+\frac{q^4}{40}-\frac{q^5}{7!}(5q^2+14)x^6+\frac{q^6}{3\cdot9!}(339q^2+280)x^8$$
$$+\frac{q^7}{5\cdot11!}(1425q^4+11436q^2+4256)x^{10}+\ldots, \tag{16}$$

where $q^2=C-1$.

Tables of the solutions of equation (15) were computed by Chandrasekhar for values of C varying from .01 to .8 over ranges varying from $x=3.5$ to $x=5.3$, together with certain auxiliary quantities connected with the solutions. Graphs of these functions for a few values of C are shown in Figure 39.

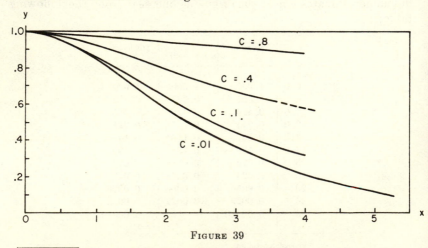

FIGURE 39

---

*For the background of this equation see Chandrasekhar: *Stellar Structure.* Chap. 11.

(c) Another equation of considerable interest, called the *Langmuir equation*, is the following:

$$3y\frac{d^2y}{dx^2}+\left(\frac{dy}{dx}\right)^2+4y\frac{dy}{dx}+y^2=1. \tag{17}$$

This equation appeared in connection with the theory of the flow of a current from a hot cathode to a positively charged anode in a high vacuum. The cathode and anode are long coaxial cylinders and the independent variable y is defined by the equation: $y=f(r/r_0)$, where r is the radius of the anode enclosing a cathode of radius r_0. The independent variable x is given by: $x=\log\ (r/r_0)$.

The current i is defined by the equation

$$i^2=\frac{8}{81}\left(\frac{e}{m}\right)\frac{V^3}{r^2y^4}, \tag{18}$$

where i is the electron current per unit length along the axis, V the voltage at any point P, r the radius at P, e and m the charge and mass respectively of an electron.

The equation is converted into a somewhat more tractable form by means of the transformation: $y=ze^{-x/2}$ and thus becomes:

$$3z\frac{d^2z}{dx^2}+\left(\frac{dz}{dx}\right)^2-e^x=0. \tag{19}$$

One now observes that equation (17) has an analytic expansion in the neighborhood of the point $x=x_0$ which assumes arbitrarily given values y_0 and y_0' provided $y_0\neq0$. But when $y_0=0$, then there exists a solution with a *movable zero*, that is to say,

$$y=(x-x_0)+A^2(x-x_0)^2+A^3(x-x_0)^3+\ .\ .\ .\ , \tag{20}$$

where A_2, A_3, etc., are fixed values, but x_0 is arbitrary.

One also observes that $y=1$ is a singular solution of the equation.

In the original solution of equation (17) it was assumed that solution (20), when x_0 is set equal to 0, approaches asymptotically the value of the singular solution. There is no reason to believe that this is, indeed, the case.

In the actual solution of the equation the coefficients A_n were computed through $n=14$ and y evaluated from $x=x_0=0$ to $x=4.2$. Beyond this point values were obtained by means of an analytic extension based upon integration.

(d) The final equation which we shall consider, due to R. E. Kidder,* is the following:

$$\sqrt{1-\alpha y}\,\frac{d^2y}{dx^2}+2x\,\frac{dy}{dx}=0, \quad 0<\alpha<1, \tag{21}$$

which appears in the problem of the unsteady flow of gas through a semi-infinite porous medium.

The origin of the equation is attractive since it appears in the one-dimensional problem obtained from the following nonlinear partial differential equation:

$$\nabla^2(P^2)=A^2\,\frac{\partial P}{\partial t}, \tag{22}$$

where A is a constant and ∇^2 is the Laplacian operator.

In terms of dimensionless variables p, ξ, and τ, the one-dimension equation obtained from (22) is the following:

$$\frac{\partial}{\partial\xi}\left(p\,\frac{\partial p}{\partial\xi}\right)=\frac{\partial p}{\partial\tau}. \tag{23}$$

Introducing the new variables y and x defined as follows:

$$p^2=1-\alpha y, \quad x=\xi/(2\sqrt{\tau}), \quad 0<\alpha<1, \tag{24}$$

we reduce the partial differential equation (23) to the ordinary nonlinear equation (21).

The boundary conditions required by the physical problem are the following:

$$y(0)=1, \quad y(x)\to0, \quad \text{as } x\to\infty. \tag{25}$$

The solution of the equation is now obtained by assuming an expansion of the form:

$$y=y^{(0)}+\alpha y^{(1)}+\alpha^2 y^{(2)}+\ \ldots\ , \tag{26}$$

where the quantities $y^{(i)}$ are functions to be determined.

The function $(1-\alpha y)^{1/2}$ is also expanded as a power series in α and it, together with (26), are introduced into (21). When the coefficients of like powers of the parameter α are equated, a series of linear differential equations in the $y^{(i)}$ are obtained for which integrals are readily computed. The first three of these are as follows:

$$\frac{d^2}{dx^2}\,y^{(i)}+2x\,\frac{d}{dx}\,y^{(i)}=F_i, \tag{27}$$

*See *Bibliography*.

where the F_i have the following values:

$$F_0 = 0, \quad F_1 = -\frac{x}{2}\frac{d}{dx}[y^{(0)}]^2, \quad F_2 = -x\frac{d}{dx}\left\{y^{(0)}y^{(1)}\frac{1}{4}[y^{(0)}]^3\right\}. \tag{28}$$

The boundary values to be satisfied by these functions are the following:

$$y^{(0)}(0) = 1, \quad y^{(0)}(x) \to 0, \quad \text{as } x \to \infty.$$
$$y^{(i)}(0) = 0, \quad y^{(i)}(x) \to 0, \quad \text{as } x \to \infty, \quad i = 1, 2, 3, \text{ etc.} \tag{29}$$

The solutions for the first two equations are as follows:*

$$y^{(0)} = 1 - erf(x), \quad y^{(1)} = -\frac{1}{2\pi}\{y^{(0)}[1 + \sqrt{\pi}\,x e^{-x^2}] - e^{-2x^2}\}, \tag{30}$$

where we use the customary abbreviation:

$$erf(x) = \frac{2}{\sqrt{\pi}}\int_0^x e^{-x^2}dx. \tag{31}$$

The numerical values of the first three coefficients are contained in the following table:

x	$y^{(0)}$	$y^{(1)}$	$y^{(2)}$
0. 0	1. 00000	0. 00000	0. 00000
0. 1	0. 88754	-0.01004	-0.00343
0. 2	0. 77730	-0.01893	-0.00611
0. 3	0. 67137	-0.02584	-0.00766
0. 4	0. 57161	-0.03037	-0.00809
0. 5	0. 47950	-0.03245	-0.00763
0. 6	0. 39614	-0.03236	-0.00661
0. 7	0. 32220	-0.03052	-0.00534
0. 8	0. 25790	-0.02748	-0.00410
0. 9	0. 20309	-0.02376	-0.00299
1. 0	0. 15730	-0.01982	-0.00210
1. 2	0. 08969	-0.01253	-0.00094
1. 5	0. 03389	-0.00514	-0.00025
2. 0	0. 00468	-0.00074	-0.00003

*The explicit value of $y^{(2)}$ is given by Kidder, but is a function of some complexity.

Chapter 13

Nonlinear Integral Equations

1. Introduction

THE DIFFICULTIES which we have encountered in the solution of nonlinear differential equations are not diminished when differential operators are replaced by integral operators. Some attempts to surmount them have been made, however, and we shall describe in this chapter the present status of a theory which must await the future efforts of mathematicians for a more satisfactory formulation.

We have already seen the necessity for such a theory in the attempts of Volterra to incorporate in his problem of the growth of populations the influences of heredity. Thus, referring to Section 6 of Chapter 5, we considered the problem of the growth of a single population (y) in which the growth was influenced (a) by a generative factor proportional to the population, (b) an inhibiting influence proportional to the square of the population, and (c) a heredity component composed of the sum of individual factors encountered in the past. This problem led to an integro-differential equation of the following form:

$$\frac{1}{y}\frac{dy}{dt}=a+by+\int_c^t K(t,s)y(s)ds. \tag{1}$$

In the case of two competing populations, one preying on the other, Volterra introduced the following system:

$$\frac{1}{x}\frac{dx}{dt}=a-by-\int_{-\infty}^t K_1(t-s)y(s)ds,$$

$$\frac{1}{y}\frac{dy}{dt}=-\alpha y+\beta x+\int_{-\infty}^t K_2(t-s)x(s)ds, \tag{2}$$

where a, b, α, and β are positive constants.

The existence theorems for equations of this type do not vary greatly, however, from those which we have introduced in Chapter 4 for the differential equation:

$$\frac{dy}{dx}=f(x,y); \tag{3}$$

and in Chapter 7 for the system:

$$\frac{dy}{dx}=f(x,y,z), \qquad \frac{dz}{dx}=g(x,y,z). \tag{4}$$

413

This follows from the fact that the existence theorem of Picard depends upon expressing equation (3) as the integral equation:

$$y=y_0+\int_{x_0}^{x} f(x,y)\,dx, \tag{5}$$

and system (4) in the following form:

$$y=y_0+\int_{x_0}^{x} f(x,y,z)\,dx, \quad z=z_0+\int_{x_0}^{x} g(x,y,z)\,dx. \tag{6}$$

A generalization of (5) can be written:

$$y(x)=f(x)+\int_{a}^{x} K[x,s,y(s)]\,ds, \tag{7}$$

which includes as a special case the linear Volterra equation of second kind, namely,

$$y(x)=f(x)+\int_{a}^{x} K(x,s)y(s)\,ds. \tag{8}$$

Existence theorems for equation (7) have been given by T. Lalesco, E. Cotton, M. Picone, and others (See *Bibliography*) in which the essential ingredient is an adaptation of a Lipschitz condition to the more general problem. G. C. Evans extended these proofs to a functional equation sufficiently general to include integro-differential equations such as equation (1) above.

Among special cases which have been studied may be mentioned the following introduced by E. Schmidt:

$$y(x)+\int_{0}^{x} K(x,s)y(s)ds+ \tag{9}$$

$$\Sigma\int_{0}^{x}\cdots\int_{0}^{s_{n-1}} K(x;s_1,s_2,\ldots,s_n)y(s_1)^{a_1}h(s_2)^{b_1}\ldots y(s_n)^{a_n}h(s_n)^{b_n}ds_1ds_2\ldots ds_n=0,$$

where the sum extends over the exponents a_i and b_i, one of which is assumed to differ from zero.

Of less generality is the equation of Lalesco:

$$y(x)=f(x)+\int_{0}^{x}[K_1(x,s)u(s)+K_2(x,s)u^2(s)+\ldots+K_n(x,s)u^n(s)]ds. \tag{10}$$

The equations which we have described above are seen to be generalizations of integral equations of Volterra type, that is to say, equations in which one of the limits of the integral is the independent variable. But some attention has been given to equations of Fredholm type, that is, equations in which the region of integration is the rectangle: $a \leqq s,\ x \leqq b$.

Lalesco has given an existence proof under general conditions for the equation

$$y(x) = f(x) + \int_a^b K[x,s,y(s)] \, ds, \qquad (11)$$

and G. Bratu has studied the following special cases:

$$y(x) = f(x) + \int_0^1 K(x,s) \, y^2(s) \, ds, \qquad (12)$$

and

$$y(x) = f(x) + \int_0^1 K(x,s) \, e^{y(s)} \, ds. \qquad (13)$$

2. An Existence Theorem for Nonlinear Integral Equations of Volterra Type

We shall consider in this section conditions under which a solution exists for the equation:

$$y(x) = f(x) + \int_a^x K[x,s,y(s)] \, ds. \qquad (1)$$

It will be convenient to make the following assumptions:

(a) The function $f(x)$ is integrable and bounded, $|f(x)| < f$, in the interval $a \leq x \leq b$.

(b) The following Lipschitz condition is satisfied by $f(x)$ in the interval (a,b):

$$|f(x) - f(x')| < k|x - x'|. \qquad (2)$$

(c) The function $K(x,y,z)$ is integrable and bounded,

$$|K(x,y,z)| < K,$$

in the domain: $a \leq x, y \leq b, |z| < c$.

(d) The following Lipschitz condition is satisfied by $K(x,y,z)$ within its domain of definition:

$$|K(x,y,z) - K(x,y,z')| < M|z - z'|. \qquad (3)$$

Employing now the method of successive approximation, we assume as the first approximation:

$$y_0(x) = f(x) - f(a),$$

from which we get

$$y_1(x) = f(x) + \int_a^x K[x,s,y_0(s)] \, ds, \qquad (4)$$

and in general,

$$y_n(x) = f(x) + \int_a^x K[x,s,y_{n-1}(s)] \, ds. \qquad (5)$$

Making use of our assumptions, we now obtain the following bound for $y_1(x)$:

$$|y_1(x)| < (f+K)|x-a|, \quad |x-a| < a'. \tag{6}$$

If x is so limited that $|x-a| < f/(f+K)$, then $|y_1| < f$.

Denoting by h the smaller of the numbers a' and $f/(f+K)$, we shall have, for each approximation, the inequality:

$$|y_n(x)| < f, \quad 1x-a| < h. \tag{7}$$

Let us now construct the series:

$$y(x) = y_0(x) + [y_1(x) - y_0(x)] + \ldots + [y_n(x) - y_{n-1}(x)] + \ldots, \tag{8}$$

which, by virtue of (5), furnishes the desired solution of the original equation, provided it converges uniformly. The uniform convergence of the series is readily established from the following considerations.

Since we have

$$y_n(x) - y_{n-1}(x) = \int_a^x \{K[x,s,y_{n-1}(s)] - K[x,s,y_{n-2}(s)]\} ds, \tag{9}$$

it follows from (3) that we have the inequality:

$$|y_n(x) - y_{n-1}(x)| < M \left| \int_a^x [y_{n-1}(s) - y_{n-2}(s)] \, ds \right|. \tag{10}$$

Letting $n = 1$, 2, 3, etc., we obtain the following sequence of inequalities:

$$|y_1(x) - y_0(x)| < M|x-a|,$$

$$|y_2(x) - y_1(x)| < M^2 \frac{|x-a|^2}{2!},$$

$$|y_3(x) - y_2(x)| < M^3 \frac{|x-a|^3}{3!},$$

and in general,

$$|y_n(x) - y_{n-1}(x)| < M^n \frac{|x-a|^n}{n!}. \tag{11}$$

Since we have $|x-a| < h$, the series

$$Y = f + Mh + \frac{(Mh)^2}{2!} + \frac{(Mh)^3}{3!} + \ldots + \frac{(Mh)^n}{n!} + \ldots, \tag{12}$$

forms a majorante for (8).

3. The Integro-Differential Problem of Volterra

We have described in Section 1 the nonlinear integro-differential equation of Volterra. Since very little progress has been made in its solution in a quantitative way, even small results may prove of interest in indicating the influence of the hereditary factor upon the stable configurations which have been found when this term is omitted.

We shall begin with the one-variable problem,

$$\frac{dy}{dt} = ay - by^2 + y \int_c^t K(t-s)y(s)ds, \tag{1}$$

which generalizes the logistic equation described earlier in Chapter 5 (Section 2). The constants a and b are assumed to be positive and c is arbitrary, although we shall find it convenient to choose $c=0$.

Since the kernel $K(t-s)$ is a function of one variable, we can write it formally as follows:

$$K(z) = K_0 + K_1 z + \frac{1}{2!} K_2 z^2 + \ \ldots \ . \tag{2}$$

Observing the formula

$$_cD_t^{-n}y(x) = \int_c^t ds \int_c^s ds \ \cdots \ \int_c^s y(s)ds = \frac{1}{\Gamma(n)} \int_c^t (t-s)^{n-1}y(s)ds, \tag{3}$$

where $_cD_t^{-n}y$ is the general integration operator, we can now write (1) as follows:

$$\frac{1}{y}\frac{dy}{dt} = a - by + \int_c^t \left[K_0 + K_1(t-s) + \frac{1}{2!} K_2(t-s)^2 + \ \ldots \ \right] y(s)ds,$$
$$= a - by + K_0 \ _cD_t^{-1}y + K_1 \ _cD_t^{-2}y + K_2 \ _cD_t^{-3}y + \ \ldots \ . \tag{4}$$

Let us now assume that all the K_i are zero except K_0, which we shall denote by K. If we now differentiate (4), we obtain the following differential equation

$$yy'' = y'^2 - by'y^2 + Ky^3. \tag{5}$$

Although this is a differential equation of second order, its solution involves the specification of only one arbitrary condition, let us say $y = y_0$ at $t = c$, since from (1) we obtain the necessary condition

$$y_0' = ay_0 - by_0^2. \tag{6}$$

Since equation (5) does not appear to be integrable in terms of known functions, its solution is obtained either by means of analytic continuation or the differential analyzer.

A somewhat simpler form can be given to equation (5) by means of the transformation: $y=au/b$, $t=\tau/a$. We thus get

$$u\frac{d^2u}{d\tau^2}=\left(\frac{du}{d\tau}\right)^2-u^2+\lambda u^3, \tag{7}$$

where $\lambda=K/ab$.

Condition (6) reduces to

$$\frac{du_0}{d\tau}=u_0-u_0^2 b, \tag{8}$$

in terms of an initial value $u=u_0$.

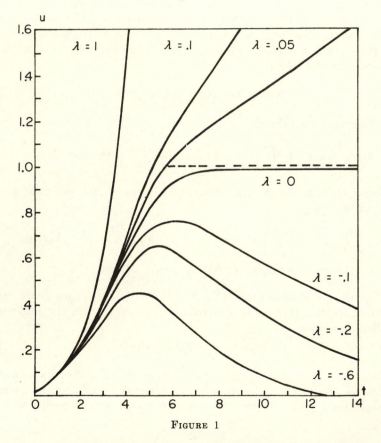

FIGURE 1

It is now possible to construct a parametric set of solutions of equation (7) in terms of λ and these are shown in Figure 1. The case where $\lambda=0$ gives us the logistic, or growth curve, which is seen to have its characteristic asymptotic approach to the line $u=1$.

If $\lambda>0$, that is, when K is positive, the growth increases without limit.

If $\lambda<0$, let us say, $\lambda=-\mu$, u attains a maximum value, and at this point the curve has the following radius of curvature:

$$R=\frac{1}{u(1+\mu u)}.$$

As μ increases the point of maximum moves toward the origin and R approaches zero, thus indicating a continuous flattening of the curve.

An associated problem of some interest, which was studied by Z. Szatrowski, is that of designing a factor of heredity which will produce an arbitrarily assumed function $y=y(t)$.

Setting $c=0$ and making the transformation: $t-s=x$, we obtain equation (1) in the following form:

$$F(y)=\int_0^t y(t-x)K(x)dx, \tag{9}$$

where $F(y)=y'/y-a+by$. If y_0 denotes the value of y when $t=0$, $F(y)$ must satisfy the boundary condition: $F(y_0)=0$.

Taking the derivative of (9) with respect to t and denoting dF/dt by $f(t)$, we obtain the following equation:

$$f(t)=y(0)K(t)+\int_0^t y'(t-x)K(x)dx. \tag{10}$$

This we observe is a Volterra integral equation of second kind in which the unknown variable $K(x)$ appears linearly and the derivative of the given function $y(t)$ is the kernel.

The Two-Variable Problem.

We consider next the problem presented by the growth of two conflicting populations both subject to the influence of a factor of heredity. The general problem is formulated in terms of the system given by equations (2) of Section 1. But for simplicity we shall consider only the following reduced system:

$$\frac{dx}{dt}=ax-bxy-K_1x\int_c^t y(s)\,ds, \quad \frac{dy}{dt}=-\alpha y+\beta xy+K_2y\int_c^t (x)\,ds, \tag{11}$$

where a, b, α, β are positive constants, but K_1 and K_2 can have either sign.

By the device employed in the one-variable problem, this system can be reduced to the following differential system of second order:

$$x\,x''=x'^2-by'x^2-K_1x^2y, \quad y\,y''=y'^2+\beta x'y^2+K_2xy^2. \tag{12}$$

If $x=x_0$ and $y=y_0$ are arbitrarily given when $t=c$, then the solution of (12) has been defined since x_0' and y_0' satisfy the following relationships:

$$x_0'=ax_0-bx_0y_0, \quad y_0'=-\alpha y_0+\beta x_0y_0. \tag{13}$$

We shall now consider three particular cases, which will reveal the effects of a constant hereditary factor upon the mutual growth of the two populations. These cases are the following:

I. $x'=2(x-xy)-0.05x\displaystyle\int_0^t y(s)ds, \quad a=b=2, \ K_1=0.05,$

$y'=-(y-xy)+0.05y\displaystyle\int_0^t x(s)ds, \ \alpha=\beta=1, \ K_2=0.05;$

II. $x'=2(x-xy)+0.05x\displaystyle\int_0^t y(s)ds. \quad a=b=2, \ K_1=-0.05,$

$y'=-(y-xy)-0.05y\displaystyle\int_0^t x(s)ds, \ \alpha=\beta=1, \ K_2=-0.05;$

III. $x'=2(x-xy)+0.05x\displaystyle\int_0^t y(s)ds, \quad a=b=2, \ K_1=-0.05,$

$y'=-(y-xy)+0.05y\displaystyle\int_0^t x(s)ds, \ \alpha=\beta=1, \ K_2=0.05. \tag{14}$

The case where $K_1=K_2=0$ has already been discussed in Section 4 of Chapter 5 where it was shown that both x and y are stable and periodic functions of t. The phase-diagram was an ovaloid figure containing the point $x=1$, $y=1$ in its interior.

But when the hereditary factor is introduced the periodicity of the solutions is destroyed and the motion may become unstable. This is the case in I where both populations, as shown in Figure 2, are observed to increase. The origin of t is assumed for the point $x=1$, $y=2$ and is designated by P in both the phase-diagram and the graphs of the functions. The number of oscillations per unit of time increases as the two populations grow in size.

Case II is illustrated in Figure 3. The motion is observed to be stable and in time both populations reach extinction. As t increases the number of oscillations per unit of time diminishes.

Figure 4 shows the behavior of the functions for Case III. The motion is not periodic, but is observed to be stable for the variable y and unstable for x. As t increases the population y ultimately reaches extinction, but the population x increases without limit. The number of oscillations per unit of time appears to diminish slowly.

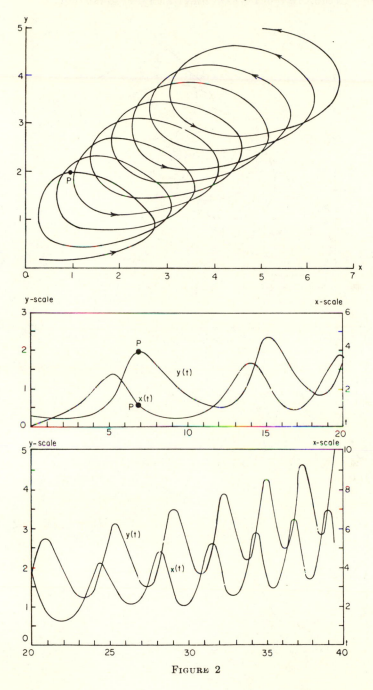

FIGURE 2

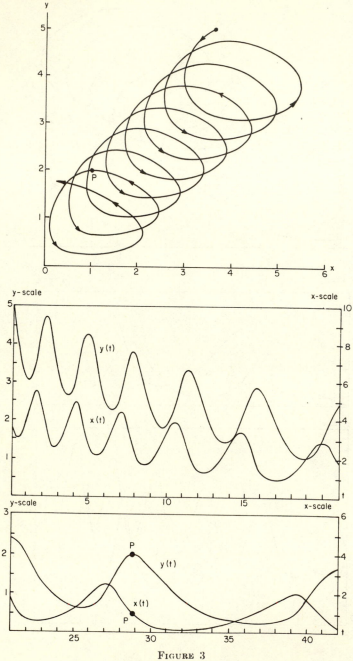

FIGURE 3

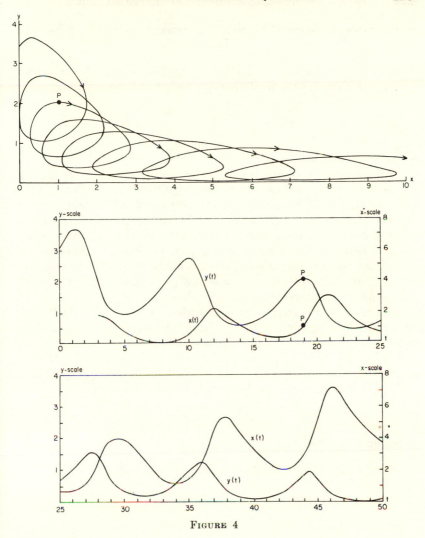

FIGURE 4

PROBLEMS

1. Show that when the hereditary function in equation (1), $c = 0$, is the following

$$K(z) = 1 + at - \log \cos z + \tan^2 z,$$

then $y = \cos t$ is a particular solution.

2. If $y = A(e^{pt} + e^{-pt})$ is a solution of equation (1), $c = 0$, show that the hereditary function is the following:

$$K(z) = -\frac{b}{a} p^2 [1 + az - \log \cosh pt - (\tanh pz)^2].$$

4. An Existence Theorem for Nonlinear Integral Equations of Fredholm Type

We now consider the problem of establishing criteria for the existence of solutions for the nonlinear equation:

$$y(x) = f(x) + \lambda \int_a^b K[x, s, y(s)] \, ds, \tag{1}$$

where λ is a parameter.

From the theory of the linear Volterra and Fredholm equations, we know that the parameter λ plays a significant role. Perhaps the most essential difference between the two equations is found in the fact that, for bounded kernels, integrable functions, and a finite range of integration, the solution in the Volterra case is an entire function of λ, while in the Fredholm case it is a meromorphic function of the parameter with singularities at the zeros of the Fredholm determinant $D(\lambda)$. This same difference is observed in general between the two cases when the integral equation is nonlinear.

In order to establish criteria under which a solution exists for (1), we make assumptions similar to those introduced in Section 2 for equations of Volterra type. These are as follows:

(a) The function $f(x)$ is bounded in the interval: $a \leq x \leq b$, that is, $|f(x)| < f$.

(b) The kernel $K(x,y,z)$ is integrable and bounded,

$$|K(x,y,z)| < K, \tag{2}$$

in the domain D: $a \leq x, y \leq b$, $|z| < c$.

(c) $K(x,y,z)$ satisfies the Lipschitz condition in D, namely,

$$|K(x,y,z) - K(x,y,z')| < M|z - z'|. \tag{3}$$

By successive approximations we now have

$$y_0(x) = f(x) - f(a),$$

$$y_1(x) = f(x) + \lambda \int_a^b K[x,s,y_0(s)] ds, \tag{4}$$

and, in general,

$$y_n(x) = f(x) + \lambda \int_a^b K[x,s,y_{n-1}(s)] ds.$$

From these we obtain

$$y_1-y_0=\lambda\int_a^b K[x,s,y_0(s)]\ ds+f(a),$$

$$y_2-y_1=\lambda\int_a^b \{K[x,s,y_1(s)]-K[x,s,y_0(s)]\}ds$$

. . . .

$$y_n-y_{n-1}=\lambda\int_a^b \{K[x,s,y_{n-1}(s)]-K[x,s,y_{n-2}(s)]\}ds.$$

From the conditions given above, we have

$$|y_1-y_0|<|\lambda|K(b-a)+|f(a)|=|\lambda|(b-a)K\left[1+\frac{f}{|\lambda|K(b-a)}\right],$$

$$=|\lambda|m(b-a),$$

where $m=K\{1+f/[|\lambda|K(b-a)]\}$.

From this inequality and the Lipschitz condition, we get

$$|y_2-y_1|<|\lambda|M\int_a^b|y_1-y_0|ds <|\lambda|^2Mm(b-a)^2 <|\lambda|^2k^2(b-a)^2,$$

where k is the larger of the two numbers M and m.
Similarly we obtain the inequalities:

$$|y_3-y_2|<|\lambda|^3k^3(b-a)^3,$$

$$|y_n-y_{n-1}|<|\lambda|^nk^n(b-a)^n.$$

A majorante for the series

$$y(x)=y_0+[y_1(x)-y_0(x)]+ \ . \ . \ . \ +[y_n(x)-y_{n-1}(x)]+ \ . \ . \ . \ , \qquad (5)$$

is furnished by the sum

$$Y=f+\sum_{n=1}^{\infty}|\lambda|^nk^n(b-a)^n,$$

and thus the series converges uniformly for all values of λ for which we have

$$|\lambda|<\frac{1}{k(b-a)}. \qquad (6)$$

Although the condition (6) is equivalent to that obtained when equation (1) is linear, the role played by λ in the case where $f(x)\equiv0$ is quite different in nonlinear equations from that which it has in the linear case. This difference will be shown in the next section.

5. A Particular Example

As a special case we shall consider the equation

$$u(x) = \lambda \int_0^1 (x-t) u^2(t) dt. \tag{1}$$

Without loss of generality we can set $\lambda = 1$, for if $v(x)$ is any solution of (1), then $u(x) = \lambda v(x)$ is a solution of the equation:

$$u(x) = \int_0^1 (x-t) u^2(t) dt. \tag{2}$$

Since (2) can be written

$$u(x) = x \int_0^1 u^2(t) dt - \int_0^1 t u^2(t) dt,$$

it is clear that its solution is a linear function of x that is

$$u(x) = px + q, \tag{3}$$

where p and q are to be determined.

Introducing this function into (2), we obtain

$$px + q = x \left(\frac{1}{3} p^2 + pq + q^2 \right) - \left(\frac{1}{4} p^2 + \frac{2}{3} pq + \frac{1}{2} q^2 \right). \tag{4}$$

Equating the coefficients of x and the constant terms and simplifying, we have the following system of equations for the determination of p and q:

$$p^2 + 3pq + 3q^2 = 3p, \tag{5}$$

$$3p^2 + 8pq + 6q^2 = -12q. \tag{6}$$

If p and q are points in a Cartesian system of coordinates, then both (5) and (6) represent ellipses which pass through the origin: $p = q = 0$. These ellipses are graphically represented in Figure 5 [$C_1 = (5)$, $C_2 = (6)$], and are observed to intersect in two real points, namely, the origin and the point $P = (-6, 12)$.

Therefore equation (2) has one real solution other than the trivial one, $u(x) = 0$, that is,

$$u_1(x) = 12x - 6. \tag{7}$$

But it also has two complex solutions obtained from the intersections of the ellipses in the points: $(3+3i, -6)$ and $(3-3i, -6)$, that is,

$$u_2(x) = -6x + 3(1+i), \tag{8}$$

$$u_3(x) = -6x + 3(1-i).$$

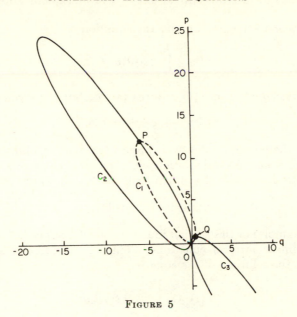

FIGURE 5

By a simple extension of this example one can derive the following theorem:

The equation

$$u(x) = \int_a^b [A(t)x + B(t)]u^2(t)\,dt, \tag{9}$$

in which $A(t)$ *and* $B(t)$ *are real functions integrable in the interval* (a,b), *has three solutions other than* $u(x) = 0$, *at least one of which is real.*

By an argument similar to that used in the example, we obtain the following system of equations defining p and q in the solution: $u(x) = px + q$:

$$p^2 \int_a^b t^2 A(t)\,dt + 2pq \int_a^b t A(t)\,dt + q^2 \int_a^b A(t)\,dt = p,$$

$$p^2 \int_a^b t^2 B(t)\,dt + 2pq \int_a^b t B(t)\,dt + q^2 \int_a^b B(t)\,dt = q. \tag{10}$$

If this system is solved for p (or q), the resulting equation which determines p is a quartic, one root of which is 0. When the quartic is reduced by the factor p, the resulting equation is a cubic with real coefficients and hence must have one real root. The theorem results from this observation.

As a second example, consider the equation

$$u(x) = \int_0^1 (x+t)u^2(t)dt. \tag{11}$$

The first equation in (10) reduces to (5), but the second becomes

$$3p^2 + 8pq + 6q^2 = 12q, \tag{12}$$

which is the ellipse C_3 shown in Figure 5. We see from the graphs that there is only one real solution, corresponding to the point Q. This solution to three decimal approximation is the following:

$$u(x) = 0.726x + 0.461. \tag{13}$$

The examples of this section illustrate the difference between linear and nonlinear integral equations. Thus, if $K(x,t)$ is a symmetric function, the solutions of the equation:

$$u(x) = \lambda \int_a^b K(x,t)u(t)dt, \tag{14}$$

will depend upon the parameter λ and will all be real. If $K(x,t)$ is a skew-symmetric function, the solutions are all complex. In the examples just given, however, one real solution and two complex solutions exist for both (1) and (11). The fact that the kernel in the first case is skew-symmetric and in the second is symmetric had nothing to do with the situation. Moreover, in the nonlinear equations the solutions were not dependent upon characteristic values of the parameter λ, a very conspicuous aspect of linear integral equations.

It should be pointed out that the method just given is also applicable to the solution of the equation:

$$u(x) = Ax + B + \lambda \int_a^b [A(t)x + B(t)]u^2(t)dt, \tag{15}$$

since the form of the solution remains unchanged. But in this case the solution depends upon λ, although not in the same critical sense as in the theory of linear integral equations. We can no longer say, however, that there exists a real solution, since the equations of system (10) are each augmented by an arbitrary constant. In this case there will be, in general, four nontrivial solutions.

PROBLEMS

1. Discuss the solution of the equation:

$$u(x) = \int_0^1 [x \sin \pi t + \cos \pi t]u^2(t)dt.$$

2. Given the equation,

$$u(x) = -\frac{7}{3}x - \frac{3}{2} + \int_0^1 (x-t)u^2(t)\,dt,$$

and the fact that $u(x) = 2x - 3$ is a solution, find the other three solutions.

6. The Equation $u(x) = \lambda \int_a^b K(x,t)u^n(t)\,dt.$

The particular equation solved in the preceding section is a special case of the following more general one:

$$u(x) = \lambda \int_a^b K(x,t)u^n(t)\,dt, \tag{1}$$

the solution of which we shall now consider.

Except in the linear case when $n=1$ the parameter λ can be set equal to unity without loss of generality. For if we make the transformation

$$u(x) = \lambda^m v(x),$$

equation (1) reduces to the following:

$$\lambda^m v(x) = \lambda^{mn+1} \int_a^b K(x,t)v^n(t)\,dt. \tag{2}$$

If m is chosen so that $mn+1=m$, that is,

$$m = -1/(n-1), \; n \neq 1.$$

we get

$$v(x) = \int_a^b K(x,t)v^n(t)\,dt. \tag{3}$$

We shall assume that the kernel has the following form:

$$K(x,t) = \sum_{i=1}^R X_i(x)Y_i(t), \tag{4}$$

when the functions: $X_1(x)$, $X_2(x)$, $X_3(x)$, . . . are assumed to form a linearly independent set.

Introducing this kernel into (1) and setting $\lambda = 1$, we get

$$u(x) = \sum_{i=1}^R X_i(x)\int_a^b Y_i(t)u^n(t)\,dt,$$

$$= \sum_{i=1}^R X_i(x)p_i, \tag{5}$$

where the p_i are constants defined by

$$p_i = \int_a^b Y_i(t) u^n(t) dt. \tag{6}$$

When (5) is introduced into (1), we have

$$\sum_{i=1}^R X_i(x) p_i = \sum_{i=1}^R X_i(x) \int_{ba}^b Y_i(t) \left[\sum_{j=1}^R X_j(t) p_j \right]^n dt.$$

Since the $X_i(x)$ are by assumption linearly independent functions of x, we can equate their coefficients and thus obtain the following system of equations for the determination of the constants p_i:

$$p_i = \int_a^b Y_i(t) \left[\sum_{j=1}^R X_j(t) p_j \right]^n dt, \quad i = 1, 2, \ldots, R. \tag{7}$$

We thus have a set of R algebraic equations of nth degree for the determination of the constants p_i. These equations will have, in general, R^n solutions, one of which is the trivial case: $p_i = 0$, since these solutions are obtained from R algebraic equations with real coefficients in each of the variables and each of degree $N = R^n$.

Since the form of these equations is as follows:

$$p_i(A_i p_i^{N-1} + B_i p_i^{N-2} + \ldots) = 0, \tag{8}$$

there will be at least one real set of values of the p_i provided $N - 1 = R^n - 1$ is an odd integer, that is to say, if R is an even integer.

We thus have the following theorem:

If in equation (1) the kernel has the form:

$$K(x,t) = \sum_{i=1}^R X_i(x) Y_i(t),$$

where the functions $X_1(x)$ *are linearly independent, there will exist in general* $N - 1 = R^n - 1$ *solutions other than the trivial one,* $u = 0$, *and of these at least one will be a real solution provided* R *is an even integer.*

When $n = 2$, equations (7) reduce to the form

$$p_i = F(p,p), \quad i = 1, 2, \ldots, R, \tag{9}$$

where $F(p,p)$ is the real quadratic form

$$F(p,p) = \sum_{jk} a_{ik}^{(i)} p_j p_k, \tag{10}$$

in which

$$a_{ik}^{(i)} = a_{mj}^{(i)} = \int_a^b X_j(t) X_k(t) Y_i(t) dt. \tag{11}$$

In this case there will exist R^2-1 solutions with at least one real solution when R is an even integer.

The actual determination of the solutions, when R is not too large, is probably most easily achieved by forming the eliminants for each p_i by means of *Sylvester's dialytic method.* Thus p_1 is eliminated between equations corresponding to $i=i$ and $i=2$, then between $i=2$ and $i=3$, and so on. In this way $n-1$ equations are obtained containing the remaining variables. From these p_2 in turn is eliminated and the process is continued until the final eliminant is obtained, which will turn out to be an equation of degree $N=R^n$ in p_R of form (8) above.

This process is illustrated readily by considering equations (5) and (6) of Section 5, namely.

$$p^2+3pq+3q^2-3p=0,$$
$$3p^2+8pq+6q^2+12q=0. \tag{12}$$

Multiplying each in turn by p, we obtain the following system of four equations in the variables p, p^2, and p^3:

$$p^2 \quad +(3q-3)p+3q^2 \quad\quad =0$$
$$p^3+(3q-3)p^2+3q^2p \quad\quad\quad\quad =0$$
$$3p^2 \quad +8qp \quad\quad +6q^2+12q=0 \tag{13}$$
$$3p^3+ \quad 8qp^2 \quad +(6q^2+12q)p \quad\quad =0$$

If these equations in p, p^2, and p^3 are to be consistent the determinant must be set equal to zero and we thus obtain the eliminant as follows:

$$\begin{vmatrix} 0 & 1 & (3q-3) & 3q^2 \\ 1 & (3q-3) & 3q^2 & 0 \\ 0 & 3 & 8q & (6q^2+12q) \\ 3 & 8q & (6q^2+12q) & 0 \end{vmatrix}=0 \tag{14}$$

which reduces to the equation:

$$3q(q^3-18q+108)=0. \tag{15}$$

A similar computation for the system corresponding to equation (11) of Section 5, namely,

$$p^2+3pq+3q^2-3p=0,$$
$$3p^2+8pq+6q^2-12q=0,$$

(16)

leads to the eliminant:

$$3q(q^3+24q^2+222q-108)=0. \tag{17}$$

7. The Equation of Bratu

The equation of Bratu is the following nonlinear integral equation:

$$y(x)=\lambda\int_a^b G(x,t)\,e^{y(t)}dt, \tag{1}$$

where the kernel is a Green's function defined as follows:

$$G(x,t)=\frac{(b-x)\,(t-a)}{b-a}, \quad t\leqq x.$$
$$=\frac{(b-t)\,(x-a)}{b-a}, \quad t\geqq x.$$

(2)

One can show that the function $y(x)$ which satisfies this equation is any solution of the differential equation

$$\frac{d^2y}{dx^2}+\lambda e^y=0, \tag{3}$$

which also satisfies the two-point boundary condition: $y(a)=y(b)=0$. It will be convenient to assume that $a=0$ and that $\lambda>0$.

We shall now prove the following theorem: (Bratu)

For every value of λ taken between 0 and λ_1, where

$$\lambda_1=\frac{(1.8745\ldots)^2}{b^2}, \tag{4}$$

equation (1) has two real and distinct solutions. These curves C_1 and C_2 are of parabolic form, concave to the x-axis and pass through the points $x=0$ and $x=b$.

As $\lambda\to\lambda_1$, the two curves tend toward a limiting curve C between them. When $\lambda=\lambda_1$, the limit curve is the unique solution of equation (1). For $\lambda>\lambda_1$, the integral equation has no real solution.

Equation (3) can be integrated and we have as a first integral

$$y'^2 - m^2 + 2\lambda(e^y - 1) = 0, \tag{5}$$

where $m = y'(0)$.

Introducing the variable

$$t = 1 + \frac{m^2}{2\lambda}, \tag{6}$$

we get $y'^2 = 2\lambda(t - e^y)$, and thus the solution of (3) in the following form:

$$x = \frac{1}{\sqrt{2\lambda}} \int_0^y \frac{dy}{\sqrt{t - e^y}}. \tag{7}$$

The curve $y = y(x)$ increases to a maximum at $y = \log t$, and hence at this point

$$x = x_1 = \frac{1}{\sqrt{2\lambda}} \int_0^{\log t} \frac{dy}{\sqrt{t - e^y}} = \frac{1}{\sqrt{2\lambda}} \int_1^t \frac{du}{u\sqrt{t - u}}. \tag{8}$$

When $x > x_1$, y diminishes, and since the curve is symmetric with respect to the line $x = x_1$, y is zero when

$$x = 2x_1 = b = \sqrt{\frac{8}{\lambda t}} \log [\sqrt{t} + \sqrt{t - 1}]. \tag{9}$$

Since $m = 0$, $b = 0$ for $t = 1$ and $m = \infty$, $b = 0$ for $t = \infty$, b passes through a maximum value as m varies between 0 and ∞. In order to find this value we compute:

$$\frac{db}{dt} = \sqrt{\frac{2}{\lambda t^2}} \left[\frac{1}{\sqrt{t - 1}} = \frac{\log (\sqrt{t} + \sqrt{t - 1})}{\sqrt{t}} \right]. \tag{10}$$

Equating db/dt to zero, we now solve the resulting transcendental equation and thus find $t = 3.2766 \ldots$. Denoting the maximum value of b by β, we now substitute t in (9) and thus obtain: $\beta = h/\sqrt{\lambda}$, where $h = 1.8745 \ldots$.

The theorem follows readily from these results. If we assume a value of b and regard λ as a variable, then for each value of β there will exist a value of λ and a value of t. If $b = \beta$, then the initial slope of $y(x)$, given by equation (6), denoted by μ, is found to be

$$\mu = 2.1338 \ldots \sqrt{\lambda}.$$

For this value, there exists one curve, namely C, which passes through $x = 0$ and $x = \beta$.

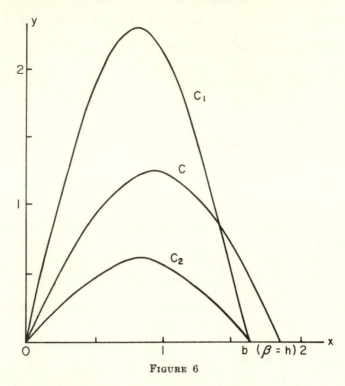

FIGURE 6

But if $b<\beta$, that is to say, if $\lambda_1>\lambda=h^2/\beta^2$, then there will be two curves, C_1 and C_2 which pass through $x=0$ and $x=b$. For the first one the slope m_1 will be greater than μ and for the second the slope m_2 will be less.

If $\lambda_1<\lambda$, then $b>\beta$, but since β is the maximum value which b can have, it is clear that there will be no solution in this case.

The situation is shown in Figure 6, where $\lambda=1$ and $\lambda_1=1.275$.

8. The Nonlinear Convolution Theorem

By a *convolution* (or *Faltung*) we shall mean an integral of the form

$$R(t)=\int_{-\infty}^{\infty} x(s)y(s+t)ds, \tag{1}$$

where $x(s)$ and $y(s)$ are functions integrable separately over the real axis, with individual convergent Fourier transforms:

$$\int_{-\infty}^{\infty} y(s)e^{\beta is}ds \text{ and } \int_{-\infty}^{\infty} x(s)e^{\beta is}ds. \tag{2}$$

If $y(s)$ and $R(t)$ are given functions, then (1) is a linear integral equation in $x(s)$. We shall proceed formally to solve for $x(s)$.

Multiplying $R(t)$ by $e^{\beta it}$ and integrating over the infinite range, we thus obtain:

$$\int_{-\infty}^{\infty} R(t)e^{\beta it}dt = \int_{-\infty}^{\infty} dt \int_{-\infty}^{\infty} x(s)y(s+t)e^{-\beta is}e^{\beta i(t+s)}ds. \qquad (3)$$

Making the transformation: $s+t=p$ and assuming such convergence as is necessary to validate the process, we can now write (3) as follows:

$$\int_{-\infty}^{\infty} R(t)e^{\beta it}dt = \int_{-\infty}^{\infty} y(p)e^{\beta ip}dp \int_{-\infty}^{\infty} x(s)e^{-\beta is}ds. \qquad (4)$$

We now denote by $r(\beta)$ and $r_1(\beta)$ the integrals

$$r(\beta) = \int_{-\infty}^{\infty} R(t)\cos\beta tdt, \qquad r_1(\beta) = \int_{-\infty}^{\infty} R(t)\sin\beta tdt, \qquad (5)$$

and by $a_x(\beta)$, $b_x(\beta)$ and $a_y(\beta)$, $b_y(\beta)$ the corresponding transforms of $x(t)$ and $y(t)$ respectively, that is,

$$a_x(\beta) = \int_{-\infty}^{\infty} x(t)\cos\beta tdt, \qquad b_x(\beta) = \int_{-\infty}^{\infty} x(t)\sin\beta tdt,$$

$$a_y(\beta) = \int_{-\infty}^{\infty} y(t)\cos\beta tdt, \qquad b_y(\beta) = \int_{-\infty}^{\infty} y(t)\sin\beta tdt. \qquad (6)$$

By equating the real and imaginary parts of (3), we obtain the following system of equations:

$$r(\beta) = a_x(\beta)a_y(\beta) + b_x(\beta)b_y(\beta),$$

$$r_1(\beta) = a_x(\beta)b_y(\beta) - a_y(\beta)b_x(\beta). \qquad (7)$$

This system can now be solved in general for $a_x(\beta)$ and $b_x(\beta)$, since its determinant is

$$\Delta = -[a_y^2(\beta) + b_y^2(\beta)_4] \neq 0.$$

The value of $x(s)$ is now obtained from the equation

$$a_x(\beta) + i\, b_x(\beta) = \int_{-\infty}^{\infty} x(s)\, e^{\beta is}\, ds, \qquad (8)$$

by means of the inverse transform:

$$x(s) = \frac{1}{2\pi}\int_{-\infty}^{\infty} [a_x(\beta) + i\, b_x(\beta)]\, e^{-\beta is}\, d\beta. \qquad (9)$$

But the problem which we have just described is essentially different if in equation (1) we assume that $x(s) = y(s) = u(s)$. The linear problem is now replaced by the following nonlinear convolution:

$$R(t) = \int_{-\infty}^{\infty} u(s) \; u(s+t) \; ds. \tag{10}$$

The uniqueness enjoyed by the linear problem has disappeared since the second equation in (7) is identically zero. We shall now prove two theorems relating to the inversion of (10).

THEOREM 1. *If the functions* u(s) *and* R(s) *exist over the infinite range, and if the integrals*

$$a(\beta) = \int_{-\infty}^{\infty} u(s) \; \cos \beta s \; ds, \; b(\beta) = \int_{-\infty}^{\infty} u(s) \; \sin \beta s \; ds,$$

$$r(\beta) = \int_{-\infty}^{\infty} R(s) \; \cos \beta s \; ds, \tag{11}$$

converge, then the functions a(β), b(β), *and* r(β) *are connected formally by the relationship:*

$$r(\beta) = a^2(\beta) + b^2(\beta). \tag{12}$$

The proof of this theorem is merely to observe that (12) is a corollary of (7). If we set $x(t) = y(t) = u(t)$, then $r_1(\beta)$ is zero and $r(\beta)$ reduces to (12).

The second theorem gives the inversion of equation (10) in the following elegant form:[*]

THEOREM 2. *If* R(t) *satisfies the conditions of Theorem 1 and if* r(β) *is defined by* (11), *then* u(s), *the solution of equation* (10), *is given by the following inversion:*

$$u(s) = \frac{1}{2\pi} \int_{-\infty}^{\infty} \sqrt{r(\beta)} \; \cos p(\beta) \; \cos \beta s \; d\beta$$

$$+ \frac{1}{2\pi} \int_{-\infty}^{\infty} \sqrt{r(\beta)} \; \sin p(\beta) \; \sin \beta s \; d\beta, \tag{13}$$

where p(β) *is an arbitrary odd function of* β, *that is,* $p(-\beta) = -p(\beta)$.

In order to prove this theorem, let us denote the first integral in (13) by $u_1(s)$ and the second by $u_2(s)$. From their definition

$$u(s) = u_1(s) + u_2(s), \quad u_1(-s) = u_1(s), \quad u_2(-s) = -u_2(s). \tag{14}$$

Employing the Fourier transform,

$$f(s) = \frac{1}{2\pi} \int_{-\infty}^{\infty} g(\beta) \, e^{\beta i s} \, ds, \quad g(\beta) = \int_{-\infty}^{\infty} f(s) \, e^{-\beta i s} \, ds,$$

and making use of (14), we now obtain

$$\sqrt{r(\beta)} \, \cos p(\beta) = \int_{-\infty}^{\infty} u_1(s) \, \cos \beta s \, ds = \int_{-\infty}^{\infty} u(s) \, \cos \beta s \, ds = a(\beta),$$

$$\sqrt{r(\beta)} \, \sin p(\beta) = \int_{-\infty}^{\infty} u_2(s) \, \sin \beta s \, ds = \int_{-\infty}^{\infty} u(s) \, \sin \beta s \, ds = b(\beta), \quad (15)$$

where $a(\beta)$ and $b(\beta)$ are defined by (11).

From these equations we then obtain the fundamental identity (12) of Theorem 1.

PROBLEMS

1. Show that if $R(t) = Ae^{-p\,t^2}$, then $u(s)$, a solution of equation (10), is a function of the same kind.

2. Find a solution of equation (10) if $R(t) = \exp(-|t|)$.

3. Show that $u(s) = $ a constant, is a solution of (10), when $R(t)$ is defined as follows: $R(t) = 0$, when $|t| > a$; $R(t) = 1 - t/a$, $0 \leqq t \leqq a$; $R(t) = 1 + t/a$, $-a \leqq t \leqq 0$.

Chapter 14

Problems From the Calculus of Variations

1. Introduction

ONE OF THE MOST FRUITFUL SOURCES of nonlinear differential equations is found in problems which center around the determination of functions that maximize or minimize certain types of integrals. This wealth of information, now so extensive that the bibliography of the subject would fill a large book, comprises what is called the *calculus of variations*.

The origin of the subject is found in antiquity. In Vergil's story of the wanderings of Aeneas, we read that Queen Dido, when she bargained with the Libyans for a site for Carthage, was offered "as much land as could be covered with a bull's hide." (Vergil's *Aeneid:* i, 368). Her crafty followers interpreted this to mean the area that could be surrounded by a cord made from a bull's hide. Desirous of obtaining as much land as possible, the Queen needed to know what shape of curve could enclose a maximum area. The answer, a circle, was known to the Greeks, although many centuries were to pass before an adequate mathematical theory was evolved to solve this and similar problems. To Zenodorus, living probably during the 2nd century B.C., is attributed the solution, although Archimedes a century earlier had considered the problem of the maximum volume enclosed by a given area and stated that the answer was the sphere. The natural generalization of *Dido's problem* led in time to the creation of the *isoperimetric problem* of the calculus of variations.

During the 18th century the subject of the maximizing and minimizing of integrals became one of great interest to the mathematicians for two reasons. The first of these was the existence of several problems such, for example, as that proposed by Newton to determine the form of a surface of revolution, which will encounter minimum resistance when moved in the direction of its axis through a resisting medium. Another equally celebrated example was that of the brachistochrone (brachistos = shortest, chronos = time), a problem due to the Bernoullis. One is required to determine a path between two points in a vertical plane along which a particle would move under gravity in the shortest time.

The second reason for the interest of the mathematicians in the subject is found in the attempts of early natural philosophers to

discover a minimizing principle in nature. The following statement
of Leonhard Euler (1707–83), which he made in 1744, is characteristic
of the philosophical origin of what has come to be called the *principle
of least action:*

"As the construction of the universe is the most perfect possible,
being the handiwork of an all-wise Maker, nothing can be met with
in the world in which some maximal or minimal property is not dis-
played. There is, consequently, no doubt but that all the effects of
the world can be derived by the method of maxima and minima
from their final causes as well as from their efficient ones."

The formulation of the equations of dynamics according to this
principle must be regarded as one of the most astonishing facts of
science. The principle of least action, where action is to be understood as the mean value of the difference between the kinetic and
potential energies of a physical system averaged over some fixed
interval of time, originated with P. L. M. de Maupertuis (1698–1759).
The general statement of Maupertuis was made in an attempt to
extend the theorem of P. Fermat (1601–65) that a ray of light, when
travelling in a homogeneous medium, will pass from one point to
another either directly or by reflection by the shortest path and in the
shortest time.

The first formulation of the equations of dynamics according to the
principles of the calculus of variations was made by J. L. Lagrange
(1736–1813) in his *Mécanique analytique*, published in 1788. This
work of Lagrange remained unchanged for nearly half a century until
Sir William R. Hamilton (1805–65) in 1834–35 produced his classical
papers on dynamics, which gave a new and very appealing form to the
equations.

The history of the calculus of variations in a modern sense began
with the work of Euler and Lagrange. The former obtained the
first necessary condition for the existence of a maximum or minimum
in the form of an equation, now known as *Euler's equation*, which
assured the vanishing of the first variation of the original integral.
Lagrange introduced the variational notation, the method now known
as that of the *Lagrange multiplier* in the isoperimetric problems,
and added numerous examples of the application of the new calculus.

Interest developed in the problem of finding a sufficient condition
for the existence of a curve which maximized or minimized the primary integral. Attention thus turned to the second variation and
in 1788, A. M. Legendre (1752–1833) published a second necessary
condition which such a curve must satisfy. But the condition thus
discovered was not also a sufficient one. Although interest in the
subject remained lively, no significant advance in its theory was
made until 1837 when C. G. J. Jacobi (1804–51) discovered a third

necessary condition associated with what has been called the *conjugate point*.

Renewed interest in the sufficiency problem, still unresolved, was awakened by the lectures of K. Weierstrass (1815–97) in which a complete reexamination of the first and second variations and the criteria of his predecessors led to the discovery of still a fourth necessary condition. This involved what is called the Weierstrass *excess function*, denoted by the symbol: $E(x,y,y',p)$. By means of this function a satisfactory sufficiency condition was finally established.

In this chapter our primary interest will be in the differential equations, principally nonlinear ones, which appear in connection with classical problems in the calculus of variations. We shall not be concerned, therefore, with the delicate problem of whether the solutions of the differential equations actually provide extremals for the primary integrals. However, some attention is paid to this question in the problems given in Section 3.

Since the differential equations generated by the calculus of variations are consequences of the first variation, we shall limit our interest to this single aspect of the subject. But this is important enough, for we shall find that from it comes the system of equations derived by Lagrange, and extended by Hamilton, which forms the basis for dynamics. The application of these equations to celestial mechanics led to many of the problems which we have already introduced in the chapter on nonlinear mechanics. We are thus, in a sense, at the fountainhead of the subject.

2. The Euler Condition

The problem of Euler concerns itself with the establishing of a necessary condition that a function, $y=y(x)$, defined within a domain R, shall maximize or minimize an integral of the following form:

$$I = \int_a^b F(x,y,y')\ dx, \tag{1}$$

in the sense that the integral will be greater or less for it than for any other function within R. Such a domain is shown in Figure 1.

It will be assumed that $y(x)$ is continuous and has a continuous derivative $y'(x)$ in the interval: $a \leqq x \leqq b$. Such a function is said to belong to class C'. We shall assume further that $F(x,y,y')$ is continuous and has continuous derivatives of first and second orders.

To obtain a necessary condition for the existence of a maximum or minimum value of (1), we replace y in (1) by $y(x)+\alpha\eta(x)$, where α

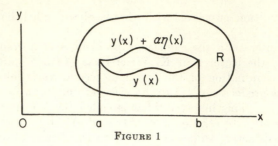

is an arbitrary constant and $\eta(x)$ is an arbitrary function belonging to class C', which vanishes at $x=a$ and $x=b$; that is,

$$\eta(a)=\eta(b)=0. \tag{2}$$

Any function lying within the domain R which gives a value to I we shall call an admissible function. We shall assume that both $y(x)$ and $y(x)+\alpha\eta(x)$ are admissible functions.

If we denote the new integral by $I(\alpha)$, that is,

$$I(\alpha)=\int_a^b F(x,y+\alpha\eta,\ y'+\alpha\eta')dx, \tag{3}$$

the integral is now a function of the parameter α. Under the assumption that F has continuous derivatives of first and second orders, $I(\alpha)$ can be expanded in a series in α at least to α^2.

We thus have

$$I(\alpha)=I(0)+\alpha\int_a^b (F_y'\eta+F_{y'}'\eta')dx \tag{4}$$

$$+\frac{\alpha^2}{2!}\int_a^b \{P\eta^2(x)+2Q\eta(x)\eta'(x)+R[\eta'(x)]^2\}dx+\ldots,$$

where we abbreviate:

$$P=\frac{\partial^2 F}{\partial y^2}, \quad Q=\frac{\partial^2 F}{\partial y\partial y'}, \quad R=\frac{\partial^2 F}{\partial y'^2}, \quad F_y'=\frac{\partial F}{\partial y}, \quad F_{y'}'=\frac{\partial F}{\partial y'}. \tag{5}$$

The coefficient of α is called the first variation and is denoted by δI, that is to say,

$$\delta I=\int_a^b (F_y'\eta+F_{y'}'\eta')dx. \tag{6}$$

It will be seen at once that if $y(x)$ is to maximize or minimize the integral, then it must be a function for which the coefficient of α in

(4) is reduced to zero. We thus obtain as a necessary condition to be satisfied by $y(x)$ the vanishing of the first variation,

$$\delta I = 0. \tag{7}$$

Integrating (6) by parts and taking account of (2), we obtain

$$I = F'_{y'}\eta(x) \Big|_a^b + \int_a^b \Big(F'_y - \frac{d}{dx} F'_{y'} \Big) \eta(x) dx,$$

$$= \int_a^b \Big(F'_y - \frac{d}{dx} F'_{y'} \Big) \eta(x) dx. \tag{8}$$

The *fundamental lemma* of the calculus of variations is now introduced. This lemma states that if in the integral

$$\int_a^b p(x)\, q(x)\, dx, \tag{9}$$

the function $p(x)$ is continuous between $x=a$ and $x=b$, and if the integral vanishes for all functions $q(x)$ of class C' which vanish at a and b, then $p(x)$ must be identically zero in the interval.*

Employing this lemma we obtain from (8) the condition which must be satisfied by $y(x)$ in the form of the following equation:

$$\frac{\partial F}{\partial y} - \frac{d}{dx} \frac{\partial F}{\partial y'} = 0. \tag{10}$$

This famous equation is called the *Euler equation*. We shall refer to any solution of it as an *extremal*. But it does not follow that an extremal is the solution of the original problem, since the condition is only a necessary and not a sufficient one.

To investigate the problem of whether or not an extremal actually provides a maximum (or a minimum) value of the integral, one is led naturally to a study of the *second variation*, namely, the coefficient of α^2 in (4). Denoting this by $\delta^2 I$, we have

$$\delta^2 I = \int_a^b \{ P\eta^2(x) + 2Q\eta(x)\eta'(x) + R[\eta'(x)]^2 \} dx, \tag{11}$$

we see that the second variation is the integral of a quadratic form in $\eta(x)$ and $\eta'(x)$. Since our interest in the calculus of variations resides mainly in its power to generate significant nonlinear equations, rather than in the integrals from which they come, we shall not attempt here to discuss the numerous problems which have been developed from a study of the second variation.

---

*Proofs of this by no means obvious lemma will be found in E. Goursat: *Cours d'analyse mathématique*, Vol. 3, 4th ed., Paris, 1924, pp. 545–547, and also in O. Bolza: *Vorlesungen über Variationsrechnung*, Berlin, 1909, pp. 25–27.

Returning to the Euler equation, we shall give a few examples of the equations which are provided in such abundance by proper specializations of the integrand of (1).

Example 1. (Minimum Surface of Revolution). If S is a surface of revolution generated by revolving about the x-axis a curve $y=y(x)$ through the points $P_0=(0,y_0)$ and $P_1=(x_1,y_1)$, the area of the surface is given by the equation:

$$S=2\pi\int_0^{x_1} y\,ds=2\pi\int_0^{x_1} y(1+y'^2)^{\frac{1}{2}}\,dx.$$

The corresponding Euler equation is readily found to be

$$y\,y''-y'^2=1,$$

which is the differential equation of the catenary and has the solution:

$$y=a\,\cosh\,(x/a+b),$$

where a and b are arbitrary.

Example 2. The following, a somewhat more complicated example,[*] requires the extremals corresponding to the integral:

$$I=\int_a^b [1+x^2+2xyy'+(1+y^2)y'^2]^{\frac{1}{2}}\,dx.$$

For this integral the Euler equation reduces to the following nonlinear equation:

$$(1+x^2+y^2)y''=(xy'-y)(1+y'^2).$$

A first integral of this equation is found to be

$$(xy'-y)^2=\frac{a^2}{1+a^2}(1+y'^2)(1+x^2+y^2),$$

where a is an arbitrary constant.

If we set $x=r\cos\theta$, $y=r\sin\theta$, this equation reduces to

$$\frac{d\theta}{dr}=\left(\frac{a}{r}\right)\left(\frac{r^2+1}{r^2-a^2}\right)^{1/2},$$

which is readily integrable and has the following solution:

$$\theta-k=\frac{1}{2}a\log\left(\frac{u+1}{u-1}\right)-\arctan(au),$$

---

[*]See A. R. Forsyth: *Calculus of Variations*, p. 46.

where we write

$$u^2 = \frac{r^2+1}{r^2-a^2} = \frac{x^2+y^2+1}{x^2+y^2-a^2}.$$

Example 3. Find the Euler equation for the integral

$$I = \int_a^b (py^2 + 2qyy' + ry'^2)dx,$$

where p, q, and r are functions of x.

The desired equation is found to be

$$\frac{d}{dx}\left(r\frac{dy}{dx}\right) + (q'-p)\,y = 0,$$

which is the general linear differential equation of second order.

Example 4. Find the Euler equation for the integral

$$I = \int_a^b \left(y'^2 + Ay^2 + \frac{1}{2}By^4\right)dx.$$

The desired equation turns out to be

$$\frac{d^2y}{dx^2} = Ay + By^3,$$

which was shown in Section 10 of Chapter 7 to have the solution:

$$y = C\mathrm{sn}(\lambda u, k), \quad u = x + p,$$

where

$$k^2 = -(\lambda^2 + A)/\lambda^2, \quad C^2 = -2(\lambda^2 + A)/B.$$

3. The Euler Condition in the Isoperimetric Case

In the *isoperimetric problem* we seek extremals for the integral

$$I = \int_a^b F(x,y,y')dx, \tag{1}$$

which at the same time give to a second integral

$$J = \int_a^b G(x,y,y')dx, \tag{2}$$

a prescribed value C. The values $y(a)=y_1$ and $y(b)=y_2$ are also prescribed.

The problem of Dido mentioned in Section 1 is the example from which this class of problems derives its name. For we are required to maximize the area integral:

$$I = \int_a^b y \, dx, \tag{3}$$

while at the same time we keep the integral of length, namely,

$$J = \int_a^b \sqrt{1 + y'^2} \, dx, \tag{4}$$

equal to a constant value. Conditions must also be imposed upon $y(x)$ so that the given length forms a closed *perimeter* for the area. Thus the areas considered are *isoperimetric*.

The Euler condition for this problem is found to be

$$\frac{\partial H}{\partial y} - \frac{d}{dx}\frac{\partial H}{\partial y'} = 0, \tag{5}$$

where $H = I + \lambda J$, in which λ is an arbitrary parameter to be evaluated from the conditions of the problem.

Thus, from (3) and (4), we get

$$H = y + \lambda\sqrt{1 + y'^2},$$

and (5) reduces to the equation:

$$\lambda \frac{d}{dx}\left[\frac{y'}{(1 + y'^2)^{1/2}}\right] = 1. \tag{6}$$

A first integral is found to be

$$\lambda y' = (x - c)(1 + y'^2)^{1/2}. \tag{7}$$

Solving for y' and integrating, we obtain

$$(x - c)^2 + (y - c')^2 = \lambda^2, \tag{8}$$

that is, the equation for a circle. The value of λ is determined from the prescribed value of J, and c and c' from the boundary conditions.

It is often convenient in isometric problems, as well as in other problems of the calculus of variations, to represent the extremals in parametric form, that is,

$$x = x(t), \; y = y(t), \tag{9}$$

In this case I and J are written

$$I=\int_a^b F(x,y; x',y') \, dt, \quad J=\int_a^b G(x,y; x',y') \, dt, \tag{10}$$

and the Euler equation is replaced by the two equations:

$$\frac{\partial H}{\partial x}-\frac{d}{dt}\frac{\partial H}{\partial x'}=0, \quad \frac{\partial H}{\partial y}-\frac{d}{dt}\frac{\partial H}{\partial y'}=0, \tag{11}$$

where $H=F+\lambda G$.

In Dido's problem considered above, when the curve is represented parametrically, equations (3) and (4) are replaced respectively by the following:

$$I=\frac{1}{2}\int_a^b (yx'-xy') \, dt, \quad J=\int_a^b \sqrt{x'^2+y'^2}dt, \tag{12}$$

from which we have:

$$H=\frac{1}{2}(yx'-xy')+\lambda(x'^2+y'^2)^{1/2}.$$

The Euler equations (11) are seen to reduce to the following:

$$x'(x'^2+y'^2)^{-1/2}=-y, \quad y'(x'^2+y'^2)^{-1/2}=x. \tag{13}$$

Dividing the second of these equations by the first, we obtain

$$y'/x'=-x/y, \tag{14}$$

which is the equation of the tangent to a circle. The solution then takes the form:

$$x-x_0=\lambda \cos t, \quad y-y_0=\lambda \sin t. \tag{15}$$

Since a closed perimeter is assumed, there must exist values of t, namely t_0 and t_1, such that: $x(t_0)=x(t_1)$, $y(t_0)=y(t_1)$. These values of t may be assumed to be 0 and 2π, respectively. When x and y as defined by (15) are substituted in the second integral in (12), we obtain for the determination of λ the equation: $J=2\pi\lambda$, where J is the length of the perimeter of the circle.

PROBLEMS

1. If the function F in equation (1) is independent of x, show that the following is a first integral of Euler's equation:

$$F-y' F'_{y'}=\text{constant}. \tag{16}$$

2. Solve the Euler equation for the function

$$F = \frac{1}{y}(1+y'^2)^{\frac{1}{2}},$$

and show that the general solution is a two-parameter family of circles with centers on the x-axis.

3. Given that

$$F = y^\alpha (1+y'^2)^{\frac{1}{2}}, \tag{17}$$

show that the solution of Euler's equation for which $y=1$, $y'=0$, when $x=0$ is given by the following parametric equations:

$$x = m \int_0^t \cos^m t \, dt, \quad y = \cos^m t, \quad m = -1/\alpha.$$

Find explicitly the solutions corresponding to $\alpha=1$ and $\alpha=-1$.

4. A bead of mass m is constrained to move under gravity on a smooth curved wire situated in a vertical plane. We shall assume that the bead starts from rest at a point h above the x-axis and moves to a point on the x-axis. The time of descent T is given by the following integral:

$$T - \frac{1}{\sqrt{2g}} \int_0^h \left(\frac{1+x'^2}{h-y} \right)^{\frac{1}{2}} dy. \tag{18}$$

The curve, $y=y(x)$, for which T is a minimum is called the *brachistochrone*, or curve of quickest descent. Show that this curve is the cycloid, defined by the following parametric equations:

$$x = a(\theta + \sin \theta), \quad y = a(1 - \cos \theta),$$

where $a=\frac{1}{2}h$.

5. (Newton's Problem). In seeking the form of a solid of revolution which experiences a minimum resistance when it moves through a fluid in the direction of the axis of revolution, we are led to minimize the following integral:

$$I = \int_a^b \frac{yy'^3}{1+y'^2} dx. \tag{19}$$

Show that the first integral of the Euler equation is

$$y = C_1(1+p^2)^2/p^3, \tag{20}$$

where $p=y'$ and C_1 is an arbitrary constant.

Observing that $dx=dy/p$, show that

$$x = C_1 \left[\log p + \frac{1}{p^2} + \frac{3}{4} p^4 \right] + C_2, \tag{21}$$

where C_2 is an arbitrary constant. Now note that equations (20) and (21) provide a solution of Euler's equation in parametric form.

6. In Problem 5, let $C_1=1$ and $C_2=0$ and graph the curve defined by (20) and (21), regarding p as a variable parameter. Show that as p varies from 0 to $\sqrt{3}$, we obtain one branch, and as p varies from $\sqrt{3}$ to ∞, we obtain a second branch of a curve, which has a cusp at the point (x,y) corresponding to $p=\sqrt{3}$.

7. Show that the Euler equation for the integral.

$$I = \int_a^b F(x, y, y', y'', \ldots, y^{(n)}) \, dx, \qquad (22)$$

is the following:

$$\frac{\partial F}{\partial y} - \frac{d}{dx}\frac{\partial F}{\partial y'} + \frac{d^2}{dx^2}\frac{\partial F}{\partial y''} + \cdots + (-1)^n \frac{\partial F}{\partial y^{(n)}} = 0. \qquad (23)$$

8. If $L > 0$ and $y(x)$ is a real-valued absolutely continuous function on $(0, L)$ which vanishes at 0 and L, and if $y'(x)$ is of integrable square on the interval, show that

$$\int_0^L \left[y'^2 - \frac{\pi^2 y^2}{L^2} \right] dx \geqq 0, \qquad (24)$$

and that the equality sign holds if and only if $y(x)$ is of the form $C \sin (\pi x/L)$.*

Hint: Observe the following identity:

$$y'^2 - \frac{\pi^2 y^2}{L^2} \equiv \frac{d}{dx}\left[\frac{\pi y^2}{L} \cot \left(\frac{\pi x}{L} \right) \right] + \left[y' - \frac{\pi y}{L} \cot \left(\frac{\pi x}{L} \right) \right]^2.$$

The following two problems are concerned with the establishing of conditions which are necessary (but not sufficient) if a solution of Euler's equation is to give a maximum or minimum value to the integral. Together these conditions are sufficient to assume that the second variation $\delta^2 I$ is of one sign.

9. If θ is an arbitrary function of class C' in (a,b), show that

$$\int_a^b (2\theta\eta\eta' + \theta'\eta^2) \, dx = 0.$$

Now add the integrand of this integral to the integrand of $\delta^2 I$ and show that, if θ is a solution of the equation:

$$R(P + \theta') = (Q + \theta)^2, \qquad (25)$$

then we can write:

$$\delta^2 I = \int_a^b R(\eta' + M\eta)^2 \, dx,$$

where $M = (Q + \theta)/R$.

From this derive the following theorem (called the *necessary condition of Legendre*): *If $\delta^2 I$ is to be of one sign, it is necessary that R does not change sign in (a,b).*

10. Observe that equation (25) is a Ricatti. Hence, let $Q + \theta = -R \, u'/u$, and thus derive the following:

$$Ru'' + Ru' + (Q' - P)u = 0, \qquad (26)$$

which is called the *equation of Jacobi*.

Now show that the second variation can be written

$$\delta^2 I = \int_a^b R \, \frac{(\eta'u - \eta u)^2}{u^2} \cdot dx.$$

---

*See W. T. Reid: *Journal of Math. and Mech.*, Vol. 8, 1959, pp. 897–906.

Observe that, when $R \neq 0$, the integral is of one sign unless the numerator of the integrand vanishes identically. Show that this is impossible unless $\eta = Cu$, where C is a constant, and that this is impossible unless u vanishes at some point in the interval (a,b).

From this derive the following *theorem of Jacobi: If $\delta^2 I$ is to be one of sign for all possible forms of the function $\eta(x)$, it is necessary that equation (26) have a solution which does not vanish in (a,b).*

11. Show that the Euler equation for the integral

$$I = \int_a^b (Pu^2 + 2Quu' + Ru'^2)\,dx$$

is Jacobi's equation.

12. If $y = y(x,\alpha,\rho)$ is a solution of Euler's equation, show that

$$y_\alpha = \frac{\partial}{\partial \alpha}\, y(x,\alpha,\beta) \text{ and } y_\beta = \frac{\partial}{\partial \beta}\, y(x,\alpha,\beta)$$

are solutions of Jacobi's equation.

13. Let $y = y(x,\alpha)$ be a one-parameter family of curves. If α is eliminated between $y = y(x,\alpha)$ and y_α as defined in Problem 12, show that the resulting equation is the envelope of the family of curves.

14. Making use of the results of Problems 12 and 13, show that Jacobi's theorem can be stated as follows: If $y = y(x,\alpha)$ is a solution of Euler's equation, then if $\delta^2 I$ is to be of one sign, it is necessary that the point of contact of y and its envelope shall lie outside of the interval (a,b).

15. Show that the envelope of the family of catenaries:

$$y = \alpha \cosh (x/\alpha),$$

is the straight line

$$y = x\, \frac{\cosh \phi}{\phi} = x \tan \psi,$$

where $\phi = 1.1997$ and $\psi = 56°28'$. (See Example 1, Section 2.)

16. Show that for Newton's problem (Problem 5), we have

$$R = \frac{2yp(3 - p^2)}{(1 + p^2)^3}.$$

4. The Euler Condition for a Double Integral

The problem which we have just considered for single integrals can be extended without difficulty to multiple integrals. We shall consider the case of a double integral, since its geometrical interpretation is readily understood. Let

$$z = z(x,y) \tag{1}$$

be the equation of a surface within a region R of space and let C be a curve upon this surface the projection of which on the xy-plane is a simply connected closed curve C' without double points. Such a configuration is shown in Figure 2.

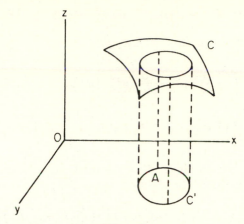

Figure 2

Let $F(x,y,z,p,q)$ be a function which is continuous in the five variables x,y,z,p,q with continuous derivatives of orders up to and including the second. We now consider the integral:

$$I = \iint_A F(x,y,z,p,q) \; dx \; dy, \tag{2}$$

evaluated over the area A inclosed by the curve C'. We seek among the surfaces of the region R which contain the curve C for that surface with minimizes (or maximizes) I. The symbols p and q have their usual significance: $p = \partial z / \partial x$ and $q = \partial z / \partial y$.

To obtain the first variation we write

$$z = z(x,y) + \alpha \eta(x,y), \tag{3}$$

where $\eta(x,y)$ is a function which is continuous with continuous derivatives in A and which vanishes along C'. The first variation is then defined by the integral

$$\delta I = \alpha \iint_A [F_z \eta(x,y) + F_p \eta_x' + F_q \eta_y'] dx \; dy. \tag{4}$$

We now make use of Green's theorem

$$\int_{C'} P \; dx + Q \; dy = \iint_A \left(\frac{\partial Q}{\partial x} - \frac{\partial P}{\partial y} \right) dx \; dy, \tag{5}$$

in which we set: $P=-\eta(x,y)\,F_q$ and $Q=\eta(x,y)\,F_p$. We thus obtain

$$\int_{C'}\eta(x,y)(-F_q\,dx+F_p\,dy)$$

$$=\iint_A\left[\frac{\partial}{\partial x}\,(\eta F_p)+\frac{\partial}{\partial y}\,(\eta F_q)\right]dx\,dy \tag{6}$$

$$=\iint_A\left[\eta_x'F_p+\eta\,\frac{\partial}{\partial x}\,F_p+\eta_y'F_q+\eta\,\frac{\partial}{\partial y}\,F_q\right]dx\,dy.$$

Since $\eta(x,y)$ is zero along C', these integrals are all zero and from the last one we obtain the identity:

$$\iint_A (F_p\eta_x'+F_q\eta_y')\,dx\,dy=-\iint_A \eta\left(\frac{\partial}{\partial x}\,F_p+\frac{\partial}{\partial y}\,F_q\right)dx\,dy. \tag{7}$$

When this is substituted in (4), the first variation reduces to

$$\delta I=\alpha\iint \eta(x,y)\left(F_z-\frac{\partial}{\partial x}\,F_p-\frac{\partial}{\partial y}\,F_q\right)dx\,dy. \tag{8}$$

It will be readily seen that the fundamental lemma of the calculus of variations, which was established in Section 2, can be extended without essential change to the case of double integrals. Applying this lemma here we derive Euler's condition in the following form:

$$\frac{\partial F}{\partial z}-\frac{\partial}{\partial x}\frac{\partial F}{\partial p}-\frac{\partial}{\partial y}\frac{\partial F}{\partial q}=0. \tag{9}$$

5. The Problem of the Minimal Surface

The problem of the minimal surface concerns itself with the determination of a surface of minimum area, which is bounded by one or more nonintersecting skew curves.

Since the area of a surface is given by the integral

$$A=\iint (1+p^2+q^2)^{1/2}\,dx\,dy, \tag{1}$$

the Euler condition, obtained from (9) in the preceding section, reduces to the following nonlinear partial differential equation of second order:

$$(1+q^2)r-2pqs+(1+p^2)t=0, \tag{2}$$

where we introduce the customary symbols:

$$r=\frac{\partial^2 z}{\partial x^2},\quad s=\frac{\partial^2 z}{\partial x\partial y},\quad t=\frac{\partial^2 z}{\partial y^2}.$$

This equation is of elliptic type, since its discriminant, namely,

$$(1+q^2)(1+p^2) - p^2q^2 = 1 + p^2 + q^2, \tag{3}$$

is greater than zero. If p and q are sufficiently small quantities, equation (2) is asymptotic to Laplace's equation, that is,

$$r + t = 0. \tag{4}$$

If k_1 and k_2 are the principal curvatures of the surface: $z = z(x,y)$, it is proved in treatises on differential geometry that $k_1 + k_2$ (the mean curvature) has the form:

$$k_1 + k_2 = \frac{1}{h^{3/2}} [(1+q^2)r - 2pqs + (1+p^2)t], \quad h = 1 + p^2 + q^2. \tag{5}$$

From this it follows that *a minimal surface can be defined as one for which the mean curvature is zero.*

If the equation of the surface is given in terms of the parameters (u,v), where

$$x = x(u,v), \quad y = y(u,v), \quad z = z(u,v), \tag{6}$$

then equation (2) can be expressed in terms of the coefficients of the first and second fundamental quadratic forms of the surface. These quadratic forms are customarily written as follows:

$$ds^2 = E\,du^2 + 2F\,du\,dv + G\,dv^2,$$

$$-d\phi^2 = D\,du^2 + 2D'\,du\,dv + D''\,dv^2, \tag{7}$$

where we use the abbreviations:

$$E = \sum (x_u)^2, \quad F = \sum (x_u x_v), \quad G = \sum (x_v)^2, \tag{8}$$

$$D = \frac{1}{\sqrt{H}} \begin{vmatrix} x_{uu} & y_{uu} & z_{uu} \\ x_u & y_u & z_u \\ x_v & y_v & z_v \end{vmatrix}, D' = \frac{1}{\sqrt{H}} \begin{vmatrix} x_{uv} & y_{uv} & z_{uv} \\ x_u & y_u & z_u \\ x_v & y_v & z_v \end{vmatrix}, D'' = \frac{1}{\sqrt{H}} \begin{vmatrix} x_{vv} & y_{vv} & z_{vv} \\ x_u & y_u & z_u \\ x_v & y_v & z_v \end{vmatrix}.$$

In these expressions the sums extend to the three variables, the subscripts denote partial derivatives, and $H = EG - F^2$.

Since the mean curvature of the surface, in terms of these coefficients, has the form

$$k_1 + k_2 = \frac{1}{H} (ED'' - 2D'F + DG), \tag{9}$$

equation (1) of the minimal surface is replaced by the following:

$$ED'' - 2D'F + DG = 0. \tag{10}$$

An example is provided by the problem of finding the minimal surface, which is a surface or revolution and which is bounded by two circles having the same axis.

The solution of this problem is immediately obtained from Example 1 of Section 2, where the surface was shown to be a *catenoid*, that is to say, one generated by revolving a catenary about an axis.

If the catenary is revolved about the z-axis, the equation of the surface assumes the form:

$$z = a + b \cosh^{-1}(\rho/b), \quad \rho^2 = x^2 + y^2. \tag{11}$$

A representation of this surface is shown in Figure 3. A study of the properties of the catenoid shows that it is not always possible to construct such a surface, which has given a directrix and at the same time passes through two preassigned points.

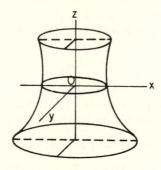

FIGURE 3

Great analytical difficulties beset the solution of equation (2), but fortunately there exists a physical analogue which makes it possible to determine the shape of minimal surfaces for many kinds of boundary curves. This analogue is due to the blind physicist Joseph Plateau (1801–83), who described it in his classical work on molecular forces in liquids published in 1873. Soap bubbles, constrained between bounding contours, will assume the shape of minimal surfaces through the action of their molecular forces. The problem of the analyst was thus transferred to the laboratory of the physicist, and we have here one of the first examples of an analogue solution of a nonlinear differential equation. Minimal surfaces obtained in this manner are shown in the accompanying plate, from experiments carried out by W. T. Reid.

The problem of determining a continuous minimal surface which passes through a given curve is called the *problem of Plateau*. It has been the subject of extensive investigations by T. Carleman, J. Douglas, A. Haar, and many others. Of special significance are the recent researches of T. Radó, who has made important extensions of the original problem.*

The problem of the minimal surface is included in the broader problem of the determination of extremals for the integral

$$\iint_A F(p,q) \; dx \; dy, \tag{12}$$

where $F(p,q)$ is an analytic function of p,q and satisfies the inequalities:

$$F_{pp}>0, \; F_{qq}>0, \; F_{pp}F_{qq}-F_{pq}^2>0, \tag{13}$$

for all values of p and q.

One observes that this problem also includes that of Dirichlet, which requires a minimum for the integral

$$\iint_A (p^2+q^2) \; dx \; dy. \tag{14}$$

The Euler condition reduces to Laplace's equation (4), which we have just seen is an approximation for the equation of the minimal surface.

Another problem closely related to that of Plateau is the determination of a surface for which the mean curvature is a constant, let us say k. Physically this corresponds to the case of a soap film in which there is a constant difference in pressure on its two sides.†

In this case equation (2) is replaced by the following:

$$(1+q^2)r-2pqs+(1+p^2)t=k(1+p^2+q^2)^{3/2}. \tag{15}$$

If p and q are sufficiently small, this equation can be replaced by the linear one

$$r+t=k. \tag{16}$$

Although this equation has the following function as its complete integral:

$$z=\phi(x+iy)+\psi(x-iy)+\frac{k}{4}(x^2+y^2), \tag{17}$$

where ϕ and ψ are arbitrary functions, this fact usually does not help in constructing a solution which also satisfies the boundary conditions, that is to say, in finding a surface which passes through one or more arbitrarily given curves.

---

*For an extensive account of the problem and its extensions, the reader is referred to the excellent volume of T. Radó: *On the Problem of Plateau*, 1932, American edition, 1951, 109 p.

†For a discussion of this problem see H. Bateman: *Bibliography*, Reference (2), pp. 169–171.

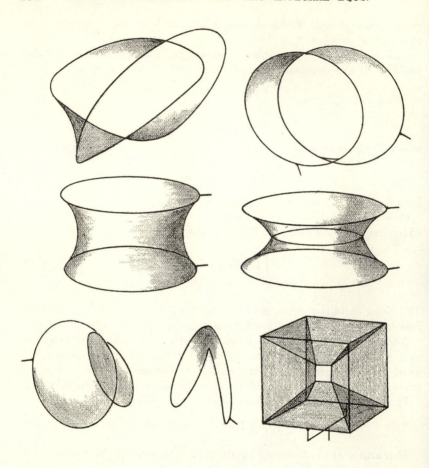

Minimal Surfaces for Various Boundary Conditions—Soap Film Pictures Obtained by W. T. Reid.

6. Hamilton's Principle—The Principle of Least Action

By means of the calculus of variations it has been possible to express in a fundamental and elegant way the general problem of dynamics. As we have already stated in Section 1 this formulation has had a long history which began with Fermat's principle. This principle, in modern terms, may be stated thus: Rays of light travel along such lines that the optical distance between any two points of the ray is a minimum. By an *optical distance* we mean the sum, $\Sigma l_i n_i$, where l_i represents the distance traveled in a medium of refractive index n_i. We have also indicated in Section 1 the metaphysical character of the arguments by means of which Maupertuis and Euler extended Fermat's theorem to the more general system of dynamics.

It was left to Hamilton to formulate the modern principle (*Hamilton's principle*). This states in the language of the calculus of variations that the first variation of the time integral of the difference between the kinetic energy (T) and the potential energy (V) of a dynamical system is zero, that is,

$$\delta \int_{t_1}^{t_2} (T-V)\,dt = 0. \tag{1}$$

Since the question is seldom asked whether or not an actual minimum is attained by the integral, it is customary to say that the motion defined by (1) is *stationary*. The equation is assumed to hold for all dynamical systems whether they are *conservative*, that is, when $T+V=C$, where C is a constant, or *nonconservative*.

The quantity $T-V$ is called the *Lagrangian* and is denoted by L. With this designation equation (1) becomes

$$\delta \int_{t_1}^{t_2} L\,dt = 0, \tag{2}$$

and Hamilton's principle then asserts that the first variation of the time-integral of the Lagrangian is zero.

If the system is conservative, V can be eliminated from (1), since $T+V=C$, and (1) then becomes

$$\delta \int_{t_1}^{t_2} 2T\,dt = 0. \tag{3}$$

Since the time-integral of $2T$ is called the *action* of the system, we have in (3) the formulation of the *principle of least action*. More properly stated, the principle of least action asserts that for any conservative system the action is stationary.

Let us first, as an example, derive Newton's laws of motion from this more general point of view. The position of a single body of mass m is given by the coordinates x, y, z, which are functions of time. If the body is constrained to move under a system of forces the respective components of which are X, Y, and Z, then the motion is defined by the following system of equations:

$$m\ddot{x}=X,\ m\ddot{y}=Y,\ m\ddot{z}=Z.* \tag{4}$$

*It will be recalled that the symbols $\dot{x}$ and $\ddot{x}$ denote respectively dx/dt and d^2x/dt^2. This notation was introduced by Newton in his theory of fluxions.

In the application of Hamilton's principle we write the kinetic energy in the form:

$$T=\frac{1}{2}m(\dot{x}^2+\dot{y}^2+\dot{z}^2),\tag{5}$$

and define W (the work) by its variation:

$$\delta W=X\delta x+Y\delta y+Z\delta z.\tag{6}$$

Then, since $\delta V=-\delta W$, we have from equation (1)

$$\delta\int_{t_1}^{t_2}(T-V)\,dt=\int_{t_1}^{t_2}(\delta T+\delta W)\,dt$$

$$=\int_{t_1}^{t_2}\left(\frac{\partial T}{\partial\dot{x}}\,\delta\dot{x}+\frac{\partial T}{\partial\dot{y}}\,\delta\dot{y}+\frac{\partial T}{\partial\dot{z}}\,\delta\dot{z}+X\delta x+Y\delta y+Z\delta z\right)dt=0.\tag{7}$$

Integrating by parts and noting that the variations vanish at $t=t_1$ and $t=t_2$, we obtain (7) in the following form:

$$\int_{t_1}^{t_2}\left[\left(-\frac{d}{dt}\frac{\partial t}{\partial\dot{x}}+X\right)\delta x+\left(-\frac{d}{dt}\frac{\partial T}{\partial\dot{y}}+Y\right)\delta y+\left(-\frac{d}{dt}\frac{\partial T}{\partial\dot{z}}+Z\right)\delta z\right]dt=0.\tag{8}$$

Since the variations δx, δy, and δz are independent of one another, their multipliers are zero and we obtain as a consequence Newton's laws of motion as given by (4).

This analysis can be extended to more complex cases, where the configuration of the dynamical system is described in terms of a set of *generalized coordinates:* q_1, q_2, q_3, . . ., q_n. For simplicity we shall consider a system of three variables, where the familiar cartesian coordinates, x, y, z, are now written as follows:

$$x=x(q_1,q_2,q_3),\ y=y(q_1,q_2,q_3),\ z=z(q_1,q_2,q_3).\tag{9}$$

Since the coordinates are functions of t, we have upon differentiation

$$\dot{x}=\frac{\partial x}{\partial q_1}\dot{q}_1+\frac{\partial x}{\partial q_2}\dot{q}_2+\frac{\partial x}{\partial q_3}\dot{q}_3,\tag{10}$$

with corresponding expressions for $\dot{y}$ and $\dot{z}$.

When these quantities are substituted in (5), the kinetic energy is then represented as the following quadratic form in the $\dot{q}_i$:

$$T=\sum A_{ij}(q_1,q_2,q_3)\dot{q}_i\dot{q}_j,\ A_{ij}=A_{ji}.\tag{11}$$

It is evident from the original form of T that this quadratic form is positive definite and thus the leading principal minors of its matrix will be positive. Moreover, if the transformation defined by (9) is orthogonal, then A_{ij} will be zero when $i \neq j$.

Since the variation x has the form

$$\delta x = \frac{\partial x}{\partial q_1} \delta q_1 + \frac{\partial x}{\partial q_2} \delta q_2 + \frac{\partial x}{\partial q_3} \delta q_3, \tag{12}$$

with similar expressions for y and z, the variation in W given by (6) assumes the following linear form in terms of the generalized variables:

$$\delta W = Q_1 \delta q_1 + Q_2 \delta q_2 + Q_3 \delta q_3, \tag{13}$$

where Q_1, Q_2, and Q_3 are functions of the q_i.

By an argument differing in no essential way from that given above, the equations of motion reduce to the following system:

$$\frac{d}{dt}\left(\frac{\partial T}{\partial \dot{q}_i}\right) - \frac{\partial T}{\partial q_i} = Q_i, \quad i = 1, 2, 3. \tag{14}$$

If the motion is defined by a potential function, V, then we have

$$Q_i = -\frac{\partial V}{\partial q_i}. \tag{15}$$

In this case equations (14) take the elegant form:

$$\frac{d}{dt}\left(\frac{\partial L}{\partial \dot{q}_i}\right) - \frac{\partial L}{\partial q_i} = 0, \tag{16}$$

where $L = T - V$ is the Lagrangian. This is the celebrated formulation of the equations of dynamics as given by Lagrange. It is clearly extended to a system of n variables without modification.

Example. A body of mass m moves under the attraction of a gravitational force the potential of which is $-k^2/r$, where r is the distance from the origin of the gravitational force to the position of the body. Show that the path of the body is a conic section.

Solution: It is convenient to use polar coordinates, that is, in terms of generalized coordinates: $q_1 = r$, $q_2 = \theta$. Equations (9) thus become

$$x = q_1 \cos q_2 = r \cos \theta, \quad y = q_1 \sin q_2 = r \sin \theta.$$

Since polar coordinates form an orthogonal system, we get

$$T = \frac{1}{2} m (\dot{x}^2 + \dot{y}^2) = \frac{1}{2} m (\dot{r}^2 + r^2 \dot{\theta}^2).$$

Introducing the potential function, $V=-mk^2/q_1=-mk^2/r$, we obtain the Lagrangian as follows:

$$L=T-V=\frac{1}{2}\,m\,(\dot{r}^2+r^2\dot{\theta}^2+2k^2/r).$$

When this is substituted in equation (16), the following nonlinear system is obtained:

$$\dot{r}-r\dot{\theta}^2=-\frac{k^2}{r^2},\quad \frac{d}{dt}\,(r^2\dot{\theta})=0. \tag{17}$$

Integrating the second equation, we get

$$r^2\dot{\theta}=h, \tag{18}$$

where h is an arbitrary constant.

Introducing $r=1/u$, and observing that

$$\dot{r}=\frac{dr}{dt}=-\frac{1}{u^2}\frac{du}{dt}=-\frac{1}{u^2}\frac{du}{d\theta}\frac{d\theta}{dt}=-h\frac{du}{d\theta},$$

$$\ddot{r}=\frac{d^2r}{dt^2}=-h\frac{d}{dt}\left(\frac{du}{d\theta}\right)=-h\frac{d^2u}{d\theta^2}\frac{d\theta}{dt}=-h^2u^2\frac{d^2u}{d\theta^2},$$

we obtain for the first equation of (17) the following:

$$\frac{d^2u}{d\theta^2}+u=\frac{k^2}{h^2}. \tag{19}$$

From the solution of this equation, namely,

$$u=-\frac{1}{p}\cos(\theta+\omega)+\frac{k^2}{h^2}, \tag{20}$$

we are able to conclude that r has the following form:

$$r=\frac{ep}{1-e\cos\theta}, \tag{21}$$

where $ep=h^2/k^2$. This is the polar equation of a conic section with origin at the focus. The constant p is the distance from the focus to the nearest directrix and e is the eccentricity.

When the orbit is an ellipse, as shown in Figure 4, we have the following relationships:

$$ep=a(1-e^2)=b^2/a,\quad a^2e^2=a^2-b^2,$$

where $2a$ and $2b$ are respectively the major and minor axes of the ellipse.

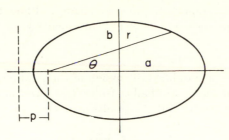

FIGURE 4.

From equation (18) we obtain:

$$\frac{1}{2}\int r^2\,d\theta = \frac{1}{2}h\int dt.$$

If the first integral is taken over a complete cycle, we obtain the area of the ellipse and the value of the second integral is the total period, denoted by T, of the motion. Since the area equals $\pi ab = \pi a^{3/2}(ep)^{1/2}$, we have

$$h^2 = \frac{4\pi^2 a^3 ep}{T^2}, \tag{22}$$

from which one derives the famous *harmonic law* of Kepler, namely, that the cubes of the mean distances of any two planets from the sun are to each other as the squares of their periods.

7. The Canonical Equations of Hamilton

The equations of dynamics as derived in the preceding section were given a very useful form by Hamilton, who introduced the new variables, p_1, p_2, . . . , p_n, defined as follows:

$$p_i = \frac{\partial T}{\partial \dot{q}_i},\ i = 1, 2, \ldots, n. \tag{1}$$

If the q_i are the familiar cartesian coordinates, x, y, z, and if T is the kinetic energy given by (5) of Section 6, then

$$p_1 = m\dot{x}, \quad p_2 = m\dot{y}, \quad p_3 = m\dot{z},$$

and their sum is the momentum of the system. Hence the p_i defined by (1) are the *generalized components of the momentum*.

It is now possible to give two forms to the kinetic energy, one as a quadratic form in terms of q_i and the other as a quadratic form in terms of p_i. The first of these we shall denote by T and the second by T'. Since both represent the same energy, although expressed in different variables, we have the obvious identity

$$T \equiv T'. \tag{2}$$

In order to give a simplified discussion of the relationships involved in these representations, we shall consider T in terms of two variables, but both the arguments and the results are extensible to n variables. We thus write,

$$T = A_{11}\dot{q}_1^2 + 2A_{12}\dot{q}_1\dot{q}_2 + A_{22}\dot{q}_2^2, \tag{3}$$

where the A_{ij} are functions of q_1 and q_2. We shall further denote the determinant of the form by D, that is, $D = A_{11}A_{22} - A_{12}^2$.

Since $p_1 = \partial T / \partial \dot{q}_i$, we have explicitly

$$p_1 = 2(A_{11}\dot{q}_1 + A_{12}\dot{q}_2), \quad p_2 = 2(A_{12}\dot{q}_1 + A_{22}\dot{q}_2). \tag{4}$$

Solving these equations for $\dot{q}_1$ and $\dot{q}_2$, ,we get

$$\dot{q}_1 = \frac{1}{2D}(A_{22}p_1 - A_{12}p_2), \quad \dot{q}_2 = \frac{1}{2D}(-A_{12}p_1 + A_{11}p_2). \tag{5}$$

When these values are substituted in (3), we obtain the explicit form of T' as follows:

$$T' = \frac{1}{4D}(A_{22}p_1^2 - 2A_{12}p_1p_2 + A_{11}p_2^2). \tag{6}$$

From this expression we get our first important result, namely,

$$\frac{\partial T'}{\partial p_1} = \frac{1}{2D}(A_{22}p_1 - A_{12}p_2) = \dot{q}_1, \text{ and } \frac{\partial T'}{\partial p_2} = \dot{q}_2. \tag{7}$$

Since this argument is readily extended to n variables, we have in the general case

$$\dot{q}_i = \frac{\partial T'}{\partial p_i}, \; i = 1, 2, 3, \ldots, n, \tag{8}$$

Somewhat less direct is the proof of the second fundamental relationship, namely, that

$$\frac{\partial T'}{\partial q_i} = -\frac{\partial T}{\partial q_i}. \tag{9}$$

To establish this we first observe that T is a homogeneous function of degree 2 in $\dot{q}_1$ and $\dot{q}_2$. Then by Euler's theorem (Section 4, Chapter 2) we have

$$2T = \dot{q}_1\frac{\partial T}{\partial \dot{q}_1} + \dot{q}_2\frac{\partial T}{\partial \dot{q}_2} = \dot{q}_1p_1 + \dot{q}_2p_2. \tag{10}$$

Since by (2) $T = T'$, we can write (10) in the form

$$T' = \dot{q}_1 p_1 + \dot{q}_2 p_2 - T. \tag{11}$$

Taking the derivative of this equation with respect to q_1 and observing (1), we get

$$\frac{\partial T'}{\partial q_1} = \frac{\partial \dot{q}_1}{\partial q_1} p_1 + \frac{\partial \dot{q}_2}{\partial q_2} p_2 - \frac{\partial T}{\partial q_1} - \frac{\partial T}{\partial \dot{q}_1} \frac{\partial \dot{q}_1}{\partial q_1} - \frac{\partial T}{\partial \dot{q}_2} \frac{\partial \dot{q}_2}{\partial q_1} = -\frac{\partial T}{\partial q_1}. \tag{12}$$

This same argument can be repeated without essential change in the general case and thus we establish (9).

The canonical equations of Hamilton are now readily derived from those of Lagrange given by (14) of Section 6, namely,

$$\frac{d}{dt} \frac{\partial T}{\partial \dot{q}_i} - \frac{\partial T}{\partial q_i} = Q_i. \tag{13}$$

To accomplish this we replace T by T' and note both the definition of p_i and equation (9). We thus obtain the desired equations as follows:

$$\dot{p}_i + \frac{\partial T'}{\partial q_i} = Q_i, \quad q_i = \frac{\partial T'}{\partial p_i}. \tag{14}$$

If the motion is determined by a potential function, V, then a very elegant form can be given to these equations by introducing the *Hamiltonian* function,

$$H = T' + V. \tag{15}$$

Since $Q = -\partial V / \partial q_i$ and since V is independent of p_i, equations (14) can be written as follows:

$$\frac{dp_i}{dt} = -\frac{\partial H}{\partial q_i}, \quad \frac{dq_i}{dt} = \frac{\partial H}{\partial p_i}. \tag{16}$$

Example 1. We shall derive the equations of motion of a body of mass m moving under the attraction of a gravitational force, using the Hamiltonian theory.

Solution: Referring to the example of Section 6, we have $q_1 = r$, $q_2 = \theta$, $V = -mk^2/q_1 = -mk^2/r$, and

$$T = \frac{1}{2} m(\dot{r}^2 + r^2 \dot{\theta}^2).$$

Computing p_1 from (1), we find $p_1 = m\dot{r}$, $p_2 = mr^2\dot{\theta}$, and thus

$$T' = \frac{1}{2m} (p_1^2 + p_2^2/r^2).$$

The Hamiltonian then becomes:

$$H = \frac{1}{2m}\,(p_1^2 + p_2^2/r^2) - mk^2/r.$$

When this is substituted in (16), the following equations result:

$$dp_1/dt = p_2^2/(mr^3) - mk^2/r^2, \quad dp_2/dt = 0,$$

which, when the values of p_i are substituted, reduce to equations (17) of Section 6.

Example 2.   The derivation of the equation of the simple pendulum by the use of the Hamiltonian is as follows:

If we set $m=1$, let L equal the length of the pendulum, and write $q = \theta$, then from Figure 5 we have as the component of the force in the

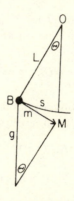

FIGURE 5.

direction BM the quantity $g \sin \theta = g \sin q$, and hence as the potential energy the function $V = -gL \cos q$.

Since $x = L \cos \theta$, $y = L \sin \theta$, we have

$$T = \frac{1}{2}(\dot{x}^2 + \dot{y}^2) = \frac{1}{2}L^2\dot\theta^2 = \frac{1}{2}L^2\dot{q}^2,$$

from which we get: $p = L^2\dot{q}$, and thus $T' = \tfrac{1}{2}p^2L^{-2}$.

The Hamiltonian then has the value

$$H = \tfrac{1}{2}p^2L^{-2} - gL \cos q.$$

When this function is introduced into (16), we obtain

$$\frac{dp}{dt}=-gL\sin q, \quad \frac{dq}{dt}=pL^{-2},$$

from which we obtain the desired equation:

$$\frac{d^2q}{dt^2}=-\frac{g}{L}\sin q, \text{ that is, } \frac{d^2\theta}{dt^2}=-\frac{g}{L}\sin\theta.$$

PROBLEMS

1. A particle acted upon by gravity is constrained to move on the surface of a fixed sphere of radius R. This is the case if the particle is suspended from a fixed point by a weightless, inextensible string and allowed to swing freely in any direction. Such a system is called a *spherical pendulum*. If the mass of the particle is m and if its position in spherical coordinates is (R, θ, ϕ), show that T and V have the following values:

$$T=\tfrac{1}{2}mR^2[\dot\theta^2+\sin^2\theta\ \dot\phi^2], \quad V=-mgR\cos\theta.$$

Show that the equations of motion are the following:

$$\ddot\theta-h^2\cot\theta\csc^2\theta+\frac{g}{r}\sin\theta=0, \quad \sin^2\dot\phi=\eta.$$

2. Given the data of Problem 1, compute the Hamiltonian and from it derive the equations of motion.

3. A particle of mass m moves on the surface of an inverted cone of revolution. If the coordinates of the particle are (r, ϕ) and α is the angle between the axis of the cone and its elements, show that

$$T=\tfrac{1}{2}m(\dot r^2+\sin^2\alpha\ r^2\dot\phi^2), \quad V=mg\cos\alpha\ r.$$

Find the equations of motion.

4. Solve Problem 3 by means of the Hamiltonian.

5. The surface of a stretched elastic membrane is defined by the equation:

$$z=z(x, y).$$

Show that the differential equation for small oscillations of the membrane is the following:

$$\frac{\partial^2 z}{\partial t^2}=k^2\left(\frac{\partial^2 z}{\partial x^2}+\frac{\partial^2 z}{\partial y^2}\right).$$

Hint: $V=a\iint(1+p^2+q^2)^{1/2}\,dx\,dy-A$, where $A=\iint dx\,dy$ is the area of the membrane at rest.

Chapter 15

The Numerical Integration of Nonlinear Equations

1. Introduction

As has been abundantly observed in preceding pages of this work, the solution of many types of nonlinear equations in a closed analytical form is not possible. The range of available functions is much too limited and many equations are intractable to the usual devices of analysis. In fact, most nonlinear equations define new functions, whose properties have not been explored nor for which tables exist. But the demands of applied science has made it necessary to obtain some insight into the nature of solutions, subject to prescribed boundary conditions. This insight usually takes the form of numerical and graphical representations of the functions.

In large laboratories at the present time there exist various machines which, with proper assistance, can achieve a graphical or a numerical description of functions defined by a broad class of nonlinear equations. One class of such machines includes what are called *analogue* computers. The mechanical type is generally referred to as a *differential analyzer*. Both types perform similar functions of graphically integrating differential equations. When applied to the solution of a differential equation of second order, for example, analogue computers will trace the solving function, such derivatives as may be desired, and also the trajectories in phase-space.

The second class of machines includes what are called *digital computers*, which are used for the solution of equations where numerical tables of the functions are desired. Ordinary desk calculators belong to this class. An astonishing development has taken place in recent years in the construction of such machines, with built-in memory units and rapid recording devices. These devices have enabled mathematicians to solve numerical problems which were hitherto far beyond their powers. Two types are in common use, the IBM computers which operate from punched cards and the UNIVAC computers which use tape. The operation of such machines, however, is very expensive and requires a staff of trained men to *program* or *code* problems submitted to them.

In the numerical solution of differential equations the method of *finite integration* in one of its several forms has been, for the most part, the favorite tool of computers. It is based essentially upon the theory of the Picard algorithm, which we have described in Chapter 4. As the reader will recall, this theory depends upon the evaluation of a sequence of integrals and thus is admirably adapted to methods of finite integration.

In this book, however, we have found the method of *continuous analytic continuation* to be admirably suited to the computation of the solution of nonlinear equations. As we have shown in Chapter 9, it is based upon the existence theory of Cauchy's calculus of limits and involves only a few derivatives of the functions involved. Its flexibility, its ready adaptation to mechanical computers, and the fact that its application is not limited to real segments, but may be extended to the evaluation of solutions over paths in the complex plane, make it a method of unusual power.

Since an adequate description of continuous analytic continuation has already been given, we shall limit our discussion in this chapter to the method of finite integration. Since this depends upon the *calculus of finite differences*, we shall give a brief description of this subject and will then describe several of the important adaptations of it to the integration of differential equations.

2. The Calculus of Finite Differences

The calculus of finite differences begins with the definition of the difference of a function $f(x)$, denoted by $\Delta f(\imath)$, or simply Δf. By this symbol we mean

$$\Delta f(x) = f(x+d) - f(x), \tag{1}$$

where d is the difference interval.

By the second difference, denoted by $\Delta^2 f(x)$, we mean the difference of $\Delta f(\imath)$, that is,

$$\Delta^2 f(x) = \Delta[\Delta f(x)] = f(x+2d) - f(x+d) + f(x); \tag{2}$$

and by the nth difference, denoted by $\Delta^n f(\imath)$, the quantity

$$\Delta^n f(x) = \Delta[\Delta^{n-1} f(x)] = f(x+nd) - {}_nC_1 f[x+(n-1)d]$$
$$+ {}_nC_2 f[x+(n-2)d] + \ldots \pm f(x), \tag{3}$$

where ${}_nC_1 = n$, ${}_nC_2 = \frac{1}{2} n(n-1)$, etc., are the binomial coefficients.

It is often useful also to employ the symbol: $E = 1 + \Delta$, since we have

$$Ef(x) = (1+\Delta)f(x) = f(x+d), \tag{4}$$

and by a simple generalization,

$$E^p f(x) = (1+\Delta)^p f(x) = f(x+pd).$$ (5)

If the term $(1+\Delta)^p$ is developed by the binomial theorem, we obtain the following important expansion of $f(x+pd)$ known as the *Gregory-Newton series:*

$$f(x+pd) = f(x) + p\Delta f(x) + \frac{p(p-1)}{2!}\Delta^2 f(x) + \frac{p(p-1)(p-2)}{3!}\Delta^3 f(x) + \cdots .$$ (6)

In applying this and other formulas to numerical approximation the following table of differences is to be used:

Argument	Tabular Value	Δ	Δ^2	Δ^3	Δ^4
$x-3d$	$f(x-3d)$		$\Delta^2 f(x-4d)$		$\Delta^4 f(x-5d)$
		$\Delta f(x-3d)$		$\Delta^3 f(x-4d)$	
$x-2d$	$f(x-2d)$		$\Delta^2 f(x-3d)$		$\Delta^4 f(x-4d)$
		$\Delta f(x-2d)$		$\Delta^3 f(x-3d)$	
$x-d$	$f(x-d)$		$\Delta^2 f(x-2d)$		$\Delta^4 f(x-3d)$
		$\Delta f(x-d)$		$\Delta^3 f(x-2d)$	
x	$f(x)$		$\Delta^2 f(x-d)$		$\Delta^4 f(x-2d)$
		$\Delta f(x)$		$\Delta^3 f(x-d)$	
$x+d$	$f(x+d)$		$\Delta^2 f(x)$		$\Delta^4 f(x-d)$
		$\Delta f(x+d)$		$\Delta^3 f(x)$	
$x+2d$	$f(x+2d)$		$\Delta^2 f(x+d)$		$\Delta^4 f(x)$
		$\Delta f(x+2d)$		$\Delta^3 f(x+d)$	
$x+3d$	$f(x+3d)$		$\Delta^2 f(x+2d)$		$\Delta^4 f(x+d)$

From this array we see that the differences which appear in expansion (6) are the diagonal differences underlined in the table. As an example of its application, let us consider the following numerical values from which we wish to compute log $\Gamma(1.0464)$:

x	log $\Gamma(x)$	Δ	Δ^2	Δ^3	Δ^4
1.046	9.989 208 037 866				
		$-218\ 546\ 226$			
1.047	9.988 989 491 640		$668\ 252$		
		$-217\ 877\ 974$		-922	
1.048	9.988 771 613 666		$667\ 330$		4
		$-217\ 210\ 644$		-918	
1.049	9.988 554 403 022		$666\ 412$		
		$-216\ 544\ 232$			
1.050	9.988 337 858 790				

Since $p=0.4$, we substitute this value and the following differences in equation (6):

$$\Delta=-0.(3)\ 212\ 546\ 226,\ \Delta^2=0.(6)\ 668\ 252,$$

$$\Delta^3=-0.(9)\ 922,\qquad\qquad \Delta^4=0.(11)\ 4,$$

where the number in parentheses indicates the number of zeros between the decimal point and the first significant figure. We thus obtain the desired value:

$$\log\ \Gamma(1.0464)=9.989\ 208\ 037\ 866-0.(4)\ 87\ 418\ 490-0.(7)\ 80\ 190$$
$$-0.(10)\ 58,$$
$$=9.989\ 120\ 539\ 128.$$

It is sometimes desirable, especially in the numerical approximation of differential equations, to obtain values from a formula expressed in *backward differences*, instead of *forward differences* as just illustrated. This is the case, for example, when one is extrapolating at the end of a table, where forward differences are not available.

A useful formula of this type is obtained by adapting the Gregory-Newton formula given in (6). We thus obtain the following:

$$f(x+pd)=f(x)+p\Delta f(x-1)+\frac{p(p+1)}{2!}\ \Delta^2 f(x-2)$$
$$+\frac{p(p+1)(p+2)}{3!}\ \Delta^3 f(x-3)+\ \ldots\ \ (7)$$

To illustrate the application of this formula, let us consider the table of $\log\ \Gamma(x)$ given above from which we are to compute the value of $\log\ \Gamma(1.051)$.

In this case $p=1$ and we have the differences:

$$\Delta=-0.(3)\ 216\ 544\ 232,\ \Delta^2=0.(6)\ 666\ 412,$$

$$\Delta^3=-0.(9)\ 918,\qquad\qquad \Delta^4=0.(11)\ 4.$$

Substituting in formula (7), we then obtain:

$$\log\ \Gamma(1.051)=9.988\ 337\ 858\ 790-0.(3)\ 216\ 544\ 232-0.(9)\ 918+0.(11)4,$$
$$=9.988\ 121\ 980\ 056,$$

which is in error by only four units in the last place.

3. Differences and Derivatives

The calculus of finite differences provides a convenient method by means of which differences of a function can be expressed in terms of its derivatives, and conversely. We shall denote successive derivatives by the following symbols: D, D^2, D^3, etc.

To establish the desired relationships, let us first write Taylor's expansion of $f(x)$ in the following way:

$$Ef(x) = f(x+d) = f(x) + df'(x) + \frac{d^2}{2!}f''(x) + \dots,$$

$$= \left(1 + D + \frac{d^2D^2}{2} + \dots\right)f(x) = e^{dD}f(x).$$

(1)

From this we get the following symbolic relationship between E and D:

$$E = e^{dD}.$$

(2)

Since we have by definition: $\Delta = E - 1$, we can then write

$$\Delta^n = (E-1)^n,$$

(3)

from which, by means of (2), we obtain the following symbolic expansion:

$$\Delta^n = (e^{dD} - 1)^n,$$

$$= \left[e^{ndD} - ne^{(n-1)dD} + \frac{n(n-1)}{2!}e^{(n-2)dD} - \dots \pm 1 \right].$$

(4)

We now let $n = 1, 2, 3$, etc. successively, and expand the resulting functions as power series in D. When we apply the resulting operators to $f(x)$, we obtain the following explicit expansions:

$$\Delta f = d\left[f' + \frac{d}{2!}f'' + \frac{d^2}{3!}f^{(3)} + \frac{d^3}{4!}f^{(4)} + \frac{d^4}{5!}f^{(5)} + \dots \right],$$

$$\Delta^2 f = d^2\left[f'' + df^{(3)} + \frac{7}{12}d^2f^{(4)} + \frac{1}{4}d^3f^{(5)} + \frac{31}{360}d^4f^{(6)} + \dots \right],$$

$$\Delta^3 f = d^3\left[f^{(3)} + \frac{3}{2}df^{(4)} + \frac{5}{4}d^2f^{(5)} + \frac{3}{4}d^3f^{(6)} + \frac{903}{2520}d^4f^{(7)} + \dots \right],$$

$$\Delta^4 f = d^4\left[f^{(4)} + 2df^{(5)} + \frac{13}{5}d^2f^{(6)} + \frac{5}{3}d^3f^{(7)} + \frac{81}{80}d^4f^{(8)} + \dots \right],$$

(5)

*   *   *   *   *   *   *

It is possible similarly to express derivatives in terms of differences. To achieve this, we solve equation (2) for dD and thus obtain

$$dD = \log(1+\Delta),$$

$$= \Delta - \frac{1}{2}\Delta^2 + \frac{1}{3}\Delta^3 - \frac{1}{4}\Delta^4 + \dots$$

(6)

The nth derivative is then given symbolically by the following equation:

$$d^n D^n = [\log (1+\Delta)]^n = \left[\Delta - \frac{1}{2}\Delta^2 + \frac{1}{3}\Delta^3 - \frac{1}{4}\Delta^4 + \cdots \right]^n. \qquad (7)$$

As before we let $n=1$, 2, 3, etc. successively. The right-hand member of (7) is expanded as a power series in Δ, and the resulting operators applied to $f(x)$. The following explicit expansions are then obtained:

$$f'(x) = \frac{1}{d}\left[\Delta f(x) - \frac{1}{2}\Delta^2 f(x) + \frac{1}{3}\Delta^3 f(x) - \frac{1}{4}\Delta^4 f(x) + \frac{1}{5}\Delta^5 f(x) - \cdots \right],$$

$$f''(x) = \frac{1}{d^2}\left[\Delta^2 f(x) - \Delta^3 f(x) + \frac{11}{12}\Delta^4 f(x) - \frac{5}{6}\Delta^5 f(x) + \cdots \right],$$

$$f^{(3)}(x) = \frac{1}{d^3}\left[\Delta^3 f(x) - \frac{3}{2}\Delta^4 f(x) + \frac{7}{4}\Delta^5 f(x) - \cdots \right], \qquad (8)$$

$$f^{(4)}(x) = \frac{1}{d^4}[\Delta^4 f(x) - 2\,\Delta^5 f(x) + \cdots],$$

*   *   *   *   *   *   *

As an example of the application of formulas (8), we shall compute the first two derivatives of $\log \Gamma$ (1.046) from the table given in Section 2. Since we have

$$\frac{d}{dx}\log_{10} \Gamma(x) = \frac{d}{dx}M\log_e \Gamma(x) = M\Psi(x),$$

where $M = 0.43429\ 44819$ is the modulus and $\Psi(x) = \Gamma'(x)/\Gamma(x)$ is the psi function, and since also

$$\frac{d^2}{dx^2}\log_{10}\Gamma(x) = M\Psi'(x),$$

the exact values to 10 places of the first two derivatives of $\log \Gamma(x)$ at $x=1.046$ are found from tables of $\Psi(x)$ and $\Psi'(x)$ to equal respectively $-0.21888\ 06595$ and $0.66917\ 53958$.

Making use of the differences in the table of $\log \Gamma(x)$, we thus compute

$$\frac{d}{dx}\log \Gamma(x) = 1000[-0.(3)\ 218\ 546\ 226 - 0.(6)\ 334\ 126 - 0.(9)\ 307$$

$$-0.(11)1],$$

$$= -0.218\ 880\ 660.$$

Similarly, for the second derivative of log $\Gamma(x)$, we obtain

$$\frac{d^2}{dx^2}\log \Gamma(x)=10^6[0.(6)668\ 252+0.(9)\ 992+0.(11)\ 4],$$

$$=0.669\ 178.$$

One should observe from these computations how rapidly the number of significant figures reduces in converting from differences to derivatives.

4. Integration Formulas

A number of integration formulas are available for the evaluation of an integral, which it will be convenient to write in the form:

$$I=\frac{1}{d}\int_x^{x+pd} f(t)dt. \tag{1}$$

The most common method of numerical integration makes use of the Euler-Maclaurin formula, which can be written as follows: *

$$\frac{1}{d}\int_x^{x+pd} f(t)dt=\sum_{n=0}^{p}f(x+nd)-\frac{1}{2}[f(x)+f(x+pd)]-\frac{d}{12}[f'(x+pd)-f'(x)]$$

$$+\frac{d^3}{720}[f^{(3)}(x+pd)-f^{(3)}(x)]-\frac{d^5}{30240}[f^{(5)}(x+pd)-f^{(5)}(x)]$$

$$+\frac{d^7}{1209600}[f^{(7)}(x+pd)-f^{(7)}(x)]-\frac{d^9}{47900160}[f^{(9)}(x+pd)$$

$$-f^{(9)}(x)]+\ ...\ +(-1)^n\frac{B_n d^{2n-1}}{(2n)!}[f^{(2n-1)}(x+pd)$$

$$-f^{(2n-1)}(x)]+R_n, \tag{2}$$

where B_n is the nth *Bernoulli number*, the first six of which are given below:

$$B_1=1/6,\ B_2=1/30,\ B_3=1/42,\ B_4=1/30,\ B_5=5/66,\ B_6=691/2730,\ \ ....$$

These numbers appear in the following expansion:

$$\frac{t}{e^t-1}=1-\frac{t}{2}+\frac{B_1t^2}{2!}-\frac{B_2t^4}{4!}+\frac{B_3t^6}{6!}-\frac{B_4t^8}{8!}+\ ...\ , \tag{3}$$

which is related to equation (1) through the operational identity

$$\Delta^{-1}=\frac{1}{e^D-1}=D^{-1}-\frac{1}{2}+\frac{B_1D}{2!}-\frac{B_2D^3}{4!}+\ ...\ , \tag{4}$$

---

*This formula was discovered independently by L. Euler and C. Maclaurin between 1730 and 1740. Euler's work was published in *Comm. Acad. Sci. Petrop.*, Vol. 6, 1738, p. 68, and Maclaurin's in his *Treatise of Fluxions*, 1742, p. 672.

where Δ^{-1} is equivalent to the summation symbol Σ and D^{-1} is equivalent to the symbol for integration.

The remainder term R_n in equation (2) is bounded by the following inequality:

$$|R_n| < p \frac{B_{n+1}}{(2n+2)!} d^{2n+2} |f^{(2n+2)}(\xi)|, \tag{5}$$

where ξ is some value between x and $x+pd$.

The derivation of formula (2) will not be given here since it can be found in any work on the calculus of finite differences.

A second very useful form of the Euler-Maclaurin formula is obtained if derivatives in (2) are replaced by differences. In this case, the resulting integration is called the Gregory formula.* It appears in the following expansion:

$$\frac{1}{d} \int_x^{x+pd} f(t)dt = \sum_{n=0}^{p} f(x+nd) - \frac{1}{2}\left[f(x)+f(x+pd)\right] - \frac{1}{12}\left\{\Delta f[x+(p-1)d]\right.$$

$$-\Delta f(x)\} - \frac{1}{24}\left\{\Delta^2 f[x+(p-2)d]+\Delta^2 f(x)\right\} - \frac{19}{720}\left\{\Delta^3 f[x+\right.$$

$$(p-3)d] - \Delta^3 f(x)\} - \frac{3}{160}\left\{\Delta^4 f[x+(p-4)d]+\Delta^4 f(x)\right\}$$

$$-\frac{863}{60480}\left\{\Delta^5 f[x+(p-5)d]-\Delta^5 f(x)\right\} - \frac{275}{24195}\left\{\Delta^6 f[x+\right.$$

$$(p-6)d] + \Delta^6 f(x)\} + R_n. \tag{6}$$

The numbers which appear as multipliers of the various terms are called *logarithmic numbers*, since they are obtained from the expansion

$$\frac{t}{\log(1+t)} = 1 + \frac{1}{2}t - \frac{1}{12}t^2 + \frac{1}{24}t^3 - \frac{19}{720}t^4 + \ldots, \tag{7}$$

which is related to formula (6) by the operational identity

$$\frac{1}{dD} = \frac{1}{\log(1+\Delta)} = \frac{1}{\Delta} + \frac{1}{2} - \frac{1}{12}\Delta + \frac{1}{24}\Delta^2 - \ldots, \tag{8}$$

where $1/D$ is equivalent to the symbol for integration and $1/\Delta$ is equivalent to the summation symbol Σ.

No proof will be given for the Gregory formula since this can be found in treatises on the calculus of finite differences. Nor will the explicit form for R_n be stated, since this is relatively complicated and

---

*This formula was announced by James Gregory (1638–75) in a letter to John Collins in 1670. See Rigaud's *Correspondence*, Vol. 2, p. 209.

in applications of (6) the convergence of the process is assured if the differences are sufficiently small. The application of the two formulas can be illustrated simply by means of the following table of values:

x	$f(x)=x^3$	Δ	Δ^2	Δ^3
1.0	1. 000	0. 331	0. 066	0. 006
1.1	1. 331	0. 397	0. 072	0. 006
1.2	1. 728	0. 469	0. 078	0. 006
1.3	2. 197	0. 547	0. 084	
1.4	2. 744	0. 631		
1.5	3. 375			

Total $(S) = 12.375$.

To compute the integral,

$$I = \int_{1.0}^{1.5} f(x)\ dx,$$

by means of (1) we first find:

$f'(1.5)=6.75,\ f''(1.5)=9.0,\ f^{(3)}(1.5)=6;\ f'(1)=3,\ f''(1)=6,\ f^{(3)}(1)=6.$

Substituting these values in (2), observing that $p=5$ and $d=0.1$, we get

$$I = \int_{1.0}^{1.5} f(x)\,dx = 0.1 \left[12.375 - \frac{1}{2}\,(1.000+3.375) - \frac{0.1}{12}\,(6.75-3.00) \right],$$

$$= 0.1\,(12.375 - 2.1875 - 0.03125) = 1.015625,$$

which is exact.

Similarly, making use of formula (6) and observing that the differences which enter the calculation are the diagonal differences at the beginning of the table, namely, 0.331, 0.066, 0.006, and the backward differences at the end of the table, namely, 0.631, 0.084, 0.006, we readily compute:

$$I = \int_{1.0}^{1.5} f(x)\ dx = 0.1 \left[12.375 - \frac{1}{2}\,(1.000+3.375) - \frac{1}{12}\,(0.631-0.331) \right.$$

$$\left. - \frac{1}{24}\,(0.084+0.066) \right] = 0.1\,(12.375 - 2.1875 - 0.025 - 0.00625)$$

$$= 1.015625.$$

Another useful integration formula is obtained by the direct integration of the Gregory-Newton series. Thus, referring to formula (6) of Section 2, we write the Gregory-Newton series with a slight change in notation as follows:

$$f(x+td)=f(x)+t\Delta f(x)+\frac{t(t-1)}{2!}\,\Delta^2 f(x)+\ \ldots\ .$$

Integrating this series with respect to t, we then obtain

$$\int_0^p f(x+td)\,dt=\frac{1}{d}\int_x^{x+pd} f(s)\,ds$$
$$=\eta_0(p)f(x)+\eta_1(p)\Delta f(x)+\frac{\eta_2(p)}{2!}\,\Delta^2 f(x)+\ \ldots\ ,\quad (9)$$

where we use the abbreviations:

$$\eta_0(p)=p,\ \eta_1(p)=\int_0^p t\,dt,\ \eta_2(p)=\int_0^p t\,(t-1)\,dt,\ \ldots\ ,$$

$$\eta_n(p)=\int_0^p t\,(t-1)\,(t-2)\,\ldots\,(t-n+1)\,dt \qquad (10)$$

A few of these polynomials are given below as follows:

$$\eta_0(p)=p,\ \eta_1(p)=\frac{1}{2}p^2,\ \eta_2(p)=\frac{1}{6}p^2(2p-3),\ \eta_3(p)=\frac{1}{4}p^2(p-2)^2,$$

$$\eta_4(p)=\frac{1}{30}p^2(6p^3-45p^2+110p-90),\ \eta_5(p)=\frac{1}{12}p^2(p-4)^2(2p^2-8p+9),$$

$$\eta_6(p)=\frac{1}{84}(12p^5-210p^4+1428p^3-4725p^2+7672p-5040).\qquad (11)$$

Another formula, similar to (9), is obtained by the direct integration of the Gregory-Newton formula expressed in terms of backward differences. Thus, referring to formula (7) of Section 2, we write it with a slight change in notation as follows:

$$f(x+td)=f(x)+t\Delta f(x-1)+\frac{t(t+1)}{2!}\,\Delta^2 f(x-2)+\ \ldots\ .$$

If we integrate this expansion, we obtain the following formula:

$$\int_0^p f(x+td)\,dt=\frac{1}{d}\int_x^{x+pd} f(s)\,ds$$
$$=-\eta_0(-p)\,f(x)+\eta_1(-p)\Delta f(x-1)$$
$$-\frac{\eta_2(-p)}{2!}\,\Delta^2 f(x-2)+\ \ldots\ ,\quad (12)$$

where $\eta_n(p)$ are the functions which appear in (9).

As an example of the application of formula (9) we shall compute the value of the integral,

$$I = \int_{1.0}^{1.5} f(x) \, dx,$$

where $f(x) = x^3$, making use of the table given earlier in this section. Since we have $d = 0.1$, $p = 5$, we fiirst compute the following values:

$$\eta_0(5) = 5, \quad \eta_1(5) = 12.5, \quad \eta_2(5) = 175/6, \quad \eta_3(5) = 225/4.$$

Substituting these values in (9), we then obtain

$$\int_{1.0}^{1.5} f(x) \, dx = 0.1 \, (5 + 4.1375 + 0.9625 + 0.05625) = 1.015625.$$

The usefulness of formula (12) in computing integrals over a range of untabulated values is illustrated by the evaluation of the integral

$$I = \int_{1.5}^{2.0} f(x) \, dx,$$

where $f(x) = x^3$, and where we make use of the values tabulated in the table given earlier in this section over the range between $x = 1.0$ and $x = 1.5$.
Observing that $d = 0.1$ and $p = 5$, we first compute,

$$\eta_0(-5), = -5, \, \eta_1(-5) = 12.5, \, \eta_2(-5) = -325/6, \, \eta_3(-5) = 1225/4.$$

When these values are substituted in (12), we find

$$\int_{1.5}^{2.0} f(x) \, dx = 0.1(16.875 + 7.8875 + 2.275 + 0.30625) = 2.734375,$$

which is the exact value of the integral.

Since it is often useful in the application of formula (12) to integrate over one interval at a time, we give the expansion for $p = 1$ below as follows:

$$\int_0^1 f(x + td) dt = \frac{1}{d} \int_x^{x+d} f(s) ds$$

$$= f(x) + \frac{1}{2} \Delta f(x-1) + \frac{5}{12} \Delta^2 f(x-2) + \frac{3}{8} \Delta^3 f(x-3)$$

$$+ \frac{251}{720} \Delta^4 f(x-4) + \frac{95}{288} \Delta^5 f(x-5) + \frac{19087}{60480} \Delta^6 f(x-6) + \cdots$$

$$(13)$$

Various relatively simple formulas can be obtained from formula (6) by specialization of the constants involved. We shall illustrate this by deriving the well-known *Simpson one-third rule.**

If we assume that $p=2$ and that third differences are constant, then formula (6) reduces as follows:

$$\frac{1}{d}\int_x^{x+2d} f(t)dt = f(x)+f(x+d)+f(x+2d)-\frac{1}{2}f[(x)+f(x+2d)]$$

$$-\frac{1}{12}[\Delta f(x+d)-\Delta f(x)]-\frac{1}{12}\Delta^2 f(x),$$

$$=f(x)+f(x+d)+f(x+2d)-\frac{1}{2}[f(x)+f(x+2d)]$$

$$-\frac{1}{12}[f(x+2d)-2f(x+d)+f(x)]$$

$$-\frac{1}{12}[f(x+2d)-2f(x+d)+f(x)].$$

$$=\frac{1}{3}f(x)+\frac{4}{3}f(x+d)+\frac{1}{3}f(x+2d). \qquad (14)$$

For example, if $d=0.2$ and $x=1.0$ we have

$$\int_{1.0}^{1.4} x^3 dx = 0.2\times\frac{1}{3}(1.000+4\times1.728+2.744)=0.7104,$$

which is exact.

5. An Illustrative Example

For purposes of illustration of the methods of solving differential equations, which are to be given in subsequent sections of this chapter, we shall consider the equation

$$\frac{dy}{dx}=xy(y-2), \qquad (1)$$

---

*This is named after Thomas Simpson (1710-61), who stated his formula in 1743, although it appears to have been given first by F. B. Cavalieri (1598-1647) in 1639 and later by James Gregory in 1668.

the solution of which is to be obtained subject to the initial conditions that $y_0=1$ when $x_0=0$.

The solution of this equation has already been developed in Chapter 4 as a power series in x by means of the calculus of limits and the method of successive approximations.

In the second method, as will be recalled from Section 3 of Chapter 4, we began with the approximate solution:

$$y_1=y_0+\int_{x_0}^{x}f(x,y_0)dx, \tag{2}$$

and obtained in succession y_2,y_3,y_4, etc. by letting $n=1$, 2, 3, etc., in the equation:

$$y_{n+1}=y_0+\int_{x_0}^{x}f(x,y_n)dx. \tag{3}$$

In this manner four approximations were obtained of which y_4 was found to have the following expansion:

$$y_4=1-\frac{x^2}{2}+\frac{x^6}{24}-\frac{x^{10}}{240}+\frac{17x^{14}}{40320}-\ \cdots \tag{4}$$

This approximation is exact to the last term as one can verify with some difficulty by expanding as a power series the solution of the differential equation, namely, the function:

$$y=\frac{2}{1+e^{x^2}}. \tag{5}$$

For purposes of illustration it will be convenient to have the accompanying table in which are recorded to ten decimal accuracy the values of the solution over the range $x=0.00$ to $x=1.00$ at intervals of 0.02, together with the corresponding values of the function: $y'=f(x,y)=xy(y-2)$ and its differences.

x	$y'=xy(y-2)$	$-\Delta$	Δ^2	$-\Delta^3$	Δ^4	$-\Delta^5$	Δ^6	$y=2/(1+e^{x^2})$
0.00	−0.00000 00000	1999 99992						1.00000 00000
0.02	−0.01999 99992	1999 99752	240					0.99980 00000
0.04	−0.03999 99744	1999 98312	1440	−1200				0.99920 00002
0.06	−0.05999 98056	1999 93752	4560	−3120	1920			0.99820 00019
0.08	−0.07999 91808	1999 83192	10560	−6000	2880	−960		0.99680 00109
0.10	−0.09999 75000	1999 62794	20398	−9838	3838	−958	−2	0.99500 00417
0.12	−0.11999 37794	1999 27759	35035	−14637	4799	−961	3	0.99280 01244
0.14	−0.13998 65553	1998 72332	55427	−20392	5755	−956	−5	0.99020 03137
0.16	−0.15997 37885	1997 89806	82526	−27099	6707	−952	−4	0.98720 06990
0.18	−0.17995 27691	1996 72522	1 17284	−34758	7659	−952	0	0.98380 14170
0.20	−0.19992 00213	1995 11882	1 60640	−43356	8598	−939	−13	0.98000 26662
0.22	−0.21987 12095	1992 98349	2 13533	−52893	9537	−939	0	0.97580 47231
0.24	−0.23980 10444	1990 21473	2 76876	−63343	10450	−913	−26	0.97120 79600
0.26	−0.25970 31917	1986 69895	3 51578	−74702	11359	−909	−4	0.96621 28656
0.28	−0.27957 01812	1982 31380	4 38515	−86937	12235	−876	−33	0.96082 00664
0.30	−0.29939 33192	1976 92838	5 38542	−100027	13090	−855	−21	0.95503 03504
0.32	−0.31916 26030	1970 40365	6 52473	−113931	13904	−814	−41	0.94884 46924
0.34	−0.33886 66395	1962 59277	7 81088	−128615	14684	−780	−34	0.94226 42810
0.36	−0.35849 25672	1953 34173	9 25104	−144016	15401	−717	−63	0.93529 05472
0.38	−0.37802 59845	1942 48987	10 85186	−160082	16066	−665	−52	0.92792 51946
0.40	−0.39745 08832	1929 87063	12 61924	−176738	16656	−590	−75	0.92017 02309
0.42	−0.41674 95895	1915 31239	14 55824	−193900	17162	−506	−84	0.91202 80002
0.44	−0.43590 27134	1898 63945	16 67294	−211470	17570	−408	−98	0.90350 12175
0.46	−0.45488 91079	1879 67306	18 96639	−229345	17875	−305	−103	0.89459 30028
0.48	−0.47368 58385	1858 23270	21 44036	−247397	18052	−177	−128	0.88530 69172
0.50	−0.49226 81655	1834 13747	24 09523	−265487	18090	−38	−139	0.87564 69982
0.52	−0.51060 95402	1807 20755	26 92992	−283469	17982	108	−146	0.86561 77965
0.54	−0.52868 16157	1777 26593	29 94162	−301170	17701	281	−173	0.85522 44116
0.56	−0.54645 42750	1744 14021	33 12572	−318410	17240	461	−180	0.84447 25276
0.58	−0.56389 56771	1707 66455	36 47566	−334994	16584	656	−195	0.83336 84486
0.60	−0.58097 23226	1667 68178	39 98277	−350711	15717	867	−211	0.82191 91319
0.62	−0.59764 91404	1624 04558	43 63620	−365343	14632	1085	−218	0.81013 22209
0.64	−0.61388 95962	1576 62287	47 42271	−378651	13308	1324	−239	0.79801 60751
0.66	−0.62965 58249	1525 29614	51 32673	−390402	11751	1557	−233	0.78557 97983
0.68	−0.64490 87863	1469 96594	55 33020	−400347	9945	1806	−249	0.77283 32636
0.70	−0.65960 84457	1410 55344	59 41250	−408230	7883	2062	−256	0.75978 71353
0.72	−0.67371 39801	1347 00275	63 55069	−413819	5589	2294	−232	0.74645 28865
0.74	−0.68718 40076	1279 28355	67 71920	−416851	3032	2557	−263	0.73284 28128
0.76	−0.69997 68431	1207 39333	71 89022	−417102	251	2781	−224	0.71897 00408
0.78	−0.71205 07764	1131 35962	76 03371	−414349	−2753	3004	−223	0.70484 85318
0.80	−0.72336 43726	1051 24204	80 11758	−408387	−5962	3209	−205	0.69049 30788
0.82	−0.73387 67930	967 13410	84 10794	−399036	−9351	3389	−180	0.67591 92983
0.84	−0.74354 81340	879 16471	87 96939	−386145	−12891	3540	−151	0.66114 36145
0.86	−0.75233 97811	787 49923	91 66548	−369609	−16536	3645	−105	0.64618 32379
0.88	−0.76021 47734	692 34047	95 15876	−349328	−20281	3745	−100	0.63105 61348
0.90	−0.76713 81781	593 92875	98 41172	−325296	−24032	3751	−6	0.61578 09914
0.92	−0.77307 74656	492 54191	101 38684	−297512	−27784	3752	−1	0.60037 71691
0.94	−0.77800 28847	388 49462	104 04729	−266045	−31467	3683	69	0.58486 46526
0.96	−0.78188 78309	282 13704	106 35758	−231029	−35016	3549	134	0.56926 39909
0.98	−0.78470 92013	173 85316	108 28388	−192630	−38399	3383	166	0.55359 62307
1.00	−0.78644 77329	64 05839	109 79477	−151089	−41541	3142	241	0.53788 28427

6. The Adams-Bashforth Method

Historically one of the first of the approximation techniques was suggested by J. C. Adams (1819–92), the astronomer, and F. Bashforth in their treatise, *Theories of Capillary Action*, 1883.

Let us consider the differential equation,

$$\frac{dy}{dx} = f(x,y), \tag{1}$$

for which we assume that initial values, x_0, y_0, are given and that the function $f(x,y)$ exists and has derivatives of all orders at (x_0, y_0).

The method assumes that an initial table of values has been provided from which a set of backward differences can be derived. This preliminary table can be computed, for example, by means of a Taylor's series as described in the calculus of limits in Chapter 4.

One then uses formula (13) of Section 4 to compute

$$\int_x^{x+d} y'(s)\,ds = y(x+d) - y(x) = \Delta y(x)$$

$$= d\left[y'(x) + \frac{1}{2}\Delta y'(x-d) + \frac{5}{12}\Delta^2 y'(x-2d) + \frac{3}{8}\Delta^3 y'(x-3d) + \dots \right]. \tag{2}$$

Since $\Delta y(x)$ has thus been found, the next value in the table is computed, a new set of differences is obtained, and the process continued to the next succeeding value.

As an example, we shall consider the equation introduced in the preceding section. We assume that values of y and y' have been computed through $x = 0.20$ and we shall use these to extend the table to $x = 0.22$. We thus have the following values available:

$$y = 0.98000\,26662, \quad y' = -0.19992\,00213, \quad \Delta y' = -1996\,72522,$$

$$\Delta^2 y' = 1\,17284, \quad \Delta^3 y' = 34758, \quad \Delta^4 y' = 7659, \quad \Delta^5 y' = 952.$$

Substituting in (2), we thus obtain:

$$\Delta y = 0.02(-19992\,00213 - 998\,36261 + 48868$$
$$+ 13034 + 2670 + 314) \times 10^{-10},$$

$$= -0.00419\,79432.$$

Hence, for $x = 0.22$ we have

$$y = 0.98000\,26662 - 0.00419\,79432 = 0.97580\,47230,$$

which differs from the recorded value by one unit in the last place.

It is clear that we can continue from this point to the next and hence construct the value of y over any range that may be desired.

7. The Runge-Kutta Method

The method that is used perhaps more than any other in the integration of differential equations is one devised by C. Runge (1856–1927)[*] in 1895 and extended by W. Kutta[†] 6 years later. Its application to the equation,

$$\frac{dy}{dx} = f(x,y), \tag{1}$$

is described as follows.

Let us assume that the initial values are x_0 and y_0. An increment of the independent variable x is then selected, let us say d, and the following values are successively computed:

$$k_1 = f(x_0, y_0)d$$

$$k_2 = f(x_0 + \frac{1}{2}d, \; y_0 + \frac{1}{2}k_1)d,$$

$$k_3 = f(x_0 + \frac{1}{2}d, \; y_0 + \frac{1}{2}k_2)d,$$

$$k_4 = f(x_0 + d, \; y_0 + k_3)d. \tag{2}$$

The new value of y which corresponds to $x_0 + d$ is then given by $y = y_0 + k$, where we write

$$k = \frac{1}{6}(k_1 + 2k_2 + 2k_3 + k_4). \tag{3}$$

In illustration of the application of this method, we shall consider the example of Section 5 subject to the conditions: $y_0 = 1$, $x_0 = 0$. If we choose $d = 0.1$, then we get the following values:

$$k_1 = 0, \; k_2 = -0.005, \; k_3 = -0.00499\,99688, \; k_4 = -0.00999\,97500,$$

from which we compute:

$$k = \frac{1}{6}(-0.02999\,96875) = -0.00499\,99479.$$

We thus obtain the approximation:

$$y_1 = y_0 + k = 0.99500\,00521,$$

which, when compared with the correct value: $y_1 = 0.99500\,00417$, shows an error of 1 in the eighth place.

[*] *Mathematische Annalen*, Vol. 46, 1895, p. 167.
[†] *Zeitschrift für Math. und Physik*, Vol. 46, 1901, p. 435.

The order of approximation achieved by this method is indicated by the following special case:

$$\frac{dy}{dx}=f(x). \tag{4}$$

Since the values of the constants are now:

$$k_1=f(x_0)d, \ k_2=f\left(x_0+\frac{1}{2}d\right)d, \ k_3=f\left(x_0+\frac{1}{2}d\right)d, \ k_4=f(x_0+d)d,$$

we obtain the solution:

$$y=y_0+\frac{d}{6}\left[f(x_0)+4f\left(x_0+\frac{1}{2}d\right)+f(x_0+d)\right]. \tag{5}$$

Observing that the solution of (4) is actually the integral

$$y=y_0+\int_{x_0}^{x}f(x)\,dx,$$

we see from (5) that the present method is equivalent in this case to Simpson's rule given by (14) in Section 4. Since the error in this formula is of the order of d^4, we can expect a similar error in the general formula.

About the order of approximation Runge and König make the following statement:[*]

"The error of the procedure one can estimate through calculation of the term of fifth order. It is very convenient, however, to apply the process a second time with twice the interval breadth. The error of the first computation amounts to about one-fifteenth of the difference of the two results."

In illustration we shall compute the value of y in the example previously used, for $x=0.20$. We shall first apply the method of this section once using $d=0.2$, and then we shall apply the method twice using $d=0.1$. The difference between the two values should then be approximately fifteen times the actual error. These computations follow:

Interval: $d=0.2$.

x	y	$y-2$	$xy(y-2)$	$k=dxy(y-2)$
0	1	-1	0	$0=k_1$
0.1	1	-1	-0.1	$-0.02=k_2$
0.1	0.99	-1.01	-0.09999	$-0.01999\ 8=k_3$
0.2	0.980002	-1.019998	$-0.19992\ 00160$	$-0.03998\ 40032=k_4$

---

[*]C. Runge and H. König: *Vorlesungen über Numerisches Rechnung.* Berlin, 1924. See p. 294.

We thus have,

$$k=\frac{1}{6}(-0.11998\ 00032)=-0.01999\ 66672,$$

$$y_2=y_0+k=0.98000\ 33328.$$

Interval: $d=0.1$.

x	y	$y-2$	$xy(y-2)$	$k=dxy(y-2)$
0	1	-1	0	$0=k_1$
0.05	1	-1	-0.05	$-0.005=k_2$
0.05	0.9975	-1.0025	$-0.04999\ 96875$	$-0.00499\ 996875=k_3$
0.1	0.99500\ 003125	$-1.00499\ 996875$	$-0.09999\ 750005$	$-0.00999\ 975001=k_4$

From this we get,

$$k=\frac{1}{6}(-0.02999\ 968751)=-0.00499\ 994792,$$

$$y_1=y_0+k=0.99500\ 005208.$$

Interval: $d=0.1$.

x	y	$y-2$	$xy(y-2)$	$k=dxy(y-2)$
0.10	0.99500\ 005208	$-1.00499\ 994792$	$-0.09999\ 750006$	$-0.00999\ 975001=k_1$
0.15	0.99000\ 017708	$-1.00999\ 982292$	$-0.14998\ 500055$	$-0.01499\ 850005=k_2$
0.15	0.98750\ 080206	$-1.01249\ 919794$	$-0.14997\ 656552$	$-0.01499\ 765655=k_3$
0.20	0.98000\ 239553	$-1.01999\ 760447$	$-0.19992\ 001906$	$-0.01999\ 200191=k_4$

We then compute,

$$k=\frac{1}{6}(-0.08998\ 406511)=-0.01499\ 734418,$$

$$y_2=y_1+k=0.98000\ 27079.$$

The difference between the two values of y_2 is thus:

$$\Delta y_2=0.00000\ 06249,$$

giving an error $$E=\frac{1}{15}\Delta y_2=0.00000\ 00417,$$

which is found to be equal to the actual error:

$$0.98000\ 27079-0.98000\ 26662=0.00000\ 00417.$$

The derivation of the Runge-Kutta equations involves a tedious computation, but the details of the argument can be inferred from the following analysis.

Referring to equation (1), we first write the expansion of its solution as the following Taylor's series:

$$y - y_0 = y_0'(x - x_0) + \frac{y_0''}{2!}(x - x_0)^2 + \dots,$$

$$= f_0(x - x_0) + \frac{1}{2!}\left(\frac{\partial f}{\partial x} + f\frac{\partial f}{\partial y}\right)_0 (x - x_0)^2 + \dots.$$

If we now write: $y - y_0 = k$, and $x - x_0 = d$, then this expansion becomes

$$k = f_0 d + \left(\frac{\partial f}{\partial x} + f\frac{\partial f}{\partial y}\right)_0 \frac{d^2}{2!} + \dots. \tag{6}$$

We next consider the following expansions:

$$k_1 = f_0 d,$$

$$k_2 = f\left(x_0 + \frac{1}{2}d, y_0 + \frac{1}{2}k_1\right)d = d\left[f_0 + \left(\frac{1}{2}d\frac{\partial}{\partial x} + \frac{1}{2}k_1\frac{\partial}{\partial y}\right)f_0 + \dots\right],$$

$$k_3 = f\left(x_0 + \frac{1}{2}d, y_0 + \frac{1}{2}k_2\right)d = d\left[f_0 + \left(\frac{1}{2}d\frac{\partial}{\partial x} + \frac{1}{2}k_2\frac{\partial}{\partial y}\right)f_0 + \dots\right],$$

$$k_4 = f(x_0 + d, y_0 + k_3)d = d\left[f_0 + \left(d\frac{\partial}{\partial x} + k_3\frac{\partial}{\partial y}\right)f_0 + \dots\right].$$

If we now write,

$$k = a_1 k_1 + a_2 k_2 + a_3 k_3 + a_4 k_4,$$

where a_1, a_2, a_3, a_4 are parameters to be determined, it is clear that we get

$$k = (a_1 + a_2 + a_3 + a_4)d \cdot f_0 + \left[\frac{d^2}{2}(a_2 + a_3 + 2a_4)\frac{\partial}{\partial x}\right.$$

$$\left. + \frac{d}{2}(a_2 k_1 + a_3 k_2 + 2a_4 k_3)\frac{\partial}{\partial y}\right]f_0 + \dots. \tag{7}$$

But we also observe that

$$a_2 k_1 + a_3 k_2 + 2a_4 k_3 = (a_2 + a_3 + 2a_4)d \cdot f_0 + \text{higher terms in } d.$$

It thus follows that when this is substituted in (7), we obtain:

$$k = (a_1 + a_2 + a_3 + a_4)d \cdot f_0 + \frac{d^2}{2!}\left[(a_2 + a_3 + 2a_4)\frac{\partial}{\partial x} + (a_2 + a_3 + 2a_4)f_0\frac{\partial}{\partial y}\right]f_0 + \dots.$$

If this expression is to be identified with (6), we see that the a_i must satisfy the following equations:

$$a_1+a_2+a_3+a_4=1, \quad a_2+a_3+2a_4=1. \tag{8}$$

Two similar equations are obtained when terms of orders d^3 and d^4 are identified in the same manner. Their derivation is somewhat involved and will not be given here. They will be found to be the following:

$$3a_2+3a_3+12a_4=4, \quad a_2+a_3+8a_4=2. \tag{9}$$

When this system of four equations is solved, the values given earlier in (3): $a_1=a_4=1/6$, $a_2=a_3=1/3$, are obtained.

8. The Milne Method

A method similar to those already described, but which recommends itself for its simplicity, is due to W. E. Milne.* This method consists in making two extrapolations from an initial set of four values. Let these values be (x_0,y_0), (x_1,y_1), (x_2,y_2) and (x_3,y_3). The corresponding values of y_0', y_1', y_2', and y_3' are then computed from the equation,

$$\frac{dy}{dx}=f(x,y). \tag{1}$$

By means of these derivatives we then obtain an approximation for y_4 from the formula:

$$y_4=y_0+\frac{4d}{3}\left(2y_3'-y_2'+2y_1'\right). \tag{2}$$

This new value is now substituted in the differential equation to obtain y_4' and a new approximation, Y_4, is obtained from the formula:

$$Y_4=y_2+\frac{d}{3}\left(y_4'+4y_3'+y_2'\right). \tag{3}$$

The error in the second value is estimated from the difference

$$\Delta=\frac{|Y_4-y_4|}{29}. \tag{4}$$

If this error is sufficiently small, so that the desired order of approximation has been attained, then one proceeds to the next value.

---

*W. E. Milne: "Numerical Integration of Ordinary Differential Equations," *American Math. Monthly*, Vol. 33, 1926, pp. 455–460.

In illustration, using the equation

$$\frac{dy}{dx} = xy(y-2), \tag{5}$$

and the initial values: $x_0=0$, $y_0=1$, we obtain the following values by one of the various methods already described:

x	y	$y'=xy(y-2)$
0. 00	1. 00000 00000	0. 00000 00000
0. 02	0. 99980 00000	−0. 01999 99992
0. 04	0. 99920 00002	−0. 03999 99744
0. 06	0. 99820 00019	−0. 05999 98056

Employing (2) and observing that $d=0.02$, we compute for $x=0.08$ the following value:

$$y_4 = 1 + \frac{0.08}{3}(-0.11999\ 96112 + 0.03999\ 99744 - 0.03999\ 99984)$$

$$= 0.99680\ 00097.$$

Substituting this value in (5) we get for $x=0.08$:

$$y'_4 = -0.07999\ 91808.$$

When this in turn is introduced into (3), we then obtain:

$$Y_4 = 0.99920\ 00002 + \frac{0.02}{3}(-0.07999\ 91808$$
$$-0.23999\ 92224 - 0.03999\ 99744)$$
$$= 0.99680\ 00110.$$

Forming the difference (4), we have $\Delta = 13/29$, which shows that the value Y_4 cannot be in error by more than one unit in the last place. Referring to the table of values given in Section 5, we see that this is, indeed, the case.

Formulas (2) and (3) are special cases of the integration formula (9) of Section 4, namely,

$$\int_x^{x+pd} f(s)\ ds = d\left[\eta_0(p)f(x) + \eta_1(p)\Delta f(x) + \frac{\eta_2(p)}{2!}\Delta^2 f(x) + \cdots\right].$$

For $p=2$, we have, to differences of fourth order,

$$I_1 = \int_x^{x+2d} f(s)\ ds = d\left[2f_0 + 2\Delta f_0 + \frac{1}{3}\Delta^2 f_0 - \frac{1}{90}\Delta^4 f_0\right],$$

$$= \frac{d}{3}(f_2 + 4f_1 + f_0) - \frac{d}{90}\Delta^4 f_0. \tag{6}$$

Similarly, for $p=4$, we have, to differences of fourth order,

$$I_2=\int_x^{x+4d} f(s)\ ds=d\left[4f_0+8\Delta f_0+\frac{20}{3}\Delta^2 f_0+\frac{8}{3}\Delta^3 f_0+\frac{14}{45}\Delta^4 f_0\right],$$

$$=\frac{4d}{3}(2f_3-f_2+2f_1)+\frac{14}{45}d\Delta^4 f_0. \tag{7}$$

Since we have

$$I_2-I_1=\frac{29}{90}d\Delta^4 f_0,$$

the error in I_1 as measured by the fourth difference is seen to be $|I_2-I_1|/29$.

Formulas (2) and (3) are obtained in an obvious manner by replacing $f(s)$ by $y'(s)$ and employing the argument given in Section 6.

9. Application to Differential Equations of Higher Order and to Systems of Equations

The methods described in the preceding sections can be applied without essential change to the solution of differential equations of higher order than the first and to systems of differential equations.

Thus, let us consider the following differential equation of second order:

$$y''=f(x,y,y'), \tag{1}$$

where $f(x,y,y')$ is a function of the three variables which is analytic in the neighborhood of the initial values: x_0,y_0,y_0'.

To reduce the problem to the cases already considered we merely write: $y'=z$, and thus (1) becomes the system:

$$z'=f(x,y,z), \quad y'=z. \tag{2}$$

From the initial values: x_0, y_0, $z_0=y_0'$, we first compute the values of y corresponding to x_0+d, x_0+2d, etc., from the series:

$$y=y_0+z_0(x-x_0)+\frac{1}{2!}z_0'(x-x_0)^2+\frac{1}{3!}z_0''(x-x_0)^3+\ldots, \tag{3}$$

where we abbreviate:

$$z'=f(x,y,z),\quad z''=\frac{\partial f}{\partial x}+\frac{\partial f}{\partial y}z+\frac{\partial f}{\partial z}z'=\frac{\partial f}{\partial x}+\frac{\partial f}{\partial y}z+\frac{\partial f}{\partial z}f,\ \text{etc.}$$

From these values a table of z is prepared with differences to a sufficiently high order to insure a satisfactory degree of approximation. The table is then extended by the integration of the equation: $y'=z$.

It is clear that the method just described can be extended to more general systems of differential equations. Thus, if we have the system:

$$\frac{dx}{dt}=f(t,x,y), \qquad \frac{dy}{dt}=g(t,x,y), \tag{4}$$

where $f(t,x,y)$ and $g(t,x,y)$ are functions analytic in the neighborhood of the initial values: x_0, y_0, $t=0$, solutions can be obtained in the form of the following series:

$$x=x_0+x_0't+\frac{1}{2!}\,x_0''t^2+\frac{1}{3!}\,x_0^{(3)}t^3+ \ldots,$$

$$y=y_0+y_0't+\frac{1}{2!}\,y_0''t^2+\frac{1}{3!}\,y_0^{(3)}t^3+ \ldots \tag{5}$$

After preliminary tables have been constructed from the series for a few values of x and y, these can be continued by the process of integration illustrated in the preceding sections.

As an example, let us consider the following equations:

$$\frac{dx}{dt}=x-xy, \quad \frac{dy}{dt}=-y+xy, \tag{6}$$

subject to the initial conditions: $x_0=-1$, $y_0=1$, when $t=0$.

We first compute the values of $x^{(n)}$ and $y^{(n)}$ for $t=0$ from the following equations:

$$x^{(n+1)}=x^{(n)}-\left[x^{(n)}y+nx^{(n-1)}y'+\frac{n(n-1)}{2!}\,x^{(n-2)}y''+\ldots+xy^{(n)}\right],$$

$$y^{(n+1)}=x^{(n)}-y^{(n)}-x^{(n-1)}.$$

We thus obtain the following sets of values:

n	$x^{(n)}$	n	$x^{(n)}$	n	$y^{(n)}$	n	$y^{(n)}$
1	0	6	-754	1	-2	6	1044
2	-2	7	5636	2	4	7	-7434
3	4	8	-47782	3	-10	8	60852
4	-22	9	453316	4	36	9	-561950
5	116	10	-4761506	5	-174	10	5776772

When these values are substituted in (5) and the coefficients reduce to decimals, the following expansions are obtained:

$$x = -1 - t^2 + 0.66666\ 66667\ t^3 - 0.91666\ 66667\ t^4 + 0.96666\ 66667\ t^5$$

$$-1.04722\ 22222\ t^6 + 1.11825\ 39683\ t^7 - 1.18506\ 94444\ t^8$$

$$+1.24921\ 73721\ t^9 - 1.31214\ 34083\ t^{10} + \dots,$$

$$y = 1 - 2t + 2t^2 - 1.66666\ 66667\ t^3 + 1.5\ t^4 - 1.45\ t^5 + 1.45\ t^6 - 1.475\ t^7$$

$$+1.50922\ 61905\ t^8 - 1.54858\ 35538\ t^9 + 1.59192\ 35001\ t^{10} + \dots.$$

We now use these expansions to compute to 10 decimal places the values of x and y from $t=0$ to $t=0.10$ at intervals of 0.01. From these quantities, the values of $x' = x - xy$ and $y' = -y + xy$ are computed over the same range and the table of their differences formed. We thus obtain the following table:

t	x	$x' = x - xy$	$\Delta x'$	$\Delta^2 x'$	$\Delta^3 x'$	$\Delta^4 x'$	$\Delta^5 x'$
0.00	$-1.00000\ 00000$	$0.00000\ 00000$	$-1980\ 36193$	$37\ 86587$	$-2\ 03504$	10205	-622
0.01	$-1.00009\ 93424$	$-0.01980\ 36193$	$-1942\ 49606$	$35\ 83083$	$-1\ 93299$	9583	-587
0.02	$-1.00039\ 48103$	$-0.03922\ 85799$	$-1906\ 66523$	$33\ 89784$	$-1\ 83716$	8996	-545
0.03	$-1.00088\ 27198$	$-0.05829\ 52322$	$-1872\ 76739$	$32\ 06068$	$-1\ 74720$	8451	-506
0.04	$-1.00155\ 95851$	$-0.07702\ 29061$	$-1840\ 70671$	$30\ 31348$	$-1\ 66269$	7945	-470
0.05	$-1.00242\ 21093$	$-0.09542\ 99732$	$-1810\ 39323$	$28\ 65079$	$-1\ 58324$	7475	-437
0.06	$-1.00346\ 71742$	$-0.11353\ 39055$	$-1781\ 74244$	$27\ 06755$	$-1\ 50849$	7038	
0.07	$-1.00469\ 18324$	$-0.13135\ 13299$	$-1754\ 67489$	$25\ 55906$	$-1\ 43811$		
0.08	$-1.00609\ 32987$	$-0.14889\ 80788$	$-1729\ 11583$	$24\ 12095$			
0.09	$-1.00766\ 89422$	$-0.16618\ 92371$	$-1704\ 99488$				
0.10	$-1.00941\ 62795$	$-0.18323\ 91859$					

t	y	$y' = -y + xy$	$\Delta y'$	$\Delta^2 y'$	$\Delta^3 y'$	$\Delta^4 y'$	$\Delta^5 y'$
0.00	$1.00000\ 00000$	$-2.00000\ 00000$	$3950\ 59287$	$-96\ 49899$	$3\ 35152$	-15458	863
0.01	$0.98019\ 83482$	$-1.96049\ 40713$	$3854\ 09388$	$-93\ 14747$	$3\ 19694$	-14595	829
0.02	$0.96078\ 69021$	$-1.92195\ 31325$	$3760\ 94641$	$-89\ 95053$	$3\ 05099$	-13766	759
0.03	$0.94175\ 61808$	$-1.88434\ 36684$	$3670\ 99588$	$-86\ 89954$	$2\ 91333$	-13007	722
0.04	$0.92309\ 70306$	$-1.84763\ 37096$	$3584\ 09634$	$-83\ 98621$	$2\ 78326$	-12285	661
0.05	$0.90480\ 06101$	$-1.81179\ 27462$	$3500\ 11013$	$-81\ 20295$	$2\ 66041$	-11624	626
0.06	$0.88685\ 83762$	$-1.77679\ 16449$	$3418\ 90718$	$-78\ 54254$	$2\ 54417$	-10998	
0.07	$0.86926\ 20706$	$-1.74260\ 25731$	$3340\ 36464$	$-75\ 99837$	$2\ 43419$		
0.08	$0.85200\ 37068$	$-1.70919\ 89267$	$3264\ 36627$	$-73\ 56418$			
0.09	$0.83507\ 55589$	$-1.67655\ 52640$	$3190\ 80209$				
0.10	$0.81847\ 01495$	$-1.64464\ 72431$					

From the values in this table it is now possible to extrapolate by any one of the methods previously described. In illustration, let us compute both x and y for $t=0.10$ by using the values corresponding to $t=0.09$. We shall employ formula (2) of Section 6 for this purpose.

Thus, observing that the differences for x' are: -1729 11583, 2555906, -150849, 7475, and -470, and that the differences for y' are: 3264 36627, -75 99837, 254417, -11624, and 661, we readily compute:

$$x(0.10) = -1.00766\ 89422 + 0.01(-16618\ 92372 - 864\ 55792$$
$$+10\ 64961 - 56568 + 2606 - 155) \times 10^{-10},$$
$$= -1.00941\ 62795;$$

$$y(0.10) = 0.83507\ 55589 + 0.01(-167655\ 52640 + 1632\ 18313$$
$$-31\ 66599 + 95406 - 4052 + 218) \times 10^{-10}$$
$$= 0.81847\ 01495.$$

The Milne method described in Section 8 is also effective in this example. Thus, if we assume as before that x and y are to be determined for $t=0.10$ from preceding values, we assume as known the values corresponding to $t=0.06$, 0.07, 0.08, and 0.09. Hence, using formula (2) of Section 8, we compute:

$$x_4 = -1.00346\ 71742 + \frac{0.04}{3}[2(-0.16618\ 92371 + 0.14889\ 80788$$
$$+2(-0.13135\ 13299)],$$
$$= -1.00941\ 62816;$$

$$y_4 = 0.88685\ 83762 + \frac{0.04}{3}[2(-1.67655\ 52640) + 1.70919\ 89267$$
$$+2(-1.74260\ 25731)],$$
$$= 0.81847\ 01529.$$

With these values the corresponding values of x_4' and y_4' are found from (6). These are respectively: $x_4' = -0.18323\ 91829$ and $y_4' = -1.64464\ 72516$.

Formula (3) of Section 8 is now employed with the following results:

$$X_4 = -1.00609.32987 + \frac{0.01}{3}[-0.18323\ 91829 + 4(-0.16618\ 92371)$$
$$-0.14889\ 80788],$$
$$= 1.00941\ 62794;$$

$$Y_4 = 0.85200\ 37068 + \frac{0.01}{3}[-1.64464\ 72516 + 4(-1.67655\ 52640)$$
$$-1.70919\ 89267],$$
$$= 0.81847\ 01492.$$

The errors are seen to be respectively 1 and 3 in the last place. Formula (4) of Section 8 estimates these errors to be of the order of 1 unit with that of Y_4 somewhat greater than that of X_4.

Appendix 1

Types of Equations with Fixed Critical Points

The following 50 types of nonlinear differential equations of second order, special cases of the general equation (1) of Section 1, Chapter 8, are taken from Gambier and Ince with some change in notation:

1. $\dfrac{d^2y}{dx^2} = 0.$

2. $\dfrac{d^2y}{dx^2} = 6y^2.$

3. $\dfrac{d^2y}{dx^2} = 6y^2 + \dfrac{1}{2}.$

4. $\dfrac{d^2y}{dx^2} = 6y^2 + x.$

5. $\dfrac{d^2y}{dx^2} = [q(x) - 2y]\dfrac{dy}{dx} + q'(x)y.$

6. $\dfrac{d^2y}{dx^2} = -[3y + q(x)]\dfrac{dy}{dx} - q(x)y^2 - y^3.$

7. $\dfrac{d^2y}{dx^2} = 2y^3.$

8. $\dfrac{d^2y}{dx^2} = a + by + 2y^3.$

9. $\dfrac{d^2y}{dx^2} = 2y^3 + xy + \mu.$

10. $\dfrac{d^2y}{dx^2} = -y\dfrac{dy}{dx} + y^3 - 12q(x)y + 12q'(x)$, where q may be an arbitrary constant;

    a solution of $q'' = 6q^2 + c$, $c = 0$ or 1; or a solution of equation 4.

11. $y\dfrac{d^2y}{dx^2} = \left(\dfrac{dy}{dx}\right)^2.$

12. $y\dfrac{d^2y}{dx^2} = \left(\dfrac{dy}{dx}\right)^2 + a + by + cy^3 + dy^4.$

13. $xy\dfrac{d^2y}{dx^2} = x\left(\dfrac{dy}{dx}\right)^2 - y\dfrac{dy}{dx} + ax + by + cy^3 + dxy^4.$

14. $y\dfrac{d^2y}{dx^2} = \left(\dfrac{dy}{dx}\right)^2 + [q(x)y^2 + r(x)]\dfrac{dy}{dx} + q'(x)y^3 - r'(x)y.$

15. $y\dfrac{d^2y}{dx^2} = \left(\dfrac{dy}{dx}\right)^2 + \dfrac{dy}{dx} + r(x)y^3 - y^2\dfrac{d}{dx}[r'(x)/r(x)].$

495

16. $y\dfrac{d^2y}{dx^2} = \left(\dfrac{dy}{dx}\right)^2 - q'(x)\dfrac{dy}{dx} + y^4 - q(x)y^3 + q''(x)y.$

17. $y\dfrac{d^2y}{dx^2} = \dfrac{m-1}{m}\left(\dfrac{dy}{dx}\right)^2.$

18. $y\dfrac{d^2y}{dx^2} = \dfrac{1}{2}\left(\dfrac{dy}{dx}\right)^2 + 4y^3.$

19. $y\dfrac{d^2y}{dx^2} = \dfrac{1}{2}\left(\dfrac{dy}{dx}\right)^2 + 4y^3 + 2y^2.$

20. $y\dfrac{d^2y}{dx^2} = \dfrac{1}{2}\left(\dfrac{dy}{dx}\right)^2 + 4y^3 + 2xy^2.$

21. $y\dfrac{d^2y}{dx^2} = \dfrac{3}{4}\left(\dfrac{dy}{dx}\right)^2 + 3y^3.$

22. $y\dfrac{d^2y}{dx^2} = \dfrac{3}{4}\left(\dfrac{dy}{dx}\right)^2 - y.$

23. $y\dfrac{d^2y}{dx^2} = \dfrac{3}{4}\left(\dfrac{dy}{dx}\right)^2 + 3y^3 + ay^2 + by.$

24. $y\dfrac{d^2y}{dx^2} = \dfrac{m-1}{m}\left(\dfrac{dy}{dx}\right)^2 + q(x)y^2\dfrac{dy}{dx} + \dfrac{mq'}{m+2}y^3 - \dfrac{mq^2}{(m+2)^2}y^4.$

25. $y\dfrac{d^2y}{dx^2} = \dfrac{3}{4}\left(\dfrac{dy}{dx}\right)^2 + \left[\dfrac{q'(x)}{2q(x)}y - \dfrac{3}{2}y^2\right]\dfrac{dy}{dx} + q(x)y + r(x)y^2 + \dfrac{q'}{2q}y^3 - \dfrac{1}{4}y^4.$

26. $y\dfrac{d^2y}{dx^2} = \dfrac{3}{4}\left(\dfrac{dy}{dx}\right)^2 + 6q'\dfrac{dy}{dx} - 36q'^2 - 12q''y + 12qy^2 + 3y^3,$

  where (1) $q'' = 6q^2$, or (2) $q'' = 6q^2 + \dfrac{1}{2}$, or (3) $q'' = 6q^2 + x.$

27. $y\dfrac{d^2y}{dx^2} = \dfrac{m-1}{m}\left(\dfrac{dy}{dx}\right)^2 + \left[-\dfrac{m-2}{m} + fy + gy^2\right]\dfrac{dy}{dx} - \dfrac{1}{m} - fy + hy^2$

$$+ \dfrac{m(g'-fg)}{m+2}y^3 - \dfrac{mg^2}{(m+2)^2}y^4,$$

  where f, g, h are definite rational functions of two arbitrary analytic functions $q(x)$ and $r(x)$ and their derivatives.

28. $y\dfrac{d^2y}{dx^2} = \dfrac{1}{2}\left(\dfrac{dy}{dx}\right)^2 + (qy - y^2)\dfrac{dy}{dx} - 72p^2 + 3\left(q' + \dfrac{1}{2}q^2\right)y^2 - 2qy^3 + \dfrac{1}{2}y^4,$

  where $p = \dfrac{1}{2}(v_2 - v_1)$ and $q = (v_2' - v_1')/(v_2 - v_1)$ and v_1, v_2 are solutions of

  (1) $v'' = 6v^2$, or (2) $v'' = 6v^2 + \dfrac{1}{2}$, or (3) $v'' = 6v^2 + x.$

29. $y\dfrac{d^2y}{dx^2} = \dfrac{1}{2}\left(\dfrac{dy}{dx}\right)^2 + \dfrac{3}{2}y^4.$

30. $y\dfrac{d^2y}{dx^2} = \dfrac{1}{2}\left(\dfrac{dy}{dx}\right)^2 - \dfrac{1}{2}a^2 + 2by^2 + 4cy^3 + \dfrac{3}{2}y^4.$

31. $y\dfrac{d^2y}{dx^2} = \dfrac{1}{2}\left(\dfrac{dy}{dx}\right)^2 - \dfrac{1}{2}a^2 + 2(x^2 - b)y^2 + 4xy^3 + \dfrac{3}{2}y^4.$

32. $y\dfrac{d^2y}{dx^2} = \dfrac{1}{2}\left(\dfrac{dy}{dx}\right)^2 - \dfrac{1}{2}.$

33. $y\dfrac{d^2y}{dx^2}=\dfrac{1}{2}\left(\dfrac{dy}{dx}\right)^2-\dfrac{1}{2}+ay^2+4y^3.$

34. $y\dfrac{d^2y}{dx^2}=\dfrac{1}{2}\left(\dfrac{dy}{dx}\right)^2-\dfrac{1}{2}-xy^2+4ay^2,\ a\neq0.$

35. $y\dfrac{d^2y}{dx^2}=\dfrac{2}{3}\left(\dfrac{dy}{dx}\right)^2+\left(\mathrm{p}+\dfrac{2}{3}qy-\dfrac{2}{3}y^2\right)\dfrac{dy}{dx}-3p^2+(2pq-3p'y)$

$$+(p+4q'+\dfrac{8}{3}q^2)y^2-\dfrac{10}{3}qy^3+\dfrac{2}{3}y^4,$$

where, if $2y^3+Su+T$ represents either $2y^3$ or $2y^3+ay+b$, or $2y^3+xy+a$, then $p=-\dfrac{1}{3}S-\dfrac{2}{3}(q'+q^2),\ q''=2q^3+Sy+T.$

36. $y\dfrac{d^2y}{dx^2}=\dfrac{4}{5}\left(\dfrac{dy}{dx}\right)^2+\left(p-\dfrac{4}{5}qy-\dfrac{2}{5}y^2\right)\dfrac{dy}{dx}-\dfrac{5}{9}p^2-\dfrac{1}{3}(pq+5p')y$

$$+\left(p-3q'+\dfrac{6}{5}q^2\right)y^2+\dfrac{11}{5}qy^3+\dfrac{4}{5}y^4,$$

where $p=\dfrac{72}{5}v_1+\dfrac{36}{5}v_2-\dfrac{9}{5}\left(\dfrac{v_2'-v_1'}{v_2-v_1}\right)^2$, $q=(v_2'-v_1')/(v_2-v_1)$, in which v_1 and v_2 are solutions of $v''=6v+S$, $S=0,\ \dfrac{1}{2},$ or x.

37. $(y-y^2)\dfrac{d^2y}{dx^2}=\dfrac{1}{2}(1-3y)\left(\dfrac{dy}{dx}\right)^2.$

38. $(y-y^2)\dfrac{d^2y}{dx^2}=\dfrac{1}{2}(1-3y)\left(\dfrac{dy}{dx}\right)^2+ay^2(1-y)^3+b(1-y)^3+cy^2(1-y)+dy^2.$

39. $x^2(y-y^2)\dfrac{d^2y}{dx^2}=\dfrac{1}{2}x^2(1-3y)\left(\dfrac{dy}{dx}\right)^2-xy(1-y)\dfrac{dy}{dx}+ay^2(1-y)^3$

$$+b(1-y)^3+cxy(1-y)+dx^2y^2(1+y).$$

40. $(y-y^2)\dfrac{d^2y}{dx^2}=(1-3y)\left(\dfrac{dy}{dx}\right)^2+2(py+qy^2)\dfrac{dy}{dx}+(1-y)^3(s^2y^2-t^2)$

$$+2y^2(1-y)[q^2-p^2+(p'+q')],$$

where $s'=-2qs$, and $t'=2pt$.

41. $(y-y^2)\dfrac{d^2y}{dx^2}=\dfrac{2}{3}(1-2y)\left(\dfrac{dy}{dx}\right)^2.$

42. $(y-y^2)\dfrac{d^2y}{dx^2}=\dfrac{2}{3}(1-2y)\left(\dfrac{dy}{dx}\right)^2+[py+q(1-y)+ry^2(1-y)$

$$-\dfrac{1}{2}(p+q+r)y(1-y)]\dfrac{dy}{dx}+(3py^2-3q^2(1-y)^2$$

$$-3r^2y^3(1-y)^2-[3r'+\dfrac{3}{2}r(p+q-r)]y^2(1-y)^2$$

$$-[3q'-\dfrac{3}{2}q(p+q-r)]y(1-y)^2+[3p'-\dfrac{3}{2}p(p+q+r)]y^2(1-y,)$$

where $3p=2v_1$, $3q=v_1'/v_1+v_1+E/v_1+2C$, $3r=v_1'/v_1-v_1+E/v_1-2C$, where v_1 is any solution of the equation: $2vv''=v'^2+3v^4+8Cv^3+4Dv^2-E^2$.

43. $(y-y^2)\dfrac{d^2y}{dx^2}=\dfrac{3}{4}(1-2y)\left(\dfrac{dy}{dx}\right)^2.$

44. $(y-y^2)\dfrac{d^2y}{dx^2}=\dfrac{3}{4}(1-2y)\left(\dfrac{dy}{dx}\right)^2+ay(1-y)^2+by^2(1-y)+2cy^2(1-y)^3.$

45. $(y-y^2)\dfrac{d^2y}{dx^2}=\dfrac{3}{4}(1-2y)\left(\dfrac{dy}{dx}\right)^2+[ay(1-y)+b(1-y)+cy]\dfrac{dy}{dx}$

$$+4d^2y^2(1-y)^2(1-2y)-b^2(1-y)^2+c^2y^2-hy(1-y)^2+ky^2(1-y),$$

where $a=(v_2'-v_1')/(v_2-v_1)$, $b-c=-\dfrac{3}{2}(v_1+v_2)$, $b+c=$

$-\dfrac{3}{2}(v_1'-v_2')/(v_2-v_1)$, $d=\dfrac{1}{2}(v_2-v_1)$, $h=2b'+ab$, $k=2c'+ac$, and v_1, v_2 are solutions of $v''=2v^3+Sv+T$, where S and T are the coefficients in equations 7, 8, or 9.

46. $(y-y^2)\dfrac{d^2y}{dx^2}=\dfrac{3}{4}(1-2y)\left(\dfrac{dy}{dx}\right)^2+\dfrac{1}{2}\dfrac{H'}{H}(y+2y^2)\dfrac{dy}{dx}+\dfrac{4a^2}{H^2}y^2(1-y)^2(1-2y)$

$$+\dfrac{3}{4}\left(\dfrac{H'}{H}\right)^2y^2(1+2y)+3\left(\dfrac{H'}{H}\right)'y^2(1-y)-Hy(1-y)^2,$$

where $H=2(v_1'+v_1^2)+b$, v_1 being any solution of: $v''=2v^3+bv+a$.

47. $(y-y^2)\dfrac{d^2y}{dx^2}=\dfrac{3}{4}(1-2y)\left(\dfrac{dy}{dx}\right)^2+\dfrac{1}{2}\dfrac{H'}{H}y(2y+1)\dfrac{dy}{dx}$

$$+y^2(1-y)^2(1-2y)\dfrac{(2a+1)^2}{H^2}+\left(\dfrac{3H'}{2H}\right)^2y^2-Hy(1-y)$$

$$+\dfrac{3}{2}\left[2\left(\dfrac{H'}{H}\right)'-\left(\dfrac{H'}{H}\right)^2\right]y^2(1-y),$$

where $H=2(v_1'+v_1^2)+x$ and v_1 is any solution of: $v''=2v^3+vx+a$.

48. $(y-y^2)\dfrac{d^2y}{dx^2}=\dfrac{1}{6}(4-7y)\left(\dfrac{dy}{dx}\right)^2+(1-y)(ay^2+by+c)\dfrac{dy}{dx}-\dfrac{3}{8}a^2y^3(1-y)^2$

$$-fy^2(1-y)^2-3c^2(1-y)^2-gy^2-hy(1-y)^2+\dfrac{1}{3}gy^2(1-y),$$

where $a=-\dfrac{10}{9}(u+w)$, $b=\dfrac{1}{9}(2u+5w)$, $c=-\dfrac{4}{9}(w-2u)$, $f=\dfrac{3}{2}(p'-pq)-\dfrac{3}{4}p^2$,

$g=-\dfrac{9}{2}z^2$, $h=3(r'-rs)-\dfrac{3}{2}r^2$, in which

$u=\dfrac{1}{2}\left[\dfrac{v_2'-v_1'}{v_2-v_1}+\dfrac{v_3'-v_1'}{v_3-v_1}\right]$, $z=\dfrac{1}{2}\left[\dfrac{v_1'-v_1'}{v_2-v_1}-\dfrac{v_3'-v_1'}{v_3-v_1}\right]$, $w=-\dfrac{z'}{z}$,

where v_1, v_2, v_3 are three particular solutions of: $v''=6v^2+S$, $S=0$, $\dfrac{1}{2}$, or x.

49. $y(1-y)(a-y)\dfrac{d^2y}{dx^2}=\dfrac{1}{2}[a-2(a+1)y+3y^2]\left(\dfrac{dy}{dx}\right)^2+by^2(1-y)^2(a-y)^2$

$$+c(1-y)^2(a-y)^2+dy^2(a-y)^2+ey^2(1-y)^2.$$

50. $y(1-y)(x-y)\dfrac{d^2y}{dx^2}=\dfrac{1}{2}[x-2(x+1)y+3y^2]\left(\dfrac{dy}{dx}\right)^2+\dfrac{y(1-y)}{x(1-x)}[x^2+(1-2x)y]\dfrac{dy}{dx}$

$$+\dfrac{1}{2x^2(1-x)^2}[ay^2(1-y)^2(x-y)^2-bx(1-y)^2(x-y)^2-c(1-x)y^2(x-y)^2$$

$$-dx(1-x)y^2(1-y)^2].$$

Appendix 2

Elements of the Linear Fractional Transformation

The following table gives the explicit values of $\psi_j^i(x)$, which are the elements of the transformation:

$$\text{T)} \quad y = \frac{a+bz}{c+dz}, \quad \Delta = ad - bc \neq 0$$

when this transformation is applied to the equation

$$L(y) \equiv (A_0 + A_1 y)\frac{d^2 y}{dx^2} + (B_0 + B_1 y)\frac{dy}{dx} + C_0\left(\frac{dy}{dx}\right)^2 + D_0 + D_1 y + D_2 y^2 + D_3 y^3 + D_4 y^4 = 0.$$

The equation, $T L(y) = M(z) = 0$, can then be written:

$$f_1(x,z)\frac{d^2 z}{dx^2} + f_2(x,z)\left(\frac{dz}{dx}\right)^2 + f_3(x,z)\,z\,\frac{dz}{dx} + f_4(x,z)\frac{dz}{dx} + f_5(x,z) = 0,$$

where we define:

$$f_i(x,z) = \psi_1^i(x) + \psi_2^i(x)\,z + \psi_3^i(x)\,z^2 + \psi_4^i(x)\,z^3 + \psi_5^i(x)\,z^4.$$

Noting the explicit values: $\Delta' = a'd + ad' - b'c - bc'$, and $\Delta'' = a''d + 2a'd' + ad'' - b''c - 2b'c' - bc''$, we obtain the following expressions for $\psi_j^i(x)$:

$$\psi_1^1(x) = -\Delta c(A_0 c + A_1 a),$$

$$\psi_2^1(x) = -\Delta[2A_0 cd + A_1(ad + bc)],$$

$$\psi_3^1(x) = -\Delta d(A_0 d + A_1 b), \qquad \psi_4^1(x) = \psi_5^1(x) = 0.$$

$$\psi_1^2(x) = \Delta[2d(A_0 c + A_1 a) + \Delta C_0],$$

$$\psi_2^2(x) = 2\Delta d(A_0 d + A_1 b),$$

$$\psi_3^2(x) = b^2 d^2(C_0 + 2A_1) + 2bd^2(A_0 d - A_1 b - C_0 b) + d^2 b(C_0 b - 2A_0 d) = 0,$$

$$\psi_4^2(x) = \psi_5^2(x) = 0.$$

$$\psi_1^3(x) = A_0(4acdd' - 2b'c^2 d - 2bc^2 d') + A_1(4a^2 dd' - 2ab'cd - 2abcd')$$
$$+ C_0(2bb'c^2 - 2ab'cd - 2abcd' + 2a^2 dd'),$$

$$\psi_2^3(x) = 4d^2 A_0(ad' - b'c) + A_1(6abdd' - 2ab'd^2 - 2bb'cd - 2b^2 cd')$$
$$+ 2C_0(bc - ad)(b'd - bd'),$$

$$\psi_3^3(x) = 2d(bd' - b'd)(A_0 d + A_1 b),$$

$$\psi_4^3(x) = \psi_5^3(x) = 0.$$

$$\psi_1^4(x) = (4c'\Delta - 2c\Delta')(A_0 c + A_1 a) + 2C_0\Delta(ac' - a'c) - c\Delta(B_0 c + B_1 a),$$

$$\psi_2^4(x) = A_0(6b'c^2 d - 4acdd' - 2bc^2 d' - 4a'cd^2) + A_1(4ab'cd + 2bb'c^2$$
$$- 2a^2 dd' - 4abcd' - 2aa'd^2 + 6abc'd - 2a'bcd - 2b^2 cc')$$
$$- 2\Delta C_0(a'd - bc') - 2\Delta B_0 cd - \Delta B_1(bc + ad),$$

$$\psi_3^4(x) = A_0(6b'cd^2 - 4bcdd' - 2ad^2d' - 2a'd^3 + 2bc'd^2) + A_1(2ab'd^2$$
$$+ 4bb'cd - 4abdd' - 2b^2cd' - 2a'bd^2 + 2b^2c'd) + B_0d^2(bc - ad)$$
$$+ B_1bd(bc - ad),$$
$$\psi_4^4(x) = 2A_0d^2(b'd - bd') + 2A_1bd(b'd - bd'),\ \psi_5^4(x) = 0.$$

From the above values, we also obtain the following sums:

$$\psi_2^4(x) + \psi_1^3(x) = 4A_0[\Delta(cd' + c'd) - cd\Delta'] + 2A_1[2\Delta(bc' + ad') - \Delta'(ad + bc)]$$
$$+ 2C_0\Delta(ad' - a'd + bc' - b'c) - 2B_0\Delta cd - B_1\Delta(bc + ad)\ ;$$
$$\psi_3^4(x) + \psi_2^3(x) = (2d'\Delta - d\Delta')(A_0d + A_1b) + 2C_0\Delta(bd' - b'd) + d\Delta(B_0d + B_1b)\ ;$$
$$\psi_4^4(x) + \psi_3^3(x) = 0.$$

$$\psi_1^5(x) = (A_0c + A_1a)[c(a'c - ac')' - 2c'(a'c - ac')] + C_0(a'c - ac')^2$$
$$+ (B_0c + B_1a)c(a'c - ac') + c^4D_0 + ac^3D_1 + a^2c^2D_2 + a^3cD_3,$$

$$\psi_2^5(x) = A_0(3a''c^2d + b''c^3 - 2acc''d - bc^2c'' - ac^2d'' - 4a'cc'd - 2b'c'c^2$$
$$- 2a'c^2d' + 2ac'^2d + 2ac'^2d + 2bcc'^2 + 4acc'd') + A_1(a''bc^2 + 2aa''cd$$
$$+ ab''c^2 - 2abcc'' - a^2c''d - a^2cd'' - 2aa'c'd - 2a'bcc' - 2ab'cc'$$
$$- 2aa'cd' + 4abc'^2 + 4a^2c'd') + 2C_0(a'b'c^2 + a'^2cd - aa'c'd - a'bcc'$$
$$- ab'cc' - aa'cd' + abc'^2 + a^2c'd') + B_0(3a'c^2d + b'c^3 - 2acc'd$$
$$- bc^2c' - ac^2d') + B_1(a'bc^2 + 2aa'cd - ab'c^2 - a^2c'd - 2abcc' - a^2cs')$$
$$+ 4c^3dD_0 + D_1(3ac^2d + bc^3) + D_2(2a^2cd + 2abc^2) + D_3(a^3d + 3a^2bc).$$

$$\psi_3^5(x) = A_0(3a''cd^2 + 3b''c^2d - 2bcc''d - ac''d^2 - 2acdd'' - bc^2d'' - 2a'c'd^2$$
$$- 4b'cc'd - 4a'cdd' + 2bc'^2d + 4ac'dd' + 4bcc'd')$$
$$+ A_1(aa''d^2 + 2ab''cd + 2a''bcd + bb''c^2 - 2abc''d - b^2cc'' - 2abcd''$$
$$- a^2dd'' - 2a'bc'd - 2ab'c'd - 2aa'dd' - 2bb'cc' - 2a'bcd' + 2b^2c'^2$$
$$+ 8abc'd') + C_0(b'^2c^2 + a'^2d^2 + 4a'b'cd - 2a'bc'd - 2ab'c'd - 2aa'dd'$$
$$- 2bb'cc' - 2a'bcd' + b^2c'^2 + 4abc'd') + B_0(3a'cd^2 + 3b'c^2d - ac'd^2$$
$$- 2bcc'd - 2acdd' - bc^2d') + B_1(aa'd^2 + 2a'bcd + 2ab'cd + bb'c^2$$
$$- 2abc'd - b^2cc' - a^2dd' - 2abcd') + 6c^2d^2D_0 + (3acd^2 + 3bc^2d)D_1$$
$$+ (a^2d^2 + 4abcd + b^2c^2)D_2 + (3ab^2c + 3a^2bd)D_3.$$

$$\psi_4^5(x) = A_0(b''cd^2 + a''d^3 + 2b''c^2d - bc''d^2 - 2bcdd'' - ad^2d'' - 2b'c'd^2$$
$$- 2a'd^2d' - 4b'cdd' + 4bc'dd' + 2add'^2 + 2bcd'^2) + A_1(ab''d^2 + a''bd^2$$
$$+ 2bb''cd - b^2c''d - 2abdd'' - b^2cd'' - 2bb'c'd - 2a'bdd' - 2ab'dd'$$
$$- 2bb'cd + 4b^2c'd' + 4abd'^2) + C_0(2b'^2cd + 2a'b'd^2 - 2bb'c'd$$
$$- 2a'bdd' - 2ab'dd' - 2bb'cd' + 2b^2c'd' + 2abd'^2) + B_0(a'd^3 + 3b'cd^2$$
$$- b'cd^2 - ad^2d' - 2bcdd') + B_1(a'bd^2 + ab'd^2 + 2bb'cd - b^2c'd$$
$$- 2abdd' - b^2cd') + 4cd^3D_0 + (ad^3 + 3bcd^2)D_1$$
$$+ (2abd^2 + 2b^2cd)D_2 + (b^3c + 3ab^2d)D_3.$$

$$\psi_5^5(x) = (A_0d + A_1b)[d(b'd - bd')' - 2d'(b'd - bd')] + C_0(b'd - bd')^2$$
$$+ B_0d^2(b'd - bd') + B_1bd(b'd - bd') + d(d^3D_0 + bd^2D_1 + b^2dD_2 + b^3D_3).$$

Appendix 3

Coefficients of the Expansions of the First Painlevé Transcendent

The following table gives the values of the coefficients of the expansion of the solution of the equation

$$\frac{d^2y}{dx^2} = 6y^2 + \lambda x, \tag{1}$$

in the neighborhood of the movable pole: $x = x_1$. The values of the coefficients are those in the solution,

$$y = \frac{a_{-2}}{v^2} + \frac{a_{-1}}{v} + a_0 + a_1 v + a_2 v^2 + \dots, \quad v = x - x_0, \tag{2}$$

and they can be computed successively from the formula:

$$a_n = \frac{6}{n^2 - n - 12} \sum_{k=-1}^{n-3} a_k a_{n-k-2}, \quad n > 4. \tag{3}$$

$a_{-2} = 1, \quad a_{-1} = a_0 = a_1 = 0, \quad a_2 = -\lambda x_1/10, \quad a_3 = -\lambda/6, \quad a_4 = h, \ a_5 = 0,$

$a_6 = \dfrac{\lambda^2 x_1^2}{300}, \quad a_7 = \dfrac{\lambda^2 x_1}{150}, \quad a_8 = -\dfrac{3\lambda h x_1}{110} + \dfrac{\lambda^2}{264}, \quad a_9 = -\dfrac{\lambda h}{30},$

$a_{10} = \dfrac{-\lambda^3 x_1^3}{19,500} + \dfrac{h^2}{13}, \quad a_{11} = \dfrac{-11\lambda^3 x_1^2}{73,500}, \quad a_{12} = \dfrac{\lambda^2 x_1}{100}\left[hx_1 \left(\dfrac{3}{110} + \dfrac{1}{30} \right) - \lambda \left(\dfrac{1}{264} + \dfrac{1}{90} \right) \right],$

$a_{13} = \dfrac{\lambda^2}{12}\left[hx_1 \left(\dfrac{1}{100} + \dfrac{1}{220} \right) - \lambda \dfrac{1}{1584} \right],$

$a_{14} = \dfrac{1}{85}\left[\lambda^4 x_1^4 \left(\dfrac{1}{30,000} + \dfrac{1}{32,500} \right) - \lambda h^2 x_1 \left(\dfrac{9}{55} + \dfrac{3}{65} \right) + \lambda^2 h \left(\dfrac{1}{30} + \dfrac{1}{44} \right) \right],$

$a_{15} = \dfrac{\lambda^4 x_1^3}{500}\left[\dfrac{1}{15 \cdot 147} + \dfrac{1}{33 \cdot 45} + \dfrac{1}{99 \cdot 39} \right] - \dfrac{2\lambda h^2}{13 \cdot 55},$

$a_{16} = \dfrac{\lambda^4 x_1^2}{3800}\left[\dfrac{1}{1320} + \dfrac{1}{450} + \dfrac{11}{2205} + \dfrac{1}{396} + \dfrac{1}{225} \right] - \dfrac{\lambda^3 h x_1^3}{3800}\left[\dfrac{1}{150} + \dfrac{3}{550} + \dfrac{2}{195} + \dfrac{1}{55} \right] + \dfrac{h^3}{247}.$

The coefficients of the expansion

$$y = y_0 + y_0'(x-x_0) + \frac{y_0''}{2!}(x-x_0)^2 + \frac{y_0^{(3)}}{3!}(x-x_0)^3 + \ldots,$$

where y is a solution of equation (1), may be obtained explicitly from the following table of values of the derivatives:

$y'' = 6y^2 + \lambda x$; $y^{(3)} = 12yy' + \lambda$; $y^{(4)} = 12[yy'' + y'^2]$;

$y^{(5)} = 12[yy^{(3)} + 3y'y'']$; $y^{(6)} = 12[yy^{(4)} + 4y'y^{(3)} + 3y''^2]$;

$y^{(7)} = 12[yy^{(5)} + 10y''y^{(3)} + 5y'y^{(4)}]$;

$y^{(8)} = 12[yy^{(6)} + 6y'y^{(5)} + 15y''y^{(4)} + 10(y^{(3)})^2$;

$y^{(9)} = 12[yy^{(7)} + 7y'y^{(6)} + 21y''y^{(5)} + 35y^{(3)}y^{(4)}]$;

$y^{(10)} = 12[yy^{(8)} + 8y'y^{(7)} + 28y''y^{(6)} + 56y^{(3)}y^{(5)} + 35(y^{(4)})^2]$;

$y^{(11)} = 12[yy^{(9)} + 9y'y^{(8)} + 36y''y^{(7)} + 84y^{(3)}y^{(6)} + 126y^{(4)}y^{(5)}]$;

$y^{(12)} = 12[yy^{(10)} + 10y'y^{(9)} + 45y''y^{(8)} + 120y^{(3)}y^{(7)} + 210y^{(4)}y^{(6)} + 126(y^{(5)})^2]$;

$y^{(13)} = 12[yy^{(11)} + 11y'y^{(10)} + 55y''y^{(9)} + 165y^{(3)}y^{(8)} + 330y^{(4)}y^{(7)} + 462y^{(5)}y^{(6)}]$;

$y^{(14)} = 12[yy^{(12)} + 12y'y^{(11)} + 66y''y^{(10)} + 220y^{(3)}y^{(9)} + 495y^{(4)}y^{(8)} + 792y^{(5)}y^{(7)} + 462(y^{(6)})^2]$;

$y^{(15)} = 12[yy^{(13)} + 13y'y^{(12)} + 78y''y^{(11)} + 286y^{(3)}y^{(10)} + 715y^{(4)}y^{(9)} + 1287y^{(5)}y^{(8)} + 1716y^{(6)}y^{(7)}]$.

Appendix 4

Coefficients of the Expansions of the Second Painlevé Transcendent

The following table gives the values of the coefficients of the expansion of the solution of the equation

$$\frac{d^2y}{dx^2} = 2y^3 + xy + \mu, \tag{1}$$

in the neighborhood of the movable pole: $x = x_1$. The values of the coefficients are those in the solution:

$$y = \frac{a_{-1}}{v} + a_0 + a_1 v + a_2 v^2 + a_3 v^3 + \ldots, \quad v = x - x_1. \tag{2}$$

They can be computed successively from formula (4) of Section 11, Chapter 8.

$$a_{-1} = 1, \quad a_0 = 0, \quad a_1 = -x_1/6, \quad a_2 = -\tfrac{1}{4}(1+\mu), \quad a_3 = h,$$

$$a_4 = \tfrac{1}{72} x_1 (1 + 3\mu), \quad a_5 = \tfrac{1}{3024}(27 + 108\mu - 216hx_1 + 81\mu^2 - 2x_1^3),$$

$$a_6 = -\tfrac{1}{864}(2x_1^3 + 3\mu x_1^2 + 72h + 108\mu h),$$

$$a_7 = \tfrac{1}{36}\big[-x_1\{a_5 + \tfrac{1}{24}(1+\mu)(1+3\mu) - \tfrac{1}{6}hx_1 + \tfrac{1}{16}(1+\mu)^2 - \tfrac{1}{72}(1+3\mu)\} + 6h^2\big],$$

$$a_8 = \tfrac{1}{50}[x_1\{-a_6 + \tfrac{1}{6}h(1+3\mu) + \tfrac{1}{432}x_1^2(1+3\mu) + \tfrac{1}{2}h(1+\mu)\} - \tfrac{1}{32}(1+\mu)^3 - a_5(2+3\mu)],$$

$$a_9 = \tfrac{1}{66}[-x_1(a_7 + h^2) + x_1^2\{\tfrac{1}{864}(1+3\mu)^2 + \tfrac{1}{6}a_5 + \tfrac{1}{144}(1+\mu)(1+3\mu)\} + h\{12a_5 + \tfrac{3}{8}(1+\mu)^2\} - (2+3\mu)a_6],$$

$$a_{10} = \tfrac{1}{84}[x_1\{-a_8 + \tfrac{1}{3}(2+3\mu)a_5 + \tfrac{1}{192}(1+\mu)^2(1+3\mu)\} + x_1^2\{\tfrac{1}{6}a_6 - \tfrac{1}{36}h(1+3\mu)\} + 12ha_6 - \tfrac{3}{2}h^2(1+\mu) - (2+3\mu)a_7].$$

$$a_{11} = \tfrac{1}{104}[x_1\{-a_9 + \tfrac{1}{3}(2+3\mu)a_6 - \tfrac{1}{24}h(1+\lambda)(1+3\lambda) - 2ha_5\} + x_1^2\{\tfrac{1}{6}a_7 - \tfrac{1}{5184}(1+3\mu)^2\} + 6a_5^2 + \tfrac{3}{8}(1+\mu)^2a_5 + 12ha_7 - (2+3\mu)a_8 + 2h^3],$$

$$a_{12} = \tfrac{1}{126}[x_1\{-a_{10} + \tfrac{1}{3}(2+3\mu)a_7 + \tfrac{1}{12}h^2(1+3\mu) - 2ha_6\} - x_1^2\{\tfrac{1}{3456}(1+\mu)(1+3\mu)^2 + \tfrac{1}{36}(1+3\mu)a_5 + \tfrac{1}{8}a_8\} - (2+3\mu)a_9 + 12ha_8 + 12a_5a_6 + \tfrac{3}{8}(1+\mu)^2a_6 - 3(1+\mu)ha_5],$$

503

$$a_{13} = \frac{1}{150}[x_1\{-a_{11}+\tfrac{1}{3}(2+3\mu)a_8-2ha_7-a_5^2-\tfrac{1}{24}(1+\mu)(1+3\mu)a_5\}$$
$$+x_1^2\{\tfrac{1}{6}a_9-\tfrac{1}{36}(1+3\mu)a_6+\tfrac{1}{864}h(1+3\mu)^2\}+12ha_9-(2+3\mu)a_{10}$$
$$+12a_5a_7+6a_6^2+\tfrac{3}{8}(1+\mu)^2a_7-3(1+\mu)ha_6+6h^2a_5],$$

$$a_{14} = \frac{1}{176}[x_1\{-a_{12}+\tfrac{1}{3}(2+3\mu)a_9-2ha_8-2a_5a_6-\tfrac{1}{24}(1+\mu)(1+3\mu)a_6$$
$$+\tfrac{1}{6}h(1+3\mu)a_5\}+x_1^2\{\tfrac{1}{6}a_{10}-\tfrac{1}{36}a_7+\tfrac{1}{(72)^3}x_1^3(1+3\mu)^3+12ha_{10}-(2+3\mu)a_{11}$$
$$+12a_5a_8+12a_6a_7+\tfrac{3}{8}(1+3\mu)^2a_8-3h(1+\mu)a_7-\tfrac{3}{2}(1+\mu)a_5^2+6h^2a_6],$$

$$a_{15} = \frac{1}{204}[x_1\{-a_{13}+\tfrac{1}{3}(2+3\mu)a_{10}-2ha_9-2a_5a_7-a_6^2-\tfrac{1}{24}(1+\mu)(1+3\mu)a_7$$
$$+\tfrac{1}{6}h(1+3\mu)a_6\}+x_1^2\{\tfrac{1}{6}a_{11}-\tfrac{1}{36}(1+3\mu)a_8+\tfrac{1}{864}(1+3\mu)^2a_5\}+12ha_{11}$$
$$-(2+3\mu)a_{12}+12a_5a_9+12a_6a_8+6a_7^2+\tfrac{3}{8}(1+\mu)^2a_9-3(1+\mu)ha_8$$
$$-3(1+\mu)a_5a_6+6h^2a_7+6ha_5^2].$$

The coefficients of the expansion

$$y=y_0+y'(x-x_0)+\frac{y_0''}{2!}(x-x_0)^2+\frac{y_0^{(3)}}{3!}(x-x_0)^3+\ \ldots,$$

where y is a solution of equation (1), may be obtained explicitly from the following table of values of the derivatives:

$$y'' = 2y^3+xy+\mu, \qquad y^{(3)}=6y^2y'+xy'+y,$$
$$y^{(4)}=6y^2y''+12yy'^2+xy''+2y',$$
$$y^{(5)}=12y'^3+36yy'y''+6y^2y^{(3)}+xy^{(3)}+3y'',$$
$$y^{(6)}=72y'^2y''+36yy''^2+48yy'y^{(3)}+6y^2y^{(4)}+xy^{(4)}+4y^{(3)},$$
$$y^{(7)}=180y'y''^2+120y'^2y^{(3)}+120yy''y^{(3)}+60yy'y^{(4)}+6y^2y^{(5)}+xy^{(5)}+5y^{(4)},$$
$$y^{(8)}=180y''^3+720y'y''y^{(3)}+180y'^2y^{(4)}+120y(y^{(3)})^2+180yy''y^{(4)}+72yy'y^{(5)}$$
$$+6y^2y^{(6)}+xy^{(6)}+6y^{(5)}\,,$$
$$y^{(9)}=1260y''^2y^{(3)}+840(y^{(3)})^2y'+1260y'y''y^{(4)}+252y'^2y^{(5)}+420yy^{(3)}y^{(4)}$$
$$+252yy''y^{(5)}+84yy'y^{(6)}+6y^2y^{(7)}+xy^{(7)}+7y^{(6)},$$
$$y^{(10)}=420\,(y^{(4)})^2+3360y''(y^{(3)})^2+2520y'^2y^{(4)}+336y'^2y^{(6)}+2016y'y''y^{(5)}$$
$$+1680y'y^{(3)}y^{(4)}+672yy^{(3)}y^{(5)}+96yy'y^{(7)}+336yy''y^{(6)}$$
$$+6y^2y^{(8)}+xy^{(8)}+8y^{(7)}.$$

TABLE I

The values tabulated are approximations of $y(x)$ and $y'(x)$, where $y(x)$ is the solution of $y''=6y^2+\lambda x$, for initial values: $y=1$, $y'=0$, $x=0$, and for $\lambda=0$, 1, 2, 3, 4, 5.

	$\lambda=0$		$\lambda=1$		$\lambda=2$	
x	y	y'	y	y'	y	y'
0. 00	1. 0000	0. 0000	1. 0000	0. 0000	1. 0000	0. 0000
0. 01	1. 0003	0. 0600	1. 0003	0. 0601	1. 0003	0. 0601
0. 02	1. 0012	0. 1201	1. 0012	0. 1203	1. 0012	0. 1205
0. 03	1. 0027	0. 1803	1. 0027	0. 1808	1. 0027	0. 1812
0. 04	1. 0048	0. 2408	1. 0048	0. 2416	1. 0048	0. 2424
0. 05	1. 0075	0. 3015	1. 0075	0. 3028	1. 0076	0. 3040
0. 06	1. 0108	0. 3626	1. 0109	0. 3644	1. 0109	0. 3662
0. 07	1. 0148	0. 4241	1. 0148	0. 4266	1. 0149	0. 4291
0. 08	1. 0193	0. 4862	1. 0194	0. 4894	1. 0195	0. 4926
0. 09	1. 0245	0. 5489	1. 0246	0. 5529	1. 0247	0. 5570
0. 10	1. 0303	0. 6122	1. 0304	0. 6172	1. 0306	0. 6223
0. 11	1. 0367	0. 6762	1. 0369	0. 6823	1. 0372	0. 6885
0. 12	1. 0438	0. 7412	1. 0441	0. 7484	1. 0444	0. 7558
0. 13	1. 0515	0. 8070	1. 0519	0. 8156	1. 0523	0. 8242
0. 14	1. 0599	0. 8739	1. 0604	0. 8338	1. 0609	0. 8939
0. 15	1. 0690	0. 9418	1. 0696	0. 9533	1. 0702	0. 9649
0. 16	1. 0788	1. 0110	1. 0795	1. 0241	1. 0802	1. 0374
0. 17	1. 0892	1. 0815	1. 0901	1. 0964	1. 0909	1. 1114
0. 18	1. 1004	1. 1534	1. 1014	1. 1702	1. 1024	1. 1870
0. 19	1. 1123	1. 2269	1. 1135	1. 2456	1. 1147	1. 2645
0. 20	1. 1249	1. 3019	1. 1263	1. 3228	1. 1277	1. 3438
0. 21	1. 1383	1. 3788	1. 1399	1. 4019	1. 1416	1. 4251
0. 22	1. 1525	1. 4575	1. 1544	1. 4830	1. 1562	1. 5086
0. 23	1. 1675	1. 5382	1. 1696	1. 5662	1. 1718	1. 5944
0. 24	1. 1833	1. 6211	1. 1857	1. 6518	1. 1881	1. 6826
0. 25	1. 1999	1. 7063	1. 2026	1. 7398	1. 2054	1. 7735
0. 26	1. 2174	1. 7939	1. 2205	1. 8304	1. 2236	1. 8671
0. 27	1. 2358	1. 8842	1. 2392	1. 9238	1. 2428	1. 9636
0. 28	1. 2551	1. 9772	1. 2589	2. 0201	1. 2629	2. 0633
0. 29	1. 2753	2. 0733	1. 2796	2. 1196	1. 2840	2. 1662
0. 30	1. 2965	2. 1725	1. 3013	2. 2225	1. 3062	2. 2728

TABLE I—Continued

	$\lambda=0$		$\lambda=1$		$\lambda=2$	
x	y	y'	$y_.$	y'	y	y'
0. 31	1. 3188	2. 2751	1. 3241	2. 3289	1. 3295	2. 3831
0. 32	1. 3420	2. 3813	1. 3480	2. 4392	1. 3539	2. 4974
0. 33	1. 3664	2. 4913	1. 3729	2. 5535	1. 3795	2. 6159
0. 34	1. 3919	2. 6054	1. 3991	2. 6721	1. 4062	2. 7390
0. 35	1. 4185	2. 7238	1. 4264	2. 7952	1. 4343	2. 8669
0. 36	1. 4464	2. 8469	1. 4550	2. 9233	1. 4636	3. 0000
0. 37	1. 4755	2. 9750	1. 4849	3. 0566	1. 4943	3. 1385
0. 38	1. 5059	3. 1083	1. 5162	3. 1954	1. 5264	3. 2829
0. 39	1. 5377	3. 2472	1. 5488	3. 3401	1. 5600	3. 4334
0. 40	1. 5709	3. 3921	1. 5830	3. 4912	1. 5951	3. 5906
0. 41	1. 6055	3. 5435	1. 6187	3. 6490	1. 6318	3. 7549
0. 42	1. 6417	3. 7016	1. 6560	3. 8139	1. 6702	3. 9267
0. 43	1. 6796	3. 8671	1. 6950	3. 9866	1. 7104	4. 1066
0. 44	1. 7191	4. 0403	1. 7357	4. 1674	1. 7524	4. 2951
0. 45	1. 7604	4. 2219	1. 7783	4. 3571	1. 7963	4. 4929
0. 46	1. 8036	4. 4124	1. 8229	4. 5561	1. 8423	4. 7006
0. 47	1. 8487	4. 6125	1. 8695	4. 7652	1. 8903	4. 9188
0. 48	1. 8958	4. 8228	1. 9182	4. 9852	1. 9407	5. 1484
0. 49	1. 9452	5. 0441	1. 9692	5. 2166	1. 9934	5. 3902
0. 50	1. 9968	5. 2772	2. 0226	5. 4606	2. 0485	5. 6452
0. 51	2. 0508	5. 523	2. 0785	5. 718	2. 1063	5. 914
0. 52	2. 1073	5. 782	2. 1370	5. 990	2. 1669	6. 198
0. 53	2. 1665	6. 056	2. 1983	6. 277	2. 2303	6. 499
0. 54	2. 2284	6. 346	2. 2626	6. 580	2. 2969	6. 817
0. 55	2. 2934	6. 653	2. 3300	6. 902	2. 3667	7. 154
0. 56	2. 3615	6. 978	2. 4007	7. 243	2. 4400	7. 512
0. 57	2. 4330	7. 322	2. 4749	7. 606	2. 5170	7. 891
0. 58	2. 5081	7. 689	2. 5529	7. 990	2. 5979	8. 295
0. 59	2. 5869	8. 078	2. 6348	8. 400	2. 6830	8. 725
0. 60	2. 6697	8. 492	2. 7210	8. 836	2. 7725	9. 183
0. 61	2. 7568	8. 934	2. 8117	9. 301	2. 8668	9. 672
0. 62	2. 8485	9. 405	2. 9071	9. 798	2. 9661	10. 19
0. 63	2. 9450	9. 909	3. 0077	10. 33	3. 0708	10. 75
0. 64	3. 0467	10. 45	3. 1138	10. 90	3. 1813	11. 35
0. 65	3. 1540	11. 02	3. 2258	11. 51	3. 2980	12. 00
0. 66	3. 2673	11. 64	3. 3441	12. 16	3. 4214	12. 69
0. 67	3. 3870	12. 31	3. 4691	12. 86	3. 5519	13. 43
0. 68	3. 5135	13. 02	3. 6015	13. 62	3. 6901	14. 23
0. 69	3. 6475	13. 79	3. 7417	14. 43	3. 8367	15. 09
0. 70	3. 7895	14. 62	3. 8904	15. 32	3. 9922	16. 02

TABLE I—Continued

x	λ=0		λ=1		λ=2	
	y	y′	y	y′	y	y′
0. 71	3. 9401	15. 51	4. 0482	16. 27	4. 1574	17. 03
0. 72	4. 1000	16. 48	4. 2160	17. 30	4. 3332	18. 13
0. 73	4. 2700	17. 53	4. 3945	18. 42	4. 5203	19. 32
0. 74	4. 4509	18. 67	4. 5847	19. 63	4. 7199	20. 61
0. 75	4. 6438	19. 91	4. 7875	20. 96	4. 9330	22. 03
0. 76	4. 8495	21. 27	5. 0042	22. 40	5. 1609	23. 57
0. 77	5. 0694	22. 74	5. 2360	23. 98	5. 4049	25. 26
0. 78	5. 3048	24. 35	5. 4844	25. 71	5. 6666	27. 11
0. 79	5. 5570	26. 12	5. 7509	27. 61	5. 9477	29. 15
0. 80	5. 8277	28. 07	6. 0373	29. 71	6. 2503	31. 40
0. 81	6. 119	30. 2	6. 346	32. 0	6. 576	33. 9
0. 82	6. 433	32. 6	6. 678	34. 6	6. 929	36. 6
0. 83	6. 771	35. 2	7. 038	37. 4	7. 310	39. 7
0. 84	7. 137	38. 1	7. 427	40. 5	7. 724	43. 1
0. 85	7. 534	41. 4	7. 850	44. 0	8. 173	46. 9
0. 86	7. 964	44. 9	8. 309	48. 0	8. 663	51. 2
0. 87	8. 433	48. 9	8. 811	52. 4	9. 198	56. 0
0. 88	8. 944	53. 5	9. 359	57. 3	9. 784	61. 4
0. 89	9. 504	58. 6	9. 959	62. 9	10. 43	67. 5
0. 90	10. 12	64. 3	10. 62	69. 3	11. 14	74. 5
0. 91	10. 79	70. 9	11. 35	76. 5	11. 92	82. 5
0. 92	11. 54	78. 4	12. 15	84. 8	12. 79	91. 7
0. 93	12. 36	86. 9	13. 05	94. 3	13. 76	102.
0. 94	13. 28	96. 8	14. 05	105.	14. 84	115.
0. 95	14. 30	108.	15. 16	118.	16. 06	129.
0. 96	15. 45	121.	16. 41	133.	17. 43	146.
0. 97	16. 74	137.	17. 83	151.	18. 98	166.
0. 98	18. 20	155.	19. 44	171.	20. 75	189.
0. 99	19. 85	177.	21. 27	196.	22. 78	218.
1. 00	21. 74	203.				

TABLE I—Continued

x	$\lambda=0$		$\lambda=1$		$\lambda=2$	
	y	y'	y	y'	y	y'
0. 00	1. 0000	0. 0000	1. 0000	0. 0000	1. 0000	0. 0000
−0. 01	1. 0003	−0. 0600	1. 0003	−0. 0600	1. 0003	−0. 0601
−0. 02	1. 0012	−0. 1201	1. 0012	−0. 1199	1. 0012	−0. 1201
−0. 03	1. 0027	−0. 1803	1. 0027	−0. 1799	1. 0027	−0. 1800
−0. 04	1. 0048	−0. 2408	1. 0048	−0. 2400	1. 0048	−0. 2400
−0. 05	1. 0075	−0. 3015	1. 0075	−0. 3002	1. 0075	−0. 3000
−0. 06	1. 0108	−0. 3626	1. 0108	−0. 3608	1. 0108	−0. 3602
−0. 07	1. 0148	−0. 4241	1. 0147	−0. 4217	1. 0147	−0. 4206
−0. 08	1. 0193	−0. 4862	1. 0192	−0. 4830	1. 0192	−0. 4814
−0. 09	1. 0245	−0. 5489	1. 0244	−0. 5447	1. 0243	−0. 5425
−0. 10	1. 0303	−0. 6122	1. 0301	−0. 6071	1. 0301	−0. 6041
−0. 11	1. 0367	−0. 6762	1. 0365	−0. 6701	1. 0364	−0. 6663
−0. 12	1. 0438	−0. 7412	1. 0435	−0. 7338	1. 0434	−0. 7290
−0. 13	1. 0515	−0. 8070	1. 0512	−0. 7984	1. 0510	−0. 7925
−0. 14	1. 0599	−0. 8739	1. 0595	−0. 8638	1. 0592	−0. 8568
−0. 15	1. 0690	−0. 9418	1. 0685	−0. 9303	1. 0681	−0. 9220
−0. 16	1. 0788	−1. 0110	1. 0781	−0. 9979	1. 0777	−0. 9882
−0. 17	1. 0892	−1. 0815	1. 0884	−1. 0666	1. 0879	−1. 0554
−0. 18	1. 1004	−1. 1534	1. 0994	−1. 1367	1. 0988	−1. 1238
−0. 19	1. 1123	−1. 2269	1. 1112	−1. 2081	1. 1104	−1. 1935
−0. 20	1. 1249	−1. 3019	1. 1236	−1. 2811	1. 1227	−1. 2646
−0. 21	1. 1383	−1. 3788	1. 1368	−1. 3556	1. 1357	−1. 3372
−0. 22	1. 1525	−1. 4575	1. 1507	−1. 4320	1. 1494	−1. 4114
−0. 23	1. 1675	−1. 5382	1. 1654	−1. 5102	1. 1639	−1. 4874
−0. 24	1. 1833	−1. 6211	1. 1809	−1. 5904	1. 1792	−1. 5652
−0. 25	1. 1999	−1. 7063	1. 1972	−1. 6728	1. 1952	−1. 6451
−0. 26	1. 2174	−1. 7939	1. 2144	−1. 7574	1. 2121	−1. 7271
−0. 27	1. 2358	−1. 8842	1. 2324	−1. 8446	1. 2298	−1. 8114
−0. 28	1. 2551	−1. 9772	1. 2513	−1. 9343	1. 2483	−1. 8982
−0. 29	1. 2753	−2. 0733	1. 2711	−2. 0269	1. 2677	−1. 9877
−0. 30	1. 2965	−2. 1725	1. 2918	−2. 1225	1. 2881	−2. 0800
−0. 31	1. 3188	−2. 2751	1. 3135	−2. 2213	1. 3094	−2. 1753
−0. 32	1. 3420	−2. 3813	1. 3363	−2. 3234	1. 3316	−2. 2738
−0. 33	1. 3664	−2. 4913	1. 3600	−2. 4292	1. 3548	−2. 3757
−0. 34	1. 3919	−2. 6054	1. 3848	−2. 5388	1. 3791	−2, 4813
−0. 35	1. 4185	−2. 7238	1. 4108	−2. 6526	1. 4045	−2. 5908
−0. 36	1. 4464	−2. 8469	1. 4379	−2. 7708	1. 4310	−2. 7045
−0. 37	1. 4755	−2. 9750	1. 4662	−2. 8936	1. 4586	−2. 8226
−0. 38	1. 5059	−3. 1083	1. 4958	−3. 0215	1. 4874	−2. 9455
−0. 39	1. 5377	−3. 2472	1. 5267	−3. 1546	1. 5175	−3. 0734
−0. 40	1. 5709	−3. 3921	1. 5589	−3. 2935	1. 5489	−3. 2068

TABLE I—Continued

	$\lambda=0$		$\lambda=1$		$\lambda=2$	
x	y	y'	y	y'	y	y'
−0. 41	1. 6055	−3. 5435	1. 5926	−3. 4384	1. 5817	−3. 3459
−0. 42	1. 6417	−3. 7016	1. 6277	−3. 5898	1. 6158	−3. 4911
−0. 43	1. 6796	−3. 8671	1. 6644	−3. 7481	1. 6515	−3. 6429
−0. 44	1. 7191	−4. 0403	1. 7027	−3. 9137	1. 6887	−3. 8018
−0. 45	1. 7604	−4. 2219	1. 7427	−4. 0873	1. 7276	−3. 9681
−0. 46	1. 8036	−4. 4124	1. 7844	−4. 2693	1. 7681	−4. 1425
−0. 47	1. 8487	−4. 6125	1. 8281	−4. 4606	1. 8104	−4. 3254
−0. 48	1. 8958	−4. 8228	1. 8737	−4. 6614	1. 8547	−4. 5176
−0. 49	1. 9452	−5. 0441	1. 9213	−4. 8725	1. 9008	−4. 7196
−0. 50	1. 9968	−5. 2772	1. 9712	−5. 0948	1. 9491	−4. 9322
−0. 51	2. 0508	−5. 523	2. 0233	−5. 329	1. 9995	−5. 156
−0. 52	2. 1073	−5. 782	2. 0778	−5. 576	2. 0522	−5. 392
−0. 53	2. 1665	−6. 056	2. 1348	−5. 837	2. 1074	−5. 641
−0. 54	2. 2284	−6. 346	2. 1946	−6. 113	2. 1651	−5. 905
−0. 55	2. 2934	−6. 653	2. 2571	−6. 405	2. 2255	−6. 183
−0. 56	2. 3615	−6. 978	2. 3227	−6. 714	2. 2888	−6. 478
−0. 57	2. 4330	−7. 322	2. 3915	−7. 041	2. 3552	−6. 790
−0. 58	2. 5081	−7. 689	2. 4636	−7, 389	2. 4247	−7. 122
−0. 59	2. 5869	−8. 078	2. 5393	−7. 759	2. 4977	−7. 473
−0. 60	2. 6697	−8. 492	2. 6188	−8. 152	2. 5742	−7. 848
−0. 61	2. 7568	−8. 934	2. 7024	−8. 570	2. 6547	−8. 246
−0. 62	2. 8485	−9. 405	2. 7903	−9. 016	2. 7392	−8. 670
−0. 63	2. 9450	−9. 909	2. 8828	−9. 493	2. 8282	−9. 122
−0. 64	3. 0467	−10. 45	2. 9802	−10. 00	2. 9218	−9. 606
−0. 65	3. 1540	−11. 02	3. 0830	−10. 55	3. 0204	−10. 12
−0. 66	3. 2673	−11. 64	3. 1913	−11. 13	3. 1244	−10. 68
−0. 67	3. 3870	−12. 31	3. 3057	−11. 76	3. 2341	−11. 27
−0. 68	3. 5135	−13. 02	3. 4266	−12. 43	3. 3499	−11. 91
−0. 69	3. 6475	−13. 79	3. 5545	−13. 15	3. 4723	−12. 59
−0. 70	3. 7895	−14. 62	3. 6899	−13. 93	3. 6019	−13. 33
−0. 71	3, 9401	−15. 51	3. 8333	−14. 78	3. 7391	−14. 12
−0. 72	4. 1000	−16. 48	3. 9855	−15. 69	3. 8845	−14. 98
−0. 73	4. 2700	−17. 53	4. 1473	−16. 67	4. 0389	−15. 91
−0. 74	4. 4509	−18. 67	4. 3192	−17. 74	4. 2029	−16. 91
−0. 75	4. 6438	−19. 91	4. 5023	−18. 90	4. 3774	−18. 00
−0. 76	4. 8495	−21. 27	4. 6974	−20. 16	4. 5632	−19. 18
−0. 77	5. 0694	−22. 74	4. 9058	−21. 53	4. 7614	−20. 47
−0. 78	5. 3048	−24. 35	5. 1285	−23. 04	4. 9730	−21. 88
−0. 79	5. 5570	−26. 12	5. 3669	−24. 68	5. 1994	−23. 41
−0. 80	5. 8277	−28. 07	5. 6225	−26. 48	5. 4418	−15. 10

TABLE I—Continued

x	$\lambda=0$		$\lambda=1$		$\lambda=2$	
	y	y'	y	y'	y	y'
−0. 81	6. 119	−30. 2	5. 897	−28. 5	5. 702	−26. 9
−0. 82	6. 433	−32. 6	6. 192	−30. 6	5. 981	−29. 0
−0. 83	6. 771	−35. 2	6. 511	−33. 1	6. 282	−31. 2
−0. 84	7. 137	−38. 1	6. 854	−35. 7	6. 606	−33. 7
−0. 85	7. 534	−41. 4	7. 226	−38. 7	6. 957	−36. 4
−0. 86	7. 964	−44. 9	7. 629	−42. 0	7. 336	−39. 5
−0. 87	8. 433	−48. 9	8. 067	−45. 7	7. 747	−42. 9
−0. 88	8. 944	−53. 5	8. 544	−49. 8	8. 194	−46. 7
−0. 89	9. 504	−58. 6	9. 065	−54. 4	8. 682	−50. 9
−0. 90	10. 12	−64. 3	9. 634	−59. 7	9. 214	−55. 7
−0. 91	10. 79	−70. 9	10. 26	−65. 6	9. 798	−61. 1
−0. 92	11. 54	−78. 4	10. 95	−72. 3	10. 44	−67. 2
−0. 93	12. 36	−86. 9	11. 71	−80. 0	11. 14	−74. 2
−0. 94	13. 28	−96. 8	12. 55	−88. 8	11. 93	−82. 1
−0. 95	14. 30	−108.	13. 49	−99. 0	12. 79	−91. 3
−0. 96	15. 45	−121.	14. 54	−111.	13. 76	−102.
−0. 97	16. 74	−137.	15. 71	−124.	14. 83	−114.
−0. 98	18. 20	−155.	17. 03	−140.	16. 04	−128.
−0. 99	19. 85	−177.	18. 53	−159.	17. 41	−145.
−1. 00	21. 74	−203.	20. 23	−182.	18. 95	−165.

TABLE I—Continued

	$\lambda=3$		$\lambda=4$		$\lambda=5$	
x	y	y'	y	y'	y	y'
0. 00	1. 0000	0. 0000	1. 0000	0. 0000	1. 0000	0. 0000
0. 01	1. 0003	0. 0602	1. 0003	0. 0602	1. 0003	0. 0603
0. 02	1. 0012	0. 1207	1. 0012	0. 1209	1. 0012	0. 1211
0. 03	1. 0027	0. 1817	1. 0027	0. 1821	1. 0027	0. 1826
0. 04	1. 0048	0. 2432	1. 0049	0. 2440	1. 0049	0. 2448
0. 05	1. 0076	0. 3053	1. 0076	0. 3065	1. 0076	0. 3078
0. 06	1. 0109	0. 3680	1. 0110	0. 3698	1. 0110	0. 3716
0. 07	1. 0149	0. 4315	1. 0150	0. 4340	1. 0151	0. 4365
0. 08	1. 0196	0. 4959	1. 0197	0. 4991	1. 0198	0. 5023
0. 09	1. 0249	0. 5611	1. 0250	0. 5652	1. 0251	0. 5693
0. 10	1. 0308	0. 6273	1. 0310	0. 6324	1. 0311	0. 6374
0. 11	1. 0374	0. 6946	1. 0376	0. 7008	1. 0379	0. 7069
0. 12	1. 0447	0. 7631	1. 0450	0. 7704	1. 0453	0. 7777
0. 13	1. 0527	0. 8328	1. 0531	0. 8414	1. 0534	0. 8500
0. 14	1. 0614	0. 9039	1. 0618	0. 9139	1. 0623	0. 9239
0. 15	1. 0708	0. 9765	1. 0713	0. 9880	1. 0719	0. 9995
0. 16	1. 0809	1. 0505	1. 0816	1. 0637	1. 0823	1. 0768
0. 17	1. 0918	1. 1263	1. 0926	1. 1412	1. 0934	1. 1561
0. 18	1. 1034	1. 2038	1. 1044	1. 2206	1. 1054	1. 2374
0. 19	1. 1159	1. 2832	1. 1170	1. 3020	1. 1182	1. 3208
0. 20	1. 1291	1. 3647	1. 1305	1. 3856	1. 1318	1. 4064
0. 21	1. 1432	1. 4483	1. 1448	1. 4714	1. 1463	1. 4945
0. 22	1. 1581	1. 5341	1. 1599	1. 5597	1. 1617	1. 5852
0. 23	1. 1739	1. 6225	1. 1760	1. 6505	1. 1781	1. 6785
0. 24	1. 1905	1. 7133	1. 1929	1. 7441	1. 1953	1. 7748
0. 25	1. 2081	1. 8070	1. 2108	1. 8405	1. 2136	1. 8741
0. 26	1. 2267	1. 9036	1. 2297	1. 9401	1. 2328	1. 9766
0. 27	1. 2462	2. 0032	1. 2497	2. 0429	1. 2531	2. 0825
0. 28	1. 2668	2. 1062	1. 2706	2. 1491	1. 2745	2. 1921
0. 29	1. 2883	2. 2127	1. 2927	2. 2591	1. 2970	2. 3055
0. 30	1. 3110	2. 3228	1. 3158	2. 3729	1. 3206	2. 4230
0. 31	1. 3348	2. 4370	1. 3401	2. 4909	1. 3454	2. 5448
0. 32	1. 3598	2. 5553	1. 3656	2. 6133	1. 3715	2. 6713
0. 33	1. 3859	2. 6782	1. 3924	2. 7404	1. 3989	2. 8027
0. 34	1. 4134	2. 8057	1. 4205	2. 8725	1. 4276	2. 9392
0. 35	1. 4421	2. 9384	1. 4499	3. 0098	1. 4577	3. 0813
0. 36	1. 4721	3. 0764	1. 4807	3. 1528	1. 4892	3. 2293
0. 37	1. 5036	3. 2201	1. 5129	3. 3018	1. 5223	3. 3836
0. 38	1. 5366	3. 3700	1. 5467	3. 4573	1. 5569	3. 5446
0. 39	1. 5710	3. 5264	1. 5821	3. 6195	1. 5932	3. 7126
0. 40	1. 6071	3. 6897	1. 6192	3. 7890	1. 6312	3. 8883

TABLE I—Continued

x	$\lambda=3$ y	y'	$\lambda=4$ y	y'	$\lambda=5$ y	y'
0. 41	1. 6449	3. 8605	1. 6579	3. 9662	1. 6710	4. 0721
0. 42	1. 6844	4. 0392	1. 6985	4. 1518	1. 7127	4. 2646
0. 43	1. 7257	4. 2263	1. 7410	4. 3462	1. 7563	4. 4663
0. 44	1. 7689	⁓ 4. 4225	1. 7855	4. 5501	1. 8020	4. 6779
0. 45	1. 8142	4. 6284	1. 8320	4. 7642	1. 8499	4. 9002
0. 46	1. 8615	4. 8447	1. 8808	4. 9891	1. 9001	5. 1339
0. 47	1. 9111	5. 0721	1. 9318	5. 2258	1. 9526	5. 3797
0. 48	1. 9630	5. 3115	1. 9853	5. 4749	2. 0077	5. 6387
0. 49	2. 0174	5. 5636	2. 0414	5. 7375	2. 0654	5. 9118
0. 50	2. 0743	5. 8296	2. 1001	6. 0145	2. 1260	6. 2000
0. 51	2. 1340	6. 110	2. 1617	6. 307	2. 1895	6. 505
0. 52	2. 1966	6. 407	2. 2263	6. 616	2. 2561	6. 827
0. 53	2. 2622	6. 721	2. 2941	6. 944	2. 3261	7. 168
0. 54	2. 3311	7. 053	2. 3653	7. 291	2. 3996	7. 529
0. 55	2. 4033	7. 406	2. 4400	7. 659	2. 4768	7. 913
0. 56	2. 4792	7. 780	2. 5185	8. 050	2. 5579	8. 321
0. 57	2. 5590	8. 178	2. 6011	8. 466	2. 6433	8. 755
0. 58	2. 6429	8. 601	2. 6879	8. 908	2. 7331	9. 217
0. 59	2. 7311	9. 051	2. 7794	9. 380	2. 8277	9. 710
0. 60	2. 8240	9. 532	2. 8756	9. 883	2. 9274	10. 24
0. 61	2. 9219	10. 05	2. 9771	10. 42	3. 0326	10. 80
0. 62	3. 0250	10. 59	3. 0842	11. 00	3. 1436	11. 40
0. 63	3. 1339	11. 18	3. 1972	11. 61	3. 2608	12. 05
0. 64	3. 2488	11. 81	3. 3166	12. 28	3. 3847	12. 74
0. 65	3. 3703	12. 49	3. 4429	12. 99	3. 5158	13. 49
0. 66	3. 4987	13. 22	3. 5765	13. 75	3. 6547	14. 29
0. 67	3. 6348	14. 00	3. 7181	14. 58	3. 8019	15. 16
0. 68	3. 7789	14. 84	3. 8682	15. 47	3. 9581	16. 10
0. 69	3. 9318	15. 75	4. 0276	16. 43	4. 1241	17. 11
0. 70	4. 0943	16. 74	4. 1971	17. 47	4. 3006	18. 21
0. 71	4. 2670	17. 81	4. 3773	18. 60	4. 4886	19. 40
0. 72	4. 4508	18. 97	4. 5694	19. 83	4. 6890	20. 71
0. 73	4. 6467	20. 23	4. 7743	21. 17	4. 9030	22. 12
0. 74	4. 8559	21. 61	4. 9932	22. 63	5. 1318	23. 66
0. 75	5. 0794	23. 11	5. 2273	24. 22	5. 3768	25. 36
0. 76	5. 3186	24. 76	5. 4781	25. 97	5. 6395	27. 21
0. 77	5. 5750	26. 56	5. 7473	27. 89	5. 9217	29. 26
0. 78	5. 8504	28. 54	6. 0366	30. 01	6. 2253	31. 51
0. 79	6. 1465	30. 72	6. 3481	32. 34	6. 5526	34. 00
0. 80	6. 4656	33. 13	6. 6842	34. 91	6. 9061	36. 75

TABLE I—Continued

	$\lambda=3$		$\lambda=4$		$\lambda=5$	
x	y	y'	y	y'	y	y'
0. 81	6. 810	35. 8	7. 047	37. 8	7. 289	39. 8
0. 82	7. 182	38. 8	7. 441	41. 0	7. 704	43. 2
0. 83	7. 586	42. 1	7. 868	44. 5	8. 154	47. 0
0. 84	8. 025	45. 7	8. 332	48. 5	8. 646	51. 3
0. 85	8. 502	49. 8	8. 839	52. 9	9. 182	56. 1
0. 86	9. 023	54. 5	9. 392	57. 9	9. 770	61. 5
0. 87	9. 594	59. 7	9. 999	63. 6	10. 42	67. 7
0. 88	10. 22	65. 6	10. 67	70. 0	11. 13	74. 7
0. 89	10. 91	72. 3	11. 40	77. 4	11. 91	82. 7
0. 90	11. 67	80. 0	12. 22	85. 8	12. 78	91. 9
0. 91	12. 51	88. 8	13. 12	95. 4	13. 75	102.
0. 92	13. 45	98. 9	14. 13	107.	14. 84	115.
0. 93	14. 50	111.	15. 26	120.	16. 06	129.
0. 94	15. 67	124.	16. 53	135.	17. 43	146.
0. 95	16. 99	140.	17. 97	153.	18. 99	166.
0. 96	18. 48	159.	19. 59	174.	20. 76	190.
0. 97	20. 18	182.	21. 45	199.		
0. 98	22. 13	208.				

TABLE I—Continued

x	$\lambda=3$		$\lambda=4$		$\lambda=5$	
	y	y'	y	y'	y	y'
0. 00	1. 0000	0. 0000	1. 0000	0. 0000	1. 0000	0. 0000
−0. 01	1. 0003	−0. 0602	1. 0003	−0. 0602	1. 0003	−0. 0603
−0. 02	1. 0012	−0. 1201	1. 0012	−0. 1201	1. 0012	−0. 1201
−0. 03	1. 0027	−0. 1799	1. 0027	−0. 1797	1. 0027	−0. 1796
−0. 04	1. 0048	−0. 2396	1. 0048	−0. 2392	1. 0048	−0. 2388
−0. 05	1. 0075	−0. 2993	1. 0075	−0. 2985	1. 0075	−0. 2978
−0. 06	1. 0108	−0. 3590	1. 0108	−0. 3578	1. 0107	−0. 3566
−0. 07	1. 0147	−0. 4189	1. 0146	−0. 4171	1. 0146	−0. 4154
−0. 08	1. 0192	−0. 4790	1. 0191	−0. 4766	1. 0190	−0. 4741
−0. 09	1. 0243	−0. 5393	1. 0242	−0. 5362	1. 0241	−0. 5330
−0. 10	1. 0299	−0. 6001	1. 0298	−0. 5961	1. 0297	−0. 5920
−0. 11	1. 0363	−0. 6613	1. 0361	−0. 6563	1. 0359	−0. 6513
−0. 12	1. 0432	−0. 7230	1. 0430	−0. 7169	1. 0427	−0. 7108
−0. 13	1. 0507	−0. 7853	1. 0504	−0. 7780	1. 0501	−0. 7708
−0. 14	1. 0589	−0. 8483	1. 0585	−0. 8397	1. 0581	−0. 8312
−0. 15	1. 0677	−0. 9121	1. 0672	−0. 9021	1. 0668	−0. 8922
−0. 16	1. 0771	−0. 9767	1. 0766	−0. 9652	1. 0760	−0. 9538
−0. 17	1. 0872	−1. 0423	1. 0865	−1. 0292	1. 0858	−1. 0161
−0. 18	1. 0980	−1. 1090	1. 0971	−1. 0941	1. 0963	−1. 0793
−0. 19	1. 1094	−1. 1768	1. 1084	−1. 1601	1. 1074	−1. 1434
−0. 20	1. 1215	−1. 2459	1. 1203	−1. 2272	1. 1192	−1. 2085
−0. 21	1. 1343	−1. 3164	1. 1330	−1. 2956	1. 1316	−1. 2747
−0. 22	1. 1478	−1. 3883	1. 1463	−1. 3653	1. 1447	−1. 3422
−0. 23	1. 1621	−1. 4619	1. 1603	−1. 4365	1. 1584	−1. 4110
−0. 24	1. 1771	−1. 5372	1. 1750	−1. 5093	1. 1729	−1. 4813
−0. 25	1. 1928	−1. 6144	1. 1905	−1. 5838	1. 1881	−1. 5531
−0. 26	1. 2094	−1. 6936	1. 2067	−1. 6602	1. 2040	−1. 6267
−0. 27	1. 2267	−1. 7750	1. 2237	−1. 7386	1. 2206	−1. 7021
−0. 28	1. 2449	−1. 8587	1. 2415	−1. 8191	1. 2380	−1. 7795
−0. 29	1. 2639	−1. 9448	1. 2601	−1. 9020	1. 2562	−1. 8591
−0. 30	1. 2838	−2. 0336	1. 2795	−1. 9873	1. 2752	−1. 9410
−0. 31	1. 3046	−2. 1253	1. 2998	−2. 0753	1. 2950	−2. 0253
−0. 32	1. 3263	−2. 2199	1. 3210	−2. 1661	1. 3157	−2. 1123
−0. 33	1. 3490	−2. 3178	1. 3431	−2. 2599	1. 3373	−2. 2021
−0. 34	1. 3727	−2. 4192	1. 3662	−2. 3570	1. 3598	−2. 2949
−0. 35	1. 3974	−2. 5242	1. 3903	−2. 4576	1. 3832	−2. 3910
−0. 36	1. 4232	−2. 6332	1. 4154	−2. 5619	1. 4076	−2. 4906
−0. 37	1. 4501	−2. 7463	1. 4415	−2. 6701	1. 4330	−2. 5939
−0. 38	1. 4781	−2. 8640	1. 4688	−2. 7825	1. 4595	−2. 7011
−0. 39	1. 5074	−2. 9864	1. 4972	−2. 8995	1. 4871	−2. 8126
−0. 40	1. 5379	−3. 1140	1. 5268	−3. 0212	1. 5158	−2. 9286

TABLE I—Continued

x	$\lambda=3$		$\lambda=4$		$\lambda=5$	
	y	y'	y	y'	y	y'
-0.41	1. 5697	-3.2469	1. 5576	-3.1481	1. 5456	-3.0494
-0.42	1. 6028	-3.3857	1. 5898	-3.2805	1. 5768	-3.1754
-0.43	1. 6374	-3.5308	1. 6233	-3.4187	1. 6092	-3.3068
-0.44	1. 6734	-3.6824	1. 6582	-3.5632	1. 6429	-3.4442
-0.45	1. 7111	-3.8412	1. 6946	-3.7144	1. 6781	-3.5879
-0.46	1. 7503	-4.0075	1. 7325	-3.8728	1. 7147	-3.7383
-0.47	1. 7912	-4.1820	1. 7720	-4.0388	1. 7529	-3.8959
-0.48	1. 8340	-4.3651	1. 8133	-4.2130	1. 7926	-4.0611
-0.49	1. 8786	-4.5576	1. 8563	-4.3959	1. 8341	-4.2347
-0.50	1. 9251	-4.7600	1. 9012	-4.5883	1. 8774	-4.4170
-0.51	1. 9738	-4.973	1. 9481	-4.791	1. 9225	-4.609
-0.52	2. 0246	-5.198	1. 9971	-5.004	1. 9696	-4.811
-0.53	2. 0778	-5.435	2. 0482	-5.229	2. 0187	-5.024
-0.54	2. 1334	-5.685	2. 1017	-5.466	2. 0701	-5.248
-0.55	2. 1915	-5.949	2. 1576	-5.717	2. 1237	-5.485
-0.56	2. 2524	-6.229	2. 2161	-5.982	2. 1798	-5.736
-0.57	2. 3162	-6.526	2. 2773	-6.262	2. 2385	-6.001
-0.58	2. 3830	-6.840	2. 3414	-6.560	2. 2999	-6.281
-0.59	2. 4530	-7.173	2. 4085	-6.875	2. 3642	-6.579
-0.60	2. 5265	-7.528	2. 4790	-7.210	2. 4315	-6.894
-0.61	2. 6037	-7.905	2. 5528	-7.566	2. 5021	-7.230
-0.62	2. 6847	-8.306	2. 6303	-7.945	2. 5762	-7.586
-0.63	2. 7699	-8.734	2. 7118	-8.348	2. 6539	-7.966
-0.64	2. 8595	-9.190	2. 7974	-8.778	2. 7356	-8.370
-0.65	2. 9538	-9.678	2. 8874	-9.237	2. 8214	-8.801
-0.66	3. 0531	-10.20	2. 9823	-9.728	2. 9117	-9.262
-0.67	3. 1579	-10.76	3. 0821	-10.25	3. 0068	-9.755
-0.68	3. 2684	-11.36	3. 1874	-10.82	3. 1069	-10.28
-0.69	3. 3852	-12.00	3. 2986	-11.42	3. 2125	-10.85
-0.70	3. 5086	-12.69	3. 4160	-12.07	3. 3240	-11.45
-0.71	3. 6393	-13.44	3. 5401	-12.77	3. 4417	-12.11
-0.72	3. 7776	-14.24	3. 6715	-13.52	3. 5663	-12.81
-0.73	3. 9243	-15.11	3. 8107	-14.33	3. 6980	-13.56
-0.74	4. 0801	-16.05	3. 9583	-15.21	3. 8377	-14.38
-0.75	4. 2456	-17.07	4. 1151	-16.15	3. 9858	-15.26
-0.76	4. 4217	-18.17	4. 2817	-17.18	4. 1431	-16.21
-0.77	4. 6093	-19.37	4. 4590	-18.30	4. 3103	-17.25
-0.78	4. 8095	-20.68	4. 6479	-19.51	4. 4883	-18.37
-0.79	5. 0233	-22.11	4. 8495	-20.83	4. 6780	-19.59
-0.80	5. 2520	-23.66	5. 0650	-22.27	4. 8805	-20.92

TABLE I—Continued

x	$\lambda=3$		$\lambda=4$		$\lambda=5$	
	y	y'	y	y'	y	y'
−0. 81	5. 497	−25. 4	5. 295	−23. 9	5. 097	−22. 4
−0. 82	5. 760	−27. 2	5. 542	−25. 6	5. 328	−24. 0
−0. 83	6. 043	−29. 3	5. 808	−27. 5	5. 577	−25. 7
−0. 84	6. 347	−31. 6	6..093	−29. 6	5. 843	−27. 7
−0. 85	6. 675	−34. 1	6. 400	−31. 9	6. 130	−29. 7
−0. 86	7. 030	−36. 9	6. 731	−34. 4	6. 438	−32. 1
−0. 87	7. 414	−40. 0	7. 089	−37. 3	6. 772	−34. 6
−0. 88	7. 831	−43. 5	7. 477	−40. 4	7. 132	−37. 5
−0. 89	8. 285	−47. 3	7. 898	−43. 9	7. 522	−40. 7
−0. 90	8. 779	−51. 7	8. 357	−47. 8	7. 946	−44. 2
−0. 91	9. 320	−56. 5	8. 857	−52. 2	8. 408	−48. 2
−0. 92	9. 912	−62. 1	9. 404	−57. 2	8. 911	−52. 7
−0. 93	10. 56	−68. 3	10. 00	−62. 8	9. 462	−57. 6
−0. 94	11. 28	−75. 4	10. 66	−69. 2	10. 07	−63. 3
−0. 95	12. 08	−83. 6	11. 39	−76. 4	10. 73	−69. 7
−0. 96	12. 96	−93. 0	12. 19	−84. 7	11. 46	−77. 1
−0. 97	13. 94	−104.	13. 09	−94. 3	12. 28	−85. 5
−0. 98	15. 04	−116.	14. 08	−105.	13. 18	−95. 1
−0. 99	16. 27	−131.	15. 20	−118.	14. 18	−106.
−1. 00	17. 67	−148.	16. 45	−133.	15. 31	−119.

TABLE II

The values tabulated are approximations of $y(x)$ and $y'(\lambda)$, where $y(x)$ is the solution of $y'' = 2y^3 + xy + \mu$, for initial values: $y = 1$, $y' = 0$, $x = 0$, and for $\mu = 0$, 1, 2, 3, 4, 5.

	$\mu = 0$		$\mu = 1$		$\mu = 2$	
x	y	y'	y	y'	y	y'
0. 00	1. 0000	0. 0000	1. 0000	0. 0000	1. 0000	0. 0000
0. 01	1. 0001	0. 0200	1. 0002	0. 0301	1. 0002	0 0401
0. 02	1. 0004	0. 0402	1. 0006	0. 0602	1. 0008	0. 0802
0. 03	1. 0009	0. 0605	1. 0014	0. 0905	1. 0018	0. 1206
0. 04	1. 0016	0. 0809	1. 0024	0. 1210	1. 0032	0. 1611
0. 05	1. 0025	0. 1015	1. 0038	0. 1516	1. 0050	0. 2018
0. 06	1. 0036	0. 1222	1. 0054	0. 1825	1. 0072	0. 2427
0. 07	1. 0050	0. 1431	1. 0074	0. 2135	1. 0099	0. 2839
0. 08	1. 0065	0. 1642	1. 0097	0. 2448	1. 0129	0. 3253
0. 09	1. 0082	0. 1855	1. 0123	0. 2763	1. 0164	0. 3671
0. 10	1. 0102	0. 2070	1. 0152	0. 3081	1. 0203	0. 4091
0. 11	1. 0124	0. 2288	1. 0185	0. 3402	1. 0246	0. 4516
0. 12	1. 0148	0. 2508	1. 0220	0. 3726	1. 0293	0. 4944
0. 13	1. 0174	0. 2730	1. 0259	0. 4054	1. 0345	0. 5377
0. 14	1. 0202	0. 2956	1. 0302	0. 4385	1. 0401	0. 5814
0. 15	1. 0233	0. 3184	1. 0347	0. 4720	1. 0461	0. 6256
0. 16	1. 0266	0. 3415	1. 0396	0. 5059	1. 0526	0. 6703
0. 17	1. 0301	0. 3649	1. 0448	0. 5403	1. 0595	0. 7156
0. 18	1. 0339	0. 3887	1. 0504	0. 5751	1. 0669	0. 7615
0. 19	1. 0379	0. 4129	1. 0563	0. 6104	1. 0747	0. 8081
0. 20	1. 0422	0. 4374	1. 0626	0. 6463	1. 0830	0. 8553
0. 21	1. 0467	0. 4623	1. 0693	0. 6827	1. 0918	0. 9032
0. 22	1. 0514	0. 4856	1. 0763	0. 7197	1. 1011	0. 9520
0. 23	1. 0564	0. 5114	1. 0837	0. 7573	1. 1109	1. 0015
0. 24	1. 0616	0. 5377	1. 0914	0. 7956	1. 1211	1. 0519
0. 25	1. 0672	0. 5644	1. 0996	0. 8345	1. 1319	1. 1033
0. 26	1. 0729	0. 5916	1. 1081	0. 8742	1. 1432	1. 1556
0. 27	1. 0790	0. 6194	1. 1171	0. 9147	1. 1550	1. 2090
0. 28	1. 0853	0. 6477	1. 1264	0. 9560	1. 1674	1. 2635
0. 29	1. 0919	0. 6766	1. 1362	0. 9982	1. 1803	1. 3192
0. 30	1. 0989	0. 7061	1. 1464	1. 0413	1. 1938	1. 3761

TABLE II—Continued

| x | $\mu=0$ | | | $\mu=1$ | | | $\mu=2$ | |
	y	y′		y	y′		y	y′
0. 31	1. 1061	0. 7363		1. 1570	1. 0854		1. 2078	1. 4344
0. 32	1. 1136	0. 7671		1. 1681	1. 1305		1. 2225	1. 4941
0. 33	1. 1214	0. 7987		1. 1796	1. 1766		1. 2377	1. 5554
0. 34	1. 1296	0. 8309		1. 1916	1. 2239		1. 2536	1. 6182
0. 35	1. 1380	0. 8640		1. 2041	1. 2724		1. 2701	1. 6827
0. 36	1. 1468	0. 8979		1. 2171	1. 3222		1. 2873	1. 7491
0. 37	1. 1560	0. 9326		1. 2306	1. 3733		1. 3051	1. 8174
0. 38	1. 1655	0. 9683		1. 2445	1. 4259		1. 3236	1. 8877
0. 39	1. 1754	1. 0048		1. 2591	1. 4799		1. 3428	1. 9602
0. 40	1. 1856	1. 0424		1. 2742	1. 5356		1. 3628	2. 0351
0. 41	1. 1962	1. 0810		1. 2898	1. 5929		1. 3836	2. 1124
0. 42	1. 2072	1. 1207		1. 3060	1. 6520		1. 4051	2. 1924
0. 43	1. 2186	1. 1615		1. 3228	1. 7130		1. 4274	2. 2752
0. 44	1. 2304	1. 2036		1. 3403	1. 7760		1. 4506	2. 3611
0. 45	1. 2427	1. 2469		1. 3584	1. 8412		1. 4746	2. 4502
0. 46	1. 2554	1. 2915		1. 3771	1. 9085		1. 4996	2. 5427
0. 47	1. 2685	1. 3376		1. 3965	1. 9783		1. 5255	2. 6389
0. 48	1. 2822	1. 3851		1. 4167	2. 0507		1. 5524	2. 7391
0. 49	1. 2962	1. 4342		1. 4376	2. 1257		1. 5803	2. 8436
0. 50	1. 3108	1. 4850		1. 4592	2. 2037		1. 6093	2. 9526
0. 51	1. 3259	1. 537		1. 4817	2. 285		1. 6394	3. 067
0. 52	1. 3416	1. 592		1. 5049	2. 369		1. 6706	3. 186
0. 53	1. 3578	1. 648		1. 5290	2. 457		1. 7031	3. 311
0. 54	1. 3745	1. 706		1. 5541	2. 548		1. 7369	3. 441
0. 55	1. 3919	1. 766		1. 5800	2. 644		1. 7720	3. 579
0. 56	1. 4099	1. 830		1. 6070	2. 744		1. 8085	3. 724
0. 57	1. 4285	1. 895		1. 6350	2. 848		1. 8465	3. 876
0. 58	1. 4478	1. 962		1. 6639	2. 957		1. 8860	4. 037
0. 59	1. 4678	2. 033		1. 6941	3. 072		1. 9272	4. 207
0. 60	1. 4884	2. 106		1. 7254	3. 192		1. 9702	4. 386
0. 61	1. 5099	2. 183		1. 7579	3. 318		2. 0150	4. 576
0. 62	1. 5321	2. 262		1. 7918	3. 451		2. 0618	4. 778
0. 63	1. 5552	2. 346		1. 8269	3. 590		2. 1106	4. 993
0. 64	1. 5791	2. 433		1. 8636	3. 738		2. 1617	5. 221
0. 65	1. 6038	2. 524		1. 9017	3. 893		2. 2151	5. 465
0. 66	1. 6296	2. 619		1. 9415	4. 058		2. 2710	5. 725
0. 67	1. 6562	2. 718		1. 9829	4. 232		2. 3296	6. 004
0. 68	1. 6839	2. 823		2. 0261	4. 416		2. 3912	6. 303
0. 69	1. 7127	2. 932		2. 0713	4. 612		2. 4558	6. 624
0. 70	1 7426	3. 047		2. 1184	4. 821		2. 5237	6. 970

TABLE II—Continued

	$\mu=0$		$\mu=1$		$\mu=2$	
x	y	y'	y	y'	y	y'
0. 71	1. 7737	3. 168	2. 1677	5. 043	2. 5953	7. 343
0. 72	1. 8060	3. 296	2. 2193	5. 279	2. 6707	7. 747
0. 73	1. 8396	3. 430	2. 2734	5. 532	2. 7503	8. 185
0. 74	1. 8746	3. 572	2. 3301	5. 803	2. 8345	8. 661
0. 75	1. 9111	3. 722	2. 3895	6. 094	2. 9237	9. 180
0. 76	1. 9491	3. 880	2. 4520	6. 405	3. 0183	9. 746
0. 77	1. 9887	4. 048	2. 5177	6. 741	3. 1188	10. 37
0. 78	2. 0301	4. 225	2. 5869	7. 104	3. 2258	11. 05
0. 79	2. 0733	4. 414	2. 6599	7. 495	3. 3400	11. 80
0. 80	2. 1184	4. 615	2. 737	7. 919	3. 4622	12. 64
0. 81	2. 166	4. 83	2. 818	8. 38	3. 593	13. 6
0. 82	2. 215	5. 06	2. 905	8. 88	3. 734	14. 6
0. 83	2. 267	5. 30	2. 996	9. 43	3. 885	15. 8
0. 84	2. 321	5. 56	3. 093	10. 0	4. 049	17. 1
0. 85	2. 378	5. 84	3. 197	10. 7	4. 227	18. 5
0. 86	2. 438	6. 14	3. 307	11. 4	4. 421	20. 2
0. 87	2. 501	6. 46	3. 426	12. 2	4. 632	22. 1
0. 88	2. 567	6. 81	3. 552	13. 1	4. 864	24. 3
0. 89	2. 637	7. 19	3. 688	14. 1	5. 119	26. 9
0. 90	2. 711	7. 59	3. 834	15. 2	5. 402	29. 8
0. 91	2. 789	8. 03	3. 992	16. 4	5. 718	33. 3
0. 92	2. 872	8. 51	4. 164	17. 8	6. 071	37. 5
0. 93	2. 960	9. 03	4. 350	19. 4	6. 471	42. 5
0. 94	3. 053	9. 61	4. 553	21. 3	6. 470	48. 6
0. 95	3. 152	10. 2	4. 776	23. 3	6. 925	56. 1
0. 96	3. 258	10. 9	5. 021	25. 7	7. 447	65. 5
0. 97	3. 371	11. 7	5. 292	28. 5	8. 053	77. 4
0. 98	3. 492	12. 5	5. 594	31. 8	8. 765	93. 0
0. 99	3. 621	13. 5	5. 931	35. 7	9. 613	114.
1. 00	3. 761	14. 5	6. 311	40. 4	10. 64	142.

TABLE II—Continued

x	$\mu=0$		$\mu=1$		$\mu=2$	
	y	y'	y	y'	y	y'
0.00	1.0000	0.0000	1.0000	0.0000	1.0000	0.0000
−0.01	1.0001	−0.0200	1.0002	−0.0301	1.0002	−0.0401
−0.02	1.0004	−0.0398	1.0006	−0.0600	1.0008	−0.0800
−0.03	1.0009	−0.0596	1.0014	−0.0899	1.0018	−0.1200
−0.04	1.0016	−0.0793	1.0024	−0.1198	1.0032	−0.1599
−0.05	1.0025	−0.0990	1.0037	−0.1496	1.0050	−0.1997
−0.06	1.0036	−0.1186	1.0054	−0.1794	1.0072	−0.2397
−0.07	1.0049	−0.1382	1.0073	−0.2093	1.0098	−0.2796
−0.08	1.0063	−0.1578	1.0096	−0.2391	1.0128	−0.3196
−0.09	1.0080	−0.1774	1.0121	−0.2690	1.0162	−0.3598
−0.10	1.0099	−0.1970	1.0150	−0.2990	1.0200	−0.4000
−0.11	1.0120	−0.2166	1.0181	−0.3290	1.0242	−0.4404
−0.12	1.0142	−0.2362	1.0215	−0.3592	1.0288	−0.4809
−0.13	1.0167	−0.2559	1.0253	−0.3894	1.0338	−0.5216
−0.14	1.0193	−0.2756	1.0293	−0.4198	1.0392	−0.5627
−0.15	1.0222	−0.2954	1.0337	−0.4504	1.0451	−0.6039
−0.16	1.0252	−0.3152	1.0383	−0.4811	1.0513	−0.6454
−0.17	1.0285	−0.3352	1.0433	−0.5121	1.0580	−0.6872
−0.18	1.0319	−0.3553	1.0486	−0.5432	1.0650	−0.7294
−0.19	1.0356	−0.3755	1.0542	−0.5746	1.0726	−0.7719
−0.20	1.0395	−0.3958	1.0601	−0.6063	1.0805	−0.8149
−0.21	1.0435	−0.4162	1.0663	−0.6382	1.0889	−0.8583
−0.22	1.0478	−0.4369	1.0728	−0.6705	1.0977	−0.9022
−0.23	1.0523	−0.4576	1.0797	−0.7031	1.1069	−0.9466
−0.24	1.0569	−0.4786	1.0869	−0.7361	1.1166	−0.9916
−0.25	1.0618	−0.4998	1.0944	−0.7695	1.1267	−1.0372
−0.26	1.0669	−0.5212	1.1023	−0.8033	1.1373	−1.0834
−0.27	1.0723	−0.5428	1.1105	−0.8376	1.1484	−1.1303
−0.28	1.0778	−0.5647	1.1190	−0.8723	1.1599	−1.1780
−0.29	1.0835	−0.5869	1.1279	−0.9076	1.1720	−1.2265
−0.30	1.0895	−0.6093	1.1372	−0.9434	1.1845	−1.2759
−0.31	1.0957	−0.6321	1.1468	−0.9798	1.1975	−1.3261
−0.32	1.1022	−0.6552	1.1568	−1.0169	1.2110	−1.3774
−0.33	1.1088	−0.6785	1.1671	−1.0546	1.2250	−1.4297
−0.34	1.1157	−0.7024	1.1779	−1.0930	1.2396	−1.4831
−0.35	1.1229	−0.7266	1.1890	−1.1322	1.2547	−1.5377
−0.36	1.1303	−0.7512	1.2005	−1.1722	1.2704	−1.5936
−0.37	1.1379	−0.7762	1.2125	−1.2130	1.2866	−1.6509
−0.38	1.1458	−0.8017	1.2248	−1.2548	1.3034	−1.7096
−0.39	1.1540	−0.8277	1.2376	−1.2975	1.3208	−1.7698
−0.40	1.1624	−0.8542	1.2507	−1.3412	1.3388	−1.8317

TABLE II—Continued

x	μ=0		μ=1		μ=2	
	y	y'	y	y'	y	y'
−0.41	1.1710	−0.8811	1.2644	−1.3860	1.3574	−1.8954
−0.42	1.1800	−0.9082	1.2785	−1.4320	1.3767	−1.9610
−0.43	1.1892	−0.9364	1.2930	−1.4791	1.3966	−2.0285
−0.44	1.1987	−0.9652	1.3081	−1.5276	1.4173	−2.0982
−0.45	1.2085	−0.9947	1.3236	−1.5775	1.4386	−2.1703
−0.46	1.2186	−1.0250	1.3396	−1.6287	1.4607	−2.2447
−0.47	1.2291	−1.0559	1.3562	−1.6816	1.4835	−2.3218
−0.48	1.2398	−1.0877	1.3733	−1.7361	1.5071	−2.4017
−0.49	1.2508	−1.1203	1.3909	−1.7923	1.5316	−2.4847
−0.50	1.2622	−1.1537	1.4091	−1.8504	1.5568	−2.5708
−0.51	1.2739	−1.188	1.4279	−1.910	1.5830	−2.660
−0.52	1.2860	−1.223	1.4473	−1.973	1.6101	−2.754
−0.53	1.2984	−1.260	1.4674	−2.037	1.6381	−2.851
−0.54	1.3111	−1.297	1.4881	−2.104	1.6671	−2.953
−0.55	1.3243	−1.336	1.5094	−2.173	1.6971	−3.059
−0.56	1.3379	−1.376	1.5315	−2.245	1.7283	−3.170
−0.57	1.3518	−1.417	1.5544	−2.320	1.7606	−3.286
−0.58	1.3662	−1.459	1.5779	−2.398	1.7940	−3.409
−0.59	1.3810	−1.503	1.6023	−2.479	1.8288	−3.537
−0.60	1.3962	−1.548	1.6275	−2.564	1.8648	−3.672
−0.61	1.4119	−1.595	1.6536	−2.653	1.9022	−3.815
−0.62	1.4281	−1.643	1.6806	−2.745	1.9411	−3.965
−0.63	1.4448	−1.694	1.7085	−2.842	1.9815	−4.124
−0.64	1.4620	−1.746	1.7375	−2.944	2.0236	−4.292
−0.65	1.4797	−1.800	1.7674	−3.050	2.0674	−4.470
−0.66	1.4980	−1.856	1.7985	−3.162	2.1131	−4.659
−0.67	1.5169	−1.915	1.8307	−3.280	2.1606	−4.860
−0.68	1.5363	−1.976	1.8641	−3.403	2.2103	−5.074
−0.69	1.5564	−2.039	1.8988	−3.534	2.2622	−5.303
−0.70	1.5771	−2.105	1.9348	−3.671	2.3164	−5.547
−0.71	1.5985	−2.174	1.9722	−3.817	2.3732	−5.809
−0.72	1.6206	−2.246	2.0112	−3.971	2.4327	−6.089
−0.73	1.6434	−2.321	2.0517	−4.134	2.4950	−6.391
−0.74	1.6670	−2.399	2.0939	−4.307	2.5605	−6.715
−0.75	1.6914	−2.481	2.1379	−4.491	2.6294	−7.066
−0.76	1.7167	−2.568	2.1837	−4.687	2.7020	−7.445
−0.77	1.7428	−2.658	2.2316	−4.895	2.7784	−7.855
−0.78	1.7698	−2.753	2.2817	−5.118	2.8592	−8.302
−0.79	1.7979	−2.852	2.3340	−5.356	2.9446	−8.788
−0.80	1.8269	−2.957	2.3888	−5.610	3.0351	−9.319

TABLE II—Continued

	$\mu=0$			$\mu=1$			$\mu=2$	
x	y	y'	y	y'		y	y'	
−0. 81	1. 857	−3. 07	2. 446	−5. 88	3. 131	−9. 90
−0. 82	1. 888	−3. 18	2. 507	−6. 18	3. 233	−10. 5
−0. 83	1. 921	−3. 31	2. 570	−6. 49	3. 342	−11. 2
−0. 84	1. 954	−3. 43	2. 637	−6 84	3. 458	−12. 0
−0. 85	1. 989	−3. 57	2. 707	−7. 20	3. 583	−12. 9
−0. 86	2. 026	−3. 72	2. 781	−7. 60	3. 716	−13. 8
−0. 87	2. 064	−3. 87	2. 859	−8. 04	3. 860	−14 9
−0. 88	2. 103	−4. 03	2. 942	−8. 51	4. 015	−16. 1
−0. 89	2. 144	−4. 20	3. 029	−9. 03	4. 183	−17. 5
−0. 90	2. 187	−4. 39	3. 122	−9. 59	4. 365	−19. 0
−0. 91	2. 232	−4. 58	3. 221	−10. 2	4. 564	−20. 8
−0. 92	2. 279	−4. 79	3. 327	−10. 9	4. 782	−22. 8
−0. 93	2. 328	−5. 02	3. 439	−11. 6	5. 022	−25. 1
−0. 94	2. 379	−5. 25	3. 560	−12. 5	5. 286	−27. 8
−0. 95	2. 433	−5. 51	3. 689	−13. 4	5. 580	−31. 0
−0. 96	2. 490	−5. 79	3. 828	−14. 4	5. 908	−34. 8
−0. 97	2. 549	−6. 08	3. 979	−15. 6	6. 278	−39. 2
−0. 98	2. 611	−6. 40	4. 141	−16. 9	6. 696	−44. 7
−0. 99	2. 677	−6. 74	4. 317	−18. 4	7. 175	−51. 3
−1. 00	2. 746	−7. 12	4. 510	−20. 1	7. 727	−59. 5

TABLE II—Continued

x	$\mu=3$		$\mu=4$		$\mu=5$	
	y	y'	y	y'	y	y'
0. 00	1. 0000	0. 0000	1. 0000	0. 0000	1. 0000	0. 0000
0. 01	1. 0003	0. 0501	1. 0003	0. 0601	1. 0004	0. 0710
0. 02	1. 0010	0. 1002	1. 0012	0. 1202	1. 0014	0. 1403
0. 03	1. 0023	0. 1506	1. 0027	0. 1806	1. 0032	0. 2106
0. 04	1. 0040	0. 2011	1. 0048	0. 2412	1. 0056	0. 2813
0. 05	1. 0063	0. 2519	1. 0075	0. 3020	1. 0088	0. 3521
0. 06	1. 0091	0. 3029	1. 0109	0. 3631	1. 0127	0. 4233
0. 07	1. 0123	0. 3542	1. 0148	0. 4246	1. 0172	0. 4949
0. 08	1. 0161	0. 4058	1. 0193	0. 4864	1. 0226	0. 5669
0. 09	1. 0205	0. 4578	1. 0245	0. 5486	1. 0286	0. 6393
0. 10	1. 0253	0. 5102	1. 0303	0. 6112	1. 0353	0. 7123
0. 11	1. 0307	0. 5630	1. 0367	0. 6744	1.·0428	0. 7858
0. 12	1. 0366	0. 6163	1. 0438	0. 7381	1. 0511	0. 8600
0. 13	1. 0430	0. 6700	1. 0515	0. 8024	1. 0600	0. 9348
0. 14	1. 0500	0. 7244	1. 0599	0. 8674	1. 0698	1. 0104
0. 15	1. 0575	0. 7793	1. 0689	0. 9330	1. 0802	1. 0868
0. 16	1. 0655	0. 8349	1. 0785	0. 9994	1. 0915	1. 1641
0. 17	1. 0742	0. 8911	1. 0889	1. 0667	1. 1035	1. 2423
0. 18	1. 0834	0. 9481	1. 0999	1. 1348	1. 1164	1. 3216
0. 19	1. 0931	1. 0059	1. 1116	1. 2039	1. 1300	1. 4020
0. 20	1. 1035	1. 0645	1. 1239	1. 2740	1. 1444	1. 4836
0. 21	1. 1144	1. 1241	1. 1370	1. 3452	1. 1596	1. 5666
0. 22	1. 1260	1. 1846	1. 1508	1. 4176	1. 1757	1. 6509
0. 23	1. 1381	1. 2461	1. 1654	1. 4912	1. 1927	1. 7368
0. 24	1. 1509	1. 3088	1. 1807	1. 5663	1. 2105	1. 8243
0. 25	1. 1643	1. 3727	1. 1967	1. 6428	1. 2292	1. 9136
0. 26	1. 1784	1. 4378	1. 2135	1. 7209	1. 2488	2. 0048
0. 27	1. 1931	1. 5043	1. 2311	1. 8006	1. 2693	2. 0980
0. 28	1. 2085	1. 5722	1. 2496	1. 8822	1. 2907	2. 1935
0. 29	1. 2245	1. 6417	1. 2688	1. 9657	1. 3131	2. 2913
0. 30	1. 2413	1. 7128	1. 2889	2. 0513	1. 3366	2. 3917
0. 31	1. 2588	1. 7857	1. 3098	2. 1391	1. 3610	2. 4949
0. 32	1. 2770	1. 8604	1. 3317	2. 2294	1. 3865	2. 6011
0. 33	1. 2960	1. 9372	1. 3544	2. 3222	1. 4130	2. 7105
0. 34	1. 3158	2. 0161	1. 3781	2. 4178	1. 4407	2. 8234
0. 35	1. 3363	2. 0973	1. 4028	2. 5163	1. 4695	2. 9400
0. 36	1. 3577	2. 1809	1. 4285	2. 6181	1. 4995	3. 0607
0. 37	1. 3800	2. 2672	1. 4552	2. 7233	1. 5307	3. 1858
0. 38	1. 4031	2. 3563	1. 4829	2. 8322	1. 5632	3. 3156
0. 39	1. 4271	2. 4484	1. 5118	2. 9451	1. 5971	3. 4506
0. 40	1. 4520	2. 5438	1. 5419	3. 0623	1. 6323	3. 5911

TABLE II—Continued

	$\mu=3$		$\mu=4$		$\mu=5$	
x	y	y'	y	y'	y	y'
0. 41	1. 4780	2. 6426	1. 5731	3. 1841	1. 6689	3. 7377
0. 42	1. 5049	2. 7451	1. 6056	3. 3110	1. 7070	3. 8909
0. 43	1. 5329	2. 8517	1. 6393	3. 4433	1. 7467	4. 0512
0. 44	1. 5620	2. 9625	1. 6744	3. 5815	1. 7881	4. 2193
0. 45	1. 5922	3. 0780	1. 7110	3. 7260	1. 8312	4. 3959
0. 46	1. 6235	3. 1984	1. 7490	3. 8774	1. 8760	4. 5817
0. 47	1. 6561	3. 3242	1. 7885	4. 0363	1. 9228	4. 7776
0. 48	1. 6900	3. 4558	1. 8297	4. 2033	1. 9716	4. 9845
0. 49	1. 7253	3. 5937	1. 8726	4. 3791	2. 0226	5. 2034
0. 50	1. 7619	3. 7383	1. 9173	4. 5646	2. 0757	5. 4356
0. 51	1. 8001	3. 890	1. 9640	4. 761	2. 1313	5. 682
0. 52	1. 8398	4. 050	2. 0126	4. 968	2. 1894	5. 945
0. 53	1. 8811	4. 219	2. 0634	5. 188	2. 2503	6. 226
0. 54	1. 9242	4. 397	2. 1164	5. 422	2. 3140	6. 525
0. 55	1. 9691	4. 585	2. 1719	5. 670	2. 3809	6. 847
0. 56	2. 0159	4. 784	2. 2299	5. 936	2. 4510	7. 192
0. 57	2. 0648	4. 995	2. 2906	6. 220	2. 5248	7. 564
0. 58	2. 1159	5. 220	2. 3543	6. 523	2. 6024	7. 966
0. 59	2. 1692	5. 459	2. 4212	6. 850	2. 6842	8. 401
0. 60	2. 2251	5. 714	2. 4914	7. 200	2. 7705	8. 872
0. 61	2. 2836	5. 987	2. 5653	7. 579	2. 8618	9. 386
0. 62	2. 3449	6. 279	2. 6431	7. 988	2. 9584	9. 947
0. 63	2. 4092	6. 593	2. 7251	8. 432	3. 0609	10. 56
0. 64	2. 4768	6. 930	2. 8118	8. 913	3. 1698	11. 23
0. 65	2. 5479	7. 293	2. 9036	9. 438	3. 2858	11. 98
0. 66	2. 6228	7. 686	3. 0008	10. 01	3. 4097	12. 80
0. 67	2. 7017	8. 111	3. 1040	10. 64	3. 5422	13. 71
0. 68	2. 7851	8. 572	3. 2138	11. 33	3. 6843	14. 73
0. 69	2. 8733	9. 074	3. 3309	12. 10	3. 8372	15. 87
0. 70	2. 9668	9. 622	3. 4560	12. 94	4. 0021	17. 15
0. 71	3. 0659	10. 22	3. 5900	13. 88	4. 1807	18. 60
0. 72	3. 1714	10. 88	3. 7339	14. 93	4. 3748	20. 24
0. 73	3. 2838	11. 61	3. 8890	16. 10	4. 5864	22. 12
0. 74	3. 4038	12. 41	4. 0565	17. 42	4. 8182	24. 29
0. 75	3. 5322	13. 30	4. 2380	18. 92	5. 0732	26. 79
0. 76	3. 6701	14. 29	4. 4356	20. 62	5. 3554	29. 71
0. 77	3. 8184	15. 40	4. 6513	22. 57	5. 6692	33. 15
0. 78	3. 9785	16. 64	4. 8880	24. 82	6. 0205	37. 23
0. 79	4. 1518	18. 05	5. 1489	27. 42	6. 4166	42. 13
0. 80	4. 3400	19. 64	5. 4380	30. 47	6. 8667	48. 09

TABLE II—Continued

	$\mu=3$		$\mu=4$		$\mu=5$	
x	y	y'	y	y'	y	y'
0. 81	4. 545	21. 5	5. 760	34. 1	7. 383	55. 4
0. 82	4. 770	23. 5	6. 121	38. 3	7. 981	64. 6
0. 83	5. 017	26. 0	6. 530	43. 5	8. 683	76. 2
0. 84	5. 291	28. 8	6. 995	49. 8	9. 517	91. 4
0. 85	5. 594	32. 1	7. 530	57. 5	10. 53	112.
0. 86	5. 934	36. 0	8. 151	67. 2	11. 77	139.
0. 87	6. 316	40. 6	8. 883	79. 7	13. 35	179.
0. 88	6. 750	46. 3	9. 757	95. 9		
0. 89	7. 246	53. 2	10. 82	118.		
0. 90	7. 820	61. 9	12. 14	148.		
0. 91	8. 491	72. 8	13. 82	191.		
0. 92	9. 286	86. 9				
0. 93	10. 24	106.				
0. 94	11. 42	131.				
0. 95	12. 90	167.				

TABLE II—Continued

x	$\mu=3$		$\mu=4$		$\mu=5$	
	y	y'	y	y'	y	y'
0. 00	1. 0000	0. 0000	1. 0000	0. 0000	1. 0000	0. 0000
−0. 01	1. 0003	−0. 0501	1. 0003	−0. 0601	1. 0004	−0. 0701
−0. 02	1. 0010	−0. 1000	1. 0012	−0. 1200	1. 0014	−0. 1401
−0. 03	1. 0023	−0. 1500	1. 0027	−0. 1800	1. 0032	−0. 2100
−0. 04	1. 0040	−0: 1999	1. 0048	−0. 2400	1. 0056	−0. 2800
−0. 05	1. 0062	−0. 2499	1. 0075	−0. 3000	1. 0088	−0. 3501
−0. 06	1. 0090	−0. 2999	1. 0108	−0. 3601	1. 0126	−0. 4203
−0. 07	1. 0122	−0. 3500	1. 0147	−0. 4203	1. 0172	−0. 4907
−0. 08	1. 0160	−0. 4002	1. 0192	−0. 4807	1. 0224	−0. 5612
−0. 09	1. 0202	−0. 4505	1. 0243	−0. 5412	1. 0284	−0. 6320
−0. 10	1. 0250	−0. 5010	1. 0300	−0. 6020	1. 0351	−0. 7031
−0. 11	1. 0303	−0. 5517	1. 0364	−0. 6631	1. 0424	−0. 7745
−0. 12	1. 0360	−0. 6027	1. 0433	−0. 7245	1. 0505	−0. 8463
−0. 13	1. 0423	−0. 6540	1. 0508	−0. 7863	1. 0594	−0. 9186
−0. 14	1. 0491	−0. 7055	1. 0590	−0. 8484	1. 0689	−0. 9914
−0. 15	1. 0564	−0. 7574	1. 0678	−0. 9110	1. 0792	−1. 0647
−0. 16	1. 0643	−0. 8097	1. 0772	−0. 9742	1. 0902	−1. 1386
−0. 17	1. 0726	−0. 8625	1. 0873	−1. 0378	1. 1020	−1. 2133
−0. 18	1. 0815	−0. 9157	1. 0980	−1. 1021	1. 1145	−1. 2887
−0. 19	1. 0909	−0. 9694	1. 1093	−1. 1671	1. 1278	−1. 3649
−0. 20	1. 1009	−1. 0237	1. 1213	−1. 2327	1. 1418	−1. 4420
−0. 21	1. 1114	−1. 0786	1. 1340	−1. 2992	1. 1566	−1. 5201
−0. 22	1. 1225	−1. 1342	1. 1473	−1. 3666	1. 1722	−1. 5993
−0. 23	1. 1341	−1. 1905	1. 1613	−1. 4348	1. 1886	−1. 6796
−0. 24	1. 1463	−1. 2476	1. 1760	−1. 5041	1. 2058	−1. 7613
−0. 25	1. 1591	−1. 3055	1. 1914	−1. 5745	1. 2238	−1. 8443
−0. 26	1. 1724	−1. 3643	1. 2075	−1. 6461	1. 2427	−1. 9287
−0. 27	1. 1864	−1. 4241	1. 2243	−1. 7189	1. 2624	−2. 0148
−0. 28	1. 2009	−1. 4850	1. 2419	−1. 7932	1. 2830	−2. 1027
−0. 29	1. 2161	−1. 5469	1. 2602	−1. 8689	1. 3045	−2. 1924
−0. 30	1. 2318	−1. 6101	1. 2793	−1. 9462	1. 3268	−2. 2842
−0. 31	1. 2483	−1. 6746	1. 2991	−2. 0253	1. 3501	−2. 3782
−0. •32	1. 2653	−1. 7405	1. 3198	−2. 1062	1. 3744	−2. 4746
−0. 33	1. 2831	−1. 8078	1. 3413	−2. 1891	1. 3996	−2. 5736
−0. 34	1. 3015	−1. 8768	1. 3636	−2. 2742	1. 4259	−2. 6754
−0. 35	1. 3206	−1. 9475	1. 3868	−2. 3616	1. 4532	−2. 7803
−0. 36	1. 3405	−2. 0200	1. 4108	−2. 4515	1. 4815	−2. 8884
−0. 37	1. 3610	−2. 0945	1. 4358	−2. 5441	1. 5109	−3. 0000
−0. 38	1. 3823	−2. 1711	1. 4617	−2. 6396	1. 5415	−3. 1156
−0. 39	1. 4044	−2. 2500	1. 4886	−2. 7383	1. 5733	−3. 2353
−0. 40	1. 4274	−2. 3313	1. 5165	−2. 8403	1. 6062	−3. 3595

TABLE II—Continued

	$\mu=3$		$\mu=4$		$\mu=5$	
x	y	y'	y	y'	y	y'
-0.41	1. 4511	-2.4152	1. 5454	-2.9460	1. 6405	-3.4886
-0.42	1. 4757	-2.5019	1. 5754	-3.0557	1. 6760	-3.6231
-0.43	1. 5011	-2.5917	1. 6066	-3.1696	1. 7130	-3.7634
-0.44	1. 5275	-2.6847	1. 6388	-3.2882	1. 7513	-3.9099
-0.45	1. 5548	-2.7812	1. 6723	-3.4117	1. 7912	-4.0634
-0.46	1. 5831	-2.8815	1. 7071	-3.5407	1. 8326	-4.2243
-0.47	1. 6125	-2.9858	1. 7432	-3.6755	1. 8757	-4.3933
-0.48	1. 6429	-3.0944	1. 7806	-3.8167	1. 9205	-4.5712
-0.49	1. 6744	-3.2078	1. 8195	-3.9648	1. 9671	-4.7589
-0.50	1. 7070	-3.3263	1. 8599	-4.1204	2. 0157	-4.9571
-0.51	1. 7409	-3.450	1. 9020	-4.284	2. 0663	-5.167
-0.52	1. 7761	-3.580	1. 9457	-4.457	2. 1191	-5.390
-0.53	1. 8125	-3.716	1. 9911	-4.639	2. 1742	-5.627
-0.54	1. 8504	-3.860	2. 0385	-4.832	2. 2317	-5.879
-0.55	1. 8898	-4.010	2. 0878	-5.037	2. 2918	-6.148
-0.56	1. 9307	-4.169	2. 1393	-5.254	2. 3547	-6.436
-0.57	1. 9732	-4.337	2. 1929	-5.485	2. 4206	-6.745
-0.58	2. 0174	-4.515	2. 2490	-5.732	2. 4897	-7.077
-0.59	2. 0635	-4.703	2. 3076	-5.995	2. 5622	-7.435
-0.60	2. 1115	-4.903	2. 3690	-6.277	2. 6385	-7.822
-0.61	2. 1616	-5.115	2. 4332	-6.580	2. 7187	-8.240
-0.62	2. 2139	-5.341	2. 5006	-6.905	2. 8034	-8.694
-0.63	2. 2685	-5.583	2. 5714	-7.256	2. 8927	-9.188
-0.64	2. 3256	-5.841	2. 6458	-7.634	2. 9873	-9.728
-0.65	2. 3854	-6.117	2. 7242	-8.044	3. 0875	-10.32
-0.66	2. 4480	-6.414	2. 8068	-8.489	3. 1939	-10.97
-0.67	2. 5137	-6.733	2. 8941	-8.974	3. 3071	-11.68
-0.68	2. 5827	-7.077	2. 9865	-9.502	3. 4278	-12.47
-0.69	2. 6553	-7.448	3. 0843	-10.08	3. 5568	-13.35
-0.70	2. 7318	-7.850	3. 1883	-10.72	3. 6952	-14.33
-0.71	2. 8124	-8.287	3. 2989	-11.42	3. 8438	-15.43
-0.72	2. 8977	-8.762	3. 4168	-12.19	4. 0041	-16.66
-0.73	2. 9878	-9.281	3. 5429	-13.05	4. 1775	-18.04
-0.74	3. 0834	-9.848	3. 6781	-14.00	4. 3656	-19.62
-0.75	3. 1850	-10.47	3. 8234	-15.07	4. 5706	-21.42
-0.76	3. 2930	-11.16	3. 9799	-16.27	4. 7949	-23.49
-0.77	3. 4083	-11.91	4. 1492	-17.62	5. 0415	-25.88
-0.78	3. 5316	-12.75	4. 3329	-19.15	5. 3138	-28.66
-0.79	3. 6637	-13.68	4. 5330	-20.90	5. 6163	-31.93
-0.80	3. 8056	-14.73	4. 7517	-22.90	5. 9545	-35.80

TABLE II—Continued

x	$\mu=3$		$\mu=4$		$\mu=5$	
	y	y'	y	y'	y	y'
−0. 81	3. 959	−15. 9	4. 992	−25. 2	6. 335	−40. 4
−0. 82	4. 124	−17. 2	5. 257	−27. 9	6. 766	−46. 0
−0. 83	4. 303	−18. 7	5. 552	−31. 1	7. 260	−52. 9
−0. 84	4. 499	−20. 4	5. 880	−34. 8	7. 831	−61. 5
−0. 85	4. 712	−22. 3	6. 249	−39. 2	8. 497	−72. 3
−0. 86	4. 947	−24. 6	6. 668	−44. 6	9. 288	−86. 3
−0. 87	5. 205	−27. 2	7. 145	−51. 1	10. 24	−105.
−0. 88	5. 492	−30. 2	7. 696	−59. 3	11. 41	−130.
−0. 89	5. 811	−33. 8	8. 338	−69. 5	12. 87	−165.
−0. 90	6. 170	−38. 1	9. 096	−82. 7	14. 77	−218.
−0. 91	6. 576	−43. 2	10. 00	−100.		
−0. 92	7. 039	−49. 5	11. 12	−123.		
−0. 93	7. 571	−57. 2	12. 50	−156.		
−0. 94	8. 190	−67. 0	14. 29	−204.		
−0. 95	8. 919	−79. 4				
−0. 96	9. 790	−95. 6				
−0. 97	10. 85	−117.				
−0. 98	12. 17	−148.				
−0. 99	13. 85	−191.				

TABLE III

The values tabulated are approximations of $y(x)$ and $y'(x)$, where $y(x)$ is the solution of Van der Pol's equation

$$\frac{d^2y}{dx^2} - \epsilon\,(1-y^2)\,\frac{dy}{dx} + y = 0,$$

for the initial conditions: $y(0)=2$, $y'(0)=0$ and for $\epsilon=0.5$, 1, 2, 3, 4, 5.

x	$\epsilon=0.5$		$\epsilon=1.0$		$\epsilon=2.0$		x
	y	y'	y	y'	y	y'	
0. 0	2. 00	0. 00	2. 00	0. 00	2. 00	0. 00	0. 0
0. 1	1. 99	−0. 19	1. 99	−0. 17	1. 99	−0. 15	0. 1
0. 2	1. 96	−0. 34	1. 97	−0. 30	1. 97	−0. 23	0. 2
0. 3	1. 92	−0. 48	1. 93	−0. 40	1. 95	−0. 28	0. 3
0. 4	1. 87	−0. 60	1. 89	−0. 47	1. 92	−0. 31	0. 4
0. 5	1. 80	−0. 71	1. 84	−0. 53	1. 88	−0. 33	0. 5
0. 6	1. 73	−0. 80	1. 78	−0. 59	1. 85	−0. 35	0. 6
0. 7	1. 64	−0. 89	1. 72	−0. 64	1. 81	−0. 36	0. 7
0. 8	1. 55	−0. 98	1. 65	−0. 68	1. 78	−0. 38	0. 8
0. 9	1. 45	−1. 07	1. 58	−0. 73	1. 74	−0. 39	0. 9
1. 0	1. 33	−1. 15	1. 51	−0. 78	1. 70	−0. 41	1. 0
1. 1	1. 21	−1. 24	1. 43	−0. 83	1. 66	−0. 43	1. 1
1. 2	1. 09	−1. 34	1. 34	−0. 89	1. 61	−0. 44	1. 2
1. 3	0. 95	−1. 44	1. 25	−0. 96	1. 57	−0. 46	1. 3
1. 4	0. 80	−1. 54	1. 15	−1. 04	1. 52	−0. 49	1. 4
1. 5	0. 64	−1. 65	1. 04	−1. 12	1. 47	−0. 51	1. 5
1. 6	0. 47	−1. 77	0. 92	−1. 23	1. 42	−0. 54	1. 6
1. 7	0. 28	−1. 88	0. 80	−1. 35	1. 36	−0. 58	1. 7
1. 8	0. 09	−2. 00	0. 65	−1. 49	1. 30	−0. 62	1. 8
1. 9	−0. 11	−2. 10	0. 50	−1. 65	1. 24	−0. 67	1. 9
2. 0	−0. 33	−2. 18	0. 32	−1. 83	1. 17	−0. 73	2. 0
2. 1	−0. 55	−2. 22	0. 13	−2. 04	1. 09	−0. 80	2. 1
2. 2	−0. 77	−2. 22	−0. 08	−2. 25	1. 01	−0. 88	2. 2
2. 3	−0. 99	−2. 15	−0. 32	−2. 46	0. 91	−0. 99	2. 3
2. 4	−1. 20	−2. 02	−0. 58	−2. 62	0. 81	−1. 13	2. 4
2. 5	−1. 39	−1. 83	−0. 84	−2. 68	0. 69	−1. 32	2. 5
2. 6	−1. 56	−1. 58	−1. 11	−2. 59	0. 54	−1. 56	2. 6
2. 7	−1. 71	−1. 29	−1. 35	−2. 34	0. 37	−1. 87	2. 7
2. 8	−1. 82	−1. 00	−1. 57	−1. 95	0. 17	−2. 28	2. 8
2. 9	−1. 91	−0. 71	−1. 74	−1. 48	−0. 09	−2. 79	2. 9

TABLE III—Continued

x	ε=0.5		ε=1.0		ε=2.0		x
	y	y′	y	y′	y	y′	
3. 0	−1. 96	−0. 43	−1. 87	−1. 02	−0. 39	−3. 34.	3. 0
3. 1	−1. 99	−0. 19	−1. 95	−0. 62	−0. 75	−3. 75	3. 1
3. 2	−2. 00	0. 02	−1. 99	−0. 29	−1. 13	−3. 74	3. 2
3. 3	−1. 99	0. 20	−2. 00	−0. 04	−1. 48	−3. 11	3. 3
3. 4	−1. 96	0. 36	−2. 00	0. 14	−1. 74	−2. 11	3. 4
3. 5	−1. 92	0. 50	−1. 98	0. 28	−1. 90	−1. 18	3. 5
3. 6	−1. 86	0. 61	−1. 95	0. 38	−1. 99	−0. 52	3. 6
3. 7	−1. 80	0. 72	−1. 91	0. 46	−2. 00	−0. 14	3. 7
3. 8	−1. 72	0. 81	−1. 86	0. 52	−2. 00	0. 08	3. 8
3. 9	−1. 63	0. 90	−1. 80	0. 58	−2. 00	0. 19	3. 9
4. 0	−1. 54	0. 99	−1. 74	0. 62	−1. 98	0. 26	4. 0
4. 1	−1. 44	1. 08	−1. 68	0. 67	−1. 95	0. 29	4. 1
4. 2	−1. 32	1. 16	−1. 61	0. 72	−1. 92	0. 32	4. 2
4. 3	−1. 20	1. 25	−1. 53	0. 77	−1. 89	0. 34	4. 3
4. 4	−1. 07	1. 35	−1. 45	0. 82	−1. 86	0. 35	4. 4
4. 5	−0. 93	1. 45	−1. 37	0. 87	−1. 82	0. 36	4. 5
4. 6	−0. 78	1. 55	−1. 28	0. 94	−1. 78	0. 38	4. 6
4. 7	−0. 62	1. 67	−1. 18	1. 01	−1. 74	0. 39	4. 7
4. 8	−0. 45	1. 78	−1. 08	1. 10	−1. 70	0. 41	4. 8
4. 9	−0. 27	1. 90	−0. 96	1. 19	−1. 66	0. 42	4. 9
5. 0	−0. 07	2. 01	−0. 84	1. 31	−1. 62	0. 44	5. 0
5. 1	0. 13	2. 11	−0. 70	1. 44	−1. 57	0. 46	5. 1
5. 2	0. 35	2. 18	−0. 55	1. 59	−1. 53	0. 48	5. 2
5. 3	0. 57	2. 22	−0. 38	1. 77	−1. 48	0. 51	5. 3
5. 4	0. 79	2. 21	−0. 19	1. 97	−1. 43	0. 54	5. 4
5. 5	1. 01	2. 14	0. 01	2. 19	−1. 37	0. 57	5. 5
5. 6	1. 22	2. 01	0. 24	2. 40	−1. 31	0. 61	5. 6
5. 7	1. 41	1. 81	0. 49	2. 58	−1. 25	0. 66	5. 7
5. 8	1. 58	1. 55	0. 75	2. 67	−1. 18	0. 72	5. 8
5. 9	1. 72	1. 27	1. 02	2. 64	−1. 10	0. 78	5. 9
6. 0	1. 83	0. 97	1. 28	2. 44	−1. 02	0. 87	6. 0
6. 1	1. 91	0. 68	1. 51	2. 08	−0. 93	0. 98	6. 1
6. 2	1. 97	0. 41	1. 69	1. 63	−0. 83	1. 11	6. 2
6. 3	2. 00	0. 17	1. 83	1. 16	−0. 71	1. 29	6. 3
6. 4	2. 00	−0. 04	1. 93	0. 73	−0. 57	1. 52	6. 4
6. 5	1. 99	−0. 22	1. 98	0. 38	−0. 40	1. 82	6. 5
6. 6	1. 96	−0. 37	2. 00	0. 11	−0. 20	2. 21	6. 6
6. 7	1. 91	−0. 51	2. 00	−0. 09	0. 05	2. 71	6. 7
6. 8	1. 86	−0. 62	1. 99	−0. 24	0. 34	3. 26	6. 8
6. 9	1. 79	−0. 73	1. 96	−0. 35	0. 69	3. 71	6. 9

TABLE III—Continued

x	$\epsilon=0.5$		$\epsilon=1.0$		$\epsilon=2.0$		x
	y	y'	y	y'	y	y'	
7. 0	1. 71	-0.82	1. 92	-0.44	1. 07	3. 78	7. 0
7. 1	1. 63	-0.91	1. 87	-0.50	1. 43	3. 24	7. 1
7. 2	1. 53	-1.00	1. 82	-0.56	1. 71	2. 28	7. 2
7. 3	1. 43	-1.09	1. 76	-0.61	1. 88	1. 30	7. 3
7. 4	1. 31	-1.17	1. 70	-0.66	1. 98	0. 60	7. 4
7. 5	1. 19	-1.26	1. 63	-0.70	2. 00	0. 18	7. 5
7. 6	1. 06	-1.36	1. 56	-0.75	2. 00	-0.05	7. 6
7. 7	0. 92	-1.46	1. 48	-0.80	2. 00	-0.18	7. 7
7. 8	0. 77	-1.56	1. 40	-0.86	1. 99	-0.25	7. 8
7. 9	0. 61	-1.68	1. 31	-0.92	1. 96	-0.29	7. 9
8. 0	0. 43	-1.79	1. 21	-0.99	1. 93	-0.32	8. 0
8. 1	0. 25	-1.91	1. 11	-1.07	1. 90	-0.33	8. 1
8. 2	0. 05	-2.02	1. 00	-1.16	1. 86	-0.35	8. 2
8. 3	-0.16	-2.12	0. 88	-1.27	1. 83	-0.36	8. 3
8. 4	-0.37	-2.19	0. 74	-1.40	1. 79	-0.38	8. 4
8. 5	-0.59	-2.23	0. 60	-1.54	1. 75	-0.39	8. 5
8. 6	-0.81	-2.21	0. 44	-1.71	1. 71	-0.40	8. 6
8. 7	-1.03	-2.13	0. 25	-1.91	1. 67	-0.42	8. 7
8. 8	-1.24	-1.99	0. 05	-2.12	1. 63	-0.44	8. 8
8. 9	-1.43	-1.78	-0.17	-2.33	1. 58	-0.46	8. 9
9. 0	-1.59	-1.53	-0.41	-2.53	1. 53	-0.48	9. 0
9. 1	-1.73	-1.24	-0.67	-2.65	1. 49	-0.50	9. 1
9. 2	-1.84	-0.94	-0.94	-2.67	1. 43	-0.53	9. 2
9. 3	-1.92	-0.65	-1.20	-2.52	1. 38	-0.57	9. 3
9. 4	-1.97	-0.38	-1.44	-2.21	1. 32	-0.61	9. 4
9. 5	-2.00	-0.15	-1.64	-1.78	1. 26	-0.65	9. 5
9. 6	-2.00	0. 06	-1.79	-1.31	1. 19	-0.71	9. 6
9. 7	-1.99	0. 24	-1.90	-0.86	1. 12	-0.77	9. 7
9. 8	-1.96	0. 39	-1.97	-0.49	1. 03	-0.86	9. 8
9. 9	-1.91	0. 52	-2.00	-0.19	0. 94	-0.96	9. 9
10. 0	-1.85	0. 63	-2.00	-0.33	0. 84	-1.09	10. 0

Table III—Continued

x	$\epsilon=3.0$		$\epsilon=4.0$		$\epsilon=5.0$		x
	y	y'	y	y'	y	y'	
0. 0	2. 00	0. 00	2. 00	0. 00	2. 00	0. 00	0. 0
0. 1	1. 99	−0. 13	1. 99	−0. 12	1. 99	−0. 10	0. 1
0. 2	1. 98	−0. 19	1. 98	−0. 15	1. 98	−0. 13	0. 2
0. 3	1. 96	−0. 21	1. 96	−0. 17	1. 97	−0. 13	0. 3
0. 4	1. 93	−0. 22	1. 95	−0. 17	1. 95	−0. 14	0. 4
0. 5	1. 91	−0. 23	1. 93	−0. 17	1. 94	−0. 14	0. 5
0. 6	1. 89	−0. 24	1. 91	−0. 18	1. 93	−0. 14	0. 6
0. 7	1. 86	−0. 24	1. 89	−0. 18	1. 91	−0. 14	0. 7
0. 8	1. 84	−0. 25	1. 88	−0. 18	1. 90	−0. 14	0. 8
0. 9	1. 81	−0. 26	1. 86	−0. 19	1. 88	−0. 15	0. 9
1. 0	1. 79	−0. 26	1. 84	−0. 19	1. 87	−0. 15	1. 0
1. 1	1. 76	−0. 27	1. 82	−0. 19	1. 85	−0. 15	1. 1
1. 2	1. 73	−0. 28	1. 80	−0. 20	1. 84	−0. 15	1. 2
1. 3	1. 71	−0. 28	1. 78	−0. 20	1. 82	−0. 15	1. 3
1. 4	1. 68	−0. 29	1. 76	−0. 20	1. 81	−0. 16	1. 4
1. 5	1. 65	−0. 30	1. 74	−0. 21	1. 79	−0. 16	1. 5
1. 6	1. 62	−0. 31	1. 72	−0. 21	1. 78	−0. 16	1. 6
1. 7	1. 59	−0. 32	1. 70	−0. 22	1. 76	−0. 17	1. 7
1. 8	1. 55	−0. 34	1. 67	−0. 22	1. 74	−0. 17	1. 8
1. 9	1. 52	−0. 35	1. 65	−0. 23	1. 73	−0. 17	1. 9
2. 0	1. 48	−0. 37	1. 63	−0. 24	1. 71	−0. 17	2. 0
2. 1	1. 44	−0. 38	1. 60	−0. 24	1. 69	−0. 18	2. 1
2. 2	1. 41	−0. 40	1. 58	−0. 25	1. 67	−0. 18	2. 2
2. 3	1. 37	−0. 43	1. 55	−0. 26	1. 66	−0. 19	2. 3
2. 4	1. 32	−0. 46	1. 53	−0. 27	1. 64	−0. 19	2. 4
2. 5	1. 27	−0. 49	1. 50	−0. 28	1. 62	−0. 19	2. 5
2. 6	1. 22	−0. 53	1. 47	−0. 29	1. 60	−0. 20	2. 6
2. 7	1. 17	−0. 58	1. 44	−0. 30	1. 58	−0. 20	2. 7
2. 8	1. 11	−0. 64	1. 41	−0. 32	1. 56	−0. 21	2. 8
2. 9	1. 04.	−0. 71	1. 38	−0. 33	1. 54	−0. 22	2. 9
3. 0	0. 96	−0. 81	1. 34	−0. 35	1. 51	−0. 22	3. 0
3. 1	0. 88	−0. 95	1. 31	−0. 38	1. 49	−0. 23	3. 1
3. 2	0. 77	−1. 13	1. 27	−0. 40	1. 47	−0. 24	3. 2
3. 3	0. 65	−1. 38	1. 23	−0. 43	1. 44	−0. 25	3. 3
3. 4	0. 49	−1. 75	1. 18	−0. 47	1. 42	−0. 26	3. 4
3. 5	0. 29	−2. 30	1. 13	−0. 52	1. 39	−0. 27	3. 5
3. 6	0. 24	−3. 10	1. 08	−0. 58	1. 36	−0. 28	3. 6
3. 7	−0. 34	−4. 12	1. 01	−0. 66	1. 33	−0. 30	3. 7
3. 8	−0. 80	−4. 98	0. 94	−0. 77	1. 30	−0. 32	3. 8
3. 9	−1. 29	−4. 71	0. 86	−0. 93	1. 27	−0. 34	3. 9

TABLE III—Continued

x	$\epsilon=3.0$		$\epsilon=4.0$		$\epsilon=5.0$		x
	y	y'	y	y'	y	y'	
4. 0	-1.69	-3.08	0. 76	-1.15	1. 24	-0.36	4. 0
4. 1	-1.91	-1.42	0. 62	-1.50	1. 20	-0.39	4. 1
4. 2	-2.00	-0.47	0. 45	-2.05	1. 16	-0.43	4. 2
4. 3	-2.00	-0.57	0. 20	-2.95	1. 11	-0.48	4. 3
4. 4	-2.00	0. 11	-0.16	-4.39	1. 06	-0.54	4. 4
4. 5	-2.00	0. 18	-0.68	-6.03	1. 00	-0.63	4. 5
4. 6	-1.98	0. 20	-1.30	-5.88	0. 93	-0.75	4. 6
4. 7	-1.96	0. 22	-1.77	-3.18	0. 85	-0.92	4. 7
4. 8	-1.94	0. 23	-1.96	-1.03	0. 74	-1.19	4. 8
4. 9	-1.92	0. 23	-2.00	-0.20	0. 60	-1.64	4. 9
5. 0	-1.90	0. 24	-2.00	0. 06	0. 40	-2.44	5. 0
5. 1	-1.87	0. 24	-2.00	0. 13	0. 09	-3.90	5. 1
5. 2	-1.85	0. 25	-2.00	0. 16	-0.41	-6.29	5. 2
5. 3	-1.82	0. 25	-1.98	0. 16	-1.13	-7.53	5. 3
5. 4	-1.80	0. 26	-1.96	0. 17	-1.74	-4.07	5. 4
5. 5	-1.77	0. 27	-1.95	0. 17	-1.97	-1.04	5. 5
5. 6	-1.74	0. 27	-1.93	0. 17	-2.00	-0.13	5. 6
5. 7	-1.72	0. 28	-1.91	0. 18	-2.00	0. 08	5. 7
5. 8	-1.69	0. 29	-1.89	0. 18	-2.00	0. 12	5. 8
5. 9	-1.66	0. 30	-1.88	0. 18	-2.00	0. 13	5. 9
6. 0	-1.63	0. 31	-1.86	0. 19	-1.98	0. 13	6. 0
6. 1	-1.60	0. 32	-1.84	0. 19	-1.97	0. 14	6. 1
6. 2	-1.56	0. 33	-1.82	0. 19	-1.96	0. 14	6. 2
6. 3	-1.53	0. 34	-1.80	0. 20	-1.94	0. 14	6. 3
6. 4	-1.49	0. 36	-1.78	0. 20	-1.93	0. 14	6. 4
6. 5	-1.46	0. 38	-1.76	0. 20	-1.91	0. 14	6. 5
6. 6	-1.42	0. 40	-1.74	0. 21	-1.90	0. 14	6. 6
6. 7	-1.38	0. 42	-1.72	0. 21	-1.89	0. 15	6. 7
6. 8	-1.33	0. 45	-1.70	0. 22	-1.87	0. 15	6. 8
6. 9	-1.29	0. 48	-1.67	0. 22	-1.86	0. 15	6. 9
7. 0	-1.24	0. 52	-1.65	0. 23	-1.84	0. 15	7. 0
7. 1	-1.18	0. 56	-1.63	0. 24	-1.82	0. 15	7. 1
7. 2	-1.13	0. 62	-1.60	0. 24	-1.81	0. 16	7. 2
7. 3	-1.06	0. 69	-1.58	0. 25	-1.79	0. 16	7. 3
7. 4	-0.99	0. 78	-1.55	0. 26	-1.78	0. 16	7. 4
7. 5	-0.90	0. 90	-1.53	0. 27	-1.76	0. 16	7. 5
7. 6	-0.81	1. 07	-1.50	0. 28	-1.74	0. 17	7. 6
7. 7	-0.69	1. 30	-1.47	0. 29	-1.73	0. 17	7. 7
7. 8	-0.54	1. 63	-1.44	0. 30	-1.71	0. 17	7. 8
7. 9	-0.36	2. 12	-1.41	0. 32	-1.69	0. 18	7. 9

TABLE III—Continued

x	$\epsilon=3.0$		$\epsilon=4.0$		$\epsilon=5.0$		x
	y	y'	y	y'	y	y'	
8. 0	−0. 11	2. 83	−1. 38	0. 33	−1. 67	0. 18	8. 0
8. 1	0. 22	3. 81	−1. 34	0. 35	−1. 65	0. 19	8. 1
8. 2	0. 65	4. 80	−1. 31	0. 37	−1. 64	0. 19	8. 2
8. 3	1. 15	4. 96	−1. 27	0. 40	−1. 62	0. 19	8. 3
8. 4	1. 59	3. 64	−1. 23	0. 43	−1. 60	0. 20	8. 4
8. 5	1. 86	1. 84	−1. 18	0. 47	−1. 58	0. 20	8. 5
8. 6	1. 98	0. 68	−1. 13	0. 52	−1. 56	0. 21	8. 6
8. 7	2. 00	0. 14	−1. 08	0. 58	−1. 54	0. 22	8. 7
8. 8	2. 00	−0. 08	−1. 02	0. 66	−1. 51	0. 22	8. 8
8. 9	2. 00	−0. 16	−0. 94	0. 77	−1. 49	0. 23	8. 9
9. 0	1. 99	−0. 20	−0. 86	0. 92	−1. 47	0. 24	9. 0
9. 1	1. 97	−0. 21	−0. 76	1. 15	−1. 44	0. 25	9. 1
9. 2	1. 95	−0. 22	−0. 63	1. 49	−1. 42	0. 26	9. 2
9. 3	1. 92	−0. 23	−0. 45	2. 03	−1. 39	0. 27	9. 3
9. 4	1. 90	−0. 24	−0. 21	2. 93	−1. 36	0. 28	9. 4
9. 5	1. 88	−0. 24	0. 15	4. 36	−1. 34	0. 30	9. 5
9. 6	1. 85	−0. 25	0. 67	6. 01	−1. 31	0. 32	9. 6
9. 7	1. 83	−0. 25	1. 29	5. 91	−1. 27	0. 34	9. 7
9. 8	1. 80	−0. 26	1. 76	3. 23	−1. 24	0. 36	9. 8
9. 9	1. 78	−0. 26	1. 96	1. 06	−1. 20	0. 39	9. 9
10. 0	1. 75	−0. 27	2. 00	0. 20	−1. 16	0. 43	10. 0
10. 1	1. 72	−0. 28	2. 00	−0. 06	−1. 11	0. 48	10. 1
10. 2	1. 70	−0. 29	2. 00	−0. 13	−1. 06	0. 54	10. 2
10. 3	1. 67	−0. 30	2. 00	−0. 16	−1. 00	0. 62	10. 3
10. 4	1. 64	−0. 31	1. 98	−0. 16	−0. 94	0. 74	10. 4
10. 5	1. 60	−0. 32	1. 96	−0. 17	−0. 85	0. 91	10. 5
10. 6	1. 57	−0. 33	1. 95	−0. 17	−0. 75	1. 17	10. 6
10. 7	1. 54	−0. 34	1. 93	−0. 17	−0. 61	1. 61	10. 7
10. 8	1. 50	−0. 36	1. 91	−0. 18	−0. 42	2. 37	10. 8
10. 9	1. 47	−0. 37	1. 89	−0. 18	−0. 12	3. 79	10. 9
11. 0	1. 43	−0. 39	1. 88	−0. 18	0. 37	6. 13	11. 0
11. 1	1. 39	−0. 41	1. 86	−0. 19	1. 08	7. 59	11. 1
11. 2	1. 35	−0. 44	1. 84	−0. 19	1. 71	4. 34	11. 2
11. 3	1. 30	−0. 47	1. 82	−0. 19	1. 96	1. 15	11. 3
11. 4	1. 25	−0. 50	1. 80	−0. 20	2. 00	0. 15	11. 4
11. 5	1. 20	−0. 55	1. 78	−0. 20	2. 00	−0. 07	11. 5
11. 6	1. 14	−0. 60	1. 76	−0. 20	2. 00	−0. 12	11. 6
11. 7	1. 09	−0. 67	1. 74	−0. 21	2. 00	−0. 13	11. 7
11. 8	1. 01	−0. 75	1. 72	−0. 21	1. 98	−0. 13	11. 8
11. 9	0. 93	−0. 86	1. 70	−0. 22	1. 97	−0. 14	11. 9
12. 0	0. 86	−1. 01	1. 67	−0. 22	1. 96	−0. 14	12. 0

TABLE IV

The values tabulated are approximations of $y(x)$ and $y'(x)$, where $y(x)$ is the solution of Volterra's equations:

$$y \frac{d^2y}{dx^2} = \left(\frac{dy}{dx}\right)^2 + (-y + y^2) \frac{dy}{dx} + 2(y^2 - y^3),$$

for the initial values:

$$y(0) = 1, \ y'(0) = -4.$$

x	y	y'	x	y	y'	x	y	y'
0. 00	1. 000	−4. 000	0. 30	0. 326	−1. 077	0. 60	0. 143	−0. 313
0. 01	0. 961	−3. 842	0. 31	0. 315	−1. 031	0. 61	0. 140	−0. 301
0. 02	0. 923	−3. 688	0. 32	0. 305	−0. 987	0. 62	0. 137	−0. 290
0. 03	0. 887	−3. 539	0. 33	0. 295	−0. 945	0. 63	0. 134	−0. 279
0. 04	0. 852	−3. 394	0. 34	0. 286	−0. 904	0. 64	0. 131	−0. 269
0. 05	0. 819	−3. 254	0. 35	0. 277	−0. 866	0. 65	0. 128	−0. 259
0. 06	0. 787	−3. 118	0. 36	0. 269	−0. 830	0. 66	0. 126	−0. 250
0. 07	0. 757	−2. 987	0. 37	0. 261	−0. 795	0. 67	0. 123	−0. 240
0. 08	0. 728	−2. 860	0. 38	0. 253	−0. 762	0. 68	0. 121	−0. 232
0. 09	0. 700	−2. 738	0. 39	0. 245	−0. 730	0. 69	0. 119	−0. 223
0. 10	0. 673	−2. 621	0. 40	0. 238	−0. 700	0. 70	0. 117	−0. 215
0. 11	0. 647	−2. 508	0. 41	0. 231	−0. 671	0. 71	0. 115	−0. 207
0. 12	0. 623	−2. 400	0. 42	0. 225	−0. 644	0. 72	0. 112	−0. 200
0. 13	0. 599	−2. 296	0. 43	0. 219	−0. 617	0. 73	0. 111	−0. 193
0. 14	0. 577	−2. 196	0. 44	0. 212	−0. 592	0. 74	0. 109	−0. 186
0. 15	0. 555	−2. 100	0. 45	0. 207	−0. 568	0. 75	0. 107	−0. 179
0. 16	0. 535	−2. 008	0. 46	0. 201	−0. 545	0. 76	0. 105	−0. 173
0. 17	0. 515	−1. 920	0. 47	0. 196	−0. 523	0. 77	0. 103	−0. 167
0. 18	0. 496	−1. 836	0. 48	0. 191	−0. 503	0. 78	0. 102	−0. 161
0. 19	0. 478	−1. 756	0. 49	0. 186	−0. 483	0. 79	0. 100	−0. 155
0. 20	0. 461	−1. 679	0. 50	0. 181	−0. 464	0. 80	0. 099	−0. 149
0. 21	0. 445	−1. 606	0. 51	0. 176	−0. 445	0. 81	0. 097	−0. 144
0. 22	0. 429	−1. 535	0. 52	0. 172	−0. 428	0. 82	0. 096	−0. 139
0. 23	0. 414	−1. 468	0. 53	0. 168	−0. 411	0. 83	0. 094	−0. 134
0. 24	0. 400	−1. 404	0. 54	0. 164	−0. 395	0. 84	0. 093	−0. 129
0. 25	0. 386	−1. 343	0. 55	0. 160	−0. 380	0. 85	0. 092	−0. 125
0. 26	0. 373	−1. 285	0. 56	0. 156	−0. 366	0. 86	0. 091	−0. 120
0. 27	0. 360	−1. 229	0. 57	0. 153	−0. 352	0. 87	0. 089	−0. 116
0. 28	0. 348	−1. 176	0. 58	0. 149	−0. 338	0. 88	0. 088	−0. 112
0. 29	0. 337	−1. 125	0. 59	0. 146	−0. 325	0. 89	0. 087	−0. 108

TABLE IV—Continued

x	y	y'	x	y	y'	x	y	y'
0. 90	0. 086	−0. 104	1. 30	0. 065	−0. 014	1. 70	0. 070	0. 033
0. 91	0. 085	−0. 100	1. 31	0. 065	−0. 013	1. 71	0. 070	0. 034
0. 92	0. 084	−0. 097	1. 32	0. 065	−0. 011	1. 72	0. 070	0. 035
0. 93	0. 083	−0. 093	1. 33	0. 065	−0. 010	1. 73	0. 071	0. 036
0. 94	0. 082	−0. 090	1. 34	0. 065	−0. 009	1. 74	0. 071	0. 038
0. 95	0. 081	−0. 087	1. 35	0. 065	−0. 007	1. 75	0. 071	0. 039
0. 96	0. 080	−0. 084	1. 36	0. 065	−0. 006	1. 76	0. 072	0. 040
0. 97	0. 080	−0. 081	1. 37	0. 065	−0. 005	1. 77	0. 072	0. 041
0. 98	0. 079	−0. 078	1. 38	0. 065	−0. 004	1. 78	0. 073	0. 042
0. 99	0. 078	−0. 075	1. 39	0. 065	−0. 002	1. 79	0. 073	0. 044
1. 00	0. 077	−0. 072	1. 40	0. 065	−0. 001	1. 80	0. 073	0. 045
1. 01	0. 077	−0. 069	1. 41	0. 065	0. 000	1. 81	0. 074	0. 046
1. 02	0. 076	−0. 066	1. 42	0. 065	0. 001	1. 82	0. 074	0. 047
1. 03	0. 075	−0. 064	1. 43	0. 065	0. 002	1. 83	0. 075	0. 049
1. 04	0. 075	−0. 061	1. 44	0. 065	0. 004	1. 84	0. 075	0. 050
1. 05	0. 074	−0. 059	1. 45	0. 065	0. 005	1. 85	0. 076	0. 051
1. 06	0. 074	−0. 057	1. 46	0. 065	0. 006	1. 86	0. 076	0. 052
1. 07	0. 073	−0. 054	1. 47	0. 065	0. 007	1. 87	0. 077	0. 054
1. 08	0. 072	−0. 052	1. 48	0. 065	0. 008	1. 88	0. 077	0. 055
1. 09	0. 072	−0. 050	1. 49	0. 065	0. 009	1. 89	0. 078	0. 056
1. 10	0. 071	−0. 048	1. 50	0. 065	0. 011	1. 90	0. 079	0. 058
1. 11	0. 071	−0. 046	1. 51	0. 065	0. 012	1. 91	0. 079	0. 059
1. 12	0. 070	−0. 044	1. 52	0. 065	0. 013	1. 92	0. 080	0. 060
1. 13	0. 070	−0. 042	1. 53	0. 066	0. 014	1. 93	0. 080	0. 062
1. 14	0. 070	−0. 040	1. 54	0. 066	0. 015	1. 94	0. 081	0. 063
1. 15	0. 069	−0. 038	1. 55	0. 066	0. 016	1. 95	0. 082	0. 064
1. 16	0. 069	−0. 036	1. 56	0. 066	0. 017	1. 96	0. 082	0. 066
1. 17	0. 069	−0. 034	1. 57	0. 066	0. 018	1. 97	0. 083	0. 067
1. 18	0. 068	−0. 033	1. 58	0. 066	0. 020	1. 98	0. 084	0. 069
1. 19	0. 068	−0. 031	1. 59	0. 067	0. 021	1. 99	0. 084	0. 070
1. 20	0. 068	−0. 029	1. 60	0. 067	0. 022	2. 00	0. 085	0. 072
1. 21	0. 067	−0. 028	1. 61	0. 067	0. 023	2. 01	0. 086	0. 073
1. 22	0. 067	−0. 026	1. 62	0. 067	0. 024	2. 02	0. 086	0. 075
1. 23	0. 067	−0. 024	1. 63	0. 067	0. 025	2. 03	0. 087	0. 076
1. 24	0. 067	−0. 023	1. 64	0. 068	0. 026	2. 04	0. 088	0. 078
1. 25	0. 066	−0. 021	1. 65	0. 068	0. 027	2. 05	0. 089	0. 079
1. 26	0. 066	−0. 020	1. 66	0. 068	0. 028	2. 06	0. 090	0. 081
1. 27	0. 066	−0. 018	1. 67	0. 069	0. 030	2. 07	0. 090	0. 083
1. 28	0. 066	−0. 017	1. 68	0. 069	0. 031	2. 08	0. 091	0. 084
1. 29	0. 066	−0. 016	1. 69	0. 069	0. 032	2. 09	0. 092	0. 086

TABLE IV—Continued

z	y	y'	z	y	y'	z	y	y'
2. 10	0. 093	0. 088	2. 50	0. 145	0. 183	2. 90	0. 251	0. 367
2. 11	0. 094	0. 090	2. 51	0. 147	0. 186	2. 91	0. 255	0. 374
2. 12	0. 095	0. 091	2. 52	0. 149	0. 189	2. 92	0. 258	0. 380
2. 13	0. 096	0. 093	2. 53	0. 151	0. 193	2. 93	0. 262	0. 387
2. 14	0. 097	0. 095	2. 54	0. 153	0. 196	2. 94	0. 266	0. 394
2. 15	0. 098	0. 097	2. 55	0. 155	0. 199	2. 95	0. 270	0. 400
2. 16	0. 099	0. 099	2. 56	0. 157	0. 203	2. 96	0. 274	0. 407
2. 17	0. 100	0. 100	2. 57	0. 159	0. 207	2. 97	0. 278	0. 415
2. 18	0. 101	0. 102	2. 58	0. 161	0. 210	2. 98	0. 282	0. 422
2. 19	0. 102	0. 104	2. 59	0. 163	0. 214	2. 99	0. 287	0. 429
2. 20	0. 103	0. 106	2. 60	0. 165	0. 218	3. 00	0. 291	0. 437
2. 21	0. 104	0. 108	2. 61	0. 167	0. 222	3. 01	0. 295	0. 444
2. 22	0. 105	0. 110	2. 62	0. 169	0. 225	3. 02	0. 300	0. 452
2. 23	0. 106	0. 112	2. 63	0. 172	0. 229	3. 03	0. 304	0. 460
2. 24	0. 107	0. 114	2. 64	0. 174	0. 233	3. 04	0. 309	0. 468
2. 25	0. 108	0. 117	2. 65	0. 176	0. 238	3. 05	0. 314	0. 476
2. 26	0. 109	0. 119	2. 66	0. 179	0. 242	3. 06	0. 319	0. 485
2. 27	0. 111	0. 121	2. 67	0. 181	0. 246	3. 07	0. 323	0. 493
2. 28	0. 112	0. 123	2. 68	0. 184	0. 250	3. 08	0. 328	0. 502
2. 29	0. 113	0. 125	2. 69	0. 186	0. 255	3. 09	0. 333	0. 511
2. 30	0. 114	0. 128	2. 70	0. 189	0. 259	3. 10	0. 339	0. 520
2. 31	0. 116	0. 130	2. 71	0. 191	0. 264	3. 11	0. 344	0. 529
2. 32	0. 117	0. 133	2. 72	0. 194	0. 268	3. 12	0. 349	0. 538
2. 33	0. 118	0. 135	2. 73	0. 197	0. 273	3. 13	0. 355	0. 547
2. 34	0. 120	0. 137	2. 74	0. 200	0. 278	3. 14	0. 360	0. 557
2. 35	0. 121	0. 140	2. 75	0. 202	0. 283	3. 15	0. 366	0. 567
2. 36	0. 122	0. 142	2. 76	0. 205	0. 288	3. 16	0. 371	0. 577
2. 37	0. 124	0. 145	2. 77	0. 208	0. 293	3. 17	0. 377	0. 587
2. 38	0. 125	0. 148	2. 78	0. 211	0. 298	3. 18	0. 383	0. 597
2. 39	0. 127	0. 150	2. 79	0. 214	0. 303	3. 19	0. 389	0. 607
2. 40	0. 128	0. 153	2. 80	0. 217	0. 309	3. 20	0. 395	0. 618
2. 41	0. 130	0. 156	2. 81	0. 220	0. 314	3. 21	0. 402	0. 629
2. 42	0. 131	0. 159	2. 82	0. 223	0. 319	3. 22	0. 408	0. 640
2. 43	0. 133	0. 161	2. 83	0. 227	0. 325	3. 23	0. 414	0. 651
2. 44	0. 135	0. 164	2. 84	0. 230	0. 331	3. 24	0. 421	0. 662
2. 45	0. 136	0. 167	2. 85	0. 233	0. 337	3. 25	0. 428	0. 674
2. 46	0. 138	0. 170	2. 86	0. 237	0. 342	3. 26	0. 434	0. 686
2. 47	0. 140	0. 173	2. 87	0. 240	0. 348	3. 27	0. 441	0. 698
2. 48	0. 141	0. 176	2. 88	0. 244	0. 355	3. 28	0. 448	0. 710
2. 49	0. 143	0. 179	2. 89	0. 247	0. 361	3. 29	0. 456	0. 722

TABLE IV—Continued

z	y	y'	z	y	y'	z	y	y'
3.30	0.463	0.735	3.70	0.885	1.452	4.10	1.700	2.726
3.31	0.470	0.747	3.71	0.899	1.477	4.11	1.728	2.766
3.32	0.478	0.760	3.72	0.914	1.501	4.12	1.756	2.805
3.33	0.485	0.774	3.73	0.930	1.527	4.13	1.784	2.845
3.34	0.493	0.787	3.74	0.945	1.552	4.14	1.812	2.884
3.35	0.501	0.801	3.75	0.961	1.578	4.15	1.841	2.924
3.36	0.509	0.815	3.76	0.976	1.604	4.16	1.871	2.964
3.37	0.517	0.829	3.77	0.993	1.631	4.17	1.901	3.005
3.38	0.526	0.843	3.78	1.009	1.658	4.18	1.931	3.045
3.39	0.534	0.858	3.79	1.026	1.685	4.19	1.962	3.086
3.40	0.543	0.873	3.80	1.043	1.713	4.20	1.993	3.126
3.41	0.552	0.888	3.81	1.060	1.741	4.21	2.024	3.167
3.42	0.561	0.903	3.82	1.078	1.770	4.22	2.056	3.207
3.43	0.570	0.919	3.83	1.095	1.799	4.23	2.088	3.247
3.44	0.579	0.935	3.84	1.114	1.828	4.24	2.121	3.288
3.45	0.589	0.951	3.85	1.132	1.858	4.25	2.154	3.328
3.46	0.598	0.967	3.86	1.151	1.888	4.26	2.188	3.368
3.47	0.608	0.984	3.87	1.170	1.918	4.27	2.221	3.408
3.48	0.618	1.001	3.88	1.189	1.949	4.28	2.256	3.447
3.49	0.628	1.018	3.89	1.209	1.981	4.29	2.290	3.486
3.50	0.638	1.036	3.90	1.229	2.013	4.30	2.325	3.525
3.51	0.649	1.054	3.91	1.249	2.045	4.31	2.361	3.563
3.52	0.659	1.072	3.92	1.270	2.077	4.32	2.397	3.601
3.53	0.670	1.090	3.93	1.291	2.110	4.33	2.433	3.638
3.54	0.681	1.109	3.94	1.312	2.143	4.34	2.469	3.675
3.55	0.692	1.128	3.95	1.333	2.177	4.35	2.506	3.710
3.56	0.704	1.147	3.96	1.355	2.211	4.36	2.544	3.745
3.57	0.715	1.167	3.97	1.378	2.246	4.37	2.581	3.779
3.58	0.727	1.187	3.98	1.400	2.281	4.38	2.619	3.812
3.59	0.739	1.207	3.99	1.423	2.316	4.39	2.658	3.844
3.60	0.751	1.228	4.00	1.447	2.352	4.40	2.696	3.875
3.61	0.763	1.249	4.01	1.470	2.388	4.41	2.735	3.904
3.62	0.776	1.270	4.02	1.494	2.424	4.42	2.774	3.932
3.63	0.789	1.292	4.03	1.519	2.461	4.43	2.814	3.958
3.64	0.802	1.313	4.04	1.544	2.498	4.44	2.853	3.982
3.65	0.815	1.336	4.05	1.569	2.535	4.45	2.893	4.005
3.66	0.829	1.358	4.06	1.594	2.573	4.46	2.933	4.026
3.67	0.842	1.381	4.07	1.620	2.611	4.47	2.974	4.044
3.68	0.856	1.405	4.08	1.646	2.649	4.48	3.014	4.061
3.69	0.870	1.428	4.09	1.673	2.688	4.49	3.055	4.074

TABLE IV—Continued

x	y	y'	x	y	y'	x	y	y'
4. 50	3. 096	4. 086	4. 90	4. 247	−0. 406	5. 30	2. 057	−7. 168
4. 51	3. 137	4. 094	4. 91	4. 241	−0. 699	5. 31	1. 986	−7. 033
4. 52	3. 178	4. 099	4. 92	4. 233	−0. 999	5. 32	1. 917	−6. 888
4. 53	3. 219	4. 101	4. 93	4. 221	−1. 306	5. 33	1. 849	−6. 735
4. 54	3. 260	4. 100	4. 94	4. 207	−1. 619	5. 34	1. 782	−6. 575
4. 55	3. 301	4. 095	4. 95	4. 189	−1. 937	5. 35	1. 717	−6. 409
4. 56	3. 342	4. 086	4. 96	4. 168	−2. 258	5. 36	1. 654	−6. 238
4. 57	3. 382	4. 073	4. 97	4. 144	−2. 583	5. 37	1. 592	−6. 063
4. 58	3. 423	4. 056	4. 98	4. 116	−2. 909	5. 38	1. 533	−5. 886
4. 59	3. 464	4. 034	4. 99	4. 085	−3. 236	5. 39	1. 475	−5. 707
4. 60	3. 504	4. 007	5. 00	4. 051	−3. 561	5. 40	1. 418	−5. 527
4. 61	3. 544	3. 975	5. 01	4. 014	−3. 884	5. 41	1. 364	−5. 346
4. 62	3. 583	3. 938	5. 02	3. 974	−4. 204	5. 42	1. 312	−5. 167
4. 63	3. 622	3. 895	5. 03	3. 930	−4. 518	5. 43	1. 261	−4. 988
4. 64	3. 661	3. 846	5. 04	3. 883	−4. 825	5. 44	1. 212	−4. 811
4. 65	3. 699	3. 791	5. 05	3. 834	−5. 123	5. 45	1. 165	−4. 637
4. 66	3. 737	3. 729	5. 06	3. 781	−5. 412	5. 46	1. 119	−4. 465
4. 67	3. 774	3. 660	5. 07	3. 725	−5. 690	5. 47	1. 075	−4. 297
4. 68	3. 810	3. 584	5. 08	3. 667	−5. 954	5. 48	1. 033	−4. 132
4. 69	3. 845	3. 501	5. 09	3. 606	−6. 205	5. 49	0. 993	−3. 970
4. 70	3. 880	3. 409	5. 10	3. 543	−6. 440	5. 50	0. 954	−3. 813
4. 71	3. 914	3. 310	5. 11	3. 478	−6. 658	5. 51	0. 916	−3. 660
4. 72	3. 946	3. 202	5. 12	3. 410	−6. 859	5. 52	0. 880	−3. 512
4. 73	3. 978	3. 085	5. 13	3. 341	−7. 041	5. 53	0. 846	−3. 367
4. 74	4. 008	2. 960	5. 14	3. 269	−7. 204	5. 54	0. 813	−3. 228
4. 75	4. 037	2. 825	5. 15	3. 197	−7. 348	5. 55	0. 781	−3. 093
4. 76	4. 064	2. 680	5. 16	3. 122	−7. 470	5. 56	0. 751	−2. 963
4. 77	4. 090	2. 526	5. 17	3. 047	−7. 572	5. 57	0. 722	−2. 837
4. 78	4. 115	2. 362	5. 18	2. 971	−7. 653	5. 58	0. 694	−2. 716
4. 79	4. 138	2. 188	5. 19	2. 894	−7. 714	5. 59	0. 668	−2. 600
4. 80	4. 159	2. 003	5. 20	2. 817	−7. 754	5. 60	0. 642	−2. 488
4. 81	4. 178	1. 808	5. 21	2. 739	−7. 774	5. 61	0. 618	−2. 380
4. 82	4. 195	1. 603	5. 22	2. 661	−7. 774	5. 62	0. 595	−2. 277
4. 83	4. 210	1. 387	5. 23	2.. 584	−7. 755	5. 63	0. 573	−2. 178
4. 84	4. 222	1. 160	5. 24	2. 506	−7. 718	5. 64	0. 551	−2. 083
4. 85	4. 233	0. 924	5. 25	2. 430	−7. 664	5. 65	0. 531	−1. 992
4. 86	4. 241	0. 677	5. 26	2. 353	−7. 593	5. 66	0. 511	−1. 904
4. 87	4. 246	0. 420	5. 27	2. 278	−7. 507	5. 67	0. 493	−1. 821
4. 88	4. 249	0. 154	5. 28	2. 203	−7. 407	5. 68	0. 475	−1. 741
4. 89	4. 249	−0. 122	5. 29	2. 130	−7. 293	5. 69	0. 458	−1. 665

TABLE IV—Continued

x	y	y'	x	y	y'	x	y	y'
5. 70	0. 442	−1. 592	5. 80	0. 313	−1. 022	5. 90	0. 230	−0. 666
5. 71	0. 426	−1. 523	5. 81	0. 303	−0. 979	5. 91	0. 224	−0. 639
5. 72	0. 411	−1. 456	5. 82	0. 294	−0. 937	5. 92	0. 217	−0. 612
5. 73	0. 397	−1. 393	5. 83	0. 284	−0. 897	5. 93	0. 211	−0. 588
5. 74	0. 383	−1. 332	5. 84	0. 276	−0. 859	5. 94	0. 206	−0. 564
5. 75	0. 370	−1. 274	5. 85	0. 267	−0. 823	5. 95	0. 200	−0. 541
5. 76	0. 358	−1. 219	5. 86	0. 259	−0. 789	5. 96	0. 195	−0. 519
5. 77	0. 346	−1. 166	5. 87	0. 251	−0. 756	5. 97	0. 190	−0. 499
5. 78	0. 335	−1. 116	5. 88	0. 244	−0. 724	5. 98	0. 185	−0. 479
5. 79	0. 324	−1. 068	5. 89	0. 237	−0. 694	5. 99	0. 180	−0. 460
6. 00	0. 176	−0. 442						

Bibliography

ABEL, N. H.: "Sur l'équation différentielles, $(y+s)$ $dy+(p+qy+ry^2)$ $dx=0$." *Werke*, Vol. 2, No. 5, pp. 26–35.

D'ADHÉMAR, R.: "Les fonctions implicites en nombre infini à l'équation intégrale non linéaire." *Bull. Soc. Math. de France*, Vol. 36, 1908, pp. 195–204.

AGNEW, R. P.: *Differential Equations*. New York, 1942, vii+341 pp.

ANDRONOW, A., and CHAIKIN, S.: *Theory of Oscillations*. Princeton, 1949, ix+358 pp.

ANDRONOW, A.: "Les cycles limites de Poincaré et le théorie des oscillations auto-entretenues." *Comptes Rendus*, Vol. 189, 1929, pp. 559–561.

ANTOSIEWICZ, H. A.: "Forced Periodic Solutions of Differential Equations." *Annals of Math.*, Vol. 57 (2), 1953, pp. 314–317.

APPELL, P.:
(1) *Traité de la mécanique rationelle*. Vol. 1, Paris, 5th ed., 1926, 619 pp.
(2) "Sur les invariants de quelques équations différentielles." *Journal de Math.*, Vol. 5 (4), 1889, pp. 361–423.

APPLETON, E. V., and VAN DER POL, B.: "On a Type of Oscillator Hysteresis in a Simple Triode Generator." *Phil. Mag.*, Vol. 42, 1921, p. 201.

APPLETON, E. V.: "The Automatic Synchronization of Triode Oscillators." *Proc. of Cambridge Phil. Soc.*, Vol. 21, 1922, p. 231

ARCHIBALD, R. C., and MANNING, H. P.: "Historical Remarks on the Pursuit Problem." *Amer. Math. Monthly*, Vol. 28, 1921, pp. 91–93.

AUTONNE, L.:
(1) "Sur la nature des intégrales algébriques de l'équation de Riccati." *Comptes Rendus*, Vol. 96, 1883, pp. 1354–1356.
(2) "Sur les intégrales algébriques de l'équation de Riccati." *Comptes Rendus*, Vol. 128, 1899, pp. 410–412.

BAROCIO, S.: "On Certain Critical Points of a Differential System in the Plane." *Contributions to Non-Linear Oscillations*, Vol. 3, Princeton, 1955.

BASS, R. W.: "On Nonlinear Repulsive Forces." *Contributions to Nonlinear Oscillations*, Vol. 4, 1958, pp. 201–211.

BATEMAN, HARRY:
(1) *Differential Equations*. London, 1918, xi+306 pp.
(2) *Partial Differential Equations of Mathematical Physics*. Cambridge, 1932, xxii+522 pp; in particular, pp. 166–169 and 491–511.

BAUTIN, N. N.: "On the Number of Limit Cycles Appearing During Changes of Coefficients From a State of Equilibrium of the Type of Focus or Center." *Mat. Sbornik*, Vol. 30, 1952, pp. 181–196.

BELLMAN, R.:
(1) "On the Boundedness of Solutions of Nonlinear Differential and Difference Equations." *Trans. Amer. Math. Soc.*, Vol. 62, 1947, pp. 357–386.
(2) *Stability Theory of Differential Equations*. New York, 1953.

BENDIXSON, I.: "Sur les courbes définies par des équations différentielles." *Acta Mathematica*, Vol. 24, 1901, pp. 1–88.

BERNHART, A.: "Curves of General Pursuit." *Scripta Mathematica*, Vol. 24, 1959, pp. 189–206.

545

BERNOULLI, DANIEL: "Solutio problematis Riccatiani proporeti in *Act. Lipo Suppl.*" Tom., viii, p. 73. *Acta Erudetorum*, 1925, pp. 473–475.

BERNOULLI, JOHN: "Methodus generalis construendi omnes aequationes differentiales primi gradus." *Acta Erudetorum*, 1694, pp. 435–437. *Opera*, Vol. 1, Lausanne and Geneva, 1742, p. 124.

BERRY, A.: "On Certain Riccati Differential Equations with Algebraic Integrals." *Quarterly Journal*, Vol. 49, 1923, pp. 303–308.

BIEBERBACH, L.: "Δu=eᵘ und die automorphen Funktionen." *Math. Annalen*, Vol. 77 1916, pp. 173–212.

BIRKHOFF, G. D.:
(1) "Quelques théorèmes générales sur le mouvement des systèmes dynamiques." *Bull. Soc. Math. de France*, Vol. 40, 1912, pp. 305–323.
(2) "Dynamical System with Two Degrees of Freedom." *Trans. Amer. Math. Soc.*, Vol. 18, 1917, pp. 199–300.
(3) "Surface Transformations and Their Dynamical Application." *Acta Mathematica*, Vol. 43, 1920, pp. 1–119.
(4) *Dynamical Systems.* Colloquium Publications, Amer. Math. Soc., Vol. 9, 1927, viii+295 pp.
(5) "Probability and Physical Systems." *Bull. Amer. Math. Soc.*, Vol. 38, 1932, pp. 361–379.
(6) "What Is the Ergodic Theorem?" *Amer. Math. Monthly*, Vol. 49, 1942, pp. 222–226.

BLASIUS, H.: "Grenzschichten in Flussigkeiten mit kleiner Reibung." *Zeitschrift für Math. und Physik*, Vol. 56, 1908, pp. 1–37.

BLOCK, H.: "Sur la solution de certaines équations intégrales." *Arkiv för Mat., Astro. cch Fys.*, Vol. 3, 1907, No. 22, 18 pp.

BOCHNER, S., and MARTIN, W. T.: *Several Complex Variables.* Princeton, 1948.

BOMPIANI, E.: "Sull'andamento degli integrali di un'equazione de Riccati." *Memorie* Bologna, Vol. 2 (8), 1925, pp. 29–33.

BOOLE, GEORGE: *Differential Equations.* London, 1859; 4th ed., 1877, xiv+496 pp.

BOUTROUX, M. P.:
(1) *Leçons sur les functions définies par les équations différentielles de premier ordre.* Paris, 1908, 187 pp.
(2) "Recherches sur les transcendants de M. Painlevé et l'étude asymptotique des équations différentielles du second ordre." *Annales de l'École Normale Supérieure*, Vol. 30 (3), 1913, pp. 255–375; Vol. 31 (3), 1914, pp. 99–159.
(3) "On Multiform Functions Defined by Differential Equations of the First Order." *Annals of Math.*, Vol. 22 (2), 1920–21, pp. 1–10.

BRASSINNE, E.:
(1) "Sur diverses équations différentielles du premier ordre analogues à l'équation de Riccati." *Mem. de l'Acad. R. des Sci. de Toulouse*, Vol. 4 (3), 1848, pp. 234–236.
(2) "Sur des équations différentielles qui se rattachant à l'équation de Riccati," *Journ. de Math.*, Vol. 16, 1851, pp. 255–256.

BRATU, G.:
(1) "Sur certaines équations intégrales non linéaires." *Comptes Rendus*, Vol. 150, 1910, pp. 896–899.
(2) "Sur les équations intégrales non linéaires." *Bull. Soc. Math. de France*, Vol. 41, 1913, pp. 346–350; Vol. 42, 1914, pp. 113–142.

BROWNELL, F. H.: "Nonlinear Difference-Differential Equations." *Contributions to Nonlinear Oscillations*, Vol. 1, Princeton, 1950, pp. 89–148.

BRUCHT, G.: "Über nichtlineare Integralgleichungen mit unverzweigten Lösung." *Arkiv för Mat., Astro. och Fys.*, Vol. 8, 1912–13, No. 8, 20 pp.

Bush, V., and Caldwell, S. H.: "Thomas-Fermi Equation Solution by the Differential Analyzer." *Physical Review*, Vol. 38 (2), 1931, pp. 1898–1901.

Bushaw, D.: "Optimal Discontinuous Forcing Terms." *Contributions to Nonlinear Oscillations*, Vol. 4, Princeton,1958, pp. 29–52.

Carleman, T.: "Application de la théorie des équations intégrales linéares aux systèmes d'équations différentielles non linéares." *Acta Mathematica*, Vol. 59, 1932, pp. 63–87.

Cartwright, Mary L.
(1) "Forced Oscillations in Nearly Sinusoidal Systems." *Journ. Inst., Elec. Eng.*, Vol. 95 (3), 1948, p. 88.
(2) "Forced Oscillations in Nonlinear Systems." *Contributions to Nonlinear Oscillations*, Vol. 1, Princeton, 1950, pp. 149–242.
(3) "Van der Pol's Equation for Relaxation Oscillations." *Contributions to Nonlinear Oscillations*, Vol. 2, Princeton, 1952, pp. 3–18.

Cartwright, M. L., and Littlewood, J. E.:
(1) "On Non-Linear Differential Equations of the Second Order." *Journ. London Math. Soc.*, Vol. 20, 1945, pp. 180–189.
(2) "On Non-Linear Differential Equations of the Second Order." *Annals of Math.*, Vol. 48 (2), 1947, pp. 472–494.

Cartwright, Mary L., Copson, E. T., and Greig, J.: "Nonlinear Vibrations." *The Advancement of Science*, Vol. 6, No. 21, 1949, 12 pp.

Cayley, A.:
(1) "On the Theory of the Singular Solution of Differential Equations of First Order." *Messenger of Math.*, Vol. 2, 1872, pp. 6–12.
(2) "On Riccati's Equations." *Phil. Mag.*, Vol. 36 (4), 1868, pp. 348–351; also *Collected Papers*, Vol. 7, 1894, pp. 9–12.

Challis, H. W.: "Extension of the Solution of Riccati's Equation." *Quarterly Journ. of Math.*, Vol. 7, 1866, pp. 51–53.

Chandrasekhar, S.: (1) "On the Radiative Equilibrium of a Stellar Atmosphere. XXII." *Astrophysical Journal*, Vol. 107, 1948, pp. 48–72.
(2) *An Introduction to the Study of Stellar Structure.* Chicago, 1939, 509 pp. In particular Ch. 11. (Dover Reprint)

Chiellini, A.: "Sulle pseudo-equazioni differenziali di Fuchs di prima specie, di ordini qualunque e su classi di equazioni di Riccati reducibili alle quadrature." *Rend. Sem. Fac. Sci. Univ. Cagliari*, Vol. 9, 1939, pp. 142–155.

Chrystal, G.: "On the *p*-discriminant of a Differential Equation of the First Order and on Certain Points in the General Theory of Envelopes Connected Therewith." *Trans. Royal Soc. of Edinburgh*, Vol. 38, 1896, pp. 803–824.

Coddington, E. A., and Levinson, N.: *Theory of Ordinary Differential Equations.* New York, 1955.

Collet, A.: "Sur les solutions approchées de certaines équations intégrales non linéaires." *Annales de Toulouse*, Vol. 4 (3), 1912, pp. 199–249.

Corbeiller, Ph. Le:
(1) *Les systemes autoentretenues et les oscillations de relaxation.* Paris, 1931.
(2) "The Nonlinear Theory of the Maintenance of Oscillations." *Journ. Inst. of Elec. Engineers*, Vol. 79, 1936, pp. 361–378.

Cotton, E.: "Equations différentielles et équations intégrales." *Bull. de la Soc. Math. de France*, Vol. 38, 1910, pp. 144–154.

Darboux, J. G.:
(1) *Leçons sur la théorie générale des surfaces.* Four volumes, Paris, 1887–96.
(2) "Sur les solutions singulières des équations aux dérivées ordinaires du premier ordre." *Bull. des Sc. Math.*, Vol. 4, 1873, pp. 158–176.
(3) "Sur l'équation de Riccati." *Chelini. Cll. Math.* In memoriam D. Chelini, Mediolani, 1881, pp. 199–205.

DAVIS, H. T.: "Studies Relating to a Non-Linear Differential Equation of Second Order, With Special Reference to the First and Second Transcendents of Painlevé." *Studies in Differential Equations*, Evanston, 1956, pp. 1–72.

DE BAGGIO, H. F.: "Dynamical Systems with Stable Structures." *Contributions to Nonlinear Oscillations*, Vol. 2, Princeton, 1952, pp. 37–60.

DE VOGELAERE, R.: "On the Structure of Symmetric Periodic Solutions of Conservative Systems, With Applications." *Contributions to Nonlinear Oscillations*, Vol. 4, Princeton, 1958, pp. 53–84.

DILIBERTO, S. P.:

(1) "On Systems of Ordinary Differential Equations." *Contributions to Nonlinear Oscillations*, Vol. 1, Princeton, 1950, pp. 1–38.

(2) "An Application of Periodic Surfaces (Solution of a Small Division Problem)." *Ibid.*, Vol. 3, 1956, pp. 257–260.

(3) "Bounds for Periods of Periodic Solutions." *Ibid.*, Vol. 3, 1956, pp. 269–276.

DILIBERTO, S. P., and HUFFORD, G.: "Perturbation Theorems for Nonlinear Ordinary Differential Equations." *Contributions to Nonlinear Oscillations*, Vol. 3, Princeton, 1956, pp. 207–236.

DILIBERTO, S. P., and MARCUS, M. D.: "A Note on the Existence of Periodic Solutions of Differential Equations." *Contributions to Nonlinear Oscillations*, Vol. 3, Princeton, 1956, pp. 237–242.

DOETSCH, G.: *Theorie und Andwendung der Laplace-Transformation.* 1937, xiii +439 pp.

DORODNICYN, A. A.: "Asymptotic Solutions of Van der Pol's Equation." *Prikl. Mat. i Mech.*, Vol. 11, 1947, pp. 313–328.

DUFF, G. F.: "Limit-cycles and Rotated Vector Fields." *Annals of Math.*, Vol. 57 (2), 1953, pp. 15–31.

DUFFIN, R. J.: "Non-Linear Networks." *Bull. Amer. Math. Soc.*, Vol. 52, 1946, p. 883; Vol. 53, 1947, p. 963; Vol. 54, 1948, p. 119.

DUFFING, G.: *Erzwungene Schwingungen bei veränderlicher Eigenfrequenz.* Braunschweig, 1918, 134 pp.

DULAC, H.: "Détermination et intégration d'une certaine class d'équations différentielles ayant pour point singulier un centre." *Bull. de Soc. Math.*, Vol. 32 (2) 1908, pp. 230–252.

DUNOYER, L.: "Sur les coubres de poursuite d'un cercle." *Nouvelles Annales de Math.*, Vol. 6 (4), 1906, pp. 193–222.

ECKWEILER, H. J.: "Non-Linear Differential Equations of the Van der Pol Type With a Variety of Periodic Solutions." From *Studies in Non-Linear Vibration Theory.* New York University, 1946.

EDDINGTON, A. S.: *The Internal Constitution of the Stars.* Cambridge, 1926, viii +407 pp. (Dover Reprint)

EMDEN, R.: *Gaskugeln.* Berlin and Leipzig, 1907, v + 497 pp.

EULER, L.: "De resolutione aequationis $dy + ayy\ dx = bx^m\ dx$." *Novi Comm. Acad. Petrop.*, IX 1762–63, pp. 154–169. Also see *Ibid.*, Vol. VIII, 1760–61, pp. 3–63.

EVANS, G. C.: "Some General Types of Functional Equations." *Fifth International Congress of Mathematicians*, Cambridge, 1912, Vol. 1, pp. 387–396.

FALKENHAGEN, J. H. M.: "Über des Verhalten den Integrale eine Riccati' schen Gleichung in der Nähe einer singulären Stelle." *Nieuw. Archief von Wiskunde*, Vol. 6 (2), 1905, pp. 209–248.

FALKNER, V. M., and Miss S. W. SKAN: "Solutions of the Boundary-layer Equations." *Phil. Mag.*, Vol. 12 (7), 1931, pp. 865–896. See also, *Phil. Mag.*, Vol. 21 (7), 1936, pp. 624–640.

FERMI, E.: "Statistical Method of Investigating Electrons in Atoms." *Zeitschrift für Physik*, Vol. 48, 1928, pp. 73–79. Also *Atti dei Lincei*, Vol. 6 (6), 1927, pp. 602–607.

FLANDERS, D., and STOKER, J.: "The Limit Case of Relaxation Oscillations." *Studies in Non-Linear Vibration Theory.* New York University, 1946.

FORD, L. E.:
(1) *Automorphic Functions.* New York, 1929; 2d ed., 1951, 333 pp.
(2) *Differential Equations.* New York, 1933, x + 263 pp.

FORSYTH, A. R.:
(1) *A Treatise on Differential Equations.* London, 1885; 4th ed, 1914, xviii + 584 pp.
(2) *Theory of Differential Equations.* Cambridge, 1900–1902. Vol. 1, exact equations and Pfaff's problem; vols. 2 and 3, ordinary equations, not linear; vol. 4, ordinary linear equations; vols. 5 and 6, partial differential equations. (Dover Reprint)
(3) "Note on the Central Differential Equation in the Relativity Theory of Gravitation." *Proc. of Royal Soc.*, Vol. 97 (A), 1920, pp. 145–151.

FRIEDRICKS, K. O.: "On Non-Linear Vibrations of Third Order." *Studies in Non-Linear Vibration Theory.* New York University, 1946.

FRIEDRICKS, K. O., and STOKER, J. J.: "Forced Vibrations of Systems With Nonlinear Restoring Force." *Quarterly Journal of Applied Math.*, Vol. 1, 1943, pp. 97–115.

FRIEDRICKS, K. O., and WASOW, W.: "Singular Perturbations of Non-Linear Oscillations." *Duke Math. Journal*, Vol. 13, pp. 367–381.

FROMMER, M.: "Über das Auftreten von Wirbeln und Strudeln (geschlossener und spiraliger Integralkurven) in der Umgebung rationaler Unbestimmtheitsstellen." *Math. Annalen*, Vol. 109, 1934, pp. 395–424.

GAMBIER, B.:
(1) "Sur les équations différentielles du second ordre dont l'intégrale générale est uniform." *Comptes Rendus*, Vol. 142, 1906, pp. 266–269, 1403–1496, 1497–1500.
(2) "Sur les équations différentielles du second ordre et du premier degré dont l'intégrale générale est à points critiques fixes." *Comptes Rendus*, Vol. 143, 1906, pp. 741–743; Vol. 144, 1907, pp. 827–830, 962–964.
(3) "Sur les équations différentielles du second ordre et du premier degré dont l'intégrale générale est à points critiques fixes." *Acta Mathematica*, Vol. 33, 1909, pp. 1–55.

GARNIER, R.: "Sur la représentation des intégrales des équations de M. Prinlevé au moyen de la théorie des équations linéaires." *Comptes Rendus*, Vol. 159, pp. 296–299; Vol. 160, 1915, pp. 795–798.

GERBER, P.: "Die räumliche und zeitliche Ausbreitung der Gravitation." *Zeitschrift für Math. und Physik*, Vol. 43, 1898, pp. 93–104.

GLAISHER, J. W. L.:
(1) "On Riccati's Equation and its Transformations and on some Definite Integrals which Satisfy Them." *Phil. Trans. of Royal Soc.*, Vol. 172, 1881, pp. 759–828.
(2) "Examples Illustrative of Cayley's Theory of Singular Solutions." *Messenger of Mathematics*, Vol. 12, 1882, pp. 1–14.

GOLDSTEIN, S.: "Concerning Some Solutions of the Boundary Layer Equations in Hydrodynamics." *Proc. Cambridge Phil. Soc.*, Vol. 26, 1930, pp. 1–30.

GOLOMB, M.: "Zur Theorie der nichtlinearen Integralgleichungen, Integralgleichungssysteme und allgemeinen Funktionalgleichungen." *Math. Zeitschrift*, Vol. 39, 1934, pp. 45–75.

GOMORY, R. E.: "Critical Points at Infinity and Forced Oscillations." *Contributions to Nonlinear Oscillations*, Vol. 3, Princeton, 1956, pp. 85–126.

GOURSAT, E.:
  (1) *Cours d'analyse mathématique.* Tome 2, Paris, 4th ed., 1924; English translation by E. R. Hedrick and O. Dunkel, Vol. 2, Part 2 (*Differential Equations*), Boston, 1917, 299 pp. (Dover Reprint)
  (2) *Leçons sur l'intégration des équations aux dérivées partielles du primier ordre.* Paris, 1891, 354 pp.
  (3) *Leçons sur l'intégration des équations au dérivées partielles du second ordre.* Paris, Vol. 1, 1896, 226 pp; Vol. 2, 1898, 344 pp.

GRAFFI, D.: "Forced Oscillations for Several Nonlinear Circuits." *Annals of Math.*, Vol. 54 (2), 1951, pp. 262–271.

GREENHILL, A. G.:
  (1) *The Applications of Elliptic Functions.* 1892, xi + 357 pp. (Dover Reprint)
  (2) "On Riccati's Equation and Bessel's Equation." *Quarterly Journ. of Math.*, Vol. 16, 1879, pp. 294–298.

GREENHILL, A. G., and HADCOCK, A. G.: "Siacci's Method of Solving Trajectories and Problems in Ballistics." *Proc. of the Royal Artillery Institution*, Vol. 17, 1890, pp. 81–108, 389–424. In particular, pp. 92 and 424.

GRONWALL, T. H.: "Note on the Derivatives With Respect to a Parameter of the Solutions of a System of Differential Equations." *Annals of Math.*, Vol. 20 (2), 1919, pp. 292–296.

GUDERMANN, C.: "Theorie der Modular-Functionen und der Modular-Integrals." *Journ. für Math.*, Vol. 19, 1839, pp. 45–83; in particular, pp. 79–81.

GUIGUE, R.: "Sur l'équation de Riccati." *Bull. de Sc. Math.*, Vol. 62 (2), 1938, pp. 166–171.

HAAG, J.:
  (1) "Étude asymptotique des oscillations de relaxation". *Annales de l'Ecole Normale Supérieure*, Vol. 60, 1943, pp. 65–111.
  (2) "Exemples concrets d'étude asymptotique d'oscillations de relaxation." *Ibid.*, Vol. 61, 1944, pp. 73–117.

HAAS, F : "On the Total Number of Singular Points and Limit Cycles of a Differential Equation." *Contributions to Nonlinear Oscillations*, Vol. 3, Princeton, 1956, pp. 137–172.

HAAS, VIOLET B.: "On a Nonlinear Differential Equation Containing a Small Parameter." *Contributions to Nonlinear Oscillations*, Vol. 3, Princeton, 1956, pp. 57–84.

HAMEL, G.: "Über erzwungene Schwingungen bei endlicher Amplituden." *Mah. Annalen*, Vol. 86, 1922, pp. 1–13.

HAMMERSTEIN, A.:
  (1) "Nichtlineare Integralgleichungen nebst Anwendung." *Acta Mathematica*, Vol. 54, 1930, pp. 117–176.
  (2) "Die erste Randwertaufgabe fur nichtlineare Differentialgleichungen zweiter Ordnung." *Sitzungsberichte der Berlin Math. Ges.*, Vol. 30, 1932, pp. 3–10.
  (3) "Über die Eigenwerte gewisse nichtlinearer Differentialgleichungen." *Journ. für. Math.*, Vol. 168, pp. 37–43.

HARGREAVE, C. J.: "On Riccati's Equation." *Quart. Journ. of Math.*, Vol. 7, 1866, pp. 256–258.

HARTOG, J. P. DEN: *Mechanical Vibrations.* 2d ed. New York, 1940, xi + 390 pp.

HARTREE, D. R.: "On an Equation Occurring in Falkner and Skan's Approximate Treatment of the Equations of the Boundary Layer." *Proc. Cambridge Phil. Soc.*, Vol. 33, 1937, pp. 223–239.

HATHAWAY, A. S.: "Pursuit in a Circle." *Amer. Math Monthly*, Vol. 28, 1921, pp. 93–97.

HILB, E.: "Nichtlineare Differentialgleichungen." *Encyclopädie der Math. Wissenschaften*, II B 6, 1921, pp. 563–603.

HILL, G. W.:
(1) "On the Part of the Motion of the Lunar Perigee Which Is a Function of the Mean Motion of the Sun and Moon." Cambridge, Mass., 1877. Reprinted in *Acta Mathematica*, Vol. 8, 1886, pp. 1–36. Also see Hill's *Collected Works*, Vol. 1, 1905, pp. 243–270.
(2) "Researches in the Lunar Theory." *American Journ. of Math.*, Vol. 1, 1878, pp. 5–26, 129–147, 245–290. Also see Hill's *Collected Works*, Vol. 1, 1905, pp. 284–335.

HILL, M. J. M.:
(1) "On the c- and p-Discriminants of Ordinary Integrable Differential Equations of First Order." *Proc. London, Math. Soc.*, Vol. 19 (1), 1888, pp. 561–589.
(2) "On Node- and Cusp-Loci Which Are Enveloped by the Tangents at the Cusps." *Ibid.*, Vol. 22 (1), 1891, pp. 216–236.
(3) "On the Singular Solutions of Ordinary Differential Equations of the First Order with Transcendental Coefficients." *Ibid.*, Vol. 17 (7), 1918, pp. 149–183.

HIRSCHMAN, I. I., and WIDDER, D. V.: *The Convolution Transform.* Princeton, 1955, x +268 pp.

HORN, J.:
(1) *"Gewöhnliche Differentialgleichungen beliebiger Ordnung.* Leipzig, 1905, x + 391 pp.
(2) *Einführung in die Theorie der partiellen Differentialgleichungen.* Leipzig, 1910, vii + 363 pp.
(3) "Laplacesche Integrale als Lösungen nict linearen Differentialgleichungen." *Journ. für Math.*, Vol. 144, 1914, pp. 167–189; Vol. 151, 1920, pp. 167–199.

HOWARTH, L.: "On the Solution of the Laminar Boundary Layer Equation." *Proc. Royal Soc. of London*, Vol. 164 (A), 1937, pp. 547–579.

HUDSON, R. W. H. T.: "A Geometrical Theory of Differential Equations of the First and Second Orders." *Proc. London Math. Soc.*, Vol. 33, 1901, pp. 380–403

HUFFORD, G.: "Banach Spaces and the Perturbation of Ordinary Differential Equations." *Contributions to Nonlinear Oscillations*, Vol. 3, Princeton, 1956, pp. 173–196.

IGLISCH, R.: "Existenz und Eindeutigkeitsatz bei Nichtlinearen Integralgleichungen." *Math. Annalen*, Vol. 108, 1933, pp. 161–189.

INCE, E. L.: *Ordinary Differential Equations.*1927,viii + 558 pp.(Dover Reprint)

ISELI, F.: "Die Riccatische Differentialgleichung." *Dissertation*, Bern,1909, 42 pp.

JACOBSEN, L. S., and AYRE, R. S.: *Engineering Vibrations.* New York, 1958, xii+564 pp.

JAHNKE, E., and EMDE, F.: *Funktionentafeln.* First ed., Leipzig, 1909, 174 pp. American reprint, New York, 1944, xv + 306 + 76 pp. (Dover Reprint)

KAKUTANI, S., and MARKUS, L.: "On the Nonlinear Difference-Differential Equation." *Contributions to Nonlinear Oscillations*, Vol. 4, Princeton, 1958, pp. 1–18.

KAMKE, E.:
(1) *Gewöhnliche Differentialgleichungen.* Leipzig, 1940, 3d. ed., 1944; American ed., New York, 1948, xxvi + 666 pp.
(2) *Partielle Differentialgleichungen erster Ordnung.* Leipzig, 1944; American ed., New York, 1946, xv + 243 pp.

KAPTEYN, W.: "Over de middelpunten de Integraalkrommer van Differentiaalvergelijkingen van de eerste Orde en den eersten Graad." *Koninkl. Nederland Akad.*, Vol. 19, 1911, pp. 1446–1457; Vol. 20, 1912, pp. 1354–1365; Vol. 21, 1912, pp. 27–33.

KIDDER, R. E.: "Unsteady Flow of Gas Through a Semi-Infinite Porous Medium" *Journal of Applied Mechanics*, Vol. 27, 1957, pp. 329–332.

KOOSIS, P.: "One-Dimensional Repeating Curves in the Nondegenerate Case." *Contributions to Nonlinear Oscillations*, Vol. 3, Princeton, 1956, pp. 277–285.

KOURENSKY, M.:
  (1) "On Lines of Electric Force Due to a Moving Point Charge." *Proc. London Math. Soc.*, Vol. 24 (2), 1926, pp. 202–210.
  (2) "Sur l'équation de Riccati." *Rendiconti Accad. dei Lincei, Roma*, Vol. 9 (6), 1929, pp. 950–957.

KRYLOFF, N., and BOGOLIUBOFF, N.: *Introduction to Non-Linear Mechanics.* Princeton, 1943, 105 pp. Trans. from Russian published in 1937.

KU, Y. H.: *Analysis and Control of Non-Linear Systems.* New York, 1958, vii + 360 pp.

KUMMER, E. E.: "Sur l'intégration de l'équation de Riccati par des intégrales définies." *Journ für Math.*, Vol. 12, 1834, pp. 144–147.

KYNER, W. T.:
  (1) "A Fixed Point Theorem." *Contributions to Nonlinear Oscillations*, Vol. 3, Princeton, 1956, pp. 195–206.
  (2) "Small Periodic Perturbations of an Autonomous System of Vector Equations." *Ibid.*, Vol. 4, Princeton, 1958, pp. 111–124.

LAGRANGE, R.: "Quelques théorèmes d'intégrabilité par quadratures de l'équation de Riccati." *Bull. Soc. Math. de France.* Vol. 66, 1938, pp. 155–163.

LALESCO, T.: *Introduction à la Théorie des Equations Intégrales.* Paris, 1912, 152 pp.

LANGENHOP, C. E.: "Note on Levinson's Existence Theorem for Forced Periodic Solutions of a Second Order Differential Equation." *Journal of Math. and Phys.*, Vol. 30, 1951, pp. 30–39.

LANGENHOP, C. E. and FARNELL, A. B.: "The Existence of Forced Periodic Solutions of Second Order Differential Equations Near Certain Equilibrium Points of the Unforced Equation." *Contributions to Nonlinear Oscillations*, Vol. 1, Princeton, 1950, pp. 313–350.

LANGMUIR, I., and BLODGETT, KATHERINE B.: "Currents Limited by Space Charge Between Coaxial Cylinders." *Phys. Review*, Vol. 22 (2), 1923, pp. 347–356.

LASALLE, J.: "Relaxation Oscillations." *Quart. Journ. of Applied Math.*, Vol. 7, 1949, pp. 1–19.

LAUE, M. VON: "The Entropy-Constant of Incandescent-Electrons." *Yahrbuch der Radioaktivitat and Elektronik*, Vol. 15, 1918, pp. 205–270.

LEFSCHETZ, S.:
  (1) *Lectures on Differential Equations.* Princeton, 1946, Annals of Math. Studies, No. 14, 209 pp.
  (2) (Editor) *Contributions to the Theory of Nonlinear Oscillations.* Princeton, Vol. 1, 1950, Annals of Math. Studies, No. 20, 350 pp.; Vol. 2, 1952, No. 29, 116 pp.; Vol. 3, 1956, No. 36, 285 pp.; Vol. 4, 1958, No. 41, 211 pp.
  (3) *Differential Equations—Geometric Theory.* New York, 1957, x + 364 pp.
  (4) "Notes on Differential Equations." *Contributions to Nonlinear Oscillations*, Vol. 2, pp. 61–74.
  (5) "On the Critical Points of a Class of Differential Equations." *Ibid.*, Vol. 4, pp. 19–28.

LEMPKE, H.: "Über die Differentialgleichunges welche den Gleichgewichtszustand eines gasförmigen Himmelskörpers bestimmen dessen Teile gegeneinander nach dem Newtonschen Gesetze gravitieren." *Journ. fur Math.* Vol. 142, 1913, pp. 118–145.

LERAY, J.: "Étude de diverses équations intégrales nonlinéaires et de quelques problèmes que pose l'hydrodynamique." *Journ. de Math.*, Vol. 12 (9), 1933, pp. 1–82.

LEVENSON, M. L.: "Harmonic and Subharmonic Response for the Duffing Equation." *Dissertation*, New York University, 1948; abridgment in *Journ. of Applied Physics*, Vol. 20, 1949, p. 1045.

LEVINSON, N.:
(1) "On a Non-Linear Differential Equation of the Second Order." *Journ. Math. and Physics*, Vol. 22, 1943, pp. 181–187.
(2) "Transformation Theory of Non-Linear Differential Equations of the Second Order." *Annals of Math.*, Vol. 45, 1944, pp. 723–737.
(3) "Perturbations of Discontinuous Solutions of Non-Linear Systems of Differential Equations." *Proc. Nat. Academy*, Vol. 33, 1947, pp. 214–218.
(4) "A Simple Second Order Differential Equation With Singular Motions." *Ibid.*, Vol. 34, 1948, p. 13.

LEVINSON, N., and SMITH, O. K.: "A General Equation for Relaxation Oscillations." *Duke Math. Journ.*, Vol. 9, 1942, pp. 382–403.

LEVY, H., and BAGGOTT, E. A.: *Numerical Studies in Differential Equations*. London, 1934; American ed., New York, 1950, viii+238 pp. (Dover Reprint)

LEVY, P.: "Sur les équations intégrales non linéaires." *Comptes Rendus*, Vol. 150 (1), 1910, pp. 899–901.

LIAPOUNOFF, A.: "Problème général de la stabilité du mouvement." *Annales de Toulouse*, Vol. 9 (2), 1907, pp. 203–474. Originally published in Russian in 1892.

LICHTENSTEIN, L.:
(1) "Intégration de l'équation $\nabla^2 u = ke^u$ sur un surface fermée." *Comptes Rendus*, Vol. 157, 1913, pp. 1508–1511.
(2) "Integration der Differentialgleichungen $\Delta_2 u = ke^u$ auf geschlossenen Flächen. Methode der unendlichvielen Variabeln." *Acta Mathematica*, Vol. 140, 1915, pp. 1–34.
(3) *Vorlesungen über einige Klassen nichtlinearer Integralgleichungen und Integrodifferentialgleichungen nebst Anwendungen.* Berlin, 1931, 164 pp.

LIÉNARD, A.: "Étude des oscillations entretenues." *Revue générale de l'electrécité*, Vol. 23, 1928, pp. 901–946.

LIOUVILLE, J.:
(1) "Sur l'équation aux différences partieiles: $\partial^2 \log\lambda / \partial u \partial v \pm \lambda/2a^2 = 0$.' *Journ. de Math.*, Vol. 18 (1), 1853, pp. 71–72.
(2) "Sur la classification des transcendents et sur l'impossibilité d'exprimer des racines de certaines équations en fonctions finies explicites des coefficients." *Journ. de Math.*, Vol. 2, 1837, pp. 56–105; Vol. 3, 1838, pp. 523–547.
(3) "Sur l'intégration d'une classe d'équations différentielles du second ordre en quantités finies explicites." *Journ. de Math.*, Vol. 4, 1839, pp. 427–456.
(4) "Remarques nouvelles sur l'équation de Riccati." *Comptes Rendus*, Vol. 11, 1840, p. 729; *Journ. de Math.*, Vol. 6, 1841, pp. 1–13.

LIOUVILLE, R.: "Sur une équation différentielle du premier ordre." *Acta Mathematica*, Vol. 27, 1903, pp. 55–78.

LOBATTO, R.: "Sur l'intégration des équations: $d^n y/dx^n - xy = 0, d^2 y/dx^2 + abx^n y = 0$." *Journ. für Math.*, Vol. 17, 1837, pp. 363–371.

LOWAN, A. N.: *Operational Approach to Problems of Stability and Convergence.* New York. 1957, 104 pp.

McCARTHY, J.: "A Method for the Calculation of Limit Cycles by Successive Approximation." *Contributions to Nonlinear Oscillations*, Vol. 2, Princeton, 1952, pp. 75–80.

McLachlan, N. W.:
(1) *Theory and Application of Mathieu Functions.* Oxford, 1947.
(2) *Ordinary Non-Linear Differential Equations in Engineering and Physical Sciences.* Oxford, 1950, vi+199 pp.

Malkin, I. G.: *Theory of Stability of Motion.* Moscow, 1952.

Marcus, M. D.:
(1) "An Invariant Surface Theorem for a Nondegenerate System." *Contributions to Nonlinear Oscillations,* Vol. 3. Princeton, 1956, pp. 243–256.
(2) "Repeating Solutions for a Degenerate System." *Ibid.,* pp. 261–268.

Markus, L.: "Asymptotic Autonomous Differential Systems." *Contributions to Nonlinear Oscillations,* Vol. 3, Princeton, 1956, pp. 17–30.

Massera, J. L.:
(1) "The Number of Subharmonic Solutions of Non-Linear Differential Equations of the Second Order." *Annals of Math.,* Vol. 50 (2), 1949, pp. 118–126.
(2) "On Liapounoff's Conditions of Stability." *Annals of Math.,* Vol. 50 (2), 1949, pp. 705–721.
(3) "Contributions to Stability Theory." *Annals of Math.,* Vol. 64 (2), 1956, pp. 182–206.

Mendelson, P.: "On Phase Portraits of Critical Points in n-Space." *Contributions to Nonlinear Oscillations,* Vol. 4, Princeton, 1958, pp. 167–200.

Minding, F.: "Bemerkungen zur Integration der Differentialgleichungen erster Ordnung zwischen zwei Veränderleicher." *Journ. für Math.,* Vol. 40, 1850, pp. 361–365.

Minetti, S.: "Sull'equazione differenziale di Riccati e su qualche resultato di geometria differenziale." *Rend. dei Lincei, Roma,* Vol. 19 (6), 1934, pp. 65–74.

Minorsky, N.: *Introduction to Non-Linear Mechanics.* Ann Arbor, 1947 xiv+447 pp.

Mitrenovitch, D.: Théorèmes relatifs à l'équation différentielle de Riccati." *Comptes Rendus,* Vol. 206, 1938, pp. 411–413.

Morley, F. V.: "A Curve of Pursuit." *Amer. Math. Monthly,* Vol. 28, 1921, pp. 54–61.

Morton, W. B.: "The Form of Planetary Orbits in the Theory of Relativity." *Phil. Mag.,* Vol. 42 (6), 1921, pp. 511–522.

Murphy, R.: "On the General Properties of Definite Integrals." *Trans. Cambridge Phil. Soc.,* Vol. 3, 1830, pp. 429–443.

Obi, C.:
(1) "Subharmonic Solutions of Non-Linear Differential Equations of the Second Order." *Journal of the London Math. Soc.,* Vol. 25, 1950, pp. 217–226.
(2) "Periodic Solutions of Non-Linear Differential Equations of Second Order." *Proc. Cambridge Phil. Soc.,* Vol. 47, 1951, pp. 741–751.
(3) "Periodic Solutions of Non-Linear Differential Equations of Order 2n." *Journal of the London Math. Soc.,* Vol. 28, 1953, pp. 163–171.
(4) "Researches on the Equation $x'' + (e_1+e_2x)x' + x + e_3x^2 = 0$." *Proc. Cambridge Phil. Soc.,* Vol. 50, 1954, pp. 26–32.

Orlando, L.:
(1) "Sopra alcune equazione integrali." *Rendiente dei Lincei,* Vol. 16 (2), 1907, pp. 601–604.
(2) "Sulle equazioni integrali." *Giornale di Battaglini,* 1908, Vol. 46, pp. 173–196.

Painlevé, P.:
(1) "Sur les équations différentielles d'ordre supérieur dont l'intégrale n'admet qu'un nombre fini déterminations." *Comptes Rendus,* Vol. 116, 1893, pp. 88–91, 173–176.

PAINLEVÉ, P.—Continued

(2) "Sur les singularités essentielles des équations différentielles d'ordre superieur." *Comptes Rendus*, Vol. 116, 1893, pp. 362–365.

(3) "Sur les transcendantes définies par les équations différentielles du second ordre." *Comptes Rendus*, Vol. 116, 1893, pp. 566–569.

(4) "Sur les équations du second degré dont l'intégrale générale est uniforme." *Comptes Rendus*, Vol. 117, 1893, pp. 211–214.

(5) "Sur les équations du second ordre à points critiques fixes et sur la correspondance univoque entre deux surfaces." *Comptes Rendus*, Vol. 117, 1893, pp. 611–614, 686–688.

(6) *Leçons sur la théorie analytique des équations différentielles.* (Lithographed.) Paris, 1897, 19+6+589 pp.

(7) "Sur la détermination explicite des équations differentielles du second ordre à points critiques fixes." *Comptes Rendus*, Vol. 126, 1898, pp. 1329–1332.

(8) "Sur les équations différentielles du second ordre à points critiques fixes." *Comptes Rendus*, Vol. 126, 1898, pp. 1185–1188, 1699–1700; Vol. 127, 1898, pp. 541–544, 945–948; Vol. 129, 1899, pp. 750–753, 949–952.

(9) "Mémoire sur les équations différentielles dont l'intégrale générale est uniforme." *Bull. Soc. Math. de France,* Vol. 28, 1900, pp. 201–261.

(10) "Sur les singularités essentielles des équations différentielles." *Comptes Rendus*, Vol. 133, 1901, pp. 910–913.

(11) "Sur les équations différentielles du second ordre et d'ordre supérieur, dont l'intégrale générale est uniforme." *Acta Mathematica*, Vol. 25, 1902, pp. 1–85.

(12) "Sur l'irreducibilité de l'équation $y''=6y^2+x$." *Comptes Rendus*, Vol. 135, 1902, pp. 411–415, 641–647, 1020–1025.

(13) "Sur les transcendantes uniformes definies par l'équation $y''=6y^2+x$." *Comptes Rendus*, Vol. 135, 1902, pp. 757–761.

PASCAL, E.: "Sulle equazioni di Riccati." *Giornale di Mat. di Ballaztina.* Vol. 62, 1924, pp. 99–110.

PEIXOTO, M. M.: "On Structural Stability." *Annals of Math.*, Vol. 59, 1959, pp. 199–222.

PERRON, O.: "Die Stabilitatsfrage bei Differentialgleichungen." *Math. Zeitschrift*, Vol. 32, 1930, pp. 703–728.

PETROVITCH, M.: "Contributions à la théorie des solutions singulières des équations différentielles du premier ordre." *Math. Annalen*, Vol. 50, 1898, pp. 103–112.

PETROVSKII, I. G. and LANDIS, E. M.: "On the Number of Limit Cycles of the Equation $dy/dx = P(x,y)/Q(x,y)$ when P and Q are Polynomials of the Second Degree." *Mat. Sbornik,* Vol. 37 (79), 1955, pp. 209–250. Translation in *American Math. Soc. Translations.* Vol. 10, 2d ser., 1958, pp. 177–221.

PICARD, E.:

(1) *Traité d'analyse.* Paris, 1896, Vols. 2 and 3.

(2) "Sur un classe d'équations différentielles." *Comptes Rendus*, Vol. 104, 1887, pp. 41–43.

(3) "Mémoire sur la théorie des fonctions algébrique de deux variables." *Journ. de Math.*, Vol. 5 (4), 1889, pp.135–318. In particular, Chap 5, pp. 263–300.

(4) "Mémoire sur la théorie des équations aux dérivées partielles et la méthode des approximations successives." *Journ. de Math.*, Vol. 6 (4), 1890, pp. 145–210.

PICARD, E.—Continued

(5) "Sur un classe d'équations différentielles dont l'intégrale générale est uniform." *Comptes Rendus,* Vol. 110, 1890, pp. 877–880.

(6) "De l'équation $\Delta u = ke^u$ sur une surface de Riemann fermée." *Journ. de Math.,* Vol. 9 (4), 1893, pp. 273–291.

(7) "Remarques sur les équations différentielles." *Acta Mathematica,* Vol. 17, 1893, pp. 297–300.

(8) "Sur l'équation $\Delta u = e^u$." *Journ. de Math.,* Vol. 4 (5), 1898, pp. 313–316.

(9) "De intégration de l'équation $\Delta u = e^u$ sur une surface de Riemann fermée." *Journ. für Math.,* Vol. 130, 1905, pp. 243–258.

(10) "Application de la théorie des complexes linéaires à l'étude des surfaces et des courbes gauches." *Annales de l'Ecole Normale Superieure,* Vol. 6 (2), 1877, pp. 329–366.

PICONE, M.: "Sulle equazione alle derivate parziali del second' ordine del tipo iperbolico in due variabili independenti." *Rendiconti di Palermo,* Vol. 30, 1910, pp. 349–376.

PINNEY, E.: "Nonlinear Differential Equations Systems." *Contributions to Nonlinear Oscillations,* Vol. 3, Princeton, 1956, pp. 31–56.

POINCARÉ, H.:

(1) "Mémoire sur les courbes définies par une équation différentielle." *Journ. de Math.,* Vol. 7 (3), 1881, pp. 375–442; Vol. 8 (3), 1882, pp. 251–296; Vol. 1 (4), 1885, pp. 167–244. *Oeuvres,* Vol. 1, Paris, 1928, pp. 3–84, 90–161.

(2) *Les méthodes nouvelles de la mécanique céleste.* Vol. 1, Paris, 1892; Vol. 2, 1893. (Dover Reprint)

(3) "Les fonctions fuchsiennes et l'équation $\Delta u = e^u$." *Journ. de Math.,* Vol. 4 (5), 1898, pp. 137–230.

POMPÉIU, D.: "Sur une solution double de l'équation de Riccati." *Comptes Rendus,* Vol. 161, 1915, pp. 235–237.

POOLE, E. G. C.: "A Simply Found Property of Riccati's Equation." *Journ. London Math. Soc.,* Vol. 1, 1926, pp. 197–208.

RAFFY, L.: "Une leçon sur l'équation de Riccati." *Nouvelles Annales de Math.,* Vol. 2 (4), 1902, pp. 529–545.

RAINVILLE, E. D.: "Necessary Conditions for Polynomial Solution of Certain Riccati Equations." *Amer. Math. Monthly,* Vol. 43, 1936, pp. 473–476.

RAUCH. L. L.: "Oscillation of a Third Order Nonlinear Autonomous System." *Contributions to Nonlinear Oscillations,* Vol. 1, Princeton, 1950, pp. 39–88.

RAUSCHER, M.: "Steady Oscillations of Systems With Non-Linear and Unsymmetrical Elasticity." *Journ. Appl. Mech.,* Vol. 5, 1938, p. 169.

RAWSON, R.:

(1) "On Cognate Riccatian Equations." *Messenger of Math.,* Vol. 7, 1878, pp. 69–72.

(2) "Note on a Transformation of Riccati's Equation." *Ibid.,* Vol. 12, 1883, pp. 34–36.

RAYLEIGH, LORD:

(1) *The Theory of Sound.* Vol. 1, London, 1877; Vol. 2, 1878; 2d ed., 1894, 1896. (Dover Reprint)

(2) "On Maintained Vibrations." *Phil. Mag.,* Vol. 15 (1), 1883, p. 229. See *The Theory of Sound,* Vol. 1, p. 81.

REID, W. T.: "A Matrix Differential Equation of Riccati Type." *Amer. Journ. of Math.,* Vol. 68, 1946, pp. 237–246; Vol. 70, 1948, p. 460.

REUTER, G. E. H.:

(1) "Subharmonics in a Non-Linear System with Unsymmetrical Restoring Force." *Quarterly Journal Mech. and Appl. Math.,* Vol. 2, 1949, pp. 198–207.

REUTER, G. E. H.—Continued
  (2) "A Boundedness Theorem for Non-Linear Differential Equations of Second Order." I. *Proc. Cambridge Phil. Soc.*, Vol. 47, 1951, pp. 49–54; II. *Journal of the London Math. Soc.*, Vol. 27, 1952, pp. 48–58.

RICHARDSON, O. W.: *The Emission of Electricity From Hot Bodies.* 1921. In particular, p. 45.

RITT, J. F.: "On the Integration in Finite Terms of Linear Differential Equations of Second Order." *Bull. Amer. Math. Soc.*, Vol. 33, 1927, pp. 51–57.

ROUTH, E. J.: *Dynamics of a Particle.* London, 1898. (Dover Reprint)

SANSONE, G.: *Equazione Differenziali nel Campo Reale.* Vols. 1 and 2. Bologna, 1948–49.

SCHÄFLI, L.: "Sulle realzioni tra diversi integrale definiti che giovano ad esprimere la soluzione generale della equazione di Riccati." *Annali di Mat.*, Vol. 1 (2) 1868, pp. 232–242.

SCHMIDT, E.: "Zur Theorie der linearen und nichtlinearen Integralgleichungen." *Math Annalen*, (I) Vol. 63, 1906–07, pp. 433–476; (II) Vol. 64, 1907–08, pp. 161–174; (III) Vol. 65, 1908, pp. 370–399.

SEIFERT, G.:
  (1) "A Rotated Vector Approach to the Problem of Stability of Solutions of Pendulum-Type Equations." *Contributions to Nonlinear Oscillations*, Vol. 3, Princeton, 1956, pp. 1–16.
  (2) "Rotated Vector Field and an Equation for Relaxation Oscillations." *Ibid.*, Vol. 4, 1958, pp. 125–140.

SHIMIZU, T.: "On Differential Equations for Non-Linear Oscillations." *Mathematica Japonica*, Vol. 2, 1951, pp. 86–96.

SHOHAT, J. A.: "On Van der Pol's and Related Non-Linear Differential Equations." *Journ. of Appl. Physics*, Vol. 15, 1944, pp. 568–574.

SIACCI, F.: "Sulla integrazione di una equazione differenziale e sulla equazione di Riccati." *Rendiconto Napoli*, Vol. 7 (3), 1901, pp. 139–143.

SLOTNICK, D. L.: "Asymptotic Behavior of Solutions of Canonical Systems Near a Closed, Unstable Orbit." *Contributions to Nonlinear Oscillations*, Vol. 4, Princeton, 1958, pp. 85–110.

SOMMERFELD, S. A.: "Integrazione asintotica dell' equazione differenziali di Thomas-Fermi." *Accad. dei Lincei, Atti-Rendiconte*, Vol. 15 (6), 1932, pp. 788–792.

STOKER, J. J.: *Nonlinear Vibrations in Mechanical and Electrical Systems.* New York, 1950, xiv + 273 pp.

SZATROWSKI, Z.: "Solution of Some Integro-Differential Equations." *Dissertation*, Northwestern University, 1942, 67 pp.

TCHACALOFF, L.:
  (1) "Le equazioni di Riccati." *Giornale di Mat.*, Vol. 63, 1925, pp. 139–179.
  (2) "Sur un théorème de M. Ernesto Pascal." *Rendiconto Napoli*, Vol. 31 (3), 1925, pp. 120–122.

THOMAS, L. H.: "The Circulation of Atomic Fields." *Proc. Cambridge Phil. Soc.*, Vol. 23, 1927, pp. 542–548.

TIMOSHENKO, S.: *Vibration Problems in Engineering.* New York, 2d ed., 1937, vi + 470pp.

TORELLI, G.: "Sulla integrazione di un'equazione di Riccati." *Rendiconto Napoli*, Vol. 33 (3), 1927, pp. 17–23, 67–73, 89–98.

TURRITTIN, H. L.: "Asymptotic Solution of Certain Ordinary Differential Equations Associated with Multiple Roots of the Characteristic Equation." *Amer. Journ. of Math.*, Vol. 58, 1936, pp. 364–376.

URABE, K.: "On the Existence of Periodic Solutions for Certain Non-Linear Differential Equations." *Mathematica Japonica*, Vol. 2, 1949, pp. 23–26.

VAN DER POL, B.:
   (1) "On Oscillation Hysteresis in a Triode Generator." *Phil. Mag.*, Vol. 43, (6), 1922, p. 700–719.
   (2) "On 'Relaxation-Oscillations.' " *Phil. Mag.*, Vol. 2 (7), 1926, pp. 978–992.
   (3) "Forced Oscillations in a Circuit With Non-Linear Resistance." *Phil. Mag.*, Vol. 3, (7), 1927, pp. 65–80.
VERGERIO, A.:
   (1) "Sulle equaxioni integrali non lineari." *Annali di Mat.*, Vol. 31, 1922, pp. 81–119.
   (2) "Sulle equazioni integrali non lineari con operazioni funzionali singolari." *Giornale di Mat.*, Vol. 59, 1921, pp. 175–214.
VESSIOT, E.: "Sur quelques équations différentielles ordinaires du second ordre." *Annales de Toulouse*, Vol. 9, 1895, No. 6, pp. 1–26.
VIGERIER, G.: "Les Chaînes de Darboux et l'équation de Fourier." *Experientia*, Vol. 5, 1949, p. 439.
VOLTERRA, V.: *Leçons sur la théorie mathématiques de la lutte pour la vie.* Paris, 1931, vi + 214 pp.
WALKER, G. W.: (Liouville's Equation). *Proc. of the Royal Soc. of London*, Vol. 91 (A), 1915, p. 410.
WALLENBERG, G.:
   (1) "Über Riccatische Differentialgleichungen höherer Ordnung." *Journ. für Math.*, Vol. 121, 1899, pp. 196–199.
   (2) "Die Differentialgleichungen, deren allgemeines Integral eine lineare gebrochene Function der willkürlichen Constanten ist." *Journ. für Math.*, 1899, Vol. 121, pp. 210–217.
   (3) "Sur l'équation différentielle de Riccati du second ordre." *Comptes Rendus*, Vol. 137, 1903, pp. 1033–1035.
WASOW, W.: "The Construction of Periodic Solutions of Singular Perturbation Problems." *Contributions to Nonlinear Oscillations*, Vol. 1, Princeton, 1950, pp. 313–350.
WATSON, G. N.: *Theory of Bessel Functions.* Cambridge, 1922; 2d ed., 1944, 804 pp.
WENDEL, J. G.: "Singular Perturbations of a Van der Pol Equation." *Contributions to Nonlinear Oscillations*, Vol. 1, Princeton, 1950, pp. 243–289.
WEYR, R.: "Zur Integration der Differentialgleichungen erster Ordnung." *Abh. böhm. Ges. Wiss.* (Prag), Vol. 7 (6), 1875–76, Math. Mem. 1.
WHITTAKER, E. T.: *Analytical Dynamics.* Cambridge, 1904; 4th ed., 1937, xiv + 456 pp.
WHITTAKER, E. T., and WATSON, G. N.: *A Course of Modern Analysis.* Cambridge, 1st ed., 1902; 4th ed., 1927, 608 pp.
WORKMAN, W. P.: "The Theory of the Singular Solutions of Integrable Differential Equations of First Order." *Quart. Journ. of Math.*, Vol. 22, 1887, pp. 175–198, 308–324.

Index of Names

Index of Subjects

Abel's equation, 74–75.
Action, definition of, 457.
Adams-Bashforth method, 481.
d'Alembert's equation, 53.
Algebraic functions, 7.
Almost periodic functions, 290–291.
Analytic continuation, 245, 246; Chap. 9.
Asymptotic series, 5.
Automorphic functions, definition of, 222.

Bendixson's theorem, 354–355.
Bernoulli's equation, 49.
Bernoulli numbers, 473.
Bessel functions, 4, 5, 68–70.
Blasius equation, 16, 400–405.
Branch points, 14, 185, 225.
Bratu's equation, 432–434.

Calculus of variations, Chap. 14; fundamental lemma of, 443; necessary conditions for extremals in, 449–450.
Calculus of limits, 79–83, 190–191.
Capture point, 117, 123–124.
Cauchy-Lipschitz, method of, 88–93.
c-discriminant, 50, 54.
Center, 322; conditions for, 350.
Cepheid variable stars, 368–371.
Characteristic equation, 311, 317.
Chrystal's equation, 51–53.
Clairaut's equation, 49–51.
Coefficient of heredity, 113.
Complementary elliptic integral, 131.
Complete elliptic integrals, definition of, 131; expansions of, 133–135; differential equations of, 137.
Continued fractions, 70–72.
Continuous analytic continuation, Chap. 9; method of, 251–256; application around a singular point, 263–266.
Convolution, nonlinear, 434–437.
Critical points, 14, 185, 225.
Curvature, method of, 29–32, 248–249.
Curves of pursuit. (See Pursuit).
Cusp-locus, 54–56.

Dido's problem, 439, 445–447.
Differential equations of first order, Chap. 2; graphical method of solution, 26–29; solution by isoclines and curvature, 29–32; homogeneous case of, 36–45; Abel's, 74–75; d'Alembert's, 53; Bernoulli's, 49; Clairaut's, 49–51; Chrystal's, 51–53; Riccati's, Chap. 3.
Differential equations of second order, Chaps. 7 and 8; classification of, 182–189; existence theorems for, 189–192; with fixed critical points, 184–185, 225–228; with periodic solutions, 186, 297–308, 339–351; of polynomial class, Chap. 8; Duffing's, 15, 180, 271, 386–400; elliptic, 14, 179, 183–184, 194–211, 270; Emden's, 15, 181, 371–377; Emden type, 381–386, 408; generalized Riccati, 76–79, 186, 218–221; Kidder's, 16, 410–411; Langmuir's, 16, 181; Painlevé's, 14, 185, 229–244, 233–234, 239–244, 258–261, 263–266, 501–504, values of, 507–530; Rayleigh's, 15, 181, 186–187; Thomas-Fermi's, 16, 405–407; Van der Pol's, 15, 17, 181, 186–187, 270, 358–368, values of, 531–536; Volterra's, 102, 180, 270, 324–327, 355, values of, 537–542; white dwarf, 16, 408.
Differential equation of third order, Blasius, 16, 400–405.
Duffing's equation, 15, 180, 271, 386–400.

Elliptic equations, 14, 179, 183–184, 194–211, 270.
Elliptic functions, Chap. 6. (See Jacobi, elliptic functions of, and Weierstrass, elliptic functions of,)
Elliptic integrals, Chap. 6; first kind, 131; second kind, 131; third kind, 132; expansions of, 135–137; computation of, 137–141; tables of, 142–143.

563

A CATALOGUE OF
SELECTED DOVER BOOKS
IN ALL FIELDS OF INTEREST

A CATALOGUE OF SELECTED DOVER
BOOKS IN ALL FIELDS OF INTEREST

RACKHAM'S COLOR ILLUSTRATIONS FOR WAGNER'S RING. Rackham's finest mature work—all 64 full-color watercolors in a faithful and lush interpretation of the *Ring*. Full-sized plates on coated stock of the paintings used by opera companies for authentic staging of Wagner. Captions aid in following complete Ring cycle. Introduction. 64 illustrations plus vignettes. 72pp. 8⅝ x 11¼. 23779-6 Pa. $6.00

CONTEMPORARY POLISH POSTERS IN FULL COLOR, edited by Joseph Czestochowski. 46 full-color examples of brilliant school of Polish graphic design, selected from world's first museum (near Warsaw) dedicated to poster art. Posters on circuses, films, plays, concerts all show cosmopolitan influences, free imagination. Introduction. 48pp. 9⅜ x 12¼. 23780-X Pa. $6.00

GRAPHIC WORKS OF EDVARD MUNCH, Edvard Munch. 90 haunting, evocative prints by first major Expressionist artist and one of the greatest graphic artists of his time: *The Scream, Anxiety, Death Chamber, The Kiss, Madonna,* etc. Introduction by Alfred Werner. 90pp. 9 x 12. 23765-6 Pa. $5.00

THE GOLDEN AGE OF THE POSTER, Hayward and Blanche Cirker. 70 extraordinary posters in full colors, from Maitres de l'Affiche, Mucha, Lautrec, Bradley, Cheret, Beardsley, many others. Total of 78pp. 9⅜ x 12¼. 22753-7 Pa. $5.95

THE NOTEBOOKS OF LEONARDO DA VINCI, edited by J. P. Richter. Extracts from manuscripts reveal great genius; on painting, sculpture, anatomy, sciences, geography, etc. Both Italian and English. 186 ms. pages reproduced, plus 500 additional drawings, including studies for *Last Supper,* Sforza monument, etc. 860pp. 7⅞ x 10¾. (Available in U.S. only) 22572-0, 22573-9 Pa., Two-vol. set $15.90

THE CODEX NUTTALL, as first edited by Zelia Nuttall. Only inexpensive edition, in full color, of a pre-Columbian Mexican (Mixtec) book. 88 color plates show kings, gods, heroes, temples, sacrifices. New explanatory, historical introduction by Arthur G. Miller. 96pp. 11⅜ x 8½. (Available in U.S. only) 23168-2 Pa. $7.95

UNE SEMAINE DE BONTÉ, A SURREALISTIC NOVEL IN COLLAGE, Max Ernst. Masterpiece created out of 19th-century periodical illustrations, explores worlds of terror and surprise. Some consider this Ernst's greatest work. 208pp. 8⅛ x 11. 23252-2 Pa. $6.00

DRAWINGS OF WILLIAM BLAKE, William Blake. 92 plates from Book of Job, *Divine Comedy, Paradise Lost,* visionary heads, mythological figures, Laocoon, etc. Selection, introduction, commentary by Sir Geoffrey Keynes. 178pp. 8⅛ x 11. 22303-5 Pa. $4.00

ENGRAVINGS OF HOGARTH, William Hogarth. 101 of Hogarth's greatest works: *Rake's Progress, Harlot's Progress, Illustrations for Hudibras, Before and After, Beer Street and Gin Lane,* many more. Full commentary. 256pp. 11 x 13¾. 22479-1 Pa. $12.95

DAUMIER: 120 GREAT LITHOGRAPHS, Honore Daumier. Wide-ranging collection of lithographs by the greatest caricaturist of the 19th century. Concentrates on eternally popular series on lawyers, on married life, on liberated women, etc. Selection, introduction, and notes on plates by Charles F. Ramus. Total of 158pp. 9⅜ x 12¼. 23512-2 Pa. $6.00

DRAWINGS OF MUCHA, Alphonse Maria Mucha. Work reveals drafts-man of highest caliber: studies for famous posters and paintings, render-ings for book illustrations and ads, etc. 70 works, 9 in color; including 6 items not drawings. Introduction. List of illustrations. 72pp. 9⅜ x 12¼. (Available in U.S. only) 23672-2 Pa. $4.00

GIOVANNI BATTISTA PIRANESI: DRAWINGS IN THE PIERPONT MORGAN LIBRARY, Giovanni Battista Piranesi. For first time ever all of Morgan Library's collection, world's largest. 167 illustrations of rare Piranesi drawings—archeological, architectural, decorative and visionary. Essay, detailed list of drawings, chronology, captions. Edited by Felice Stampfle. 144pp. 9⅜ x 12¼. 23714-1 Pa. $7.50

NEW YORK ETCHINGS (1905-1949), John Sloan. All of important American artist's N.Y. life etchings. 67 works include some of his best art; also lively historical record—Greenwich Village, tenement scenes. Edited by Sloan's widow. Introduction and captions. 79pp. 8⅜ x 11¼. 23651-X Pa. $4.00

CHINESE PAINTING AND CALLIGRAPHY: A PICTORIAL SURVEY, Wan-go Weng. 69 fine examples from John M. Crawford's matchless private collection: landscapes, birds, flowers, human figures, etc., plus calligraphy. Every basic form included: hanging scrolls, handscrolls, album leaves, fans, etc. 109 illustrations. Introduction. Captions. 192pp. 8⅞ x 11¾. 23707-9 Pa. $7.95

DRAWINGS OF REMBRANDT, edited by Seymour Slive. Updated Lipp-mann, Hofstede de Groot edition, with definitive scholarly apparatus. All portraits, biblical sketches, landscapes, nudes, Oriental figures, classical studies, together with selection of work by followers. 550 illustrations. Total of 630pp. 9⅛ x 12¼. 21485-0, 21486-9 Pa., Two-vol. set $15.00

THE DISASTERS OF WAR, Francisco Goya. 83 etchings record horrors of Napoleonic wars in Spain and war in general. Reprint of 1st edition, plus 3 additional plates. Introduction by Philip Hofer. 97pp. 9⅜ x 8¼. 21872-4 Pa. $4.00

THE EARLY WORK OF AUBREY BEARDSLEY, Aubrey Beardsley. 157 plates, 2 in color: *Manon Lescaut, Madame Bovary, Morte Darthur, Salome,* other. Introduction by H. Marillier. 182pp. 8⅛ x 11. 21816-3 Pa. $4.50

THE LATER WORK OF AUBREY BEARDSLEY, Aubrey Beardsley. Exotic masterpieces of full maturity: *Venus and Tannhauser, Lysistrata, Rape of the Lock, Volpone,* Savoy material, etc. 174 plates, 2 in color. 186pp. 8⅛ x 11. 21817-1 Pa. $5.95

THOMAS NAST'S CHRISTMAS DRAWINGS, Thomas Nast. Almost all Christmas drawings by creator of image of Santa Claus as we know it, and one of America's foremost illustrators and political cartoonists. 66 illustrations. 3 illustrations in color on covers. 96pp. 8⅜ x 11¼. 23660-9 Pa. $3.50

THE DORÉ ILLUSTRATIONS FOR DANTE'S DIVINE COMEDY, Gustave Doré. All 135 plates from Inferno, Purgatory, Paradise; fantastic tortures, infernal landscapes, celestial wonders. Each plate with appropriate (translated) verses. 141pp. 9 x 12. 23231-X Pa. $4.50

DORÉ'S ILLUSTRATIONS FOR RABELAIS, Gustave Doré. 252 striking illustrations of *Gargantua and Pantagruel* books by foremost 19th-century illustrator. Including 60 plates, 192 delightful smaller illustrations. 153pp. 9 x 12. 23656-0 Pa. $5.00

LONDON: A PILGRIMAGE, Gustave Doré, Blanchard Jerrold. Squalor, riches, misery, beauty of mid-Victorian metropolis; 55 wonderful plates, 125 other illustrations, full social, cultural text by Jerrold. 191pp. of text. 9⅜ x 12¼. 22306-X Pa. $7.00

THE RIME OF THE ANCIENT MARINER, Gustave Doré, S. T. Coleridge. Dore's finest work, 34 plates capture moods, subtleties of poem. Full text. Introduction by Millicent Rose. 77pp. 9¼ x 12. 22305-1 Pa. $3.50

THE DORE BIBLE ILLUSTRATIONS, Gustave Doré. All wonderful, detailed plates: Adam and Eve, Flood, Babylon, Life of Jesus, etc. Brief King James text with each plate. Introduction by Millicent Rose. 241 plates. 241pp. 9 x 12. 23004-X Pa. $6.00

THE COMPLETE ENGRAVINGS, ETCHINGS AND DRYPOINTS OF ALBRECHT DURER. "Knight, Death and Devil"; "Melencolia," and more—all Dürer's known works in all three media, including 6 works formerly attributed to him. 120 plates. 235pp. 8⅜ x 11¼. 22851-7 Pa. $6.50

MECHANICK EXERCISES ON THE WHOLE ART OF PRINTING, Joseph Moxon. First complete book (1683-4) ever written about typography, a compendium of everything known about printing at the latter part of 17th century. Reprint of 2nd (1962) Oxford Univ. Press edition. 74 illustrations. Total of 550pp. 6⅛ x 9¼. 23617-X Pa. $7.95

THE COMPLETE WOODCUTS OF ALBRECHT DURER, edited by Dr. W. Kurth. 346 in all: "Old Testament," "St. Jerome," "Passion," "Life of Virgin," Apocalypse," many others. Introduction by Campbell Dodgson. 285pp. 8½ x 12¼. 21097-9 Pa. $7.50

DRAWINGS OF ALBRECHT DURER, edited by Heinrich Wölfflin. 81 plates show development from youth to full style. Many favorites; many new. Introduction by Alfred Werner. 96pp. 8⅛ x 11. 22352-3 Pa. $5.00

THE HUMAN FIGURE, Albrecht Dürer. Experiments in various techniques—stereometric, progressive proportional, and others. Also life studies that rank among finest ever done. Complete reprinting of *Dresden Sketchbook*. 170 plates. 355pp. 8⅜ x 11¼. 21042-1 Pa. $7.95

OF THE JUST SHAPING OF LETTERS, Albrecht Dürer. Renaissance artist explains design of Roman majuscules by geometry, also Gothic lower and capitals. Grolier Club edition. 43pp. 7⅞ x 10¾ 21306-4 Pa. $3.00

TEN BOOKS ON ARCHITECTURE, Vitruvius. The most important book ever written on architecture. Early Roman aesthetics, technology, classical orders, site selection, all other aspects. Stands behind everything since. Morgan translation. 331pp. 5⅜ x 8½. 20645-9 Pa. $4.50

THE FOUR BOOKS OF ARCHITECTURE, Andrea Palladio. 16th-century classic responsible for Palladian movement and style. Covers classical architectural remains, Renaissance revivals, classical orders, etc. 1738 Ware English edition. Introduction by A. Placzek. 216 plates. 110pp. of text. 9½ x 12¾. 21308-0 Pa. $10.00

HORIZONS, Norman Bel Geddes. Great industrialist stage designer, "father of streamlining," on application of aesthetics to transportation, amusement, architecture, etc. 1932 prophetic account; function, theory, specific projects. 222 illustrations. 312pp. 7⅞ x 10¾. 23514-9 Pa. $6.95

FRANK LLOYD WRIGHT'S FALLINGWATER, Donald Hoffmann. Full, illustrated story of conception and building of Wright's masterwork at Bear Run, Pa. 100 photographs of site, construction, and details of completed structure. 112pp. 9¼ x 10. 23671-4 Pa. $5.50

THE ELEMENTS OF DRAWING, John Ruskin. Timeless classic by great Viltorian; starts with basic ideas, works through more difficult. Many practical exercises. 48 illustrations. Introduction by Lawrence Campbell. 228pp. 5⅜ x 8½. 22730-8 Pa. $3.75

GIST OF ART, John Sloan. Greatest modern American teacher, Art Students League, offers innumerable hints, instructions, guided comments to help you in painting. Not a formal course. 46 illustrations. Introduction by Helen Sloan. 200pp. 5⅜ x 8½. 23435-5 Pa. $4.00

THE ANATOMY OF THE HORSE, George Stubbs. Often considered the great masterpiece of animal anatomy. Full reproduction of 1766 edition, plus prospectus; original text and modernized text. 36 plates. Introduction by Eleanor Garvey. 121pp. 11 x 14¾. 23402-9 Pa. $6.00

BRIDGMAN'S LIFE DRAWING, George B. Bridgman. More than 500 illustrative drawings and text teach you to abstract the body into its major masses, use light and shade, proportion; as well as specific areas of anatomy, of which Bridgman is master. 192pp. 6½ x 9¼. (Available in U.S. only) 22710-3 Pa. $3.50

ART NOUVEAU DESIGNS IN COLOR, Alphonse Mucha, Maurice Verneuil, Georges Auriol. Full-color reproduction of *Combinaisons ornementales* (c. 1900) by Art Nouveau masters. Floral, animal, geometric, interlacings, swashes—borders, frames, spots—all incredibly beautiful. 60 plates, hundreds of designs. 9⅜ x 8-1/16. 22885-1 Pa. $4.00

FULL-COLOR FLORAL DESIGNS IN THE ART NOUVEAU STYLE, E. A. Seguy. 166 motifs, on 40 plates, from *Les fleurs et leurs applications decoratives* (1902): borders, circular designs, repeats, allovers, "spots." All in authentic Art Nouveau colors. 48pp. 9⅜ x 12¼. 23439-8 Pa. $5.00

A DIDEROT PICTORIAL ENCYCLOPEDIA OF TRADES AND INDUSTRY, edited by Charles C. Gillispie. 485 most interesting plates from the great French Encyclopedia of the 18th century show hundreds of working figures, artifacts, process, land and cityscapes; glassmaking, papermaking, metal extraction, construction, weaving, making furniture, clothing, wigs, dozens of other activities. Plates fully explained. 920pp. 9 x 12. 22284-5, 22285-3 Clothbd., Two-vol. set $40.00

HANDBOOK OF EARLY ADVERTISING ART, Clarence P. Hornung. Largest collection of copyright-free early and antique advertising art ever compiled. Over 6,000 illustrations, from Franklin's time to the 1890's for special effects, novelty. Valuable source, almost inexhaustible.
Pictorial Volume. Agriculture, the zodiac, animals, autos, birds, Christmas, fire engines, flowers, trees, musical instruments, ships, games and sports, much more. Arranged by subject matter and use. 237 plates. 288pp. 9 x 12. 20122-8 Clothbd. $14.50

Typographical Volume. Roman and Gothic faces ranging from 10 point to 300 point, "Barnum," German and Old English faces, script, logotypes, scrolls and flourishes, 1115 ornamental initials, 67 complete alphabets, more. 310 plates. 320pp. 9 x 12. 20123-6 Clothbd. $15.00

CALLIGRAPHY (CALLIGRAPHIA LATINA), J. G. Schwandner. High point of 18th-century ornamental calligraphy. Very ornate initials, scrolls, borders, cherubs, birds, lettered examples. 172pp. 9 x 13. 20475-8 Pa. $7.00

ART FORMS IN NATURE, Ernst Haeckel. Multitude of strangely beautiful natural forms: Radiolaria, Foraminifera, jellyfishes, fungi, turtles, bats, etc. All 100 plates of the 19th-century evolutionist's *Kunstformen der Natur* (1904). 100pp. 9⅜ x 12¼. 22987-4 Pa. $5.00

CHILDREN: A PICTORIAL ARCHIVE FROM NINETEENTH-CENTURY SOURCES, edited by Carol Belanger Grafton. 242 rare, copyright-free wood engravings for artists and designers. Widest such selection available. All illustrations in line. 119pp. 8⅜ x 11¼.
23694-3 Pa. $4.00

WOMEN: A PICTORIAL ARCHIVE FROM NINETEENTH-CENTURY SOURCES, edited by Jim Harter. 391 copyright-free wood engravings for artists and designers selected from rare periodicals. Most extensive such collection available. All illustrations in line. 128pp. 9 x 12.
23703-6 Pa. $4.50

ARABIC ART IN COLOR, Prisse d'Avennes. From the greatest ornamentalists of all time—50 plates in color, rarely seen outside the Near East, rich in suggestion and stimulus. Includes 4 plates on covers. 46pp. 9⅜ x 12¼. 23658-7 Pa. $6.00

AUTHENTIC ALGERIAN CARPET DESIGNS AND MOTIFS, edited by June Beveridge. Algerian carpets are world famous. Dozens of geometrical motifs are charted on grids, color-coded, for weavers, needleworkers, craftsmen, designers. 53 illustrations plus 4 in color. 48pp. 8¼ x 11. (Available in U.S. only) 23650-1 Pa. $1.75

DICTIONARY OF AMERICAN PORTRAITS, edited by Hayward and Blanche Cirker. 4000 important Americans, earliest times to 1905, mostly in clear line. Politicians, writers, soldiers, scientists, inventors, industrialists, Indians, Blacks, women, outlaws, etc. Identificatory information. 756pp. 9¼ x 12¾. 21823-6 Clothbd. $40.00

HOW THE OTHER HALF LIVES, Jacob A. Riis. Journalistic record of filth, degradation, upward drive in New York immigrant slums, shops, around 1900. New edition includes 100 original Riis photos, monuments of early photography. 233pp. 10 x 7⅞. 22012-5 Pa. $7.00

NEW YORK IN THE THIRTIES, Berenice Abbott. Noted photographer's fascinating study of city shows new buildings that have become famous and old sights that have disappeared forever. Insightful commentary. 97 photographs. 97pp. 11⅜ x 10. 22967-X Pa. $5.00

MEN AT WORK, Lewis W. Hine. Famous photographic studies of construction workers, railroad men, factory workers and coal miners. New supplement of 18 photos on Empire State building construction. New introduction by Jonathan L. Doherty. Total of 69 photos. 63pp. 8 x 10¾.
23475-4 Pa. $3.00

THE DEPRESSION YEARS AS PHOTOGRAPHED BY ARTHUR ROTH-STEIN, Arthur Rothstein. First collection devoted entirely to the work of outstanding 1930s photographer: famous dust storm photo, ragged children, unemployed, etc. 120 photographs. Captions. 119pp. 9¼ x 10¾.
23590-4 Pa. $5.00

CAMERA WORK: A PICTORIAL GUIDE, Alfred Stieglitz. All 559 illustrations and plates from the most important periodical in the history of art photography, Camera Work (1903-17). Presented four to a page, reduced in size but still clear, in strict chronological order, with complete captions. Three indexes. Glossary. Bibliography. 176pp. 8⅜ x 11¼.
23591-2 Pa. $6.95

ALVIN LANGDON COBURN, PHOTOGRAPHER, Alvin L. Coburn. Revealing autobiography by one of greatest photographers of 20th century gives insider's version of Photo-Secession, plus comments on his own work. 77 photographs by Coburn. Edited by Helmut and Alison Gernsheim. 160pp. 8⅛ x 11.
23685-4 Pa. $6.00

NEW YORK IN THE FORTIES, Andreas Feininger. 162 brilliant photographs by the well-known photographer, formerly with Life magazine, show commuters, shoppers, Times Square at night, Harlem nightclub, Lower East Side, etc. Introduction and full captions by John von Hartz. 181pp. 9¼ x 10¾.
23585-8 Pa. $6.95

GREAT NEWS PHOTOS AND THE STORIES BEHIND THEM, John Faber. Dramatic volume of 140 great news photos, 1855 through 1976, and revealing stories behind them, with both historical and technical information. Hindenburg disaster, shooting of Oswald, nomination of Jimmy Carter, etc. 160pp. 8¼ x 11.
23667-6 Pa. $5.00

THE ART OF THE CINEMATOGRAPHER, Leonard Maltin. Survey of American cinematography history and anecdotal interviews with 5 masters—Arthur Miller, Hal Mohr, Hal Rosson, Lucien Ballard, and Conrad Hall. Very large selection of behind-the-scenes production photos. 105 photographs. Filmographies. Index. Originally Behind the Camera. 144pp. 8¼ x 11.
23686-2 Pa. $5.00

DESIGNS FOR THE THREE-CORNERED HAT (LE TRICORNE), Pablo Picasso. 32 fabulously rare drawings—including 31 color illustrations of costumes and accessories—for 1919 production of famous ballet. Edited by Parmenia Migel, who has written new introduction. 48pp. 9⅜ x 12¼.
(Available in U.S. only)
23709-5 Pa. $5.00

NOTES OF A FILM DIRECTOR, Sergei Eisenstein. Greatest Russian filmmaker explains montage, making of Alexander Nevsky, aesthetics; comments on self, associates, great rivals (Chaplin), similar material. 78 illustrations. 240pp. 5⅜ x 8½.
22392-2 Pa. $4.50

HOLLYWOOD GLAMOUR PORTRAITS, edited by John Kobal. 145 photos capture the stars from 1926-49, the high point in portrait photography. Gable, Harlow, Bogart, Bacall, Hedy Lamarr, Marlene Dietrich, Robert Montgomery, Marlon Brando, Veronica Lake; 94 stars in all. Full background on photographers, technical aspects, much more. Total of 160pp. 8⅜ x 11¼.   23352-9 Pa. $6.00

THE NEW YORK STAGE: FAMOUS PRODUCTIONS IN PHOTO-GRAPHS, edited by Stanley Appelbaum. 148 photographs from Museum of City of New York show 142 plays, 1883-1939. *Peter Pan, The Front Page, Dead End, Our Town,* O'Neill, hundreds of actors and actresses, etc. Full indexes. 154pp. 9½ x 10.   23241-7 Pa. $6.00

DIALOGUES CONCERNING TWO NEW SCIENCES, Galileo Galilei. Encompassing 30 years of experiment and thought, these dialogues deal with geometric demonstrations of fracture of solid bodies, cohesion, leverage, speed of light and sound, pendulums, falling bodies, accelerated motion, etc. 300pp. 5⅜ x 8½.   60099-8 Pa. $4.00

THE GREAT OPERA STARS IN HISTORIC PHOTOGRAPHS, edited by James Camner. 343 portraits from the 1850s to the 1940s: Tamburini, Mario, Caliapin, Jeritza, Melchior, Melba, Patti, Pinza, Schipa, Caruso, Farrar, Steber, Gobbi, and many more—270 performers in all. Index. 199pp. 8⅜ x 11¼.   23575-0 Pa. $7.50

J. S. BACH, Albert Schweitzer. Great full-length study of Bach, life, background to music, music, by foremost modern scholar. Ernest Newman translation. 650 musical examples. Total of 928pp. 5⅜ x 8½. (Available in U.S. only)   21631-4, 21632-2 Pa., Two-vol. set $11.00

COMPLETE PIANO SONATAS, Ludwig van Beethoven. All sonatas in the fine Schenker edition, with fingering, analytical material. One of best modern editions. Total of 615pp. 9 x 12. (Available in U.S. only)
                           23134-8, 23135-6 Pa., Two-vol. set $15.50

KEYBOARD MUSIC, J. S. Bach. Bach-Gesellschaft edition. For harpsichord, piano, other keyboard instruments. English Suites, French Suites, Six Partitas, Goldberg Variations, Two-Part Inventions, Three-Part Sinfonias. 312pp. 8⅛ x 11. (Available in U.S. only)   22360-4 Pa. $6.95

FOUR SYMPHONIES IN FULL SCORE, Franz Schubert. Schubert's four most popular symphonies: No. 4 in C Minor ("Tragic"); No. 5 in B-flat Major; No. 8 in B Minor ("Unfinished"); No. 9 in C Major ("Great"). Breitkopf & Hartel edition. Study score. 261pp. 9⅜ x 12¼.
                                                23681-1 Pa. $6.50

THE AUTHENTIC GILBERT & SULLIVAN SONGBOOK, W. S. Gilbert, A. S. Sullivan. Largest selection available; 92 songs, uncut, original keys, in piano rendering approved by Sullivan. Favorites and lesser-known fine numbers. Edited with plot synopses by James Spero. 3 illustrations. 399pp. 9 x 12.   23482-7 Pa. $9.95

PRINCIPLES OF ORCHESTRATION, Nikolay Rimsky-Korsakov. Great classical orchestrator provides fundamentals of tonal resonance, progression of parts, voice and orchestra, tutti effects, much else in major document. 330pp. of musical excerpts. 489pp. 6½ x 9¼.   21266-1 Pa. $7.50

TRISTAN UND ISOLDE, Richard Wagner. Full orchestral score with complete instrumentation. Do not confuse with piano reduction. Commentary by Felix Mottl, great Wagnerian conductor and scholar. Study score. 655pp. 8⅛ x 11.   22915-7 Pa. $13.95

REQUIEM IN FULL SCORE, Giuseppe Verdi. Immensely popular with choral groups and music lovers. Republication of edition published by C. F. Peters, Leipzig, n. d. German frontmaker in English translation. Glossary. Text in Latin. Study score. 204pp. 9⅜ x 12¼.
23682-X Pa. $6.00

COMPLETE CHAMBER MUSIC FOR STRINGS, Felix Mendelssohn. All of Mendelssohn's chamber music: Octet, 2 Quintets, 6 Quartets, and Four Pieces for String Quartet. (Nothing with piano is included). Complete works edition (1874-7). Study score. 283 pp. 9⅜ x 12¼.
23679-X Pa. $7.50

POPULAR SONGS OF NINETEENTH-CENTURY AMERICA, edited by Richard Jackson. 64 most important songs: "Old Oaken Bucket," "Arkansas Traveler," "Yellow Rose of Texas," etc. Authentic original sheet music, full introduction and commentaries. 290pp. 9 x 12.   23270-0 Pa. $7.95

COLLECTED PIANO WORKS, Scott Joplin. Edited by Vera Brodsky Lawrence. Practically all of Joplin's piano works—rags, two-steps, marches, waltzes, etc., 51 works in all. Extensive introduction by Rudi Blesh. Total of 345pp. 9 x 12.   23106-2 Pa. $14.95

BASIC PRINCIPLES OF CLASSICAL BALLET, Agrippina Vaganova. Great Russian theoretician, teacher explains methods for teaching classical ballet; incorporates best from French, Italian, Russian schools. 118 illustrations. 175pp. 5⅜ x 8½.   22036-2 Pa. $2.50

CHINESE CHARACTERS, L. Wieger. Rich analysis of 2300 characters according to traditional systems into primitives. Historical-semantic analysis to phonetics (Classical Mandarin) and radicals. 820pp. 6⅛ x 9¼.
21321-8 Pa. $10.00

EGYPTIAN LANGUAGE: EASY LESSONS IN EGYPTIAN HIERO-GLYPHICS, E. A. Wallis Budge. Foremost Egyptologist offers Egyptian grammar, explanation of hieroglyphics, many reading texts, dictionary of symbols. 246pp. 5 x 7½. (Available in U.S. only)
21394-3 Clothbd. $7.50

AN ETYMOLOGICAL DICTIONARY OF MODERN ENGLISH, Ernest Weekley. Richest, fullest work, by foremost British lexicographer. Detailed word histories. Inexhaustible. Do not confuse this with *Concise Etymological Dictionary*, which is abridged. Total of 856pp. 6½ x 9¼.
21873-2, 21874-0 Pa., Two-vol. set $12.00

A MAYA GRAMMAR, Alfred M. Tozzer. Practical, useful English-language grammar by the Harvard anthropologist who was one of the three greatest American scholars in the area of Maya culture. Phonetics, grammatical processes, syntax, more. 301pp. 5⅜ x 8½.   23465-7 Pa. $4.00

THE JOURNAL OF HENRY D. THOREAU, edited by Bradford Torrey, F. H. Allen. Complete reprinting of 14 volumes, 1837-61, over two million words; the sourcebooks for *Walden*, etc. Definitive. All original sketches, plus 75 photographs. Introduction by Walter Harding. Total of 1804pp. 8½ x 12¼.   20312-3, 20313-1 Clothbd., Two-vol. set $70.00

CLASSIC GHOST STORIES, Charles Dickens and others. 18 wonderful stories you've wanted to reread: "The Monkey's Paw," "The House and the Brain," "The Upper Berth," "The Signalman," "Dracula's Guest," "The Tapestried Chamber," etc. Dickens, Scott, Mary Shelley, Stoker, etc. 330pp. 5⅜ x 8½.   20735-8 Pa. $4.50

SEVEN SCIENCE FICTION NOVELS, H. G. Wells. Full novels. *First Men in the Moon, Island of Dr. Moreau, War of the Worlds, Food of the Gods, Invisible Man, Time Machine, In the Days of the Comet.* A basic science-fiction library. 1015pp. 5⅜ x 8½. (Available in U.S. only)
20264-X Clothbd. $8.95

ARMADALE, Wilkie Collins. Third great mystery novel by the author of *The Woman in White* and *The Moonstone.* Ingeniously plotted narrative shows an exceptional command of character, incident and mood. Original magazine version with 40 illustrations. 597pp. 5⅜ x 8½.
23429-0 Pa. $6.00

MASTERS OF MYSTERY, H. Douglas Thomson. The first book in English (1931) devoted to history and aesthetics of detective story. Poe, Doyle, LeFanu, Dickens, many others, up to 1930. New introduction and notes by E. F. Bleiler. 288pp. 5⅜ x 8½. (Available in U.S. only)
23606-4 Pa. $4.00

FLATLAND, E. A. Abbott. Science-fiction classic explores life of 2-D being in 3-D world. Read also as introduction to thought about hyperspace. Introduction by Banesh Hoffmann. 16 illustrations. 103pp. 5⅜ x 8½.
20001-9 Pa. $2.00

THREE SUPERNATURAL NOVELS OF THE VICTORIAN PERIOD, edited, with an introduction, by E. F. Bleiler. Reprinted complete and unabridged, three great classics of the supernatural: *The Haunted Hotel* by Wilkie Collins, *The Haunted House at Latchford* by Mrs. J. H. Riddell, and *The Lost Stradivarious* by J. Meade Falkner. 325pp. 5⅜ x 8½.
22571-2 Pa. $4.00

AYESHA: THE RETURN OF "SHE," H. Rider Haggard. Virtuoso sequel featuring the great mythic creation, Ayesha, in an adventure that is fully as good as the first book, *She.* Original magazine version, with 47 original illustrations by Maurice Greiffenhagen. 189pp. 6½ x 9¼.
23649-8 Pa. $3.50

UNCLE SILAS, J. Sheridan LeFanu. Victorian Gothic mystery novel, considered by many best of period, even better than Collins or Dickens. Wonderful psychological terror. Introduction by Frederick Shroyer. 436pp. 5⅜ x 8½. 21715-9 Pa. $6.00

JURGEN, James Branch Cabell. The great erotic fantasy of the 1920's that delighted thousands, shocked thousands more. Full final text, Lane edition with 13 plates by Frank Pape. 346pp. 5⅜ x 8½.
23507-6 Pa. $4.50

THE CLAVERINGS, Anthony Trollope. Major novel, chronicling aspects of British Victorian society, personalities. Reprint of Cornhill serialization, 16 plates by M. Edwards; first reprint of full text. Introduction by Norman Donaldson. 412pp. 5⅜ x 8½. 23464-9 Pa. $5.00

KEPT IN THE DARK, Anthony Trollope. Unusual short novel about Victorian morality and abnormal psychology by the great English author. Probably the first American publication. Frontispiece by Sir John Millais. 92pp. 6½ x 9¼. 23609-9 Pa. $2.50

RALPH THE HEIR, Anthony Trollope. Forgotten tale of illegitimacy, inheritance. Master novel of Trollope's later years. Victorian country estates, clubs, Parliament, fox hunting, world of fully realized characters. Reprint of 1871 edition. 12 illustrations by F. A. Faser. 434pp. of text. 5⅜ x 8½. 23642-0 Pa. $5.00

YEKL and THE IMPORTED BRIDEGROOM AND OTHER STORIES OF THE NEW YORK GHETTO, Abraham Cahan. Film Hester Street based on Yekl (1896). Novel, other stories among first about Jewish immigrants of N.Y.'s East Side. Highly praised by W. D. Howells—Cahan "a new star of realism." New introduction by Bernard G. Richards. 240pp. 5⅜ x 8½. 22427-9 Pa. $3.50

THE HIGH PLACE, James Branch Cabell. Great fantasy writer's enchanting comedy of disenchantment set in 18th-century France. Considered by some critics to be even better than his famous Jurgen. 10 illustrations and numerous vignettes by noted fantasy artist Frank C. Pape. 320pp. 5⅜ x 8½. 23670-6 Pa. $4.00

ALICE'S ADVENTURES UNDER GROUND, Lewis Carroll. Facsimile of ms. Carroll gave Alice Liddell in 1864. Different in many ways from final Alice. Handlettered, illustrated by Carroll. Introduction by Martin Gardner. 128pp. 5⅜ x 8½. 21482-6 Pa. $2.50

FAVORITE ANDREW LANG FAIRY TALE BOOKS IN MANY COLORS, Andrew Lang. The four Lang favorites in a boxed set—the complete Red, Green, Yellow and Blue Fairy Books. 164 stories; 439 illustrations by Lancelot Speed, Henry Ford and G. P. Jacomb Hood. Total of about 1500pp. 5⅜ x 8½. 23407-X Boxed set, Pa. $15.95

HOUSEHOLD STORIES BY THE BROTHERS GRIMM. All the great Grimm stories: "Rumpelstiltskin," "Snow White," "Hansel and Gretel," etc., with 114 illustrations by Walter Crane. 269pp. 5⅜ x 8½.
21080-4 Pa. $3.50

SLEEPING BEAUTY, illustrated by Arthur Rackham. Perhaps the fullest, most delightful version ever, told by C. S. Evans. Rackham's best work. 49 illustrations. 110pp. 7⅞ x 10¾. 22756-1 Pa. $2.50

AMERICAN FAIRY TALES, L. Frank Baum. Young cowboy lassoes Father Time; dummy in Mr. Floman's department store window comes to life; and 10 other fairy tales. 41 illustrations by N. P. Hall, Harry Kennedy, Ike Morgan, and Ralph Gardner. 209pp. 5⅜ x 8½. 23643-9 Pa. $3.00

THE WONDERFUL WIZARD OF OZ, L. Frank Baum. Facsimile in full color of America's finest children's classic. Introduction by Martin Gardner. 143 illustrations by W. W. Denslow. 267pp. 5⅜ x 8½.
20691-2 Pa. $3.50

THE TALE OF PETER RABBIT, Beatrix Potter. The inimitable Peter's terrifying adventure in Mr. McGregor's garden, with all 27 wonderful, full-color Potter illustrations. 55pp. 4¼ x 5½. (Available in U.S. only)
22827-4 Pa. $1.25

THE STORY OF KING ARTHUR AND HIS KNIGHTS, Howard Pyle. Finest children's version of life of King Arthur. 48 illustrations by Pyle. 131pp. 6⅛ x 9¼. 21445-1 Pa. $4.95

CARUSO'S CARICATURES, Enrico Caruso. Great tenor's remarkable caricatures of self, fellow musicians, composers, others. Toscanini, Puccini, Farrar, etc. Impish, cutting, insightful. 473 illustrations. Preface by M. Sisca. 217pp. 8⅜ x 11¼. 23528-9 Pa. $6.95

PERSONAL NARRATIVE OF A PILGRIMAGE TO ALMADINAH AND MECCAH, Richard Burton. Great travel classic by remarkably colorful personality. Burton, disguised as a Moroccan, visited sacred shrines of Islam, narrowly escaping death. Wonderful observations of Islamic life, customs, personalities. 47 illustrations. Total of 959pp. 5⅜ x 8½.
21217-3, 21218-1 Pa., Two-vol. set $12.00

INCIDENTS OF TRAVEL IN YUCATAN, John L. Stephens. Classic (1843) exploration of jungles of Yucatan, looking for evidences of Maya civilization. Travel adventures, Mexican and Indian culture, etc. Total of 669pp. 5⅜ x 8½. 20926-1, 20927-X Pa., Two-vol. set $7.90

AMERICAN LITERARY AUTOGRAPHS FROM WASHINGTON IRVING TO HENRY JAMES, Herbert Cahoon, et al. Letters, poems, manuscripts of Hawthorne, Thoreau, Twain, Alcott, Whitman, 67 other prominent American authors. Reproductions, full transcripts and commentary. Plus checklist of all American Literary Autographs in The Pierpont Morgan Library. Printed on exceptionally high-quality paper. 136 illustrations. 212pp. 9⅛ x 12¼. 23548-3 Pa. $12.50

AN AUTOBIOGRAPHY, Margaret Sanger. Exciting personal account of hard-fought battle for woman's right to birth control, against prejudice, church, law. Foremost feminist document. 504pp. 5⅜ x 8½.

20470-7 Pa. $5.50

MY BONDAGE AND MY FREEDOM, Frederick Douglass. Born as a slave, Douglass became outspoken force in antislavery movement. The best of Douglass's autobiographies. Graphic description of slave life. Introduction by P. Foner. 464pp. 5⅜ x 8½. 22457-0 Pa. $5.50

LIVING MY LIFE, Emma Goldman. Candid, no holds barred account by foremost American anarchist: her own life, anarchist movement, famous contemporaries, ideas and their impact. Struggles and confrontations in America, plus deportation to U.S.S.R. Shocking inside account of persecution of anarchists under Lenin. 13 plates. Total of 944pp. 5⅜ x 8½.

22543-7, 22544-5 Pa., Two-vol. set $12.00

LETTERS AND NOTES ON THE MANNERS, CUSTOMS AND CONDITIONS OF THE NORTH AMERICAN INDIANS, George Catlin. Classic account of life among Plains Indians: ceremonies, hunt, warfare, etc. Dover edition reproduces for first time all original paintings. 312 plates. 572pp. of text. 6⅛ x 9¼. 22118-0, 22119-9 Pa.. Two-vol. set $12.00

THE MAYA AND THEIR NEIGHBORS, edited by Clarence L. Hay, others. Synoptic view of Maya civilization in broadest sense, together with Northern, Southern neighbors. Integrates much background, valuable detail not elsewhere. Prepared by greatest scholars: Kroeber, Morley, Thompson, Spinden, Vaillant, many others. Sometimes called Tozzer Memorial Volume. 60 illustrations, linguistic map. 634pp. 5⅜ x 8½.

23510-6 Pa. $10.00

HANDBOOK OF THE INDIANS OF CALIFORNIA, A. L. Kroeber. Foremost American anthropologist offers complete ethnographic study of each group. Monumental classic. 459 illustrations, maps. 995pp. 5⅜ x 8½.

23368-5 Pa. $13.00

SHAKTI AND SHAKTA, Arthur Avalon. First book to give clear, cohesive analysis of Shakta doctrine, Shakta ritual and Kundalini Shakti (yoga). Important work by one of world's foremost students of Shaktic and Tantric thought. 732pp. 5⅜ x 8½. (Available in U.S. only)

23645-5 Pa. $7.95

AN INTRODUCTION TO THE STUDY OF THE MAYA HIEROGLYPHS, Syvanus Griswold Morley. Classic study by one of the truly great figures in hieroglyph research. Still the best introduction for the student for reading Maya hieroglyphs. New introduction by J. Eric S. Thompson. 117 illustrations. 284pp. 5⅜ x 8½. 23108-9 Pa. $4.00

A STUDY OF MAYA ART, Herbert J. Spinden. Landmark classic interprets Maya symbolism, estimates styles, covers ceramics, architecture, murals, stone carvings as artforms. Still a basic book in area. New introduction by J. Eric Thompson. Over 750 illustrations. 341pp. 8⅜ x 11¼.

21235-1 Pa. $6.95

GEOMETRY, RELATIVITY AND THE FOURTH DIMENSION, Rudolf Rucker. Exposition of fourth dimension, means of visualization, concepts of relativity as Flatland characters continue adventures. Popular, easily followed yet accurate, profound. 141 illustrations. 133pp. 5⅜ x 8½.
23400-2 Pa. $2.75

THE ORIGIN OF LIFE, A. I. Oparin. Modern classic in biochemistry, the first rigorous examination of possible evolution of life from nitrocarbon compounds. Non-technical, easily followed. Total of 295pp. 5⅜ x 8½.
60213-3 Pa. $4.00

PLANETS, STARS AND GALAXIES, A. E. Fanning. Comprehensive introductory survey: the sun, solar system, stars, galaxies, universe, cosmology; quasars, radio stars, etc. 24pp. of photographs. 189pp. 5⅜ x 8½. (Available in U.S. only)
21680-2 Pa. $3.75

THE THIRTEEN BOOKS OF EUCLID'S ELEMENTS, translated with introduction and commentary by Sir Thomas L. Heath. Definitive edition. Textual and linguistic notes, mathematical analysis, 2500 years of critical commentary. Do not confuse with abridged school editions. Total of 1414pp. 5⅜ x 8½. 60088-2, 60089-0, 60090-4 Pa., Three-vol. set $18.50

Prices subject to change without notice.

Available at your book dealer or write for free catalogue to Dept. GI, Dover Publications, Inc., 180 Varick St., N.Y., N.Y. 10014. Dover publishes more than 175 books each year on science, elementary and advanced mathematics, biology, music, art, literary history, social sciences and other areas.